Voltammetry for Sensing Applications

Edited by

J.G. Manjunatha

Department of Chemistry,
FMKMC College, Madikeri,
Constituent College of Mangalore University,
Karnataka, India

Voltammetry for Sensing Applications

Editor: J.G. Manjunatha

ISBN (Online): 978-981-5039-71-9

ISBN (Print): 978-981-5039-72-6

ISBN (Paperback): 978-981-5039-73-3

need for a court order if at any point you breach any terms of this License Agreement. In no event will any delay or failure by Bentham Science Publishers in enforcing your compliance with this License Agreement constitute a waiver of any of its rights.

3. You acknowledge that you have read this License Agreement, and agree to be bound by its terms and conditions. To the extent that any other terms and conditions presented on any website of Bentham Science Publishers conflict with, or are inconsistent with, the terms and conditions set out in this License Agreement, you acknowledge that the terms and conditions set out in this License Agreement shall prevail.

Bentham Science Publishers Pte. Ltd.
80 Robinson Road #02-00
Singapore 068898
Singapore
Email: subscriptions@benthamscience.net

CONTENTS

PREFACE

History has revealed that innovations in "Voltammetry for Sensing Applications" have been the significant approach in the advancement of electrochemical analysis in various platforms such as drug testing and analysis, sensors for point-of-care devices, sensors for diverse analysis, advanced energy storage devices, clinical sample analysis, sensors for the detection of heavy metals, nanomaterials, disease detection, immune sensors, food sample analysis, and anti-inflammatory and anticancer drug detection. The high significance, stability, repeatability, reproducibility, high performance, inexpensive, less time consuming, lower detection limit and quantification, and so on are the most appropriate applications for the sensing tools and methodologies and which portray an imperative character in the environment, biological, medicinal, and food safety-related analysis. Recently, a new era was commenced in voltammetry for sensing applications through the expansion of large-scale, sensitive, selective, and lower concentration level detection to create new sensing devices for giving a kinetic and electrochemical reaction phenomenon. Current advancements in sensing technology will authorize an advanced control in material physical and chemical characteristics and behavior. Hereby contributing an opportunity for new sensory materials in voltammetric analysis with progressive characteristics, such as greater reliability, low-cost, and improved steadiness.

J.G. Manjunatha

Department of Chemistry, FMKMC College, Madikeri,
Constituent College of Mangalore
University, Karnataka, India

List of Contributors

A.H. Sneharani	DoS in Biochemistry, Jnana Kaveri P.G. Center, Mangalore University, Chikka Aluvara, Kodagu
A S Santhosh	Department of Chemistry (UG), NMKRV College for Women, Jayanagar, Bengaluru – 560011, India
Aarti S. Bhatt	Department of Chemistry, N.M.A.M. Institute of Technology, Visvesvaraya Technological University, Belgavi, Nitte, 574110, Udupi District, Karnataka, India
Agnes Chinecherem Nkele	Department of Physics and Astronomy, University of Nigeria, Nsukka, Enugu, Nigeria
Akshatha Nemumoolya	Department of Chemistry, FMKMC College, Madikeri, Constituent College of Mangalore University, Karnataka, India
Amrutha B. Monnappa	Department of Chemistry, FMKMC College, Madikeri, Constituent College of Mangalore University, Karnataka, India Department of Chemistry, N.M.A.M. Institute of Technology, Visvesvaraya Technological University, Belgavi, Nitte, 574110, Udupi District, Karnataka, India
Anup Pandith	Department of Chemistry, School of Natural Sciences, KyungHee University, Seoul, South Korea
Aruna Kumar D B	Department of Studies and Research in Organic Chemistry , Tumkur University, Tumakuru-572 103, Karnataka, India
B P Sanjay	Department of Chemistry, JSS Science and Technology University, Mysuru- 570006, Karnataka, India
B S Surendra	Department of Science, East West Institute of Technology, Bengaluru 560 091, Karnataka, India
B. P. Nandeshwarappa	Department of PG Studies and Research in Chemistry, Shivagangothri, Davangere University, Davanagere, Karnataka - 577 007, India
Basappa C Yallur	Department of Chemistry, M. S. Ramaiah Institute of Technology, Bangalore, Karnataka, India
Bruna Coldibeli	Laboratório de Eletroanalítica e Sensores, Departamento de Química, Universidade Estadual de Londrina (UEL), Rodovia Celso Garcia Cid, PR 445 Km 380, Londrina – PR, C.P. 10.011, 86057-970, Brazil
C S. Karthik	Department of Chemistry, JSS Science and Technology University, Mysuru- 570006, Karnataka, India
Carlos Alberto Rossi Salamanca-Neto	Laboratório de Eletroanalítica e Sensores, Departamento de Química, Universidade Estadual de Londrina (UEL), Rodovia Celso Garcia Cid, PR 445 Km 380, Londrina – PR, C.P. 10.011, 86057-970, Brazil
D N Varun	Department of Chemistry, JSS Science and Technology University, Mysuru- 570006, Karnataka, India

Débora Nobile Clausen

Laboratório de Eletroanalítica e Sensores, Departamento de Química, Universidade Estadual de Londrina (UEL), Rodovia Celso Garcia Cid, PR 445 Km 380, Londrina – PR, C.P. 10.011, 86057-970, Brazil
Centro Universitário Cesumar (UniCesumar), Departamento de Biomedicina, Avenida Santa Mônica, 450, Londrina – PR, 86027-610, Brazil

Debdas Bhowmik

High Energy Materials Research Laboratory, Defence Research and Development Organization, Ministry of Defence, Government of India, Sutarwadi, Pune, India

Deepadarshan Urs

Department of Studies and Research in Biochemistry, Jnana Kaveri Post Graduate Centre, Mangalore University, Chikka Aluvara, Kodagu, Karnataka, India

Elen Romão Sartori

Laboratório de Eletroanalítica e Sensores, Departamento de Química, Universidade Estadual de Londrina (UEL), Rodovia Celso Garcia Cid, PR 445 Km 380, Londrina – PR, C.P. 10.011, 86057-970, Brazil

Erison Pereira de Abreu

Centro Universitário Cesumar (UniCesumar), Departamento de Biomedicina, Avenida Santa Mônica, 450, Londrina – PR, 86027-610, Brazil

Fabian I. Ezema

Department of Physics and Astronomy, University of Nigeria, Nsukka, Enugu, Nigeria
Nanosciences African Network (NANOAFNET), iThemba LABS-National Research Foundation,, 1 Old Faure road, Somerset West 7129, P.O. Box 722, Somerset West, Western Cape Province, South Africa
UNESCO-UNISA Africa Chair in Nanosciences/Nanotechnology, College of Graduate Studies, University of South Africa (UNISA), Muckleneuk Ridge, P.O. Box 392,, Pretoria, South Africa

Fatih ŞEN

Sen Research Group, Department of Biochemistry, University of Dumlupinar, 43000 Kütahya, Turkey

G Krishnaswamy

Department of Studies and Research in Organic Chemistry , Tumkur University, Tumakuru-572 103, Karnataka, India

G Shivaraja

Department of Studies and Research in Organic Chemistry , Tumkur University, Tumakuru-572 103, Karnataka, India

Gabriel Junquetti Mattos

Laboratório de Eletroanalítica e Sensores, Departamento de Química, Universidade Estadual de Londrina (UEL), Rodovia Celso Garcia Cid, PR 445 Km 380, Londrina – PR, C.P. 10.011, 86057-970, Brazil

Gabriel Rainer Pontes Manrique

Laboratório de Eletroanalítica e Sensores, Departamento de Química, Universidade Estadual de Londrina (UEL), Rodovia Celso Garcia Cid, PR 445 Km 380, Londrina – PR, C.P. 10.011, 86057-970, Brazil

Geethanjali N. Karthammaiah

Department of Chemistry, FMKMC College, Madikeri, Constituent College of Mangalore University, Karnataka, India

Gnanesh Rao

Department of Biochemistry, Bangalore University, Bangalore, Karnataka, India

Gururaj Kudur Jayaprakash

School of Advanced School of Chemical Science, Shoolini University, Bajhol, Himachal Pradesh 173229, India

H P Nagaswarupa

Department of Studies in Chemistry, Shivagangothri, Davangere University, Davangere - 577 007, Karnataka,, India

Hanaa S. El-Desoky

Analytical and Electrochemistry Research Unit, Department of Chemistry, Faculty of Science, Tanta University, 31527 Tanta, Egypt

J Shankar

Department of Studies and Research in Food Technology, Davangere University, Shivagangothri, Davangere, Karnataka, India

Jamballi G. Manjunatha

Department of Chemistry, FMKMC College, Madikeri, Constituent College of Mangalore University, Karnataka, India

Jessica Scremin

Laboratório de Eletroanalítica e Sensores, Departamento de Química, Universidade Estadual de Londrina (UEL), Rodovia Celso Garcia Cid, PR 445 Km 380, Londrina – PR, C.P. 10.011, 86057-970, Brazil

K S Nithin

Department of Chemistry, The National Institute of Engineering, Mysuru, India

K. K. Dharmappa

Department of Studies and Research in Biochemistry, Jnana Kaveri Post Graduate Centre, Mangalore University, Chikka Aluvara, Kodagu, Karnataka, India

Kiran Kumar Mudnakudu-Nagaraju

Department of Biotechnology & Bioinformatics, Faculty of Life Sciences, JSS Academy of Higher Education and Research, Mysore 570015, Karnataka, India

Kumara Swamy N k

Department of Chemistry, Sri Jayachamarajendra College of Engineering, JSS Science and Technology University, Mysuru, India

M H Chethana

Department of Chemistry, Sri Jayachamarajendra College of Engineering, JSS Science and Technology University, Mysuru, India

M. B Siddesh

Department of Chemistry, KLE'S S. K. Arts College and H. S. K. Science Institute, Hubballi, India

Monima Sarma

Department of Chemistry, KL Deemed to be University (KLEF), Greenfields, Vaddeswaram, Andhra Pradesh 522502, India

Muhammed BEKMEZCİ

Sen Research Group, Department of Biochemistry, University of Dumlupinar, 43000 Kütahya, Turkey
Department of Materials Science & Engineering, Faculty of Engineering, Dumlupınar University, Evliya Çelebi Campus, 43100 Kutahya, Turkey

N Raghavendra

Department of Science, East West Institute of Technology, Bengaluru 560 091, Karnataka, India

Natalia Sayuri Matunaga Campos

Laboratório de Eletroanalítica e Sensores, Departamento de Química, Universidade Estadual de Londrina (UEL), Rodovia Celso Garcia Cid, PR 445 Km 380, Londrina – PR, C.P. 10.011, 86057-970, Brazil

Nagaraja Sreeharsha

Department of Pharmaceutical Sciences, College of Clinical Pharmacy, King Faisal University,, Al- Ahsa-31982, Saudi Arabia
Department of Pharmaceutics, Vidya Siri College of Pharmacy, Off Sarjapura Road, Bengaluru - 560035, Karnataka, India

P Mallu

Department of Chemistry, Sri Jayachamarajendra College of Engineering, JSS Science and Technology University, Mysuru, India

P. Mallu	Department of Chemistry, JSS Science and Technology University, Mysuru- 570006, Karnataka, India
P Manikanta	Department of Chemistry, Sri Jayachamarajendra College of Engineering, JSS Science and Technology University, Mysuru, India
Raghu Ningegowda	Jyoti Nivas College Autonomous, Department of Studies in Chemistry, Bangalore-560095, India
Rajkumar S. Meti	Department of Studies and Research in Biochemistry, Jnana Kaveri Post Graduate Centre, Mangalore University, Chikka Aluvara, Kodagu, Karnataka, India
Ramazan BAYAT	Sen Research Group, Department of Biochemistry, University of Dumlupinar, 43000 Kütahya, Turkey Department of Materials Science & Engineering, Faculty of Engineering, Dumlupınar University, Evliya Çelebi Campus, 43100 Kutahya, Turkey
S C Prashantha	Department of Science, East West Institute of Technology, Bengaluru 560 091, Karnataka, India
S Sreenivasa	Department of Studies and Research in Chemistry, University College of Science, Tumkur University, Tumakuru-572 103, Karnataka, India Deputy Adviser, National Assessment and Accreditation Council, Bengaluru-560 072, Karnataka, India
S Sandeep	Department of Chemistry, JSS Science and Technology University, Mysuru- 570006, Karnataka, India
Sandeep Chandrashekharappa	Institute for Stem Cell Science and Regenerative Medicine, NCBS, TIFR, GKVK-Campus Bellary road, Bengaluru 560065, Karnataka, India Department of Medicinal Chemistry, National Institute of Pharmaceutical Education and Research (NIPER) Raebareli, Lucknow (UP)-226002, India
Santosh S Nandi	Chemistry Section, Department of Engineering Science and Humanities, KLE Dr. M.S. Sheshgiri College of Engineering & Technology, Udhyambagh, Belagavi-590008, Karnataka, India
Shankar Ashok Itagi	Department of Chemistry, Sri Jayachamarajendra College of Engineering, JSS Science and Technology University, Mysuru, India
Shankramma Kalikeri	Division of Nanoscience and Technology, Department of Water and Health (Faculty of life sciences) JSS Academy of Higher Education & Research (Deemed to be University), Mysore-570015, India
Sharmila B. Medappa	Department of Chemistry, FMKMC College, Madikeri, Constituent College of Mangalore University, Karnataka, India
Shefali Sharma	School of Advanced School of Chemical Science, Shoolini University, Bajhol, Himachal Pradesh 173229, India
Sophiya P	Department of Studies and Research in Biochemistry, Jnana Kaveri Post Graduate Centre, Mangalore University, Chikka Aluvara, Kodagu, Karnataka, India
T R Shashi Shekhar	Department of Civil, East West Institute of Technology, Bengaluru 560 091, Karnataka, India

İsmail Mert ALKAÇ   Sen Research Group, Department of Biochemistry, University of Dumlupinar, 43000 Kütahya, Turkey
Department of Materials Science & Engineering, Faculty of Engineering, Dumlupınar University, Evliya Çelebi Campus, 43100 Kutahya, Turkey

Vinayak Adimule   Department of Chemistry, Angadi Institute of Technology and Management (AITM), Savagaon Road, Belagavi-5800321, Karnataka, India

CHAPTER 1

Advanced Sensor Materials for Drug Analysis

Hanaa S. El-Desoky[1,*]

[1] Analytical and Electrochemistry Research Unit, Department of Chemistry, Faculty of Science, Tanta University, 31527 Tanta, Egypt

Abstract: Nanomaterials play an important role in the fabrication of many devices and modified materials, due to their unique properties, such as large surface area/volume ratio, conductivity and high mechanical strength. In the present chapter, the applicability of nanomaterials in drug analysis is well investigated. The recent trends in the development of the electrochemical sensor platforms based on state-of-the-art nanomaterials such as metal nanoparticles, metal oxide nanoparticles, carbon nanomaterials, conducting polymer and nanocomposites are discussed. The unique synthetic approaches, properties, integration, strategies, selected sensing applications and future prospects of these nanostructured materials for the design of advanced sensor platforms are also highlighted. Various kinds of functional nanocomposites have led to the enhancement in voltammetric response due to drug - nanomaterials interaction at the modified electrode surface. So, different mechanisms for the extraordinary and unique electrocatalytic activities of such nanomaterials will be highlighted. Potential applications of electrochemical sensor platforms based on advanced functional nanomaterials for drug analysis are presented. High sensitivity and selectivity, fast response, and excellent durability in biological media are all critical aspects which will also be addressed. It is expected that the chemically modified electrodes with various nanomaterials can be easily miniaturized and used as wearable, portable and user friendly devices. This will pave the way for in-vivo onsite real monitoring of single as well as multi-component pharmaceutical compounds. The significant development of the nanomaterials based electrochemical sensor platforms is giving rise to a new impetus of generating novel technologies for securing human and environmental safety.

Keywords: Analysis of drug, Biological fluids, Carbon nanotubes, Conducting polymer, Electrochemical sensor, Graphene, Hybrid nanostructure, Imprinted polymers, Metal nanoparticles, Metal oxide nanoparticles.

---

* **Corresponding author Hanaa S. El-Desoky:** Analytical and Electrochemistry Research Unit, Department of Chemistry, Faculty of Science, Tanta University, 31527 Tanta, Egypt; Tel: +201098846641; E-mails: hseldesoky@hotmail.com, hanaa_eldesoky@Science.tanta.edu.eg

J.G. Manjunatha (Ed.)
All rights reserved-© 2022 Bentham Science Publishers

INTRODUCTION

Nanotechnology involves the synthesis and application of materials having one of the dimensions in the range of 1–100 nm. The recent accomplishments in nanotechnology mainly nano-material-based electrochemical systems have led to the development of unique platforms that have significantly improved the sensory characteristics of conventional electrochemical systems. The combination of nano-materials of distinct nature and exceptional properties has notably contributed to fundamental biological research, environmental monitoring, drug and food safety, pharmaceutical procedures, healthcare diagnostics, and drug quality control. The interdisciplinary feature of such a synergic platform has not only extended the scope of sensor systems but has opened new pathways for the development of flexible and portable personal care and field applicable devices. Superior surface area to volume ratio and higher active site availability allow higher sensing response and catalysis as well as better magnetic, optical and electrical properties for biological, pharmaceutical and biomedical applications. This chapter mainly focuses on the modern advances in the growth of nanomaterials based electrochemical sensor platforms for the detection of potent biological analytes such as drugs and their ability for analysing complex samples such as urine, blood and pharmaceutical preparations.

NANOMATERIALS APPLIED FOR NANOSENSORS

Nanomaterials have unique physical and chemical properties as compared to their bulk materials due to their high surface area and electronic properties as well as the controlled morphology. The commonly used nanomaterials in electrochemical nanosensors are mainly carbon-based nanomaterials and metal oxide nanoparticles. Meanwhile, many emerging materials are explored to modify the surface of the working electrodes, such as conducting polymers [1], metal-based nanomaterials [2 - 4], carbon nanotubes [5 - 7], graphene [8 - 11], and metal-organic framework nanomaterials [12].

This leads to the development of electrodes with good stability, huge specific area, improved redox performance, and recyclability. The fabrication of the nanocomposites with many combinations such as; metal nanoparticles, metal oxide nanoparticles, carbon nanotubes (CNTs), graphene (GR), quantum dots, and conducting polymer further improve the electrochemical sensing properties of such electrodes [13]. Fig. (1) shows the schematic representation of the most important nanomaterials employed for biological and biomedical applications, especially drug analysis.

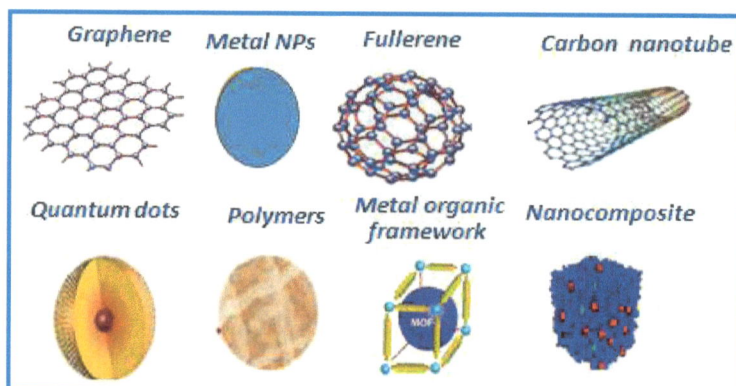

Fig. (1). Nanomaterials based electrochemical sensor platforms for drug analysis application.

CLASSIFICATION OF NANOMATERIALS

A simple classification of nano-materials based on their structures includes zero, one, two, and three dimensions. Fig. (**2**) presents some examples of various morphological structures of nano-materials. These nano-materials have many applications in electrochemistry, photochemistry, and biomedicine [14]. Nano-materials have many functional platforms which can be utilized for therapeutic functions.

Fig. (2). Nanomaterials with various morphologies.

NANOPARTICLES SYNTHESIS

Several methods have been used for the synthesis of nanoparticles (NPs), including physical, chemical and biological methods [2, 4, 15 - 18] (Fig. **3**). There are two different approaches for preparing the NPs; the bottom-up approach and the top-down approach. In the bottom-up approach, the atoms are assembled in nuclei and then grown into NPs. The top-down approach starts with bulk material at the macroscopic level, followed by trimming the material to the desired NPs. Biological and chemical methods which are used for NPs synthesis are considered bottom-up approaches. The selection of any of these methods in terms of scalability, costs, particle sizes, and size distribution should be considered.

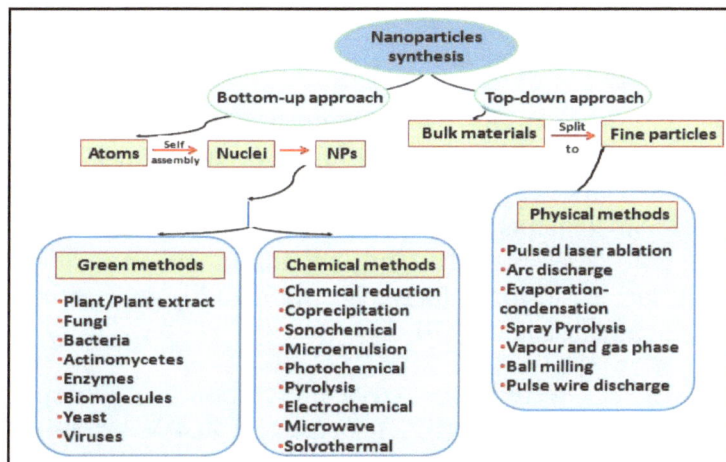

Fig. (3). Flowchart of different approaches for nanoparticles synthesis.

The most openly used physical methods for the inexpensive synthesis of NPs are wet and dry mechanical grindings. The former is preferable because it allows more options to control the NPs size. The physical methods are generally required to have the raw material to grind, surfactant to cover the particle surface and prevent their aggregation, overheating during grinding and fluid carrier where both raw material and surfactant are mixed with a fluid carrier.

Chemical methods generally provide an effective approach to synthesize NPs. The most widely used chemical methods are sol-gel technique, solvothermal method, hydrothermal method, microwave, microemulsion, and electrochemical reduction [2, 4, 15, 16]. The main components in the chemical approach are the metallic precursors, stabilizing agents and reducing agents (inorganic or organic). Chemical reducing agents such as sodium citrate, hydrazine, ascorbate, sodium borohydride ($NaBH_4$), elemental hydrogen, tollens reagent, polyol process, N,N-dimethylformamide (DMF) and poly(ethylene glycol)-block copolymers are used [17]. The various chemical methods often need various treating steps, controlled pH and temperature, much expensive equipment and toxic chemicals. Further, these methods also generate several by-products which are toxic to ecosystems. Therefore, the requirement of generating an eco-friendly method using biological (green) synthesis approaches is urgently recommended [18].

Biological methods for nanoparticles synthesis are proposed as green "eco-friendly" alternatives to existing physical and chemical methods. Biosynthesis of various types of very small nanoparticles (5-10 nm) is available [18]. Green chemistry has appeared as a novel concept for the development and implementation of chemical processes to decrease or remove the use of hazardous substances.

Metal Nanostructures in Sensors

Metal nanoparticles (MNPs) have unique physical and chemical properties which make them extremely suitable for designing novel and improved electrochemical sensors and biosensors [19]. MNPs can be used as analytical transducers and signal amplification elements in various sensing devices [20]. Various MNPs such as silver (Ag), gold (Au), platinum (Pt), palladium (Pd), cobalt (Co) and copper (Cu), including rare earth metals have been utilized in fabricating biosensors as well as electrochemical sensors [19].

For example, Au NPs were deposited at the carbon paste electrode (CPE) and screen-printed carbon electrode (SPE) surfaces at -0.4 V for 300 s for designing an effective electrochemical sensor for *Moxifloxacin Hydrochloride* (*Moxi*) drug [21]. Both electrodes gave rise to the largest current responses compared to graphene oxide (GO), Ag NPs, nano-Co (II, III) oxide, CNTs and Zeolite [21]. The SPE support was preferred over the CPE for its ability to be used as a disposable single-use sensor enabling the circumvention of the problems of electrode surface fouling. Scan electron spectroscopy (SEM) and Transimision electron spectroscopy (TEM) indicate the successful deposition of Au NPs (sizes of 13–58 nm), which dispersed well onto the electrode surface. Differential pulse voltammetry (DPV) was applied to *Moxi* detection which gave rise to an accessible concentration window ranging between 8 μM and 0.48 mM, and a detection limit (LOD) of 11.6 μM using AuNPs modified SPE. It was also practiced in a human baby urine sample with excellent recoveries (R%) between 99.8% and 101.6% and relative standard deviations (RSDs%) of 1.1–3.4%.

Pd is also abundant over other noble metals such as Pt and Au, and this is making it a cheaper alternative for developing a number of sensors [22]. Pd, in combination with other materials such as GR or GO to form nanocomposites has improved the mass diffusion of analytes. Nanocomposites normally offer electron tunneling which enables electron transfer between the active site and the electrode [22, 23]. Ex-situ decoration of graphene oxide with palladium NPs was prepared for sensitive electrochemical determination of antibiotic drug *Chloramphenicol* (*CPL*) in food and biological samples [23]. Pd NPs/GO nanocomposite modified glassy carbon electrode (Pd NPs/GO/GCE) exhibits wide linear range (0.01 to 102.68 μM), high sensitivity (3.048 μA μM^{-1} cm^2), and low limit of detection

(LOD = 0.001 µM) towards *CPL* determination in bulk form. This sensor exhibits an excellent selectivity for the *CPL* sensing in the presence of different interfering compounds. It is applied to the food and biological samples for the determination of *CPL* with good (R±RSD%) values of (97.88 ± 1.05%) and (99.52±2.05%), respectively.

A reproducible method for simultaneous determination of *Entacapone (EN)*, *Levodopa (LD)* and *Carbidopa (CD)* drugs is also described utilizing Pd NPs [24]. It is based on electrodeposition of Pd NPs on a methionine modified CPE in the presence of sodium dodecyl sulphate (Met/Pd/CPE/SDS). Chrono-amperometry (CA), electrochemical impedance spectroscopy (EIS), cyclic voltammetry (CV) and DPV techniques were used to characterize the properties of the sensor. The voltammetric results showed well-defined anodic peaks at potentials of 650, 488 and 320 mV, corresponding to the oxidation of EN, LD, and CD, respectively, indicating that the simultaneous determination of these compounds is feasible. The respective linear ranges were (2.0×10^{-8} to 0.8×10^{-3} M), (0.5×10^{-5} to 60.0×10^{-3} M) and (0.3×10^{-5} to 15×10^{-3} M) for *EN, LD* and *CD,* respectively, using DPV. Validation of this sensor for determination of *EN* in urine sample was examined using DPV. A wide linear dynamic range of (2.0×10^{-7} to 0.4×10^{-4} M) and LOD value of 1.87×10^{-8} M were achieved. Also, the recovery (99.88 – 100.1) and standard deviation (0.55×10^{-7} - 2.5×10^{-7}) were calculated.

Cu has fascinated many researchers as an ideal material for use in sensors's construction, since it has good stability, outstanding electrical conductivity, electrocatalytic properties, and low cost when compared with Pt, Au and Ag [25]. Cu-based nanostructures have many exceptional properties such as high mass-transport rate, surface to volume ratio and sensitivity in electroanalytical measurements. Cu nanoclusters prepared *via* a simple one-step electrodeposition process at -0.7 V on Pt electrode were utilized to determine nitrates within the concentration range from 0.1 – 4 mM [25]. Moreover, synthesized Au NPs were used as a chemosensor for *Piroxicam* in a concentration range of 20–60 µM [26].

Metal Oxide Nanostructures in Sensors

Semiconductor metal oxide-based nanostructured materials have been expansively utilized as sensors in numerous applications. Their small crystallite size can enhance the sensor's performance. Metal oxide nanoparticles (MONPs)-based sensors are robust, inexpensive and easy to produce [27]. Among the various porous nanomaterials, metal oxides have attracted increasing industrial and technological interest [28].

NiO NPs have received considerable attention due to its lower cost, outstanding high specific capacitance, catalytic and electrical properties. NiO is p-type metal oxide semiconductor; it has wide range of applications in various fields [28 - 30]. NiO nano-flakes (NiO NFs) were synthesized using ammonia precipitation or precursor method [29, 30] and characterized (as shown in Fig. (**4**)) for *Ledipasvir* drug determination [4]. Thermal gravimetric analysis (TGA) depicted that the precursor $Ni(OH)_2$ decomposed completely at 400°C which was applied during calcinations process for nano-scale NiO production. The Energy dispersive X-ray (EDX) spectrum of NiO shows that O/Ni molar ratio equals nearly 1:1. SEM reveals the formation of ultra-thin NiO nano-flakes (17.60±4.20 nm) with a diameter ≈ 190 nm on average while TEM shows average crystallite size of NiO NFs of (17.83 ±4.15 nm). From X-ray diffraction (XRD) profile of NiO NFs, the interplanar spacing of 0.21 nm and the crystalline size of 14.31 nm confirm formation of the nano-crystalline structures. The N_2 adsorption–desorption isotherm of NiO NFs indicates its mesoporous nature. The calculated BET specific surface area was 11.56 ± 0.81 m^2 g^{-1}.

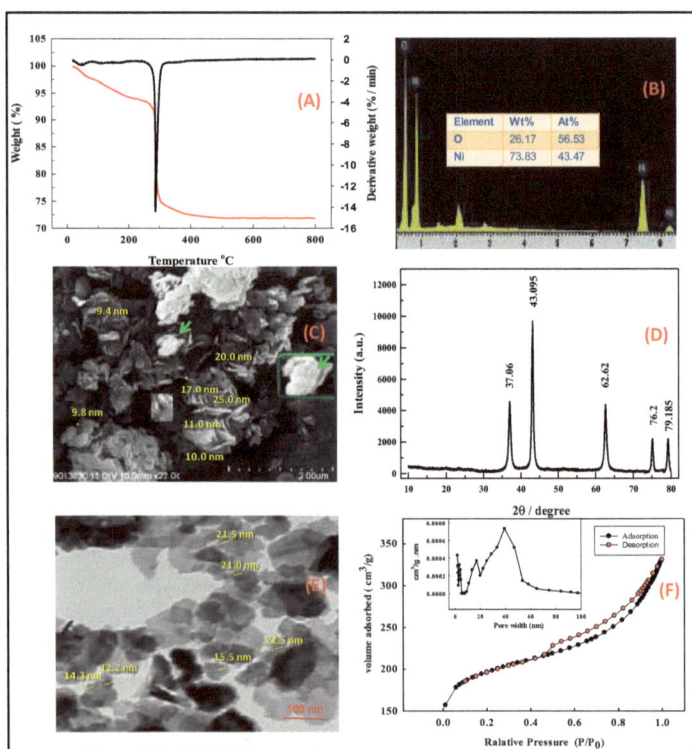

Fig. (4). A) TGA and DTA of $Ni(OH)_2$ precursor, **B)** EDX analysis, **C)** SEM image, **D)** XRD pattern, **E)** TEM image and **F)** N_2 adsorption isotherm of the NiO NFs (© The Electrochemical Society-Permission of IOP Publishing, [DOI: 10.1149/1945-7111/ab9e86]).

The synergetic benefit of [NiO NFs – activated charcoal] platform was examined for the first time for ultra trace determination of *Ledipasvir drug*. It improved the performance of CPE sensor in terms of linear dynamic range ($3.0 \times 10^{-9} - 1.5 \times 10^{-6}$ M), LOD (5.49×10^{-10} M), R% (98.33 -102.03) and RSD% (1.82-2.80) in human plasma samples. This sensor is adequate for clinical pharmacokinetic study since it offers high selectivity, stability, accuracy and precision values as well as its wide concentration range covers the C_{max} and C_{min} of the Ledipasvir in its pharmacokinetic study [4].

CuO nanostructure as an active transducing material is utilizing in direct electrochemical determination of *N-acetyl-l-cysteine* (*NAC*) (which is used in the treatment of chronic respiratory diseases) [31]. The CuO NPs were produced using the hydrothermal method with the assistance of templates (NAC itself, adipic acid and citric acid). The electrode modification was achieved by casting the decided volume of the nano-dispersion over a pre-polished electrode then the modified dry layer was followed by a layer of Nafion to prevent the surface erosion. The highest current with low-over potential value was achieved using citric acid as an active template due to favorable interaction that perceived between the surface-bound functionality (carbonyl) and hydroxyl moiety of NAC. Such sensor exhibited excellent linearity in the concentration range of 0.1 to 5.0 µM and LOD was as low as 0.01 µM.

Further, a disposable CuO NPs modified screen-printed carbon electrode (CuO NPs/SPCE) was studied for sensitive determination of *Mirtazapine (MZ) drug* [32]. The linear response was in the concentration range of 66.62– 662.25 ng mL^{-1} using DPV. LOD value was 4.49 ng mL^{-1}. This sensor was applied to the determination of *MZ* drug in tablets and spiked plasma samples. Satisfactory R±RSD% of (100.02±0.04) and (101.15±2.05) were found for *MZ* in tablets and serum, respectively. This indicates that the suggested sensor is highly suitable for clinical analysis, quality control determination of *MZ* in pharmaceutical formulations and spiked serum.

Also, SiO$_2$ and TiO$_2$ NPs have been employed in the modification of CPE [33, 34] for the determination of *Gallic acid*. Voltammetric studies show that the SiO$_2$ NPs modified CPE is sensitive to *Gallic acid* in the concentration range of (8.0×0^{-7} - 1.0×10^{-4} M). LOD and sensitivity were calculated as 2.5×10^{-7} M and 1790.7 µAmM^{-1}, respectively [33]. However, TiO$_2$ NPs/CPE offers a linear dynamic range of (2.5×10^{-6} to 1.5×10^{-4} M) with LOD of 9.4×10^{-7} M [34]. Finally, the proposed modified electrodes were successfully used in real sample analysis.

Magnetite nanoparticles (Fe$_3$O$_4$ NPs) have attracted also great attention for its high sensitivity response toward drugs determination [35]. With ongoing explorations, a typical bimetallic iron-based oxide, spinel ferrite with the general formula of MFe_2O_4 (M = Fe, Ni, Co, Mn, Zn), has attracted much attention [36]. The excellent electrochemical properties which originate from electrons hopping between Fe^{2+} and Fe^{3+} ions render it useful in several applications [37]. It was reported that doping Mn in Fe_3O_4 NPs could provide a synergistic effect with Fe species for higher electrochemical reactivity [35]. For example, Mn ferrites NPs modified CPE was developed for sensitive and selective voltammetric determination of a new antiplatelet agent, *Ticagrelor hydrochloride (TIC.HCl)* drug in formulations and human blood samples (Fig. **5**) [38]. First, a series of Mn ferrites NPs $\{Mn_xFe_{3-x}O_4\ (x = 0.2 - 1)\}$ was easily synthesized using the co-precipitation method and characterized using different techniques (Figs. **6** and **7**).

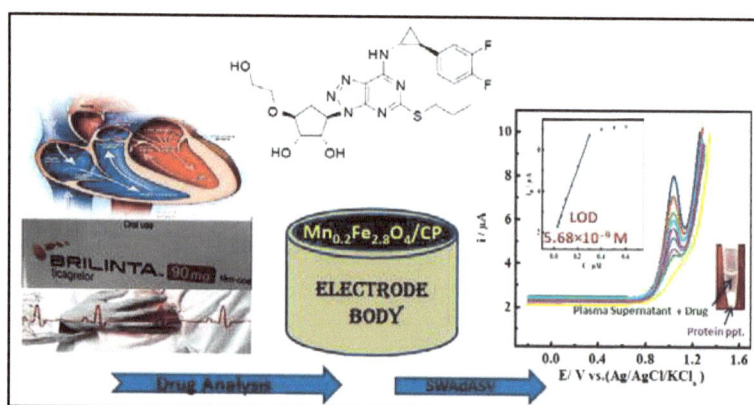

Fig. (5). Symbolizes the determination of a new antiplatelet agent, *Ticagrelor hydrochloride (TIC•HCl)* drug in human blood at $Mn_{0.2}Fe_{2.8}O_4$ NPs modified carbon paste electrode (© The Electrochemical Society-Permission of IOP Publishing, [DOI: 10.1149/1945-7111/ab7e21]).

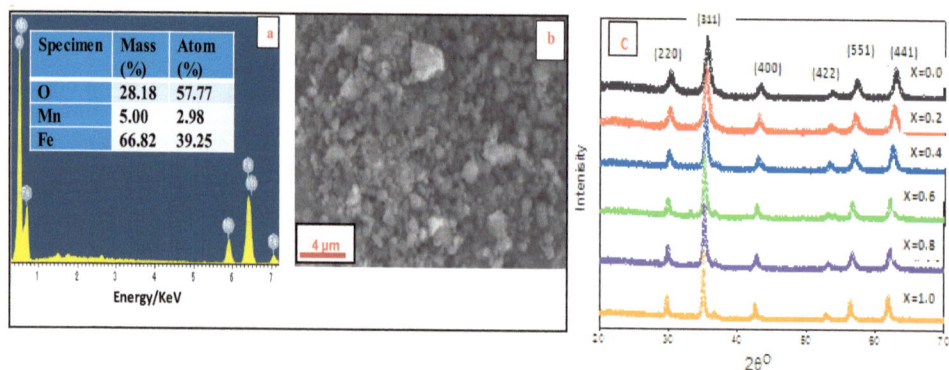

Fig. (6). EDX spectrum (a), SEM image (b) and X-ray diffraction patterns (c) of the as-prepared $Mn_{0.2}Fe_{2.8}O_4$ NPs (© The Electrochemical Society-Permission of IOP Publishing, [DOI: 10.1149/1945-7111/ab7e21]).

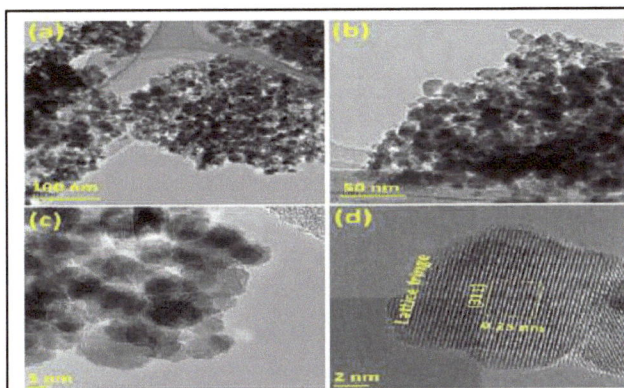

Fig. (7). TEM (a-c) and HRTEM (d) images of $Mn_{0.2}Fe_{2.8}O_4$ nanoparticles (© The Electrochemical Society - Permission of IOP Publishing, [DOI: 10.1149/1945-7111/ab7e21]).

EDX spectroscopy indicated that the composition of $Mn_{0.2}Fe_{2.8}O_4$ was consistent with their estimated molar ratios. X-ray diffraction pattern reveals the phase purity and the formation of crystalline nanoparticles with cubic inverse spinel structures. $Mn_{0.2}Fe_{2.8}O_4$ sample was found to have the smallest particles size (10.72 nm), lattice parameter (0.84 nm), crystal volume (58.70 nm) and highest BET surface area (79.54 m^2/g) compared to the higher Mn^{2+} content in the other samples ($0.2 < x \leq 1$). Fig. (**8**) illustrates square-wave adsorptive anodic stripping voltammetry (SW-AdASV) at different %compositions of $Mn_{0.2}Fe_{2.8}O_4$ modified carbon paste electrode. At 2%(w/w) $Mn_{0.2}Fe_{2.8}O_4$/CPE, the electro-chemical behavior of *TIC.HCl* was investigated and the electrode reaction mechanism was suggested, (Fig. **8**). The total number of electrons exchanged per molecule was found to be 2e$^-$. The oxidation of *TIC.HCl* occurs first at S atom with the removal of 1e$^-$, leading to the formation of a cationic radical. After that formation of sulfoxide species can be took place *via* losing of another 1e$^-$ and nucleophilic attack by water. $Mn_{0.2}Fe_{2.8}O_4$/CPE offered $\approx$ one order of magnitude improvement in LOD value (1.39×10^{-9} M) compared to the bare CPE (1.53×10^{-8} M). Assay of *TIC.HCl* in its dosage forms (Thrombolinta and Brilinta® tablets) with excellent percent recovery values of (%R = 99.21- 99.72%) and in human plasma sample with very low LOD value (5.68×10^{-9} M) was performed. Reproducibility of the method was evaluated by 3 successive determinations of *TIC.HCl* with 3 different modified electrodes. The RSD% value of less than 3.0% was obtained for 8×10^{-8} M of *TIC.HCl* indicating a good reproducibility. This approach has high sensitivity, stability and good reproducibility. Acquired results demonstrate that proposed strategy can be effortlessly applied for routine examination of *TIC.HCl* in its formulations and in human plasma samples.

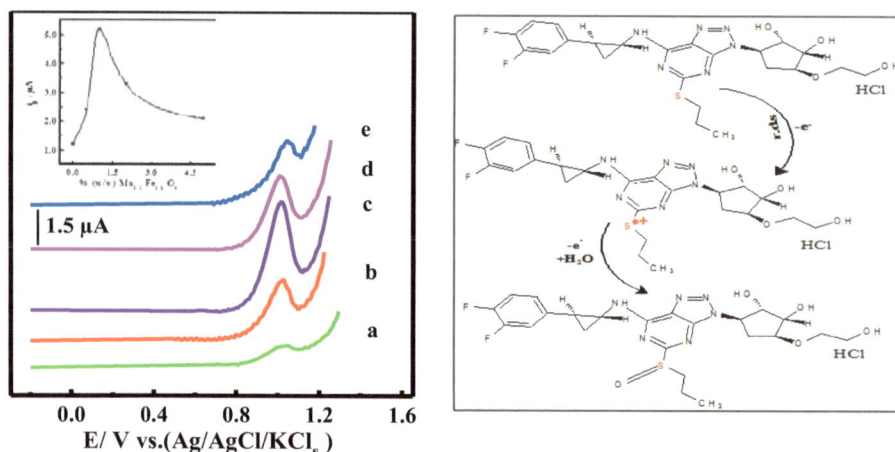

Fig. (8). SW-AdAS voltammograms of 1.0×10^{-7} M *TIC•HCl* in B-R universal buffer solution of pH = 2 recorded at E_{acc}= -0.2 V for 80s onto modified CPE with various % (w/w) $Mn_{0.2}Fe_{2.8}O_4$: a) 0.5%, b) 1%, c) 2%, d) 5% and e) 10% (w/w) $Mn_{0.2}Fe_{2.8}O_4$ and the suggested oxidation mechanism of *TIC•HCl* at CPE (© The Electrochemical Society - Permission of IOP Publishing, [DOI: 10.1149/1945-7111/ab7e21]).

Carbonaceous Nanostructures

One of the most currently used materials in the nanotechnology field is the carbon-based one due to its remarkable properties. Carbonaceous structures present numerous advantages compared to other usually employed materials, especially their extraordinary physical-chemical properties. Carbon offers matchless versatility among the elements of the periodic table. Relying on its hybridization state and atomic arrangement, carbon forms the layered semiconductor graphite, the insulator diamond with its surpassing hardness, the high surface area amorphous carbons, and the nanoscale forms of carbon with various shapes including ball shapes such as fullerenes (C60), wires such as carbon nanotubes (CNTs), sheets such as graphene (GR), *etc* [39]. By combining the advantages of carbon materials with those of nanostructured materials, carbon-based nanoscale materials have been widely used (as the nano-electrocatalysts) in the design of advanced electrochemical sensors. The abilities of carbon based nano electrocatalyst electrodes to enhance electron transfer reactions and to provide resistance to surface fouling have been documented in connection with a plenty of species [2, 3, 6 - 9, 11, 40, 41]. This would be most likely attributed to the presence of edge plane like sites on nanostructured carbon materials [42]. Among those the carbon nanomaterials, graphene (GR) is considered as the basic building block for graphitic materials of all other dimensionalities. GR is an individual graphite layer. It is a two-dimensional (2-D) monolayer of carbon atoms parked into a dense hexagonal network structure. GR can be wrapped up into 0-D C60, rolled into 1-D CNTs, or stacked into 3-D graphite, (Fig. **9**) [43].

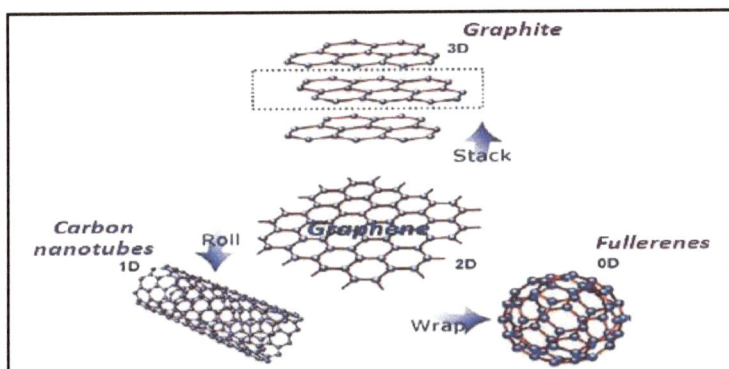

Fig (9). Carbon nanomaterials, including graphene (a 2-D building material for carbon materials of all other dimensionalities) which can be wrapped up into 0-D buckyballs, rolled into 1-D nanotubes or stacked into 3-D graphite.

Carbon Nanotubes in Sensors

The 1-D CNTs can be described as a 2-D GR sheet rolled up into a nanoscale hollow tube (which are single-wall CNTs), or with additional GR sheets around the core of a single-wall CNTs (called multi-wall CNTs). The desirable properties of CNTs are referred to as their unparalleled sp^2 structures. CNTs have diameters in the range between fractions of nanometers (nm) and tens of nm, lengths > hundred nm and extremely high surface area (For one side of GR sheet, the value obtained is 1315 m^2g^{-1} while using different multi-walled geometries and nanotubes bundles the value decreases to 50 m^2g^{-1} [44]). The GR layers of CNT themselves are weakly bound to each other (weak long-range Van der Walls-type interaction and interlayer distance of 0.34 nm). CNTs can be synthesized *via* laser ablation, chemical vapour deposition or arc discharge [45].

The functionalization of CNTs (f-CNTs) depends on the attachment of inorganic or organic moieties to their tubular structure. The functionalization of CNTs allows the modification of the structural framework and the creation of supramolecular complexes [45]. By this process, it is possible to modulate their physicochemical properties, increasing their ease of dispersion, reactivity, manipulation, biocompatibility and processability. Functionalized CNTs have several remarkable mechanical, thermal, electrical, and adsorption properties, which make them optimal in manufacturing electrochemical sensors and biosensors [45]. The different approaches for the modification of CNTs can be classified in four main groups (Fig. **10**).

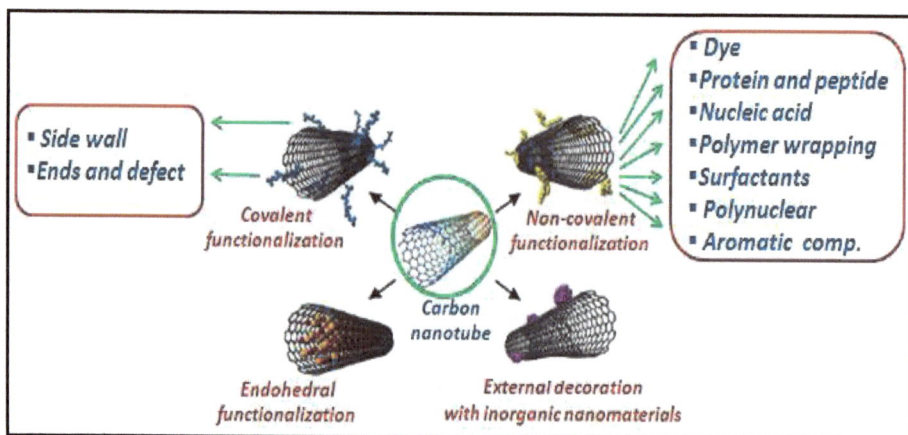

Fig. (10). The different approaches for the functionalization of CNTs.

The covalent functionalization of CNTs has two strategies which are direct sidewall and defect group functionalization. The sidewall functionalization of CNTs is based on the rehybridization of a sp^2 carbon atom into a sp^3 configuration and forms a covalent bond between the attacking species and a carbon atom of the CNT scaffold. However, defect group functionalization is relied on anchoring the desired functionalities through appropriate chemical groups introduced in intentionally created or preexisting defects on the CNT scaffold. Both approaches create a disturbance of the CNT tubular structure [45].

Noncovalent functionalization of carbon nanotubes depends on the wrapping or adsorption of different functional molecules on the tubular surface of the CNTs. It is based on π-π stacking, van der Waals or charge-transfer interactions and so it preserves the extended π-network of the carbon tubes. A wide range of compounds have been used for the noncovalent functionalization of CNTs (Fig. **10**).

Generally, pristine MWCNTs and the various types of the functionlized MWCNTs (f-MWCNTs) are extensively used in modification of carbon electrodes, especially CPE for developing electrochemical sensors for drug analysis. With this respect, covalent f-MWCNTs were prepared by a simple surface oxidization of the pristine MWCNTs using $KMnO_4$ for determination of *Domperdon* drug [7].

The f-MWCNTs were characterized using FTIR spectra, TEM, BET N_2 sorption isotherms and TGA [7]. FTIR spectra reveals the introducing surface functional groups such as –OH, >C=O and –COOH by chemical oxidation of pristine MWCNTs using $KMnO_4$. TEM images (Fig. **11**) show that the overall level of amorphous carbon and catalyst impurities was reduced and a few defects were generated in the outer most graphene sheets of the MWCNT. The end caps of the MWCNTs are opened and the diameters begin to narrow. Van der Waals interactions between different CNTs are decreased and nanotube bundles are separated into individual tubes.

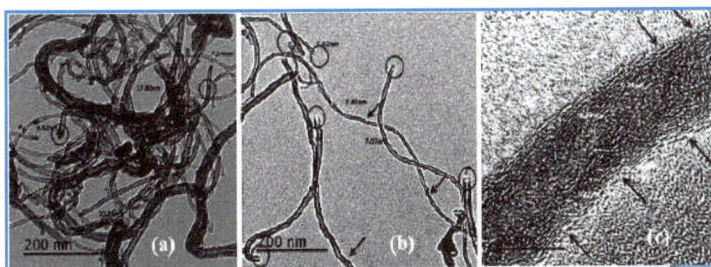

Fig. (11). TEM images of **(a)**: pristine MWCNTs; amorphous carbon and/or metallic impurities were marked by circles and **(b - c)**: f-MWCNTs opening tips were marked by circles and side wall defects were marked by arrows (© The Electrochemical Society-Permission of IOP Publishing, [DOI: 10.1149/2.1091714jes]).

The electrochemical sensor (f-MWCNTs/CPE) exhibited excellent electrocatalytic behavior for oxidation of *Domperidone* in comparison with a bare CPE (Fig. 12). The 1st quasi-reversible oxidation process was occurring on the −NH groups of the two amide moieties *via* 1e− and 1H+ for each, forming stable free radical species (II) and the 2nd irreversible oxidation step was located on the piperidine ring, which represented a typical redox system with 2e−. Neutral *Domperidone* looses an e− to form a cation radical, which on loosing 1H+ and 1e− in subsequent steps forms a quaternary Schiff base. The resulted quaternary Schiff base was rapidly hydrolyzed to the aldehydic derivative (III) and secondary amine (IV) Fig. (12).

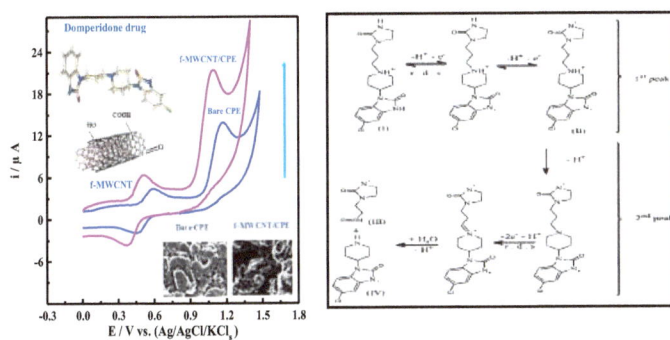

Fig. (12). Cyclic voltammograms of 1.0×10^{-5} M *Domperidone* recorded at bare CPE (violet) and 20%(w/w) f-MWCNT/CPE (pink) {Inset is SEM images of bare CPE and 20%(w/w) f-MWCNTs/CPE} and the corresponding reaction mechanism of *Domperidone* (© The Electrochemical Society-Permission of IOP Publishing, [DOI: 10.1149/2.1091714jes]).

Limit of quantitation (LOQ) value of 6.23×10^{-11} M was achieved for assay of *Domperidone* in bulk form (which was much lower than that reported (1.43×10^{-9} – 4.10×10^{-8} M) using the previous voltammetric assays using different types of the modified electrodes [7]. For intra-day-assay, the achieved SD, RSD%, R% and relative error (RE%) were in the range of (0.02–0.05), (0.52–0.91%), (98.75–99.83%) and (-1.17 – -1.25%), respectively indicating high precision and accuracy of the proposed assay procedure for *Domperidone*. The results of R±RSD% (97.25±2.02 – 98.50±1.57%) obtained due to (Lab.-to-Lab.) and even (day-to-day) using the same and different prepared electrodes were also found reproducible indicating the stability of the proposed sensor.

The determination of *Domperidone* in biological systems has been considered as useful indicator of problem related to arrhythmias, hyper-prolactinemia, sudden death and cardiac arrest. So, its detection in body fluids is of great essential in the field of clinical diagnostics. Direct assay of *Domperidone* spiked in human plasma samples was carried out successfully by the described SW-AdASV method. The average LOD (4.68×10^{-11} M) indicates the sensitivity of the developed electrode for assay of *Domperidone* in plasma samples without interferences from some foreign organic and inorganic species. The calculated mean %R (98.06–101.49%) and %RSD (1.79–2.68%) using five determinations of various concentrations of *Domperidone* elucidate insignificant differences between the spiked and the detected amounts of *Domperidone* in plasma samples. The simplicity, accuracy, precision, and sensitivity of the developed method offer the possibility to assay the drug in real plasma samples at various therapeutic dose levels for pharmacokinetic studies.

Moreover, cyclic voltammetry and molecular docking were used to determine the interactions of *Domperidone* with *ds-DNA*. The decrease in the peak current of *DNA* and the positive shift in its peak potential in a successive addition of *Domperidone* are a good indication for *drug-DNA* interaction (Fig. **13**). Molecular docking confirmed that *Domperidone* binds to *DNA* by groove binding mode which constitutes an important class in anticancer therapy (Fig. **13**). The higher binding constant (8.77×10^{4} M^{-1}) might be sufficient to interfere with *DNA* replication. Thus, *Domperidone* can be used as an anticancer therapy.

Similarly, HNO_3 acid oxidation of pristine MWCNTs [6] helps in decreases diameters, opens up tube ends, thus increases BET surface area. The carboxylic groups functionalization of MWCNTs was confirmed using FTIR spectroscopy (due to the appearance of a band at 3428 cm^{-1} for stretching of bending -OH group of carboxylic group, a band for -COOH group at 1646 cm^{-1}, a band at 1431 cm^{-1} for vibration of carbonyl and carboxylic groups in addition to that ascribed to stretching vibration C-OH at 1051 cm^{-1}) (Fig. **14A**).

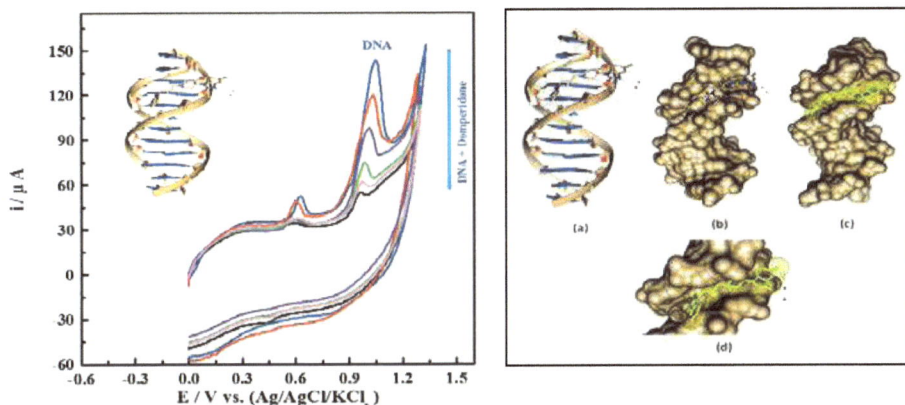

Fig. (13). Cyclic voltammograms of 5×10^{-4} M *DNA* in the absence (blue) and presence of 3×10^{-5} (red), 4×10^{-5} (violet), 5×10^{-5} (green), 6×10^{-5} (orange) and 7×10^{-5} M (black) *Domperidone* and 3D representation of *Domperidone* showing its interaction with *DNA* (1BNA): **(a)** The ribbon structure of the B-*DNA* dodecamer interacting with *Domperidone* and **(b-d)** The hydrophobicity structure of the B-*DNA* with *Domperidone* docked in the minor groove (© The Electrochemical Society-Permission of IOP Publishing, [DOI: 10.1149/2.1091714jes]).

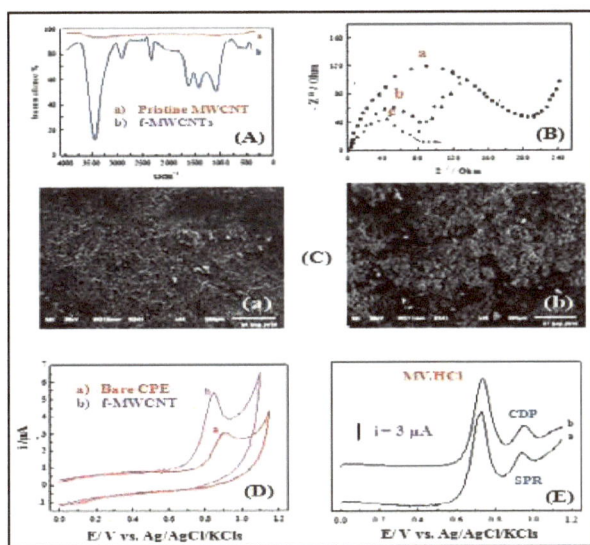

Fig. (14). A): FTIR spectra of pristine and f-MWCNTs samples, **B)** Nyquist plots of the EIS for the bare CPE (a), 5%(w/w) f-MWCNTs/CPE **(b)** and 7%(w/w) f-MWCNTs/CPE **(c)** for 1.0×10^{-6} M *MV.HCl*, **C**): SEM images of bare CPE (a) and 7%(w/w) f-MWCNTs/CPE (b) and **D):** CV of 1.0×10^{-5} M *MV.HCl* at scan rate = 300 mVs^{-1} and **E**): SW-AdAS voltammograms for 1.0×10^{-6} M of *MV.HCl* in its formulations: (a) Colona tablets® {100 mg *mebeverine* (*MV.HCl*) + 25 mg *Sulpiride* (*SPR*)} and (b) Coloverin A® tablets {135 mg *mebeverine* (*MV.HCl*) + 5 mg *chloridazepoxide* (*CDP*)}, (© The Electrochemical Society-Permission of IOP Publishing, [DOI: 10.1149/2.0941706jes]).

EIS and SEM indicate that the lowest charge transfer resistances and the heighest surface area are for 7%(w/w) f-MWCNTs/CPE, (Fig. **14B** and **14C**), respectively which was applied in determination of *Mebeverine Hydrochloride (MV.HCl)* drug using SWAdASV method. This electrode exhibited higher electrocatalytic activities towards oxidation of *MV.HCl* compared to bare CPE (Fig. **14D**). The oxidation process of *MV.HCl* was occurring on the methoxybenzene groups *via* 2e⁻ according to (Fig. **15**). 7%(w/w) f-MWCNTs/CPE was applied for trace determination of *MV.HCl* in different pharmaceutical preparation samples; "Coloverin A®", "Colona®" and "Colofac®" tablets. Excellent mean (R%) and (RSD%) of (99.53±1.02%), (99.25±0.72%) and (99.10±0.99%), were obtained for analysis of *MV.HCl* in "Coloverin A", "Colona" and "Colofac" tablets, respectively, indicating that there were no interferences from excipients and co-formulated *Sulpiride (SPR)* or *Chloridazepoxide (CDP)* drugs (Fig. **14E**). The average LOD value of *MV.HCl* spiked in six human serum samples of three healthy volunteers was 2.0×10^{-10} M and a wide concentration range of (8.0×10^{-10} to 2.0×10^{-8} M) was achieved. Satisfactory mean %R (99.2 to 101.00) and RSD% (0.25 to 1.12) and relative error% (−0.80 to 1.00) were also achieved.

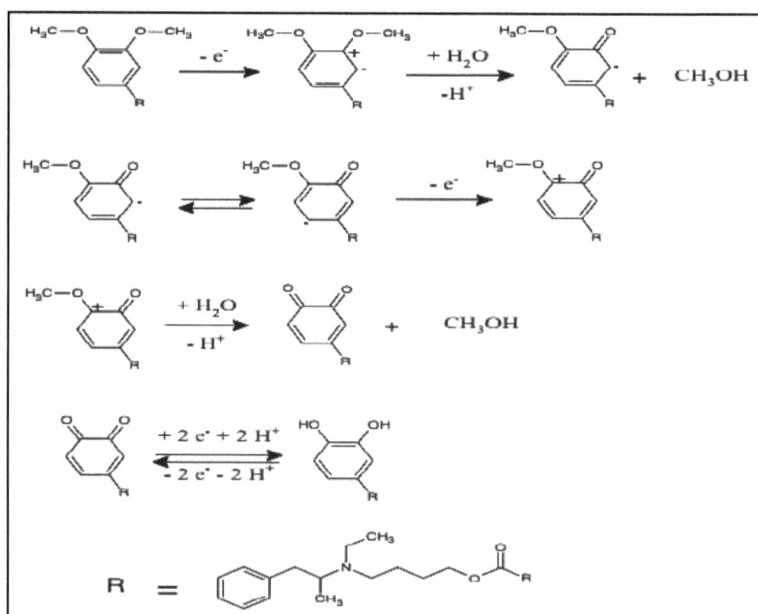

Fig. (15). Electrode reaction mechanism of *Mebeverine hydrochloride,* (© The Electrochemical Society-Permission of IOP Publishing, [DOI: 10.1149/2.0941706jes]).

MWCNTs-TiO$_2$NPs/GCE was prepared [46]. The enzyme horseradish peroxidase (HRP) was then immobilized to enhance the sensing ability of GCE. The proposed (MWCNTs-TiO$_2$NPs-HRP)/GCE biosensor was used for the determination of *Isoniazid* in various pharmaceutical samples using DPV. The increment of anodic peak currents for the enzyme-induced sensor was almost 8-fold greater than that of a bare GCE. The enzyme horseradish peroxidase (HRP) shows greater affinity for coupling with the nanocomposite for electrochemical transduction due to the presence of amino groups in the HRP enzyme. The DPV technique exhibited good LOD value of 0.034 µM. The stability study was carried out for 40 days with the same electrode where the electrochemical signal implies only 4.39% of the electrochemical signal decreased. This result indicates that the fabricated electrode showed good repeatability and long-term stability. Moreover, the real sample (Commercially-available *INZ* tablets /100 mg) analysis gave good RSD% values (1.69 - 1.98%) with an excellent R% (98.9 - 99.2%). The devolped sensor may have scope for use in the pharmaceutical industries in the near future.

A mixture of bimetallic Au–Pt NPs was electrodeposited on MWCNT/GCE to construct a sensitive voltammetric sensor for *Cefotaxime* (*CFX*) drug [47]. A remarkable enhancement in the peak current was observed by a factor of 3.53, 13.07, 20.00 at the surface of Au–PtNPs/GCE, MWCNTs/GCE and Au–PtNPs/MWCNTs/GCE, respectively, compared to bare CPE. Using linear sweep voltammetry (LSV), LOD of (1.0 nM) was achieved at Au–PtNPs/MWCNTs/GCE. To study the reproducibility of the electrode preparation procedure, 5 electrodes were prepared. The average RSD% for the electrodes' peak currents of 3 determinations on each electrode was 3.96%. An amount of 483.87 mg with a good accuracy of 96.77% and RSD of 3.86% was found for the analysis of drug pharmaceutical sample (500 mg *CFX* per ampoule). R% evaluation of 96.28% *CFX* was found by spiking of its standard solutions in the range of 0.01– 4.00 µM into the diluted plasma samples. This sensor was thus validated for *CFX* detection in pharmaceutical and clinical preparations.

A GCE was modified with a TiO$_2$-Au NPs hybrid integrated with MWCNTs in a dihexadecylphosphate film (TiO$_2$-Au NP-MWCNT-DHP/GCE) and applied to amperometric determination of *ascorbic acid* at 0.4 V [48]. A statistical linear concentration range for the acid from 5.0 to 51 µM, with a LOD of 1.2 µM was obtained. It was applied to its determination in pharmaceutical (500 mg *ascorbic acid* per tablet) and fruit juice samples. Excellent R% values ranging from (97.70 – 104.00%) and (96.30 to 105.00%), respectively, for the pharmaceutical and fruit juice indicate that this method does not suffer from any significant effects of matrix interference.

A highly sensitive method was developed for simultaneous determination of *warfarin* and *mycophenolic acid* using CPE modified by β-cyclodextrin/mult--walled carbon nanotubes/cobalt oxide nanoparticles (β-CD/MWCNTs/CoONPs/CPE) [49]. The oxidation peaks of *warfarin* and *mycophenolic acid* drugs at 0.65

V and 0.86 V, respectively, were separated enough using the constructed electrode. CV, DPV and EIS were utilized for study the electrochemical response of the fabricated electrode. The stripping voltammetric responses were linear in the concentration ranges (0.05-150 μM) and (0.5-200 μM) and LOD values were 0.02 and 0.03 μM for *warfarin* and *mycophenolic acid*, respectively. This electrode was applied for simultaneous determination of these drugs in urine and human serum samples.

An electrochemical sensor based on carboxylated-MWCNTs, polythionine and Pt NPs nanocomposite modified GCE (cMWCNT@pTh@@Pt/GCE) was described [50] for simultaneous determination of *Myricetin* and *Rutin* by DPV. *Myricetin* and *Rutin* oxidation peaks appeared at 0.16 and 0.34 V *vs.* SCE, respectively. Based on a synergistic effect among cMWCNT, pTh and Pt, the modified GCE has wide linear response in the range of 0.01-15 μM *Myricetin* and *Rutin*. LOD of 3 nM and 1.7 nM were achieved for *Myricetin* and *Rutin*, respectively. This sensor was also applied for simultaneous determination of *myricetin and rutin* in spiked juice samples, and satisfactory results of (98.55±2.00%) and (98.87±1.55%) were obtained, respectively.

Epirubicin antibiotic was detected at GCE modified with Ag decorated MWCNTs composite (Ag-MWCNTs/GCE) [51]. SWV detects *Epirubicin* with a LOD of $1.0×10^{-9}$ M. Recently, a nanocomposite from nitrogen decorated reduced graphene oxide and single-walled carbon nanotubes is loaded with Pt NPs and is then used to modify a GCE (N-rGO-SWCNTs-Pt/GCE) [52]. This electrode achieved LOD of $5.7×10^{-9}$ M for *Daunorubicin* drug.

Graphene in Sensors

Graphene (GR) is comprised of sp^2-hybridized carbon atoms packed in a hexagonal network structure to form a flat, 2-D sheet [53]. This structure is responsible for GR and CNTs unique properties such as their high thermal conductivity and unique electronic properties. This structure is also like those found in bulk graphite, but the layer thickness should be below (10 – 12) layers and the inter-planar distance should be around 0.34 nm [53]. GR is an ideal material for electroanalytical applications [2, 8, 9, 11, 54] because of its outstanding properties which involve high electrical conductivity, fast adsorption kinetics and large surface area (2630 m²/g for single-layer graphene, which is double that of SWCNTs). Further, its subtle electronic properties suggest that it has the ability to promote electron transfer when used as a modifier of working electrode [2, 8, 9, 11, 53, 55]. This behavior may arise from delocalization of π orbital through the structure. In particular, the reported carrier mobility in graphene can reach up to 10^6 cm²/Vs [56]. This can enable devices with higher operating frequencies, and ultimately superior performance. Consequently, GR provides great domain in the analysis field.

The GR synthesis can be performed by bottom-up and top down strategies. The direct fabrication of GR on a large scale using present various methods such as chemical-vapor deposition [57], graphite micro-mechanical exfoliation [58], ultrasound [59], and graphite liquid-phase exfoliation [60] are still unavailing. The main drawbacks of these methods are the difficulties in further proceeding of the carbon nanomaterial and the low amount of material which can be synthesized.

Functionalizing graphene (chemically modified graphene) with other species can have a signicant impact on its properties [61]. The carbon atoms in pristine GR adopt a sp^2-hybridized scheme, with three σ bonds in plane and a conjugated π orbital out of plane. Upon covalently binding molecules or elements to the GR surface, the three bonds can be converted to a tetragonal sp^3 configuration and form so-called chair/herringbone and boat configurations. Edge passivation is believed to have a significant impact on the electronic properties of graphene, particularly for graphene nanostructures.

Pristine GR was utilized for producing GR-based sensors, capable of nano determination of a wide range of drugs [2, 8]. An effective GR-based sensor was constructed for determination of *Dapoxetine* drug (which has medical features for the treatment of premature ejaculation in men) in real samples such as in "Joypox® tablets" and in spiked human plasma [8]. Pristine GR was characterized using XRD, FTIR, SEM and TEM techniques (Fig. **16**). The FTIR spectrum of pristine GR shows some residual oxygen containing groups such as OH groups remained during its production process. Its XRD pattern showed a peak at 2θ = 25.5, referring to (002) reflection, (Fig. **16A**). This peak is intense and sharp confirming the crystalline plane of GR. The number of GR layers and the average interlayer distance were found to be 11.00 and 0.35 nm, respectively. SEM image is featured with a layer-by- thin layer structure, (Fig. **16B**). TEM image indicates a wide area of transparent GR layers which is a reflection of a few layers of GR. Further, the obvious wrinkles and folds revealed the flexibility of graphene nanosheets, (Fig. **16C**). The inter lattice spacing/lattice fringe measured from the fringe pattern, as indicated from the high resolution TEM (HR-TEM) image, was about 0.22 nm, (Fig. **16D**).

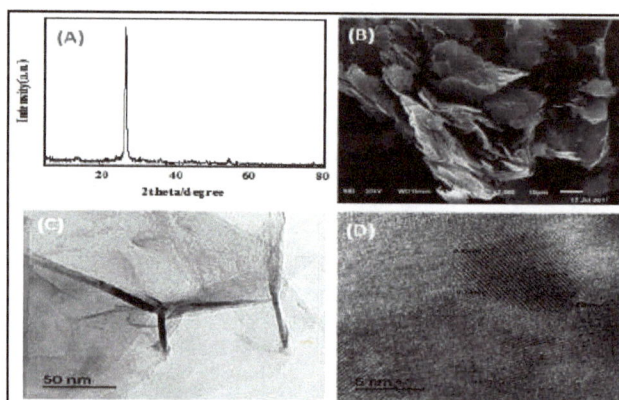

Fig. (16). A) XRD spectra, **B)** SEM image, **C)** TEM image and **D)** HRTEM image correspond to atomic spacing of the pristine graphene samples (© The Electrochemical Society-Permission of IOP Publishing, [DOI: 10.1149/2.0491803jes]).

Three times enhancement in the oxidation response of *Dapoxetine* at GR/CPE surface compared to that at bare CPE was observed. This can be due to the unique sp^2-hybridized single-atom-layer structure of GR. π-π interactions and H-bond may also be responsible for the excellent adsorption of *Dapoxetine* on GR sheets (Fig. **17A**) [8]. The number of electrons exchanged per *Dapoxetine* molecule was found to be 2e⁻. Tertiary amine group can be oxidized to give the corresponding ketone and amine derivatives as in suggested reaction mechanism shown in Fig. **(17B)**.

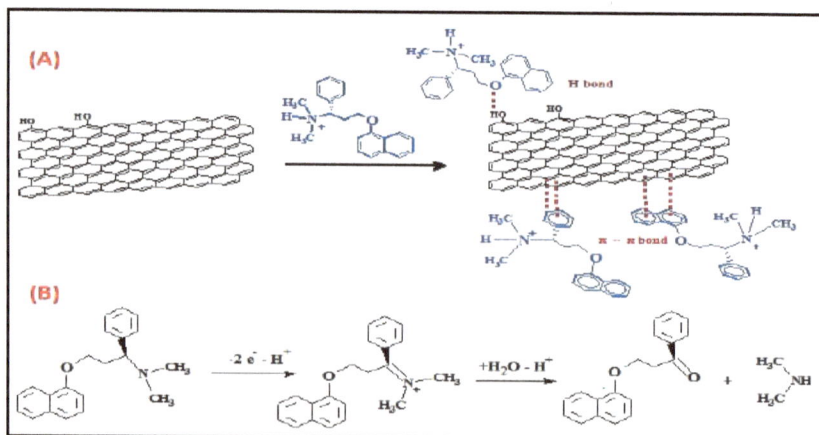

Fig. (17). A) The schematic of π-π interaction and hydrogen-bonding between *Dapoxetine* and graphene and **B)** the suggested electrode reaction mechanism of *Dapoxetine* (© The Electrochemical Society-Permission of IOP Publishing, [DOI: 10.1149/2.0491803jes]).

Direct determination of *Dapoxetine* drug in human plasma as well as pharmaceutical samples was possible without interference from endogenous human plasma constituents or excipients which are present in its Jaypox tablets [8]. A wide concentration range of $(8.0 \times 10^{-9}$ to 1.1×10^{-7} M$)$ with low LOQ value of 2.25×10^{-9} M was achieved in human plasma. Mean % R (98.22 – 99.10) and RSD % (1.05 – 2.68), reveal that the proposed method is highly precise and accurate for determination of *Dapoxetine* in spiked human plasma samples. No siginificant intereference was observed in presence of 120-fold concentration of (ibuprofen, paracetamol, aspirin, sucrose, glucose and starch), 100-fold concentration of (urea, uric and ascorbic acids), and 200-fold concentration of $(Na^+, K^+, Ca^{2+}, Mg^{2+}, Zn^{2+}, Co^{2+}, Cu^{2+}$ and Fe^{3+} at tolerance limit corresponding to a relative error of 5% in the analytical signal of *Dapoxetine*. GR/CPE response to *Dapoxetine* lost by (0.40 – 0.55%) only of its original response after storage for 60 days indicates long-term stability of the electrode. The LOQ value is well below the reported concentration level encountered in human plasma specimens $(3.27 \times 10^{-9}$ M$)$ after therapeutic dosing. So, the developed GR sensor can be applied for quality control analysis and assay of *Dapoxetine* in clinical samples.

An economical GR/CPE in situ modified with various surfactants was used for trace voltammetrical determination of *Itraconazole* drug (an inhibitor of the cytochrome P-450 14 α-demethylase enzyme) in formulation and human plasma samples [11]. First, there is an obvious enhanced response of $K_4[Fe(CN)_6]$ (as a redox probe) at GR/CPE compared to that at bare CPE (Fig. **18A**) which can rely on the high surface area, the high conductivity and the favorable electronic properties of GR. *Itraconazole* drug exhibits a main irreversible anodic peak at potential of 920 mV at bare CPE and 890 mV at GR/CPE, (Fig. **18B**). This peak is also more developed at GR/CPE indicating the strong electrocatalytic effect of the GR towards *Itraconazole* oxidation. Further, anionic (sodium dodecyl sulfate, SDS) exhibited better improvement in the electrical properties of the electrode-solution interface compared to cationic (cetyltrimethyl ammonium bromide, CTAB) and non-ionic (Triton X-100), (Fig. **18B**).

The adsorbed-SDS at GR/CPE interacts with itraconazole molecules through hydrophobic and electrostatic attraction which encourages the electron transmission between the drug and electrode surface, as shown in graphic scheme (Fig. **19**). The oxidation of *Itraconazole* follows ECE reaction pathway [11]. It involves loosing of $1e^-$ and $1H^+$ first to form a radical cation which stabilizes through deprotonation (chemical reaction step) and another $1e^-$ oxidation to form a quaternary Schiff base, (Fig. **20A**).

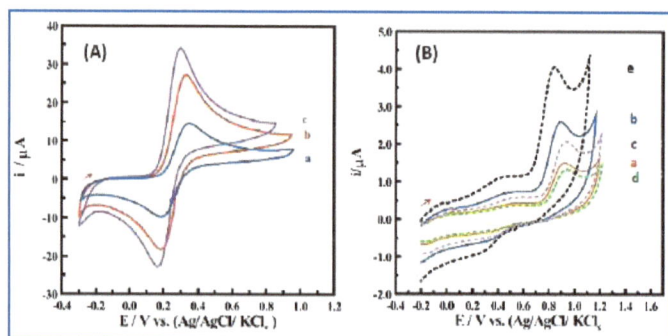

Fig. (18). A) CV characterizations of bare CPE (a), GR/CPE (b) and SDS-GR/CPE (c) using the redox probe $K_4[Fe(CN)_6]$ at scan rate of 100 mV/s. **B)** CV of 2×10^{-5} M *Itraconazole* recorded at scan rate $v = 300$ mV/s in the universal buffer of pH 2 in absence of the surfactant at: (a) bare CPE and (b) 3%(w/w) GR/CPE and at 3%(w/w) GR/CPE in presence of 0.5 mM of: (c) Triton X-100, (d) CTAB and (e) SDS (© The Electrochemical Society-Permission of IOP Publishing, [DOI: 10.1149/1945-7111/abbdd5]).

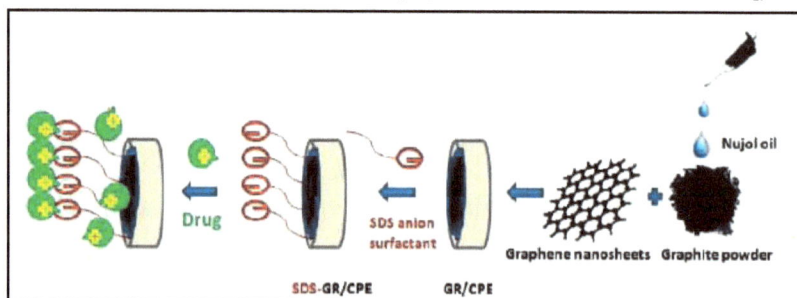

Fig. (19). SDS anion surfactant-functionalized graphene modified carbon paste electrode surface for determination of *Itraconazole* drug (© The Electrochemical Society-Permission of IOP Publishing, [DOI: 10.1149/1945-7111/abbdd5]).

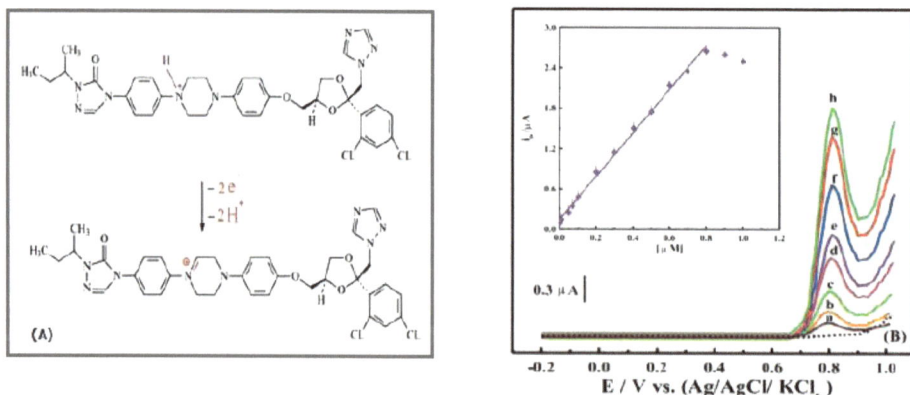

Fig. (20). A) Scheme represents the proposed electrode reaction mechanism of *Itraconazole* at carbon past electrode and **B)** SW-AdAS voltammograms at SDS-GR/CPE for various concentrations of *Itraconazole* in spiked human plasma: (a) 5×10^{-9}, (b) 5×10^{-8}, (c) 1×10^{-7}, (d) 2×10^{-7}, (e) 3×10^{-7}, (f) 5×10^{-7}, (g) 7×10^{-7} and (h) 8×10^{-7} M in the universal buffer of pH 2, $E_{acc} = -0.2$ V, $t_{acc} = 10$ s (© The Electrochemical Society-Permission of IOP Publishing, [DOI: 10.1149/1945-7111/abbdd5]).

SDS-GR/CPE has also been favorably applied to analyze *Itraconazole* in its pharmaceutical product "*Itrafungex capsules; 100 mg/Itraconazole*" and in human plasma samples (*e.g.*, Fig. **20B**) [11]. Higher %R (98.94 – 99.82) and small RSD% (1.76 – 2.26) were obtained indicating the accuracy and precision of the proposed method for assay the *Itraconazole* in its formulation sample. It presents adequately LOD value (1.36×10^{-9} M) for detection of *Itraconazole* in plasma. This value was comparable to the LOD values of plasma concentration-time pharmacokinetic curve. The obtained values of %R (98.52 – 102.13), RSD% (1.58 – 2.48) and RE% (-0.23 – 2.13) of various concentrations of *Itraconazole* spiked in human plasma samples confirmed the accuracy and precision of the proposed method. Besides, SDS-GR/CPE avoids interference from important common substances in biological and pharmaceutical samples as well as some co-administrated drugs. Signal changes still below 5% in presence of 1400-fold of some metal ions (*e.g.*, Fe^{2+}, Cd^{2+}, Mg^{2+}, Zn^{2+}, Pb^{2+} and Ca^{2+}), 1000 fold of vitamins A and E, ketoprofen, Ketorolac, and ibuprofen, 800-fold of ascorbic acid and 600-fold of uric acid. The achieved %R and %RSD were (97.06%–98.00%) and (0.60%–1.02%) after 2 weeks storage of the proposed electrode. The fabricated sensor has desirable stability and reproducibility that can be used in routine quality control and pharmacokinetic study.

Reduced Graphene Oxide

Graphite oxide is a precursor for the inexpensive mass production of graphene oxide (GO) and reduced graphene oxide (rGO). Graphite oxide has layered structure similar to that of graphite, but its plane of C atoms is decorated by several oxygen-containing functional groups [9]. If the exfoliated sheets of graphite oxide contain only one or even few layers of carbon atoms like graphene, these sheets are donated graphene oxide (GO), (Fig. **21**). GO nanosheets can be produced by treating graphite with strong acids using Hummers' method [62].

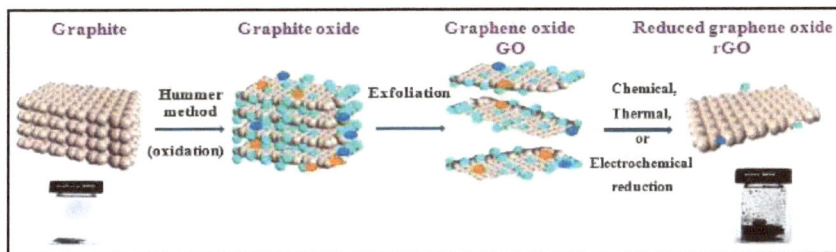

Fig. (21). Graphical scheme for synthesis of graphene oxide and reduced graphene oxide from precursor graphite oxide.

Graphene oxide (GO) can be partly reduced chemically, thermally or electrochemically [63 - 65] to GR-like sheets (more precisely named reduced graphene oxide (rGO)) (Fig. **21**), which has a partially restored sp^2 lattice and also containing oxygen-bearing groups [9, 55]. GO films reduced *via* chemical methods typically use highly toxic substances [63]. It is well also known that microwave irradiation method elicited as a green alternative source for speedy heating, increasing reaction rate, decreasing reaction time and offering good yield as compared to the classical synthesis methods.

With this respect, a direct "*green microwave irradiation*" method was applied for only 30 s for synthesis of rGO nanosheets by simultaneous exfoliation and reduction of as-prepared graphite oxide in the absence of any hazard reducing, chemical activating or intercalation agents [9]. Compared to graphite which appears as a large thick dark flakes and its edge showed several layers (Fig. **22A**), TEM image of rGO exhibited a transparent few-layered structure with a typical wrinkled and crumpled GR structure, (Fig. **22B**). The silk-like parts and the restacked parts can also be seen in Fig. (**22B**). A number of graphene layers of 3.0 and interlayer spacing of 0.38 nm were calculated from XRD pattern of rGO, [Fig. (**22C**); curve b].

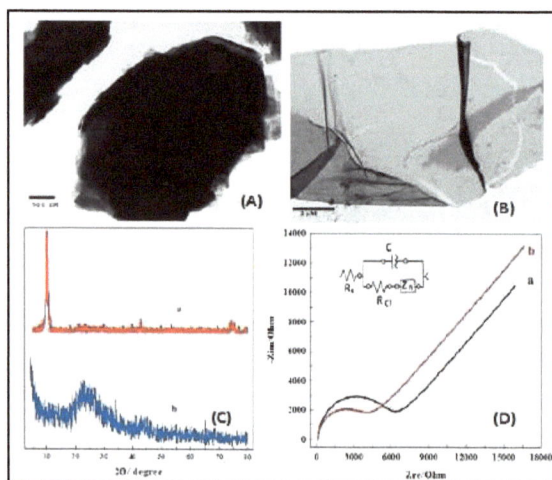

Fig. (22). A) TEM image of graphite, **B)** TEM image of reduced graphene oxide, **C)** XRD for a) graphite oxide and b) reduced graphene oxide and **D)** Nyquist plots of the EIS for the (a) bare CPE and (b) 0.4% (w/w) GR/CPE in 0.1 M KCl containing 1×10^{-3} M $K_4Fe(CN)_6$. Inset: is the corresponding equivalent circuit (© The Electrochemical Society-Permission of IOP Publishing, [DOI: 10.1149/2.0391814jes]).

Reduced graphene oxide modified carbon pste electrode (rGO/CPE) as an electrochemical sensor has been employed for detection of a potent antibacterial and antifungal agent; *Chloroxylenol* in its formulations as well as in tap and underground waters. A charge transfer resistance (R_{CT}) of 4000 Ω was achieved for 0.4% (w/w) rGO/CPE while it is found to be 7560 Ω for bare CPE revealing the fast electron transfer at rGO/CPE, (Fig. **22D**). Morphology of 0.4% (w/w) rGO/CPE also reflects its highest surface area, (Fig. **23**). This sensor exhibited LOD of 1.37×10^{-9} M in bulk form. R% values (98.00 − 102.29) and RE% (-2.00− 2.29) indicate the accuracy of the method. The estimated RSD% values (0.53–1.08%) were highly precise. The average (R±RSD%) value of *Chloroxylenol* in real samples *"Rosa Clean Lotion"* is (98.78±1.02%) and in *"Dettol Antiseptic Liquid"* is (102.50 ± 1.11%) using 0.4%(w/w) GR/CPE. The RSD values were less than 2%, giving good evidence for the validity of the method.

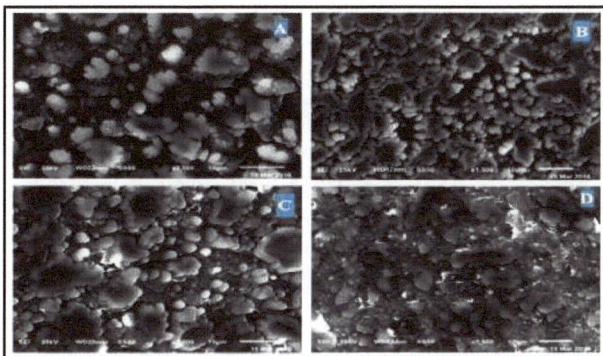

Fig. (23). SEM images for CPE with various percentages of GR: **A)** bare CPE, **B)** 0.4%, **C)** 1% and **D)** 2.5% (w/w) GR/CPE (© The Electrochemical Society-Permission of IOP Publishing, DOI: 10.1149/2.0391814jes]).

Metal-oxide-doped Reduced Graphene Oxide Composites

The agglomeration of nano-sized carbon is one of the primary obstacles that limit the applications of GR or rGO due to the effects of π-π adhesion and van der Waals interactions (which may cause restacking of graphene sheets to form graphite). Surface modification can be introduced to reduce agglomeration and allow the implementation of the inherent properties of graphene [45]. Introduction of metal oxide NPs (MONPs) onto GO or rGO sheets prevents them from undergoing restacking and provides an innovative way for developing advanced composite materials that are useful in a myriad of applications. Recently, various metal-oxide-doped rGO composites have been widely used in electrochemical devices and electrocatalysis.

Zirconium oxide nanoparticles (ZrO_2 NPs) show excellent properties including thermal stability, nontoxicity, wide band gap and good electrical and surface properties [66]. Therefore, dedicated extensive efforts have been giving to synthesize ZrO_2 decorated on rGO sheets. For example, the detection of an anticancer drug (*Regorafenib*) was demonstrated based on a zirconia-nanoparticle-decorated reduced graphene oxide composite (ZrO_2/rGO) [66]. Reduction of the GO supports of the Zr^{2+} ions with hydrazine hydrate exhibited an excellent electrocatalytic response of the nanostructure ZrO_2/rGO-based electrochemical sensor. The (ZrO_2/rGO)/GCE showed a linear response in the dynamic range of (11−343 nM) with lower LOD of 17 nM. It was also used for the determination of *Regorafenib* in both pharmaceutical formulations and serum samples with satisfactory results. (Pt/ZrO_2-rGO)/GCE was used for simultaneous determination of *Catechol* and *Hydroquinone* [67].

Magnetic Fe_3O_4 NPs has been considered one of the most promising electrode materials because of its elegant properties involving excellent magnetic, electrical and catalytic properties, low toxicity, great biocompatibility, extremely small size and high surface-area-to-volume ratio [68]. Fe_3O_4 has a broad potential electrochemical window up to 1.2 V, which is conducive to improve the energy density. Fe_3O_4 with oxygen forming face centred cubic has a cubic inverse spinal structure and in the interstitial tetrahedral sites and octahedral sites are occupied by iron (Fe) cations. At the room temperature, Fe^{+2} and Fe^{+3} ions flip between themselves in the octahedral sites giving rise to a class called half-metallic materials [69]. Fe_3O_4 NPs can therefore effective acceleration agents for electron transfer between electrode and detection molecules.

Recently, growing efforts have also been pointed toward tailoring the properties of magnetite–graphene [Fe_3O_4-GR] based nanocomposites for promoting more efficient electrochemical sensors [70]. The large surface area of GR as well as uniformly distributed active sites of nanoparticles makes this structure appropriate for sensing applications. For example, Fe_3O_4/rGO composite exhibited high specific surface area, and its morphology consists of very fine spherical particles in nanoscales [71]. So, it was used as an electrode modifier for the determination of *Paracetamol* by DPV. A LOD value (0.72×10^{-6} M) was achieved. This sensor was applied to the real samples with satisfactory results.

Recently, a very simple and economic viable setup for synthesis of Fe_3O_4 NPs and [Fe_3O_4-GR] plateform was used [2]. Fe_3O_4 NPs synthesis is based on a reduction-precipitation method utilizing $FeCl_3$ as a starting-material, which was partially reduced to $FeCl_2$ by Na_2SO_3 before alkalinizing with ammonia. Then, Fe_3O_4 NPs supported on GR was simply synthesized using *Ex situ* method in 0.5 M DMF (pH 6.7) which based on electrostatic interaction as the main driving force for self-assembling Fe_3O_4 NPs on GR sheets, (Fig. **24**).

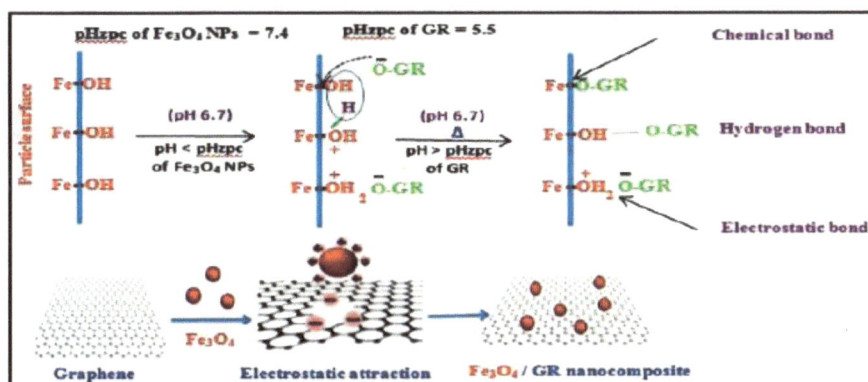

Fig. (24). The possible pathways of interaction between Fe_3O_4 NPs and GR (© The Electrochemical Society-Permission of IOP Publishing, [DOI: 10.1149/1945-7111/ab8366]).

Fe_3O_4 NPs and [Fe_3O_4-GR] nanocomposites were characterized using XRD, TEM, SEM, FTIR, vibrating sample magnetometer (VSM) and Brunauer-Emmett–Teller (BET) surface area measurements (Figs. **25** and **26**). XRD and FTIR confirm the formation of the Fe_3O_4-GR composite (Figs. **25A** and **25B**). The average particles size was found to be 16 nm. The values of saturation magnetization Fe_3O_4 NPs and [Fe_3O_4-GR] were found to be 61.50 and 30 emu/g, respectively, (Fig. **25C**). TEM images reveal that the GR nanosheets are decorated by Fe_3O_4 NPs, (Fig. **26**).

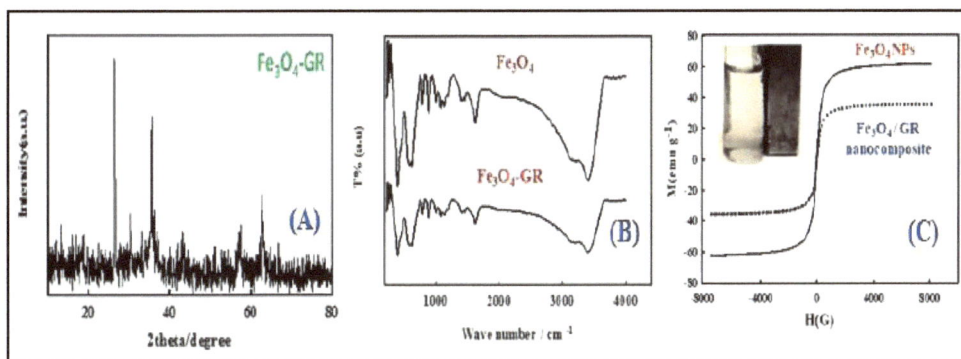

Fig. (25). A) XRD of synthesized [Fe_3O_4-GR] nanocomposite, **B)** FTIR spectra of synthesized Fe_3O_4 NPs and [Fe_3O_4-GR] nanocomposite and **C)** Magnetic hysteresis loops of Fe_3O_4 NPs and Fe_3O_4-GR composite (© The Electrochemical Society-Permission of IOP Publishing, [DOI: 10.1149/1945-7111/ab8366]).

Fig. (26). TEM images: **(a)** and **(b)** synthesized Fe_3O_4 NPs at different magnification, **(c)** graphene and **(d)** synthesized Fe $_3O_4$-GR nanocomposite (© The Electrochemical Society-Permission of IOP Publishing, [DOI: 10.1149/1945-7111/ab8366]).

Electrochemical oxidation of *Olanzapine* drug at Fe_3O_4-GR/CPE takes places *via* $1e^-$ transfer. The change in current with the scan rate is consistent with E_rC_{irr} mechanism (*i.e.*, the anodic charge transfer is followed by a homogenous irreversible chemical reaction), as shown in Fig. (**27**). It was suggested that this chemical reaction may be attributed to radical–radical coupling reaction (dimer formation) and the whole mechanism of oxidation of *Olanzapine* was proposed as in Fig. (**28**) [2].

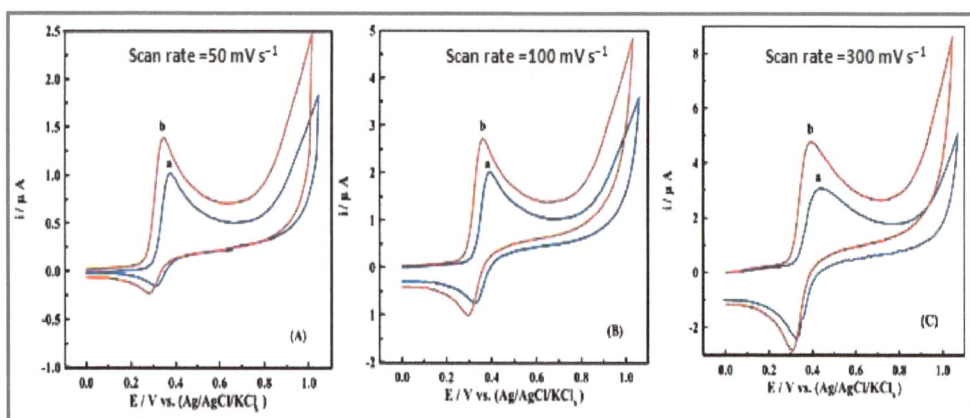

Fig. (27). Cyclic voltammograms of 1×10^{-4} M *Olanzapine* recored in B-R buffer of pH = 5 at different scan rates for bare CPE (curve a) and 5%(w/w) Fe_3O_4-GR /CPE (curve b), (© The Electrochemical Society-Permission of IOP Publishing, [DOI: 10.1149/1945-7111/ab8366]).

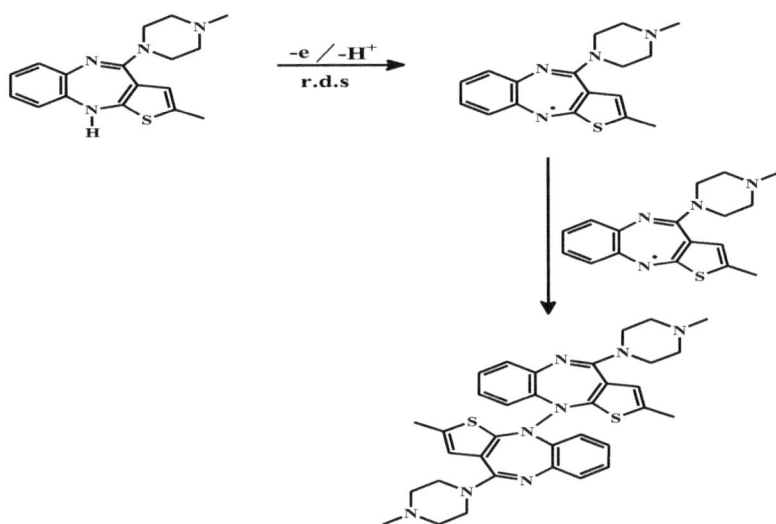

Fig. (28). Mechanism for the oxidation of *Olanzapine* at CPE (© The Electrochemical Society-Permission of IOP Publishing, [DOI: 10.1149/1945-7111/ab8366]).

Fe_3O_4-GR/CPE was also applied for the determination of *Olanzapine* drug in its "*Olazine®* *tablets*" and in human plasma [2]. Higher %R (98.97– 100.17) and small %RSD (0.77 - 1.05) were obtained revealing the accuracy and precision of the proposed method for assay the *Olanzapine* in real sample. The LOD of *Olanzapine* in spiked human plasma sample was found to be 7.29×10^{-11} M. *Olanzapine* can detected in plasma in presence of 1000–1200-fold of Na^+, K^+, Ca^{2+}, Mg^{2+}, Zn^{2+}, Cu^{2+}, Fe^{3+} metal ions, 950–1050-fold of vitamins C and E, paracatmol, aspirin, and ibuprofen administrated drugs and 1100–1300-fold of some excipients that are ordinarily exist in its formulation (*e.g.*, glucose, sucrose, starch, gelatin, lactose). This assay in biological fluids such as human plasma samples has been considered as a useful indicator of the therapeutic dose level for pharmacokinetic studies as well as therapeutic drug monitoring

Heteroatom-doped Graphene as Sensing Material

Graphene derivatives doped with heteroatoms are highly promising materials for applications in fuel cells, energy storage, electrocatalysis, sensing, *etc*. Chemical doping is usually of two types [72]. The first one, known as surface transfer doping, is the adsorption of foreign agents onto the surface of GR (which do not cause sp^3 defects in the GR cross-section). The second type, known as substitutional doping, is the disruption of the sp^2 network by foreign agents (which create sp^3 defect regions through covalent bonding with GR). The chemical doping of heteroatoms such as nitrogen (N), boron (B), phosphorus (P), silicon (Si), sulfur (S), fluorine (F), chlorine (Cl), bromine (Br) and iodine (I) into GR will cause structural and electronic alterations. This lead to modification in the properties of GR (such as Fermi level, bandgap, localized electronic state, charge transport, spin density, optical characteristics, thermal stability and magnetic properties). The chemically active sites induced by doped heteroatoms are favourable for the adsorption, anchoring of functional moieties or molecules, and accelerating the charge transfer between analyte and electrode [73].

Recently, a novel electrochemical sensor was designed for the detection of of *Flunitrazepam.* It based on electropolymerized b-cyclodextrin (EbCD)/boron-doped reduced graphene oxide (B–rGO) composite on GCE. DPV confirmed two linear ranges of *Flunitrazepam* between (2.0 nM - 0.5 μM) and (0.5 μM - 20.0 μM) with LOD of 0.6 nM [74].

Electrochemical sensors based on nitrogen-doped graphene sheets (NGS) were used as anticancer drug sensors. NGS doped in 1-methyl-3-octylimidazolium chloride (MOICl) was used for simultaneous determination of anticancer agent's *Doxorubicin* and *Topotecan.* The LOD of NGS/MOICl/CPE for them was found to be as 3.1 nM and 0.27 μM, respectively [75].

An effective electrochemical sensor based on chlorine-doped reduced GO (Cl–rGO) was established [76] for the detection of *Chloramphenicol* (*CAP*), a veterinary drug in milk, water, calf plasma and pharmaceutical samples. The sensor showed a linear relation between current intensity and *CAP* concentration over a range of (2 - 35 μM), with LOD of 1 μM.

Phosphorus-doped graphene-based sensor was used for the detection of *Acetaminophen* in pharmaceuticals as real samples [77]. An extraordinary electrocatalytic activity was shown by a P– rGO-coated GCE. Low LOD (0.36 μM) and wide linear range (1.5 - 120 μM) for *Acetaminophen* were exhibited by this sensor.

Hybrid Material Based on Mesoporous Silica and Graphene

Presently, silica, including mesoporous silica (SiO$_2$) is vastly used as a substrate for the development of electrochemical sensors. This can be attributed to its characteristic high surface area, chemical and thermal resistance, good chemical stability, easy modification, low cytotoxicity, and the presence of reactive silanol groups (Si–OH) and pores with diameters of 2 to 50 nm. Therefore, the mass transfer process is favoured by its mesoporous structure, providing electrochemical sensors with high sensitivity and low detection limits [78]. Hybrid materials based on the mesoporous silica / graphene combination

(SiO$_2$/GR) have attracted the attention because of the possibility of integrating the main features of silica and carbon-based materials in a single material, (Fig. **29**).

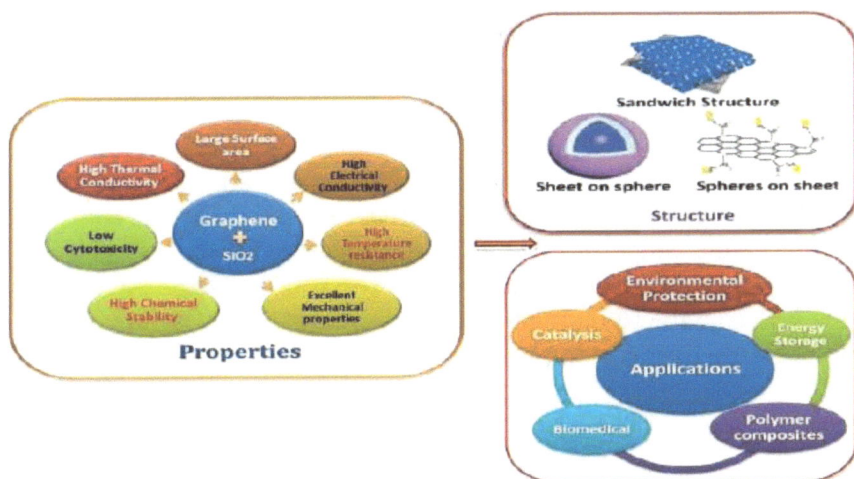

Fig. (29). The properties, structures and the applications of silica-graphene composite.

SiO$_2$/GR nanocomposites could be prepared *via* the noncovalent bonding between GO and silica. GO contains (-COOH), (-OH) and epoxy (-O(C)O-) on the surface, thus making it electronegative and allowing its combination with positively charged amino-modified silica into composites *via* electrostatic interaction [79]. Further, several -OH groups on the SiO$_2$ surface can be easily reacted with the functional groups on the GO surface. The SiO$_2$/GR composites can also be fabricated *via* covalent bonding between GO and silica by removing a molecule of water to form amide bond (-CO-NH-), silicone ester bond (-COOSi-), and carboxysilicon (-C-O-Si-). The morphology of the composites indicates that the fabrication strategy contains three categories [78]: Growing silica films on GR sheets in situ to form sandwich structures, grafting silica spheres on GR sheets and coating/ wrapping GR sheets on silica sphere, (Fig. **29**). Designing different SiO$_2$/GR composite materials is vital for its electrochemical application.

For example, a hybrid material of (SiO$_2$/GO) and decorated with AgNPs of a size of > 20 nm, was prepared and characterized [80]. A GCE modified with (AgNPs/SiO$_2$/GO) was used as an electrochemical sensor for the simultaneous determination of *Epinephrine* and *Dopamine* using SWV method. Well-separated reduction peaks were observed with no significant interference from uric and ascorbic acid. (AgNP/SiO$_2$/GO)/GCE is highly sensitive for the simultaneous determination of *Epinephrine* and *Dopamine*, with LODs being 0.26 and 0.27 μM, respectively. It is also used to detect them in human urine samples.

Further, the synergistic effect of GO and SiO_2 NPs has been used to modify GCE for the determination of *Gallic acid* [81]. DPV indicated that the nano-GO-SiO$_2$-GCE was sensitive in a concentration range of 6.25×10^{-6}-1.0×10^{-3} M with a LOD of 2.09×10^{-6} M. It was also successfully utilized for the determination of anti-carcinogenic *Gallic acid* in white wine, red wine and juice samples.

The functional nanocomposites (F-SiO$_2$/GO) obtained by surface functionalization with NH_2 group were subsequently employed as a support for loading AgNPs to synthesize AgNPs-decorated F-SiO$_2$/GO nanosheets (AgNPs/F-SiO$_2$/GO) [82]. AgNPs/F-SiO$_2$/GO exhibits remarkable catalytic performance for H_2O_2 reduction. This sensor has a fast amperometric response, linear range from $(1 \times 10^{-4}$ - 0.26 M) and LOD of 4 µM. Also, a glucose biosensor was prepared by immobilizing glucose oxidase enzyme (GOD) into AgNPs/F-SiO$_2$/GO nanocomposite-modified GCE for glucose detection in human blood serum.

Activated Charcoal in Sensors

As discussed recently by our research group [4], activated charcoal (ACH) holds great promise for sensing applications that it can be promising used as an alternative material of graphene. The term ACH defines a group of natural materials with highly developed internal surface area and porosity, and consequently a large capacity for adsorbing various chemicals [83]. According to IUPAC, charcoal is a traditional term for a char obtained from wood, peat, coal and some other related natural organic materials. However; the precise structure of ACH has yet to be fully understood because of its heterogeneity and complexity. Although GR, CNTs or rGO are extensively used as signal amplifier materials in the electrochemical sensors, the fully characterization of ACH [4] will open up a new application of ACH in electroanalytical chemistry that it can be used as an alternative GR. This is due to its low cost, non toxic, natural carbon material, extensive sources, nano and few layer structure, edge-plane-like defective sites, porous structure, high surface area and few oxygenated functional groups.

EDX spectrum of the ACH (Fig. **30A**) shows only C and O with relative composition of 95.84 and 4.16%, respectively. Comared to graphit, the Raman spectrum of the ACH shows a typical spectroscopic profile (turbostratic structure) [84] of charred and coal materials, (Fig. **30B**). It shows a band at ≈ 1592 cm^{-1} (G-band) assigned to in-plane stretching motions of carbon sp^2 atoms originated in the crystalline carbon of the graphite and another band at ≈ 1333 cm^{-1} (D-band) assigned to carbons with sp^3 configuration. The D-band represents discontinuities in crystallites [84, 85] and disordered carbon configurations arising from defects in the lattice (mainly edges in graphene-like structures) [85]. So, it was assumed that ACH contains two types of carbon structures, *viz.*, crystalline carbon (condensed aromatic rings) and amorphous carbon (refers to any non-aromatic carbon which is grafted on the edges of crystallites) [86]. The edge plane like defective sites may provide many active sites for electron transfer to electro-active substances [84, 85]. The fully Raman spectrum of ACH may also be similar to that shown for rGO due to broaden of D band compared to G band [85]. XRD pattern of the ACH is consistent with the suggested structures of coal, activated carbon [86, 87] and rGO [88]. Compared to XRD of graphite (Fig. **30C**), ACH exhibited two broad peaks at 2Θ of 24° and 43.32° corresponding to the (002) and (100) planes, respectively, (Fig. **30C**); inset). The asymmetrical and broad nature of (002) band at 24° indicates the existence of amorphous carbon (*i.e.*, aliphatic side chains on the edges of the crystallites of ACH) [86, 87]. However, the sharp peak at 26.50° is related to the (002) graphitic plane (graphitic crystalline structure). This peak indirectly confirms the presence of in-plane conductivity in ACH which required for electrochemical applications. Therefore, ACH has partially graphitic structure (*i.e.*, intermediate structures between graphite and amorphous state called turbostratic structure) [86, 87].

Table **1** shows crystallite dimensional of ACH. Its interplanar distances (d_{200} and d_{001}) are very similar to the reported data of coal, activated carbon, rGO [87]. d_{200} of ACH (3.70 Å) is higher than the typical graphitic dimensions of (3.35 Å). Lc crystallite size of ACH is much smaller than that for graphite [86].

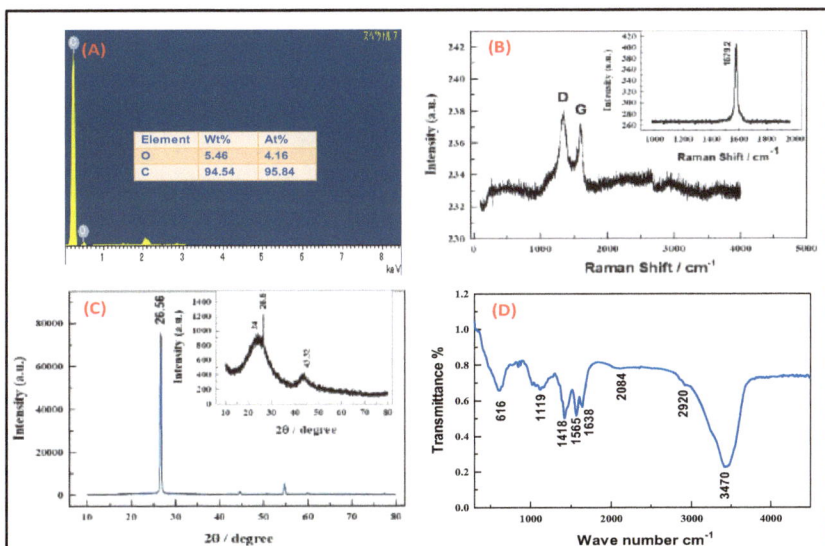

Fig. (**30**). **A)** EDX and **B)** Raman spectrum of activated charcoal (Inset; spectrum of graphite), **C)** XRD patterns of graphite (Inset; activated charcoal) and **D)** FTIR of activated charcoal (© The Electrochemical Society-Permission of IOP Publishing, [DOI: 10.1149/1945-7111/ab9e86]).

Table 1. The crystallite dimensional parameters and number of layers of activated charcoal (at the angular values of 24° and 43.32°) compared to graphite (at the angular value of 26.56°) (© The Electrochemical Society-Permission of IOP Publishing, [DOI: 10.1149/1945-7111/ab9e86]).

	Inter planar distance (d_{002}) A°	Stacking height (Lc_{002}) A°	Layer plane width (La_{002}) A°	Inter planar distance (d_{100}) A°	Stacking height (Lc_{100}) A°	Layer plane width (La_{100}) A°	Aromatic layers number (N) [002] [100]
Graphite	3.35	282.32	615	--	--	--	≈ 84
ACH	3.70	9.00	19.62	2.09	14.57	31.76	$\approx 3 \approx 8$

This confirms the disturbance of graphitic stacked ordering of the ACH or presence of graphene planes which stacked turbostraticaly [87]. The smaller value of crystallite diameter (La) of ACH is also consistent with the smaller size of coal crystallites [87]. This may be due to the constraints imposed by side chains, which prevent the adjacent crystallites from merging together during carbonization and/or activation processes [86]. It is also found that ACH crystallites consist of 3–8 aromatic layers on average. This small number of stacked layers of ACH referred to graphene nano-platelets. The FTIR spectroscopy provides that ACH sample consists of well-known carbonaceous bands of aromatic and aliphatic structures with few oxygenated functional groups representing in bands at 1119 cm^{-1} (assigned to the C-O stretching vibration) and at 3470 cm^{-1} (assigned to O–H stretching vibrations of OH⁻ groups or due to chemisorbed water), Fig. (**30D**).

SEM images of the ACH (Fig. **31**) clearly show its porous structure with lots of edge-plane-like defective sites which increase the surface area and may also provide many active sites for electro-active compounds. The BET surface area and pore volume are (630.43 ± 13.79) m^2/g and 0.20 cm^3/g, respectively. The mesopores structure of ACH may provide more reactive sites and better penetration of the electroactive species into the whole electrode materials.

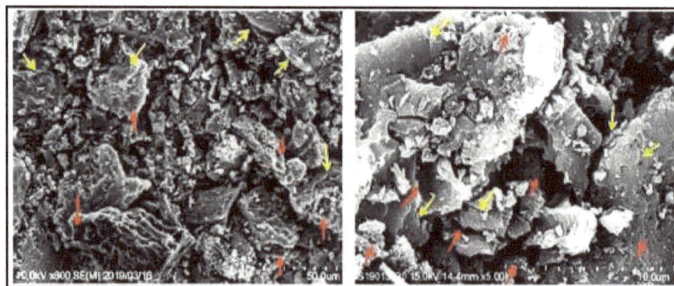

Fig. (31). SEM images of the activated charcoal of porous structure with cracks and crevices of different sizes (as marked by red arrow) and edge-plane-like defective sites (as marked by yellow arrow), (© The Electrochemical Society-Permission of IOP Publishing, [DOI: 10.1149/1945-7111/ab9e86]).

Furthermore, the performance of the modified CPE at various %compositions of ACH in comparison with GR was examined towards electrochemical sensing of *chronic hepatitis C virus inhibitor Ledipasvir* using square-wave adsorptive anodic stripping voltammetry (SW-AdASV) [4]. About two-fold enhancement of the current value for 1%(w/w) ACH/CPE (18.4 µA; Curve a) compared to unmodified CPE (9.6 µA) and even to various % composition of graphene, (curve b) was observed (Fig. **32A**).

Therefore, ACH will be a good alternative material for GR and an excellent model for constructing a promising electrochemical sensing platform in the future. The number of electron transfer for oxidation process is found to be 2e⁻ and the reaction mechanism of *Ledipasvir* at the ACH/CPE was suggested as shown in Fig. (**33**). It involves the oxidation of each of the imidazole rings to the corresponding radical.

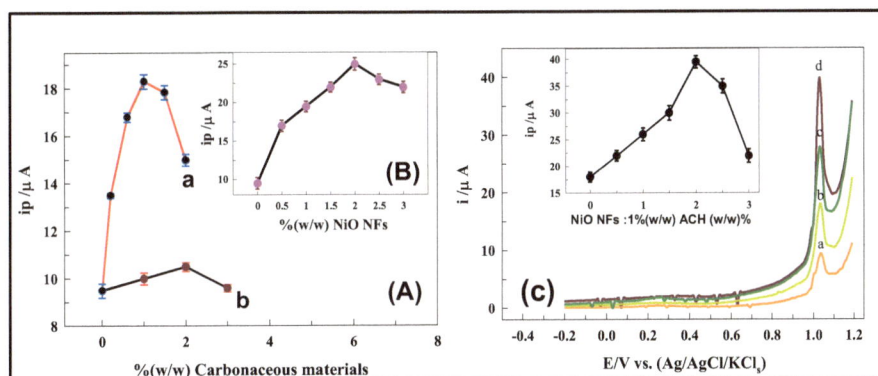

Fig. (32). Effect of % composition of: A) the carbonaceous materials [a) activated charcoal and b) graphene] and B): NiO NFs. C): SW-AdAS voltammograms recorded for 1.0×10⁻⁶ M *Ledipasvir* at E_acc = -0.2 V for 15s (pH = 2) at: a) bare CPE, b) 1%(w/w) ACH/CPE, c) 2%(w/w) NiO NFs/CPE and d) [2%(w/w) NiO NFs-1%(w/w) ACH]/CPE; (Inset: Effect of % composition of NiO NFs: 1% (w/w) ACH in carbon past matrix), (© The Electrochemical Society-Permission of IOP Publishing, [DOI: 10.1149/1945-7111/ab9e86]).

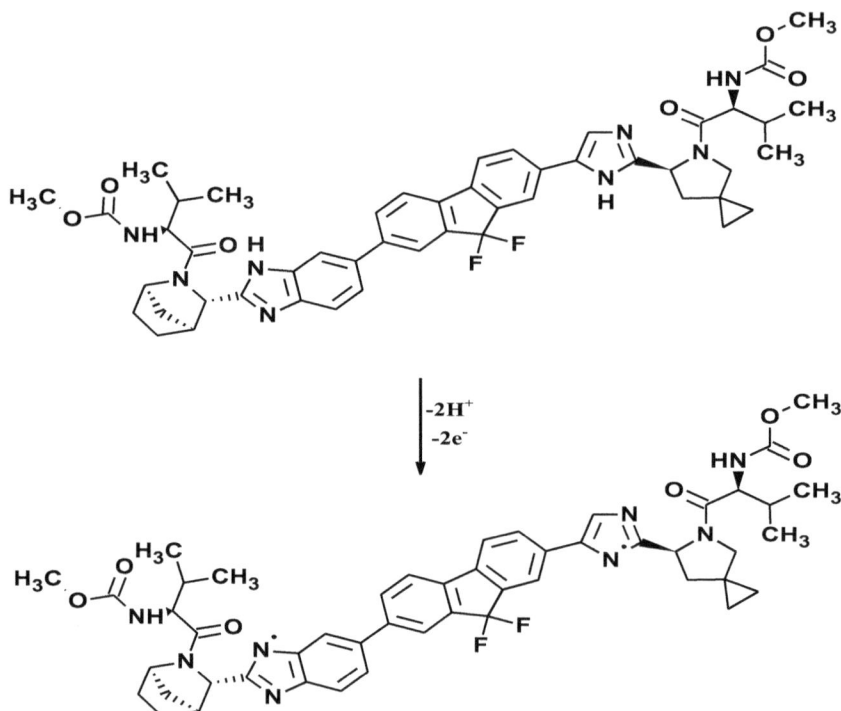

Fig. (33). Electrode reaction mechanism of *Ledipasvir* at modified CPE (© The Electrochemical Society- Permission of IOP Publishing, [DOI: 10.1149/1945-7111/ab9e86]).

Another new electrochemical sensor based on NiO nano-flakes (NiO NFs) and ACH was constructed [4]. Among bare CPE, ACH/CPE, NiO NFs/CPE and [NiO NFs–ACH]/CPE, the electrochemical properties of [2% (w/w) NiO NFs – 2% (w/w) ACH] nanocomposite were superior to the individual ACH and NiO NFs as platforms, (Fig. **32B** and **C**). [NiO NFs–ACH]/CPE was then applied for determination of *Ledipasvir* in human plasma samples with wide linear range of (3 nM - 1.5 µM) and very low LOD value of (0.55 nM), (Fig. **34A**). Upon storing the developed [NiO NFs – ACH]/CPE for 3 weeks, the deviation in peak current (i_p) was 1.35 to 2.30% and in E_p was < 0.01 V revealing its perfect stability. Besides, reasonable %R and %RSD values of (97.88 - 99.22%) and (1.69 - 2.47%) were got, respectively, for 3 assays of *Ledipasvir* utilizing set of the developed [NiO NFs – ACH]/CPEs indicates their good repeatability and reproducibility.

There are also negligible interferences on quantification of *Ledipasvir* in real samples from several foreign species such as its co-formulated drug (*Sofosbuvir*), excipients (gelatin, starch, sucrose and glucose), some co-administered drugs (paracetamol and ibuprofen) and common metal ions (Zn^{2+}, Cu^{2+}, Mg^{2+}, Co^{2+}, Ca^{2+}

and Fe^{3+}). Although vitamin C is electroactive under the applied potential range, its peak at peak potential ($E_p = 0.48$ V) was well potentially separated from that of *Ledipasvir* ($E_p = 1.0$ V), (Fig. **34B**). The current of 1.2×10^{-6} M *Ledipasvir* was not influenced by addition of 585 fold of *vitamin C* indicating the reliability of the developed sensor for trace assay of *Ledipasvir* in human plasma samples.

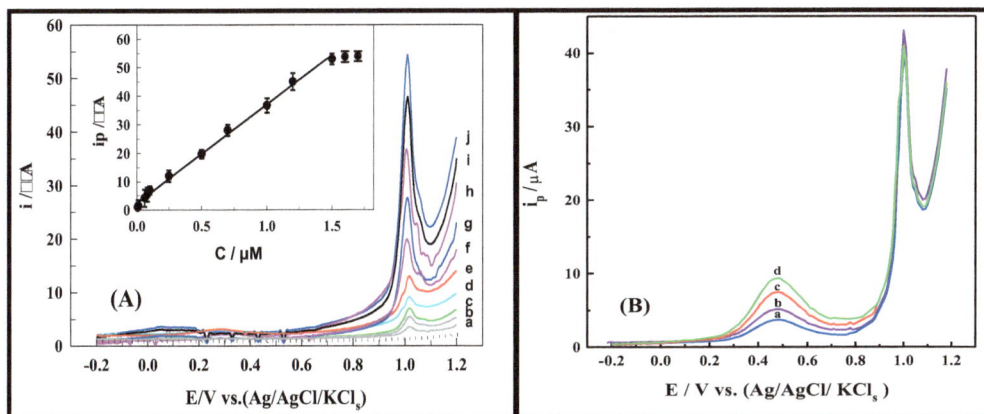

Fig. (34). A): SW-AdAS voltammograms recorded for various concentrations of *Ledipasvir* spiked in human plasma: a) 5×10^{-9}, b) 1×10^{-8}, c) 6×10^{-8}, d) 1×10^{-7}, e) 2.5×10^{-7}, f) 5×10^{-7}, g) 7×10^{-7}, h) 1×10^{-6}, i) 1.2×10^{-6} and j) 1.5×10^{-6} M and its calibration plot; (Dotted line is the blank). B): SW-AdAS voltammograms recorded for various concentrations of *Ascorbic acid*: a) 1×10^{-4}, b) 3×10^{-4}, c) 5×10^{-4}, d) 7×10^{-4} at constant concentration (1.2×10^{-6} M) of *Ledipasvir* spiked in human plasma (at pH 2, $E_{acc} = -0.2$ V, $t_{acc} = 15$ s, $f = 60$, $\Delta Es = 10$ mV and $a = 25$ mV) at [2% (w/w) NiO NFs – 2% (w/w) ACH]/ CPE), (© The Electrochemical Society-Permission of IOP Publishing, [DOI: 10.1149/1945-7111/ab9e86]).

Metal–organic Frameworks Based Nanosensors

Generally, metal-organic frameworks (MOFs) are chemical systems comprising organic molecules coordinated to metal ions or clusters to form 1-, 2-, or 3-D structural arrangement (*e.g.*, Fig. **35**)). Such open frameworks show exceptional feature of permanent porosity, pore volume, stable framework, and high surface area and bear also the ability to systematically vary and functionalize their pore structure. The organic units (linkers/bridging ligands) consist of carboxylates, or anions, such as phosphonate, sulfonate, and heterocyclic compounds. MOFs geometry is determined by the coordination geometry of the metal ions, coordination number and the nature of the functional groups. Based on the nature of the system used, discrete-closed oligomeric or infinite-extended polymeric structures can arise [89, 90].

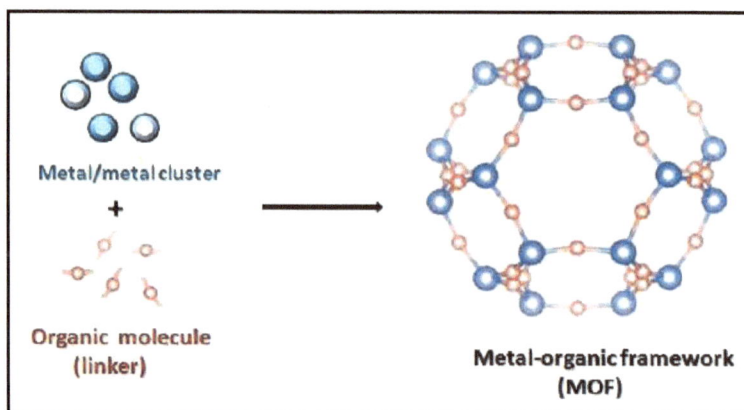

Fig. (35). Generalized scheme for metal-organic framework construction.

The application of MOFs as an electrode surface modifier was reported in some electrochemical sensors because of the MOFs' high surface areas, uncharged active sites, various pore sizes, and specific adsorption affinity [89, 90]. These materials can be considered as a bridge between organic chemistry and inorganic chemistry, which owes their flexibility to the organic ligand and their regular crystalline structure due to the metal ions.

Although the inherent micropores in MOFs (pore sizes below 2 nm) [91] are desirable for reactions due to their strong affinity for guest molecules, but they can also hinder the mass transfer and fast diffusion of reactants or products, especially large molecules which therefore limits their application. To solve the diffusion and accessibility problems that guest species face in conventional MOFs with micropores, hierarchical porous MOF materials (HP-MOFs; MOFs with micro- and mesopores, or MOFs with micro-, meso-, and macropores) with tunable porosity, such as copper benzenetricarboxylic acid (Cu–BTC) [92], zeolitic imidazolate framework (ZIF-8) [93] and zinc(II) 2,5-dihydroxy-1-4-benzenedicarboxylate (Zn–MOF-74) have been successfully synthesized [94].

Several MOF-based electrochemical functional frameworks with good electrocatalytic activities have also been reported in the electroanalysis [95 - 103]. For example, MOF as a porous matrix to encapsulate Cu nanoparticles (NPs) was applied for nonenzymatic glucose sensing in alkaline media. The hybrid of Cu NPs encapsulated in ZIF-8 (Cu-in-ZIF-8) was further modified onto screen-printed electrodes for nonenzymatic sensing of glucose in alkaline medium [95].

Cu-MOF-SWCNT modified electrode offers very promising results for simultaneous detection of *Hydroquinone (HQ)* and *Catechol (CT)* with respect to the extended linear ranges and lower LODs [96]. The composite electrode exhibited an excellent electrocatalytic activity towards the oxidation of *HQ* and *CT*. The linear ranges were from (0.10 - 1453 µM) for *HQ* and (0.1-1150 µM) for *CT*. LODs were as low as 0.08 and 0.10 µM, respectively. The analytical performance of such sensor for the simultaneous detection of *HQ* and *CT* had been evaluated in practical samples with satisfying results.

Cu-, Co- and Cr-based MOFs as modifiers were used in some electro-chemical sensors due to their electrocatalytic performances [97, 98]. The combination of carbon nanomaterials with MOFs (CdS quantum dots-encapsulated metal-organic frameworks) [99], cerium metal-organic frameworks [100]) and self-immobilization with biomolecules [101] are different ways that MOFs can be used in the modification of electrodes.

For example, Zn-MOF; [Zn4(1,4-BDC)4(BPDA)4]5DMF3H$_2$O, was synthesized from a mixture of BPDA (bis-(4-pyridinyl)-1,4-benzene dicarboxamide), NH$_2$BDC (2-amino-1,4-benzene dicarboxylic acid), ZnCl$_2$, water and DMF [12]. Zn-based MOF catalyzed the oxidation of *Fentanyl* (a potent analgesic and narcotic drug) on the screen-printed carbon electrode. It showed a LOD value of 0.3 µM in the concentration range of (1–100 µM) in an aqueous solution. It was also used to determine the low levels of *Fentanyl* in urine and plasma as the real samples. R% values were in the range of (99.00–104.00%), indicating satisfactory accuracy values and low matrix effect and the applicability of the Zn-MOF/ SPCE in real samples.

One of the most interesting MOFs for practical application is [Cu$_3$(BTC)$_2$] MOF (BTC = 1,3,5 benzentricarboxilate) that is known as HKUST-1. The HKUST-1 metal-organic framework-carbon nanofiber composite (HKUST-CNF) electrode has been prepared by film casting method and characterized morphologically and electrically. This electrode was tested for selective detection of *Ibuprofen* (*IBP*) and/or *Diclofenac* (*DCF*) simultaneously [102]. The *DCF* and *IBP* oxidation process occurred at +0.75 V and +1.25 V *vs.* SCE, respectively. The stability of the electrode was investigated by cyclic voltammetry at the time intervals of 10, 20, and 30 days and the peak currents were changed by 1.5% and 2.1% for *DCF* and *IBP*, respectively, indicating high stability of the electrode. The reproducibility was tested for 3 replicates and RSD of 2.5% for *IBP* and 2.6% for *DCF* showed good reproducibility. The analysis of *DCF* and *IBP* in water exhibited R% of 99.5% for *DCF* and 99.0% *IBP*, and RSD% of 2.8% for *IBP* and 3.5% for *DCF*, indicating adequate precision and accuracy of this method. The HKUST-CNF electrode exhibited good stability and reproducibility for selective detect such drugs.

A voltammetric sensor was developed using a ZIF-8 derived nitrogen-doped porous carbon and nitrogen-doped graphene nanocomposite modified GCE for the simultaneous determination of *Dopamine (DA)*, *Uric acid (UA)* and *Ascrobic acid (AA)* [103]. The sensor demonstrated a linear plot of (0.01–15.0 µM) for *AA*, (0.08–350.0 µM) for *DA*, and (0.5–100.0 µM) for *UA*, with LODs of 1.99 µM, 0.011 µM and 0.088 µM, respectively. The established sensor was applied for their determination in real urine samples.

Polymer-based Nanosensors

The nanostructure conductive polymers are among the most commonly used nanomaterials to modify electrode surfaces owing to their essential role in enhancing the electrocatalytic activity and sensitivity, protecting the electrode surface and reducing its fouling. Conducting polymers have been successfully applied in chemical sensing, biosensing, gas sensing, super-capacitors, *etc.* due to their unique electronic properties. The nanostructure polyaniline (PANI), polypyrrole (Ppy), poly-(3,4-ethylenedioxythiophene (PEDOT), (Fig. **36**) and their functionalized derivatives have been studied intensely for sensing and biosensing applications because of their intrinsic conductivity [1].

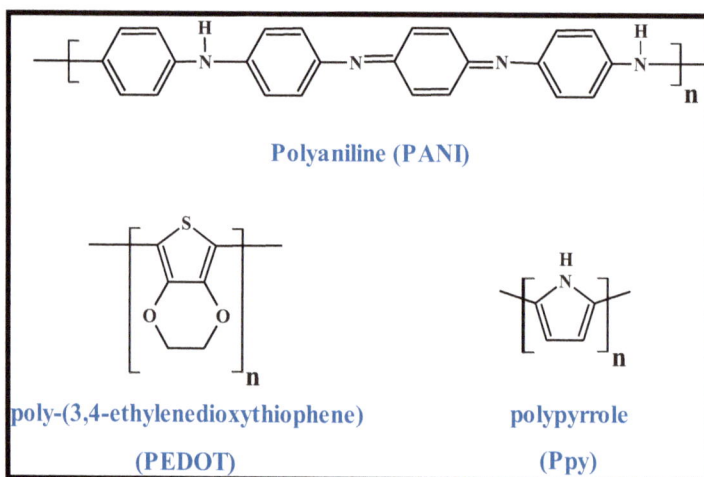

Polyaniline (PANI)

poly-(3,4-ethylenedioxythiophene)   polypyrrole

(PEDOT)   (Ppy)

Fig. (36). Chemical structure of some conducing polymers.

Because of the delocalization of π-bonded electrons over polymeric backbone, the conducting polymers exhibit unique electronic properties (such as low ionization potentials and high electron affinities) [104]. Owing to excellent electronic conductivity, operability and low cost, conducting polymers are considered to be one of the most attractive materials to modify the electrode surface. The electrode modification can be done either by direct electro-polymerization of the monomer or by simple adsorption of the polymer onto the surface of the working electrodes.

Such organic polymers are characterized by an alternating single (σ) and double (π) bonds and by the presence of π electrons delocalized across their entire conjugated structure; polymers can be easily oxidized or reduced. This doping, that can be performed upon oxidation (p-doping) or reduction (n-doping), increases their conductivity significantly (this conductivity can vary from $< 10^{-6}$ S/cm in the neutral state to $> 10^5$ S/cm in the doped state) [104].

Further, polymer nanocomposites have improved electrochemical properties and can be used as transducers in the development of electrochemical sensors and biosensors. Polyaniline (PANI) is a very excellent material for sensing applications because of its high conductivity and redox reversibility character. In the following, several examples are shown that illustrate analysis of some drugs employing many nanocomposites based- conducting polymer.

A novel composite for electrochemical determination of *Tramadol* was proposed by inclusion of a metallocene mediator (*e.g.*, ferrocene carboxylic acid) between two layers of conducting poly(3,4-ethylenedioxythiophene) (PEDOT) polymer [105]. In the presence of SDS, an enhancement of the accumulation of *Tramadol* at the interface was observed which leads to faster electron transfer. The practical application of the proposed composite for the determination of *Tramadol* in real urine and serum samples was successfully achieved with adequate recovery results. Very low LOD values of (18.6 nM) and (16 nM) and very wide linear dynamic ranges of (7–300 µM) and (5–280 µM) were obtained in urine and serum samples, respectively. The simultaneous determination of *Tramadol* with common interfering species, paracetamol, morphine, dopamine, ascorbic and uric acids showed good resolution. The good stability of the electrode and high reproducibility, sensitivity, selectivity, anti-interference ability enhanced its application in the determination of other narcotic drugs.

Electrochemical determination of *Paracetamol* using PEDOT {Poly-(3,--ethylenedioxythiophene)} modified electrode prepared by electropolymerization on SPE was also investigated [106]. An irreversible oxidation peak is noticed at the bare SPE at +0.65 V which shifted to +0.54 V at the electrochemically pretreated SPE. Oxidation of *Paracetamol* at GCE is irreversible, involving $2e^-$ and $2H^+$ transfer process to produce the unstable, oxidized product N-acetyl-p-quinoneimine. The hydrolysis of this product is responsible for the disappearing of cathodic wave in cyclic voltammograms. The achieved LOD values were 3.71, 1.39 and 0.16 µM, using CV, DPV and flow-injection amperometry respectively. The RSD is 1.81, 0.57, 0.36% for 3 runs of 40, 200, 800 mM *Paracetamol* standards, respectively. A loss in the current signal of 10.30% was only observed after 20 days suggesting that the PEDOT film is highly stable and no loss of film during the detection. This sensor was also successfully implemented in the detection of *Paracetamol* in commercial tablet samples, with excellent recovery values (98.04 – 106.10%).

Electro-polymerized Poly-4-amino-6-hydroxy-2-mercaptopyrimidine modified glassy carbon electrode (Poly-AHMP/GCE) has also been investigated for selective and simultaneous determination of *Paracetamol* and *Dopamine* [107]. Their anodic peak currents increased linearly within the concentration ranges 2.0–20.0 µM and 2.5–25 µM with correlation coefficients of 0.991 and 0.999, respectively. LOD values were found to be 0.20 and 0.15 µM, respectively by DPV. This electrode sensor was successfully utilized to determine *Dopamine* and *Paracetamol* in their injection and tablets samples

Similarly, the detection of *Serotonin* in banana was performed by SWV at a SPCE modified with polypyrrole doped with green iron (III) oxide nanoparticles (SPCE-PPy-Fe_3O_4 NPs). This modified electrode demonstrated good electrocatalytic activity towards simultaneous detection of *Serotonin* and *Ascorbic acid*. Their oxidation peaks were observed at 0.1259 V and 0.1099 V, respectively. Electrodes were repeatedly scanned (12 scans) with an only current drop of $\approx 9.0\%$ observed. Thus, this developed sensor succeeded in simultaneous detection of 100 mM *Ascorbic acid* and 0.1 mM *Serotonin* [108]. Moreover, detection of *Serotonin* in banana samples offers good R±RSD% (95.70±1.07 – 114.40±0.37).

Based on the excellent properties of the polymer nanocomposites, they have been used in the development of other electrochemical sensors and biosensors for various analytes, some of which are also summarized in Table **2**.

Table 2. Sensors based on polymer Nanocomposites.

Polymer Nanocomposite	Analyte	LOD	Refs.
$NiCo_2O_4$-polyaniline (Ni Co_2O_4-PANI)	Glucose	0.38 µM	[109]
Graphene polyaniline nanocomposite- (GR–PANI-) modified glassy carbon electrode (GR-PANI-GCE)	4-Aminophenol	0.07 µM	[110]
Graphene-polyaniline-horseradish peroxidase (Gr-p-PANI-HRP)	Artesunate	0.012 ng/mL	[111]
Polyaniline and zirconia nanocomposite film (PANI-ZrO_2)	Esomeprazole	97.21 ng/mL	[112]
Gold-polyaniline-graphene nanocomposites (Au-Grp-PANI)	Nitrite	0.01 µM	[113]
Graphene-polyaniline-Bi_2O_3 Grp-PANI- Bi_2O_3) composite	Etodolac	10.03 ng/mL	[114]
Zinc oxide nanoparticles intercalated into polypyrrole (ZnO NPs-PPy)	Xanthine	0.80 µM	[115]

(Table 2) cont.....

Polymer Nanocomposite	Analyte	LOD	Refs.
Polypyrrole nanosheets decorated with platinum nanoparticles (PPy-Pt NPs)	Hydrogen peroxide	0.60 μM	[116]
Polypyrrole/graphene nanocomposite (PPy-Grp-GCE)	Adenine, guanine	0.02 μM, 0.01 μM	[117]
Polypyrrole/graphene oxide nanosheets (PPy-GrpO)	Dopamine	73.30 nM	[118]
Praphene-poly(3,4-ethylenedioxythiophene) (graphene–PEDOT) nanocomposite film with ascorbate oxidase (AO) entrapped (Grp-PEDOT-AO)	Ascorbic acid	2.00 μM	[119]

Molecularly Imprinted Polymers

A new trend in the area of electrochemical sensors and biosensors concerns the applying of molecularly imprinted polymers (MIPs). MIPs-based sensor platforms have been used for a variety of target chemical and biological molecules. The imprinting of organic or biomolecules, including pharmaceuticals, amino acids, pesticides, peptides, steroids and sugars, even metal and other ions are well demonstrated to favor the selective organization of functional groups in the imprinting network.

A variety of methods have been employed for the production of nanostructured MIPs (such as suspension, dispersion, precipitation and emulsion seeded polymerization). MIPs technology is used to synthesize highly crosslinked polymers capable of selective recognition. MIP is in creating selective recognition sites for the target molecule on the polymer matrix [120]. This approach practically stands on the molecular detection, identification, capturing, and/or recognition events occurring around the target molecule.

As shown in Fig. (37), in MIP process, the target (template) molecule, crosslinker(s), functional monomers, initiator, and progenic solvent are used. Monomer polymerization takes place in the presence of this target molecule incorporated into the polymer matrix. The process begins with the dissolution of the target, crosslinker, functional monomer, and initiator in the respective solvent. Functional monomers having the ability to interact with the template molecules are preferred. A successful MIP and recognition require the formation of a stable target molecule-function monomer complex. After the monomers are placed around the target molecule, the position is fixed by co-polymerization of the crosslinked monomers (Functional monomers and crosslinkers are used for co-polymerization with the target molecule).

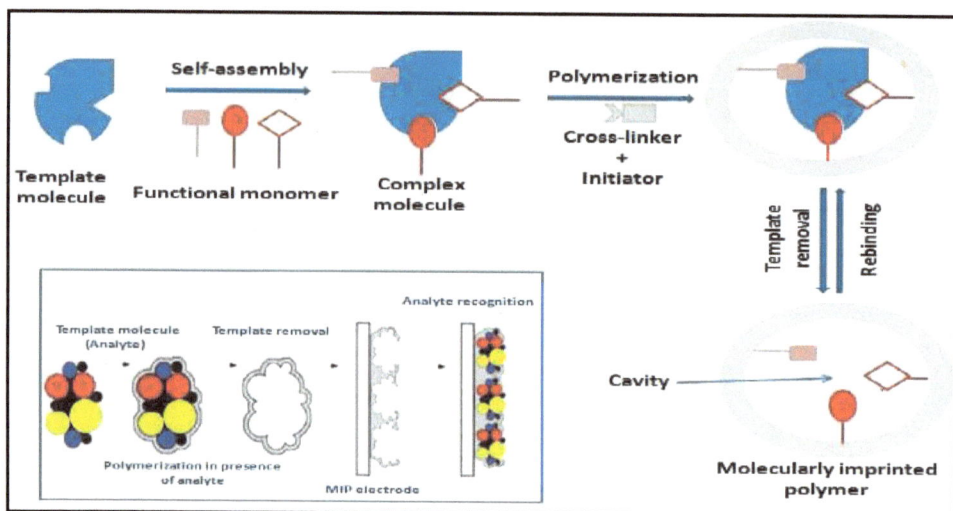

Fig. (37). Simple schematic presentation of the preparation of MIPs.

The resulting polymer is a porous matrix with microcavities containing three-dimensional structure complementary to the target molecule. Thus, by removing the target molecules from the polymer by washing them with solvent, binding sites are formed which recognize the target molecule shape (*i.e.*, when the template is removed, cavities remain which resemble the template in shape, size and surface chemistry. Since it fits into the cavity, the template will be incorporated preferentially). Consequently, polymers with high affinity and selectivity for the target molecule selectively recognize and bind the molecule [120].

Several electrochemical sensors based on MIP were reported in literature. The formation of a MIP film on the surface of electrodeposited hollow nickel nanospheres (hNiNS) was reported [121]. This nanocomposite is utilized in an effective electrochemical sensor for *Dopamine* detection. The use of the 3D hNiNS as a support material enlarges the sensing area and conductivity, while the MIP film warrants improved selectivity for *Dopamine*. CV, EIS as well as SEM were employed to characterize the sensor materials. This sensor has fairly low LOD of 1.7×10^{-14} M, and superb selectivity.

A 3D imprinted CNTs@Cu NPs hybrid nanostructure was applied to fabricate an electrochemical sensor to monitor a model veterinary drug, *Chloramphenicol* [13]. Cu NPs are deposited on CNTs and then a hybrid structure is formed by coating MIP on 3D CNTs@Cu NPs. SEM revealed the presence of Cu NPs (100–500 nm) anchored along the whole length of CNTs, topped with imprinted layer. This sensor exhibited sensitivity with LOD of 10 μM.

MWCNTs coated with MIPs poly-(3-aminophenylboronic acid) (PAPBA) was prepared *via* potentiodynamic electropolymerization and utilized as an effective electrochemical material for *Epinephrine* detection [122]. Compared with MWCNTs or non-imprinted polymers PAPBA modified MWCNTs electrodes, the PAPBA(MIPs)/MWCNTs electrode exhibited a lower charge transfer resistance and enhanced electrochemical performance for *Epinephrine* detection. This is due to a large amount of specific imprinted cavities with boric acid group which can selectively adsorb *Epinephrine* molecules. This sensor could effectively recognize *Epinephrine* from many possible interferents within a linear range of 0.2-800 nM and low LOD of 35 nM. The detection of *Epinephrine* in human serum and real injection samples also gave satisfactory results.

A poly(sulfosalicylic acid) film was electropolymerized on a Au electrode, which was modified by 2-mercaptobenzothiazole self-assembled monolayer and MWCNTs (PSSA/CNTs/MBT/Au) for determining *Rutin* [123]. The developed sensor displayed excellent electrocatalytic activity towards its oxidation due to the presence of the PSSA/MWCNTs nanocomposite (LOD = 1.8 nM). This electrode was also applied for *Rutin* determination in orange, red apple, red onion, strawberry, oat and salvia samples.

Several other drug sensors based on MIP [124 - 132] are reported in Table **3**. A sensor based on PANI/GO/GNU {an aniline MIP film and gold nano urchins (GNU) /graphene oxide}-modified GCE was designed for assaying *Azithromycin* [124]. *Calycosin* was determined using Polyaniline/graphene quantum dots (PANI /GQD) nanocomposite modified GCE [125]. *5-Fluorouracil (5-FU)* was determined using Ag NPs polyaniline nanotube nanocomposite (AgNP@PANINT) modified pencil graphite [126]. *Diclofenac* was detected using PANI/reduced graphene oxide nanocomposite (PANI/rGO) [127]. *Bezafibrate* was determined based carbon nanofibers (CNF) /PANI/Pt NPs composite onto the CPE surface [128]. A binary nanocomposite composed of a copper tungstate $(CuWO_4)$ and PANI synthesized by chemical oxidative single step polymerization on the surface of a GCE was used to determine *Quercetin* [129]. Cefixime was determined utilizing Au nanowires/graphene oxide/electropolymerized MIP (PANI/GO/AuNW) [130]. *Dexamethasone* was detected using Fe_3O_4/PANI–CuII microsphere modified carbon ionic liquid electrode (Fe_3O_4/PANI–CuII/ CILE) [131]. *Nifedipine* was determined using ITO electrode modified with nano-hybrid uniformly distributed film composed of PANI NPs and carboxylic acid functionalized MWCNTs (MWNTs-COOH) *via* a layer-by-layer method [132].

Table 3. Application of different molecularly imprinted polymer (MIP) electrochemical sensors in drug analysis.

Drug	Nanocomposite	Technique	Linear range	LOD	Refs.
Azithromycin	MIP/GNU/ GO	DPV	0.3 – 920 nM	0.10 nM	[124]
Calycosin	PANI/GQD	DPV	1.1×10^{-5}–3.5×10^{-4} M	9.80 µM	[125]
5-Fluorouracil	AgNPs@PANINT	DPV	1.0 to 300.0 µM	0.06 µM	[126]
Diclofenace	PANI/rGO	CV	5 – 80 mg/mL	1.10 mg/ml	[127]
Bezafibrate	CNF /PANI/Pt NPs	DPV	0.025 – 100 mM	2.46 nM	[128]
Quercetin	$CuWO_4$@PANI	DPV	0.001–0.500 µM	1.20 nM	[129]
Cefixime	PANI/GO/AuNW	DPV	20.0 – 950.0 nM	7.10 nM.	[130]
Dexamethasone	Fe_3O_4/PANI–CuII/ CILE	DPV	0.05 – 30 mM	3.00 nM	[131]
Nifedipine	MWNT-COOH/PANINP	CV	1.10×10^{-6}–1.1×10^{-4} M	10.00 µM	[132]

Green Synthesis of Nanoparticles

Nanoparticles produced by green technology are far superior to those manufactured with chemical and even with physical methods. Green techniques eliminate the use of expensive chemicals, consume less energy, and generate environmentally benign products and byproducts [18]. The green synthesis of nanoparticles reflects a bottom-up approach. Some of the sensing and biosensing applications of green synthesis metal and metal oxide nanoparticles by biomass filtrate obtained from various biological systems such as bacteria, yeast, virus, actinomycetes, algae, fungi and plant extract [133 - 148] are given in Table **4**.

Table 4. Synthesis of some metal and metal oxide NPs from various biological species for sensing applications.

Reducing/Stabilizing agent	NPs	Size, nm	Morphology	Application	Refs.
Glucose	Ag NPs/GO	30-80	--	Electrochemical sensor	[133]
Bacteria					
Bacillus megaterium D01	Au NPs	<2.5	spherical	Biosensing, Catalysis,	[134]
E. coli DH 5α	Au NPs	8-25	spherical	Biosensing	[135]
Bacillus subtilis	Se NPs	50–400		H_2O_2 sensoristic device	[136]
Virus					
Bacteriophage	Au NPs	20–50	..	Biosensor electrode	[137]
M13 virus	Au-DNs	-	-	Biosensor platform	[138]

(Table 4) cont.....

Reducing/Stabilizing agent	NPs	Size, nm	Morphology	Application	Ref.
M13 virus	TiO$_2$ NPs	20–40		Photo-electrochemical properties	[139]
Fungs					
Verticillium luteoalbum	Au NPs	< 10	Triangular, hexagonal	Sensor, Optics, coatings	[140]
Sargassum muticum	Ag NPs	5-35	spherical	Sensors, biolabeling, drug delivery	[141]
Aspergillus terreus	ZnO NPs	8	spherical	Biosensing, Catalysis, drug delivery, molecular diagnostics, solar cell, optoelectronics, and imaging	[142]
Plant origin					
Camellia sinensis (black tea leaf extracts)	Au, Ag NPs	20	spherrical, prism	Sensors, catalysts,	[143]
Ocimum sanctum (tulsi; leaf extract)	Au, Ag NPs	30 and 10–20	Crystalline, hexagonal, triangular, spherical	Biosensor, biolabeling,	[144]
Pear fruit extract	Au NPs	200-500	Triangular, hexagonal	Biosensor, Catalysis	[145]
Tanacetum vulgare (tansy fruit)	Au, Ag NPs	11, 12	Triangular, spherical	Sensor, Antibacterial	[146]
Onion extract	Ag NPs	6.0	mainly spherical	Electrochemical sensor	[147]
Piper betle leaves extract	Ag NPs	20	spherical	Electrochemical sensor	[148]

Several reports have also appeared that metal NPs, such as gold, silver, silver–gold alloy, platinum, titanium, tellurium, zinc, copper, magnesium, selenium, palladium, zirconium, silica, quantum dots, and magnetite, can be biosynthesized by various microorganisms, especially fungi and bacteria [134 - 142, 149, 150]. However, the majority of plants have features as renewable and sustainable resources compared with enzymes or microbes. Plants have chemicals like antioxidants and biomolecules (like carbohydrates, proteins, and coenzyme) with exemplary potential to reduce metal salt into nanoparticles [143 - 148, 151 - 153]. Hence, plants with various reducing agents are favorable candidates for the manufacture of noble MNPs, (Table **4**).

As in Fig. (**38**), for NPs synthesis mediated by plant leaf extract, the extract is mixed with metal precursor solutions at different reaction conditions. The types of phytochemicals, phytochemical concentration, metal salt concentration, pH, and temperature) controlled the rate of nanoparticle formation as well as their yield and stability [154]. Plant leaf extract plays a dual role by acting as both reducing and stabilizing agents in nanoparticles synthesis process to facilitate NPs synthesis [153]. The main role of stabilizing the NPs is to prevent further growth and agglomeration. The main phytochemicals present in plants are flavones, terpenoids, sugars, ketones, aldehydes, carboxylic acids, phenol, and amides, which are responsible for bioreduction of NPs.

However, when fungi are exposed to metal salts (such as $AgNO_3$ or $AuCl_4^-$), they produce enzymes and metabolites to protect themself from unwanted foreign matters and in doing so, the metal ions are reduced to metal NPs. The fungi also produce anthraquinones and napthoquinones, which act as reducing agents. Thus, a specific enzyme can act on a specific metal [154]. Ag NPs possess large specific surface area as well as the ability of quick electron transfer and strong antibacterial activity, so they have been extremely applied in the preparation of biosensors and bactericidal agents.

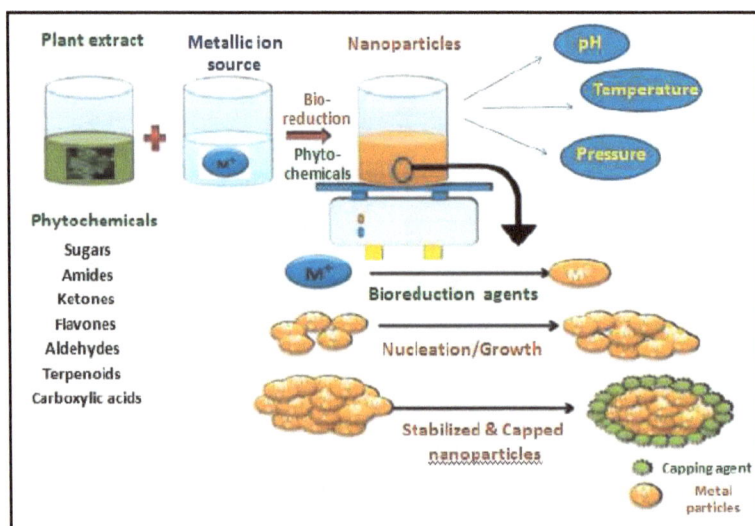

Fig. (38). Biological synthesis of metal nanoparticles using plant extracts.

With this respect, a nanocomposite based on (AgNPs/GO) was obtained through a green synthesis approach using glucose as a reducing and stabilizing agent [133]. AgNPs/GO was prepared by reducing Ag^+ ions directly on GO with glucose. The AgNPs/GO/GCE exhibited higher activity for the electro-oxidation of *Tryptophan* drug than GO film with 10 fold enhancement of peak current. At AgNPs/GO/GCE, the currents were proportional to the concentrations over the range of (0.01 to 50.00 μM) with LOD was 2.0 nM. The proposed method is free of interference from tyrosine and other coexisting species.

A high-quality method for one-pot biosynthesis of AgNPs using onion extracts as reductant and stabilizer was also reported [147]. AgNPs reduce charge transfer resistance. The effect of synthesized NPs on *Ascorbic acid* signal was investigated by SWV. The current increased linearly with its concentration in the range of (0.4 - 450.0) μM with LOD of 0.1 μM.

The detection of *nitrite* (NO_2^-) and its compounds are substantial, owing to its extensive utilization as additives, therapeutics, fertilizers and polymer inhibitors and its widely used in food science, pharmaceutical, agriculture and rubber industry fields [155]. An environmentally benign method exploiting Piper betle biomass (dried leaves extract) as a reducing and stabilizing agent was exploited for the green preparation of Ag NPs [148]. The fabricated Ag NPs/GCE sensor exhibited a high electrocatalytic activity toward NO_2^- oxidation with a lower LOD of 0.05 μM.

Besides, in the cases of the biometallic alloys Ag, Au, and Au/Ag, it was shown that polyphenols and polyols that carry antioxidant properties present in plants have an important role in the formation of such nanostructures [143]. Further, *Quercetin*; a flavonol with high antioxidant activity appeared as the main component in the formation of metallic nanostructures [156].

The suggested reaction mechanism for the green synthesis of AgNPs due to the flavanoids in plant extract is shown in Fig. (**39**).

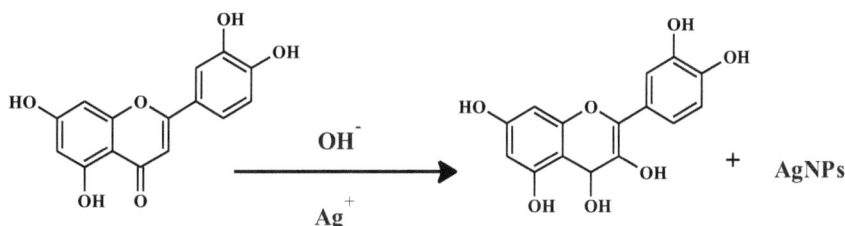

Fig. (**39**). Suggested reaction mechanism for the green synthesis of AgNPs due to the flavanoids in plant extract.

Furthermore, Au NPs were synthesized with dicarboxylic acids (oxalic, malonic, succinic, glutaric, and adipic) as reducing agents of $HAuCl_4$, without the presence of any other surfactant agents [157]. Additionally, pimelic dicarboxylic acid is used as a nucleating agent for the synthesis of TiO_2 nanoparticles [158].

CONCLUSION

Pharmaceutical and biomedical analysis are among the most important branches of applied analytical chemistry. Pharmaceutical drugs play an important role in human life. They are commonly employed to diagnose, treat or prevent and cure a disease *via* a biological effect on the human body. Drug analysis should be performed during all steps of pharmaceutical development, from the stage of synthesis, formulation, stability testing and quality control to toxicological and pharmacological investigations in humans or animals, including preclinical and clinical trials. Electrochemical sensors based on various nanomaterials are promising candidates for drug quality control and in a clinical laboratory. This chapter signifies the *via*ble integration of numerous types of nanomaterials and electrochemical systems, which have led to the unparalleled maturity of electrode-based sensor devices. We have discussed strategies and applications in drug analysis in this chapter that are evident of growing efforts toward the increased sensitivity, selectivity, speedy analysis with field portability of the sensors.

CONSENT FOR PUBLICATION

Not applicable.

CONFLICT OF INTEREST

The author declares no conflict of interest, financial or otherwise.

ACKNOWLEDGEMENT

Declared none.

REFERENCE

[1]   N. Shoaie, M. Daneshpour, M. Azimzadeh, S. Mahshid, S.M. Khoshfetrat, F. Jahanpeyma, A. Gholaminejad, K. Omidfar, and M. Foruzandeh, "Electrochemical sensors and biosensors based on the use of polyaniline and its nanocomposites: a review on recent advances", *Mikrochim. Acta,* vol. 186, no. 7, pp. 465-494, 2019.
[http://dx.doi.org/10.1007/s00604-019-3588-1] [PMID: 31236681]

[2]   H.S. El-Desoky, A. Khalifa, and M.M. Abdel-Galeil, "An advanced and facile synthesized graphene/magnetic Fe3O4 nanoparticles platform for subnanomolar voltammetric determination of antipsychotic olanzapine Drug in human plasma", *J. Electrochem. Soc.,* vol. 167, pp. 067527-067540, 2020.
[http://dx.doi.org/10.1149/1945-7111/ab8366]

[3]   Z. Li, C. Liu, V. Sarpong, and Z. Gu, "Multisegment nanowire/nanoparticle hybrid arrays as electrochemical biosensors for simultaneous detection of antibiotics", *Biosens. Bioelectron.,* vol. 126, pp. 632-639, 2019.
[http://dx.doi.org/10.1016/j.bios.2018.10.025] [PMID: 30513482]

[4]   H.S. El-Desoky, M.M. Ghoneim, M.A. Momtaz, and M.M. Abdel-Galeil, "Utilizing the synergetic benefit of synthesized NiO nano flakes and natural activated charcoal (an alternative of graphene) as a novel platform for voltammetric determination of chronic hepatitis C virus inhibitor ledipasvir", *J. Electrochem. Soc.,* vol. 167, pp. 117504-117518, 2020.
[http://dx.doi.org/10.1149/1945-7111/ab9e86]

[5]   A.U. Alam, M.M.R. Howlader, N-X. Hu, and M.J. Deen, "Electrochemical sensing of lead in drinking water using β-cyclodextrin-modified MWCNTs", *Sens. Actuators B Chem.*, vol. 296, p. 126632, 2019.
[http://dx.doi.org/10.1016/j.snb.2019.126632]

[6]   H.S. El-Desoky, M.M. Ghoneim, and F.M. El-badawy, "F. M. Carbon nanotubes modified electrode for enhanced voltammetric sensing of mebeverine hydrochloride in formulations and human serum samples", *J. Electrochem. Soc.*, vol. 164, pp. B212-B222, 2017.
[http://dx.doi.org/10.1149/2.0941706jes]

[7]   H. El-Desoky, M. Ghoneim, M. Abdel-Galeil, and A. Khalifa, "Electrochemical sensor based on functionalized multiwalled carbon nanotubes, domperidone determination, DNA binding and molecular docking", *J. Electrochem. Soc.*, vol. 164, pp. H1133-H1147, 2017.
[http://dx.doi.org/10.1149/2.1091714jes]

[8]   A. Khalifa, H. El-Desoky, and M. Abdel-Galeil, "Graphene-based sensor for voltammetric quantification of dapoxetine hydrochloride: A drug for premature ejaculation in formulation and human plasma", *J. Electrochem. Soc.*, vol. 165, pp. H128-H140, 2018.
[http://dx.doi.org/10.1149/2.0491803jes]

[9]   F.M. El-Badawy, and H.S. El-Desoky, "Quantification of chloroxylenol, a potent antimicrobial agent in various formulations and water samples: environmental friendly electrochemical sensor based on microwave synthesis of grapheme", *J. Electrochem. Soc.*, vol. 165, pp. B694-B707, 2018.
[http://dx.doi.org/10.1149/2.0391814jes]

[10]   W. Yi, Z. He, J. Fei, and X. He, "Sensitive electrochemical sensor based on poly(l-glutamic acid)/graphene oxide composite material for simultaneous detection of heavy metal ions", *RSC Advances*, vol. 9, pp. 17325-17334, 2019.
[http://dx.doi.org/10.1039/C9RA01891C]

[11]   H. El-Desoky, "khattab, A.; A simple approach for trace voltammetrical determination of itraconazole an inhibitor of the cytochrome P - 450 14 α-demethylase enzyme in formulation and human plasma utilizing graphene-carbon paste electrode in situ modified with an anionic surfactant", *J. Electrochem. Soc.*, vol. 167, pp. 147502-147517, 2020.
[http://dx.doi.org/10.1149/1945-7111/abbdd5]

[12]   E. Naghian, E.M. Khosrowshahi, E. Sohouli, F. Ahmadi, M. Rahimi-Nasrabadi, and V. Safarifard, "A new electrochemical sensor for the detection of fentanyl lethal drug by a screen-printed carbon electrode modified with the open-ended channels of Zn(II)-MOF", *New J. Chem.*, vol. 44, pp. 9271-9277, 2020.
[http://dx.doi.org/10.1039/D0NJ01322F]

[13]   A. Munawar, M.A. Tahir, A. Shaheen, P.A. Lieberzeit, W.S. Khan, and S.Z. Bajwa, "Investigating nanohybrid material based on 3D CNTs@Cu nanoparticle composite and imprinted polymer for highly selective detection of chloramphenicol", *J. Hazard. Mater.*, vol. 342, pp. 96-106, 2018.
[http://dx.doi.org/10.1016/j.jhazmat.2017.08.014] [PMID: 28823921]

[14]   T.A. Saleh, "Detection: from electrochemistry to spectroscopy with chromatographic techniques, recent trends with nanotechnology", *Detection*, vol. 2, pp. 27-32, 2014.
[http://dx.doi.org/10.4236/detection.2014.24005]

[15]   I.A. Wani, S. Khatoon, A. Ganguly, J. Ahmed, A.K. Ganguli, and T. Ahmad, "Silver nanoparticles: large scale solvothermal synthesis and optical properties", *Mater. Res. Bull.*, vol. 45, pp. 1033-1038, 2010.
[http://dx.doi.org/10.1016/j.materresbull.2010.03.028]

[16]   M.I. Dar, A.K. Chandiran, M. Grätzel, M.K. Nazeeruddin, and S.A. Shivashankar, "Controlled synthesis of TiO2 nanoparticles and nanospheres using a microwave assisted approach for their application in dye-sensitized solar cells", *J. Mater. Chem. A Mater. Energy Sustain.*, vol. 2, pp. 1662-1667, 2014.
[http://dx.doi.org/10.1039/C3TA14130F]

[17]   X-F. Zhang, Z-G. Liu, W. Shen, and S. Gurunathan, "Silver nanoparticles: synthesis, characterization, properties, applications, and therapeutic approaches", *Int. J. Mol. Sci.,* vol. 17, no. 9, pp. 13-17, 2016.
[http://dx.doi.org/10.3390/ijms17091534] [PMID: 27649147]

[18]   K.B. Narayanan, and N. Sakthivel, "Biological synthesis of metal nanoparticles by microbes", *Adv. Colloid Interface Sci.,* vol. 156, no. 1-2, pp. 1-13, 2010.
[http://dx.doi.org/10.1016/j.cis.2010.02.001] [PMID: 20181326]

[19]   X. Luo, A. Morrin, A.J. Killard, and M.R. Smyth, "Application of nanoparticles in electrochemical sensors and biosensors", *Electroanalysis,* vol. 18, pp. 319-326, 2006.
[http://dx.doi.org/10.1002/elan.200503415]

[20]   R.M. Pallares, N.T.K. Thanh, and X. Su, "Sensing of circulating cancer biomarkers with metal nanoparticles", *Nanoscale,* vol. 11, no. 46, pp. 22152-22171, 2019.
[http://dx.doi.org/10.1039/C9NR03040A] [PMID: 31555790]

[21]   M. Shehata, A.M. Fekry, and A. Walcarius, "Moxifloxacin hydrochloride electrochemical detection at gold nanoparticles modified screen-printed electrode", *Sensors (Basel),* vol. 20, no. 10, pp. 2797-2814, 2020.
[http://dx.doi.org/10.3390/s20102797] [PMID: 32423013]

[22]   R. Abdel-Karim, Y. Reda, and A. Abdel-Fattah, "Review—nanostructured materials-based nanosensors", *J. Electrochem. Soc.,* vol. 167, pp. 037554-0375566, 2020.
[http://dx.doi.org/10.1149/1945-7111/ab67aa]

[23]   T. Kokulnathan, T.S.K. Sharm, C. Shen-Ming, C. Tse-Wei, and B. Dinesh, "Ex-situ decoration of graphene oxide with palladium nanoparticles for the highly sensitive and selective electrochemical determination of chloramphenicol in food and biological samples", *J. Taiwan Inst. Chem. Eng.,* vol. 89, pp. 26-38, 2018.
[http://dx.doi.org/10.1016/j.jtice.2018.04.030]

[24]   N.N. Salama, S.M. Azab, M.A. Mohamed, and A.M. Fekry, "A novel methionine/palladium nanoparticle modified carbon paste electrode for simultaneous determination of three antiparkinson drugs", *RSC Advances,* vol. 5, pp. 14187-14195, 2015.
[http://dx.doi.org/10.1039/C4RA15909H]

[25]   Y. Li, J.Z. Sun, C. Bian, J.H. Tong, H.P. Dong, H. Zhang, and S.H. Xia, "Copper nano-clusters prepared by one-step electrodeposition and its application on nitrate sensing", *AIP Adv.,* vol. 5, pp. 041312-041318, 2015.
[http://dx.doi.org/10.1063/1.4905712]

[26]   M. Ateeq, M.R. Shah, N.U. Ain, S. Bano, I. Anis, Lubna, S. Faizi, M.F. Bertino, and S. Sohaila Naz, "Green synthesis and molecular recognition ability of patuletin coated gold nanoparticles", *Biosens. Bioelectron.,* vol. 63, pp. 499-505, 2015.
[http://dx.doi.org/10.1016/j.bios.2014.07.076] [PMID: 25129513]

[27]   A. Staerz, U. Weimar, and N. Barsan, "Understanding the potential of WO3 based sensors for breath analysis", *Sensors (Basel),* vol. 16, no. 11, pp. 1815-1832, 2016.
[http://dx.doi.org/10.3390/s16111815] [PMID: 27801881]

[28]   G. Niu, C. Zhao, H. Gong, Z. Yang, X. Leng, and F. Wang, "NiO nanoparticle-decorated SnO2 nanosheets for ethanol sensing with enhanced moisture resistance", *Microsyst. Nanoeng.,* vol. 5, p. 21, 2019.
[http://dx.doi.org/10.1038/s41378-019-0060-7] [PMID: 31123595]

[29]   S. Reddy, B.K. Swamy, S. Ramakrishana, L. He, and H. Jayadevappa, "NiO nanoparticles based carbon paste as a sensor for detection of dopamine", *Int. J. Electrochem. Sci.,* vol. 13, pp. 5748-5761, 2018.
[http://dx.doi.org/10.20964/2018.06.06]

[30]   J. Wang, Y. Zhou, M. He, P. Wangyang, Y. Lu, and L. Gu, "Electrolytic approach towards the

controllable synthesis of NiO nanocrystalline and self-assembly mechanism of Ni(OH)$_2$ precursor under electric, temperature and magnetic fields.", *CrystEngComm,* vol. 20, pp. 2384-2395, 2018.
[http://dx.doi.org/10.1039/C8CE00263K]

[31] M.M. Tunesi, R.A. Soomro, and R. Ozturk, "CuO nanostructures for highly sensitive shape dependent electrocatalytic oxidation of N-acetyl-L-cysteine", *J. Electroanal. Chem. (Lausanne),* vol. 777, pp. 40-47, 2016.
[http://dx.doi.org/10.1016/j.jelechem.2016.07.034]

[32] M. ElShal, "A copper oxide nanoparticle modified screen-printed electrode for determination of mirtazapine", *Egypt. J. Chem.,* vol. 62, pp. 1739-1748, 2019.
[http://dx.doi.org/10.21608/ejchem.2019.10062.1663]

[33] J. Tashkhourian, and S.F. Nami-Ana, "A sensitive electrochemical sensor for determination of gallic acid based on SiO2 nanoparticle modified carbon paste electrode", *Mater. Sci. Eng. C,* vol. 52, pp. 103-110, 2015.
[http://dx.doi.org/10.1016/j.msec.2015.03.017] [PMID: 25953546]

[34] J. Tashkhourian, S.F.N. Ana, S. Hashemnia, and M.R. Hormozi-Nezhad, Construction of a modified carbon paste electrode based on TiO$_2$ nanoparticles for the determination of gallic acid., *J. Solid State Electrochem.,* vol. 17, pp. 157-165, 2013.
[http://dx.doi.org/10.1007/s10008-012-1860-y]

[35] T.A. Rocha-Santos, "Sensors and biosensors based on magnetic nanoparticles, TrAC", *Trends Analyt. Chem.,* vol. 62, pp. 28-36, 2014.
[http://dx.doi.org/10.1016/j.trac.2014.06.016]

[36] A. Lassoued, Ben hassine, M.; Karolak, F.; Dkhil, B.; Ammar, S.; Gadri, A. Synthesis and magnetic characterization of spinel ferrites MFe2O4 (M = Ni, Co, Zn and Cu) *via* chemical co-precipitation method., *J. Mater. Sci. Mater. Electron.,* vol. 28, pp. 18857-18864, 2017.
[http://dx.doi.org/10.1007/s10854-017-7837-y]

[37] P. Garcia-Muñoz, and F. Fresno, "A de la Peña O'Shea, V.; Keller, N. Ferrite materials for photoassisted environmental and solar fuels applications", *Top. Curr. Chem. (Cham),* vol. 378, pp. 107-1062, 2020.
[http://dx.doi.org/10.1007/s41061-019-0270-3]

[38] H. El-Desoky, M. Ghoneim, M. Gado, and M. Abdel-Galeil, Development of Feasible and Economic Electrochemical Sensor Based on Manganese Ferrite Nanoparticles (Mn$_{0.2}$Fe$_{2.8}$O$_4$) for Determination of a Platelet Aggregation Inhibitor Ticagrelor Drug in Formulations and Human Blood., *J. Electrochem. Soc.,* vol. 167, pp. 067510-067521, 2020.
[http://dx.doi.org/10.1149/1945-7111/ab7e21]

[39] P.S. Karthik, "P. S.; A. L. Himaja, Singh, S. P.; Carbon-allotropes: synthesis methods, applications and future perspectives", *Carbon Letters,* vol. 15, pp. 219-237, 2014.
[http://dx.doi.org/10.5714/CL.2014.15.4.219]

[40] A. Walcarius, "Mesoporous materials and electrochemistry", *Chem. Soc. Rev.,* vol. 42, no. 9, pp. 4098-4140, 2013.
[http://dx.doi.org/10.1039/c2cs35322a] [PMID: 23334166]

[41] A. Walcarius, "Electrocatalysis, sensors and biosensors in analytical chemistry based on ordered mesoporous and macroporous carbon-modified electrodes", *Trends Analyt. Chem.,* vol. 38, pp. 79-97, 2012.
[http://dx.doi.org/10.1016/j.trac.2012.05.003]

[42] M. Zhou, L-P. Guo, Y. Hou, and X-J. Peng, "Peng, immobilization of nafion-ordered mesoporous carbon on a glassy carbon electrode: Application to the detection of epinephrine", *Electrochim. Acta,* vol. 53, pp. 4176-4184, 2008.
[http://dx.doi.org/10.1016/j.electacta.2007.12.077]

[43] A.K. Geim, and K.S. Novoselov, "The rise of graphene", *Nat. Mater.,* vol. 6, no. 3, pp. 183-191, 2007.

[http://dx.doi.org/10.1038/nmat1849] [PMID: 17330084]

[44] A. Peigney, C. Laurent, E. Flahaut, R. Bacsa, and A. Rousset, "Specific surface area of carbon nanotubes and bundles of carbon nanotubes", *Carbon,* vol. 39, pp. 507-514, 2001.
[http://dx.doi.org/10.1016/S0008-6223(00)00155-X]

[45] D.M. Guldi, and N. Martın, "Carbon Nanotubes and Related Structures: Synthesis, Characterization, Functionalization and Applications", In: *John Wiley & Sons* Wiley-VCH: Weinheim, 2010, p. 562.
[http://dx.doi.org/10.1002/9783527629930]

[46] R. Chokkareddy, N.K. Bhajanthri, and G.G. Redhi, "An Enzyme-induced novel biosensor for the sensitive electrochemical determination of isoniazid", *Biosensors (Basel),* vol. 7, no. 2, pp. 21-32, 2017.
[http://dx.doi.org/10.3390/bios7020021] [PMID: 28587260]

[47] S. Shahrokhian, and S. Rastgar, "Construction of an electrochemical sensor based on the electrodeposition of Au-Pt nanoparticles mixtures on multi-walled carbon nanotubes film for voltammetric determination of cefotaxime", *Analyst (Lond.),* vol. 137, no. 11, pp. 2706-2715, 2012.
[http://dx.doi.org/10.1039/c2an35182j] [PMID: 22543355]

[48] J. Scremin, E.C.M. Barbosa, C.A.R. Salamanca-Neto, P.H.C. Camargo, and E.R. Sartori, "Amperometric determination of ascorbic acid with a glassy carbon electrode modified with TiO2-gold nanoparticles integrated into carbon nanotubes", *Mikrochim. Acta,* vol. 185, no. 5, pp. 251-262, 2018.
[http://dx.doi.org/10.1007/s00604-018-2785-7] [PMID: 29651559]

[49] M-B. Gholivand, and M. Solgi, "Simultaneous electrochemical sensing of warfarin and maycophenolic acid in biological samples", *Anal. Chim. Acta,* vol. 1034, pp. 46-55, 2018.
[http://dx.doi.org/10.1016/j.aca.2018.06.045] [PMID: 30193639]

[50] C. Liu, J. Huang, and L. Wang, "Electrochemical synthesis of a nanocomposite consisting of carboxy-modified multi-walled carbon nanotubes, polythionine and platinum nanoparticles for simultaneous voltammetric determination of myricetin and rutin", *Mikrochim. Acta,* vol. 185, no. 9, pp. 414-423, 2018.
[http://dx.doi.org/10.1007/s00604-018-2947-7] [PMID: 30116901]

[51] A. Shams, and A. Yari, "A new sensor consisting of Ag-MWCNT nanocomposite as the sensing element for electrochemical determination of Epirubicin", *Sens. Actuators B Chem.,* vol. 286, pp. 131-138, 2019.
[http://dx.doi.org/10.1016/j.snb.2019.01.128]

[52] F-Y. Kong, R-F. Li, L. Yao, Z-X. Wang, W-X. Lv, and W. Wang, "An electrochemical daunorubicin sensor based on the use of platinum nanoparticles loaded onto a nanocomposite prepared from nitrogen decorated reduced graphene oxide and single-walled carbon nanotubes", *Mikrochim. Acta,* vol. 186, no. 5, p. 321, 2019.
[http://dx.doi.org/10.1007/s00604-019-3456-z] [PMID: 31049702]

[53] N.O. Weiss, H. Zhou, L. Liao, Y. Liu, S. Jiang, Y. Huang, and X. Duan, "Graphene: an emerging electronic material", *Adv. Mater.,* vol. 24, no. 43, pp. 5782-5825, 2012.
[http://dx.doi.org/10.1002/adma.201201482] [PMID: 22930422]

[54] S. Kochmann, T. Hirsch, and O.S. Wolfbeis, "Graphenes in chemical sensors and biosensors. TrAC", *Trends Analyt. Chem.,* vol. 39, pp. 87-113, 2012.
[http://dx.doi.org/10.1016/j.trac.2012.06.004]

[55] Y. Liu, X. Dong, and P. Chen, "Biological and chemical sensors based on graphene materials", *Chem. Soc. Rev.,* vol. 41, no. 6, pp. 2283-2307, 2012.
[http://dx.doi.org/10.1039/C1CS15270J] [PMID: 22143223]

[56] E.V. Castro, H. Ochoa, M.I. Katsnelson, R.V. Gorbachev, D.C. Elias, K.S. Novoselov, A.K. Geim, and F. Guinea, "Limits on charge carrier mobility in suspended graphene due to flexural phonons", *Phys. Rev. Lett.,* vol. 105, no. 26, pp. 266601-266604, 2010.

[http://dx.doi.org/10.1103/PhysRevLett.105.266601] [PMID: 21231692]

[57]   K.S. Kim, Y. Zhao, H. Jang, S.Y. Lee, J.M. Kim, K.S. Kim, J.H. Ahn, P. Kim, J.Y. Choi, and B.H. Hong, "Large-scale pattern growth of graphene films for stretchable transparent electrodes", *Nature*, vol. 457, no. 7230, pp. 706-710, 2009.
[http://dx.doi.org/10.1038/nature07719] [PMID: 19145232]

[58]   K.S. Novoselov, A.K. Geim, S.V. Morozov, D. Jiang, Y. Zhang, S.V. Dubonos, I.V. Grigorieva, and A.A. Firsov, "Electric field effect in atomically thin carbon films", *Science*, vol. 306, no. 5696, pp. 666-669, 2004.
[http://dx.doi.org/10.1126/science.1102896] [PMID: 15499015]

[59]   H. Gao, K. Zhu, G. Hu, and C. Xue, "Large-scale graphene production by ultrasound-assisted exfoliation of natural graphite in supercritical CO_2/H_2O medium", *Chem. Eng. J.*, vol. 308, pp. 872-879, 2017.
[http://dx.doi.org/10.1016/j.cej.2016.09.132]

[60]   A.B. Bourlinos, V. Georgakilas, R. Zboril, T.A. Steriotis, and A.K. Stubos, "Liquid-phase exfoliation of graphite towards solubilized graphenes", *Small*, vol. 5, no. 16, pp. 1841-1845, 2009.
[http://dx.doi.org/10.1002/smll.200900242] [PMID: 19408256]

[61]   L. Yan, Y.B. Zheng, F. Zhao, S. Li, X. Gao, B. Xu, P.S. Weiss, and Y. Zhao, "Chemistry and physics of a single atomic layer: strategies and challenges for functionalization of graphene and graphene-based materials", *Chem. Soc. Rev.*, vol. 41, no. 1, pp. 97-114, 2012.
[http://dx.doi.org/10.1039/C1CS15193B] [PMID: 22086617]

[62]   Y. Zhu, S. Murali, W. Cai, X. Li, J.W. Suk, J.R. Potts, and R.S. Ruoff, "Graphene and graphene oxide: synthesis, properties, and applications", *Adv. Mater.*, vol. 22, no. 35, pp. 3906-3924, 2010.
[http://dx.doi.org/10.1002/adma.201001068] [PMID: 20706983]

[63]   S. Park, J. An, I. Jung, R.D. Piner, S.J. An, X. Li, A. Velamakanni, and R.S. Ruoff, "Colloidal suspensions of highly reduced graphene oxide in a wide variety of organic solvents", *Nano Lett.*, vol. 9, no. 4, pp. 1593-1597, 2009.
[http://dx.doi.org/10.1021/nl803798y] [PMID: 19265429]

[64]   M.A. Rafiee, J. Rafiee, Z. Wang, H. Song, Z-Z. Yu, and N. Koratkar, "Enhanced mechanical properties of nanocomposites at low graphene content", *ACS Nano*, vol. 3, no. 12, pp. 3884-3890, 2009.
[http://dx.doi.org/10.1021/nn9010472] [PMID: 19957928]

[65]   Y. Shao, J. Wang, M. Engelhard, C. Wang, and Y. Lin, "Facile and controllable electrochemical reduction of graphene oxide and its applications", *J. Mater. Chem.*, vol. 20, pp. 743-748, 2010.
[http://dx.doi.org/10.1039/B917975E]

[66]   M. Venu, S. Venkateswarlu, Y.V.M. Reddy, A. Seshadri Reddy, V.K. Gupta, M. Yoon, and G. Madhavi, "Highly Sensitive Electrochemical Sensor for Anticancer Drug by a Zirconia Nanoparticle-Decorated Reduced Graphene Oxide Nanocomposite", *ACS Omega*, vol. 3, no. 11, pp. 14597-14605, 2018.
[http://dx.doi.org/10.1021/acsomega.8b02129] [PMID: 30555980]

[67]   A.T.E. Vilian, S.M. Chen, L.H. Huang, M.A. Ali, and F.M.A. Al-Hemaid, "Simultaneous determination of catechol and hydroquinone using a Pt/ZrO2-rGO/GCE composite modified glassy carbon electrode", *Electrochim. Acta*, vol. 125, pp. 503-509, 2014.
[http://dx.doi.org/10.1016/j.electacta.2014.01.092]

[68]   N. Zhang, X. Yan, Y. Huang, J. Li, J. Ma, and D.H.L. Ng, "Electrostatically Assembled Magnetite Nanoparticles/Graphene Foam as a Binder-Free Anode for Lithium Ion Battery", *Langmuir*, vol. 33, no. 36, pp. 8899-8905, 2017.
[http://dx.doi.org/10.1021/acs.langmuir.7b01519] [PMID: 28768104]

[69]   R.M. Cornell, and U. Schwertmann, "The Iron Oxides: Structure, Properties, Reactions, Occurrences and Uses", In: *John Wiley & Sons* VCH: New York, 2003, p. 664.

[70]   J. Salamon, Y. Sathishkumar, K. Ramachandran, Y.S. Lee, D.J. Yoo, A.R. Kim, and G. Gnana Kumar, "One-pot synthesis of magnetite nanorods/graphene composites and its catalytic activity toward electrochemical detection of dopamine", *Biosens. Bioelectron.,* vol. 64, pp. 269-276, 2015.
[http://dx.doi.org/10.1016/j.bios.2014.08.085] [PMID: 25240127]

[71]   N.T.A. Thu, H.V. Duc, N.H. Phong, N.D. Cuong, N.T. Hoan, and D.Q. Khieu, "Electrochemical determination of paracetamol using Fe3O4/reduced graphene-oxide-based electrode", *J. Nanomater.,* vol. •••, p. 7619419, 2018.
[http://dx.doi.org/10.1155/2018/7619419]

[72]   V. Georgakilas, M. Otyepka, A.B. Bourlinos, V. Chandra, N. Kim, K.C. Kemp, P. Hobza, R. Zboril, and K.S. Kim, "Functionalization of graphene: covalent and non-covalent approaches, derivatives and applications", *Chem. Rev.,* vol. 112, no. 11, pp. 6156-6214, 2012.
[http://dx.doi.org/10.1021/cr3000412] [PMID: 23009634]

[73]   G. Chen, Y. Liu, Y. Liu, Y. Tian, and X. Zhang, "Zhang, Nitrogen and sulfur dual-doped graphene for glucose biosensor application", *J. Electroanal. Chem. (Lausanne),* vol. 738, pp. 100-107, 2015.
[http://dx.doi.org/10.1016/j.jelechem.2014.11.020]

[74]   M.H. Ghanbari, Z. Norouzi, and M.M. Ghanbari, "Using a nanocomposite consist of boron-doped reduced graphene oxide and electropolymerized b-cyclodextrin for flunitrazepam electrochemical sensor", *Microchem,* vol. 156, pp. 104994-1041004, 2020.
[http://dx.doi.org/10.1016/j.microc.2020.104994]

[75]   A. Mohammadian, M. Ebrahimi, and H. Karimi-Maleh, "Synergic effect of 2D nitrogen doped reduced graphene nano-sheet and ionic liquid as a new approach for fabrication of anticancer drug sensor in analysis of doxorubicin and topotecan", *J. Mol. Liq.,* vol. 265, pp. 727-732, 2018.
[http://dx.doi.org/10.1016/j.molliq.2018.07.026]

[76]   K-P. Wang, Y-C. Zhang, X. Zhang, and L. Shen, "Green preparation of chlorine-doped graphene and its application in electrochemical sensor for chloramphenicol detection", In: *SN Appl. Sci,* 2019, p. 157.

[77]   X. Zhang, K-P. Wang, L-N. Zhang, Y-C. Zhang, and L. Shen, "Phosphorus-doped graphene-based electrochemical sensor for sensitive detection of acetaminophen", *Anal. Chim. Acta,* vol. 1036, pp. 26-32, 2018.
[http://dx.doi.org/10.1016/j.aca.2018.06.079] [PMID: 30253834]

[78]   M. Ma, H. Li, Y. Xiong, and F. Dong, "Rational design, synthesis, and application of silica/graphene-based nanocomposite: A review", *Mater. Des.,* vol. 198, pp. 109367-109385, 2021.
[http://dx.doi.org/10.1016/j.matdes.2020.109367]

[79]   Y. Du, L. Liu, Y. Xiang, and Q. Zhang, "Enhanced electrochemical capacitance and oilabsorbability of N-doped graphene aerogel by using amino-functionalized silica as template and doping agent", *J. Power Sources,* vol. 379, pp. 240-248, 2018.
[http://dx.doi.org/10.1016/j.jpowsour.2018.01.047]

[80]   F.H. Cincotto, T.C. Canevari, A.M. Campos, R. Landers, and S.A. Machado, "Simultaneous determination of epinephrine and dopamine by electrochemical reduction on the hybrid material SiO2/graphene oxide decorated with Ag nanoparticles", *Analyst (Lond.),* vol. 139, no. 18, pp. 4634-4640, 2014.
[http://dx.doi.org/10.1039/C4AN00580E] [PMID: 25050410]

[81]   C.O. Chikere, N.H. Faisal, P.K.T. Lin, and C. Fernandez, "The synergistic effect between graphene oxide nanocolloids and silicon dioxide nanoparticles for gallic acid sensing", *J. Solid State Electrochem.,* vol. 23, pp. 1795-1809, 2019.
[http://dx.doi.org/10.1007/s10008-019-04267-9]

[82]   W. Lu, Y. Luo, G. Chang, and X. Sun, "Synthesis of functional SiO2-coated graphene oxide nanosheets decorated with Ag nanoparticles for H2O2 and glucose detection", *Biosens. Bioelectron.,* vol. 26, no. 12, pp. 4791-4797, 2011.

[http://dx.doi.org/10.1016/j.bios.2011.06.008] [PMID: 21733668]

[83]   J. Pastor-Villegas, J.F. Pastor-Valle, J.M.M. Rodríguez, and M.G. García, "Study of commercial wood charcoals for the preparation of carbon adsorbents", *J. Anal. Appl. Pyrolysis,* vol. 76, pp. 103-108, 2006.
[http://dx.doi.org/10.1016/j.jaap.2005.08.002]

[84]   J. Huang, Q. Zeng, S. Bai, and L. Wang, "Application of coal in electrochemical sensing", *Anal. Chem.,* vol. 89, no. 16, pp. 8358-8365, 2017.
[http://dx.doi.org/10.1021/acs.analchem.7b01612] [PMID: 28700826]

[85]   J. McDonald-Wharry, M. Manley-Harris, and K. Pickering, "Carbonisation of biomass-derived chars and the thermal reduction of a graphene oxide sample studied using Raman spectroscopy", *Carbon,* vol. 59, pp. 383-405, 2013.
[http://dx.doi.org/10.1016/j.carbon.2013.03.033]

[86]   L. Lu, V. Sahajwalla, C. Kong, and D. Harris, "Quantitative X-Ray diffraction analysis and its application to various coals", *Carbon,* vol. 39, pp. 1821-1833, 2001.
[http://dx.doi.org/10.1016/S0008-6223(00)00318-3]

[87]   B. Manoj, and A.G. Kunjomana, "Study of stacking structure of amorphous carbon by X-Ray diffraction technique", *Int. J. Electrochem. Sci.,* vol. 7, pp. 3127-3134, 2012.

[88]   S. Park, J. An, J.R. Potts, A. Velamakanni, S. Murali, and R.S. Ruoff, "Hydrazine-reduction of graphite- and graphene oxide", *Carbon,* vol. 49, pp. 3019-3023, 2011.
[http://dx.doi.org/10.1016/j.carbon.2011.02.071]

[89]   H. Furukawa, K.E. Cordova, M. O'Keeffe, and O.M. Yaghi, "The chemistry and applications of metal-organic frameworks", *Science,* vol. 341, no. 6149, p. 1230444, 2013.
[http://dx.doi.org/10.1126/science.1230444] [PMID: 23990564]

[90]   W. Lu, Z. Wei, Z-Y. Gu, T-F. Liu, J. Park, J. Park, J. Tian, M. Zhang, Q. Zhang, T. Gentle III, M. Bosch, and H-C. Zhou, "Tuning the structure and function of metal-organic frameworks via linker design", *Chem. Soc. Rev.,* vol. 43, no. 16, pp. 5561-5593, 2014.
[http://dx.doi.org/10.1039/C4CS00003J] [PMID: 24604071]

[91]   W. Xuan, C. Zhu, Y. Liu, and Y. Cui, "Mesoporous metal-organic framework materials", *Chem. Soc. Rev.,* vol. 41, no. 5, pp. 1677-1695, 2012.
[http://dx.doi.org/10.1039/C1CS15196G] [PMID: 22008884]

[92]   L.G. Qiu, T. Xu, Z.Q. Li, W. Wang, Y. Wu, X. Jiang, X.Y. Tian, and L.D. Zhang, "Hierarchically micro- and mesoporous metal-organic frameworks with tunable porosity", *Angew. Chem. Int. Ed. Engl.,* vol. 47, no. 49, pp. 9487-9491, 2008.
[http://dx.doi.org/10.1002/anie.200803640] [PMID: 18972472]

[93]   P.R. Jothi, R.R. Salunkhe, M. Pramanik, S. Kannan, and Y. Yamauchi, "Surfactant-assisted synthesis of nanoporous nickel sulfide flakes and their hybridization with reduced graphene oxides for supercapacitor applications", *RSC Advances,* vol. 6, pp. 21246-21253, 2016.
[http://dx.doi.org/10.1039/C5RA26946F]

[94]   Y. Yue, Z-A. Qiao, P.F. Fulvio, A.J. Binder, C. Tian, J. Chen, K.M. Nelson, X. Zhu, and S. Dai, "Template-free synthesis of hierarchical porous metal-organic frameworks", *J. Am. Chem. Soc.,* vol. 135, no. 26, pp. 9572-9575, 2013.
[http://dx.doi.org/10.1021/ja402694f] [PMID: 23796254]

[95]   L. Shi, X. Zhu, T. Liu, H. Zhao, and M. Lan, "Encapsulating Cu nanoparticles into metal-organic frameworks fornonenzymatic glucose sensing", *Sens. Actuators B Chem.,* vol. 227, pp. 583-590, 2016.
[http://dx.doi.org/10.1016/j.snb.2015.12.092]

[96]   J. Zhou, X. Li, L. Yang, S. Yan, M. Wang, D. Cheng, Q. Chen, Y. Dong, P. Liu, W. Cai, and C. Zhang, "The Cu-MOF-199/single-walled carbon nanotubes modified electrode for simultaneous determination of hydroquinone and catechol with extended linear ranges and lower detection limits",

Anal. Chim. Acta, vol. 899, pp. 57-65, 2015.
[http://dx.doi.org/10.1016/j.aca.2015.09.054] [PMID: 26547493]

[97]   D. Zhang, J. Zhang, H. Shi, X. Guo, Y. Guo, R. Zhang, and B. Yuan, Redox-active microsized metal-organic framework for efficient nonenzymatic H₂O₂ sensing., *Sens. Actuators B Chem.,* vol. 221, pp. 224-229, 2015.
[http://dx.doi.org/10.1016/j.snb.2015.06.079]

[98]   Y. Li, C. Huangfu, H. Du, W. Liu, Y. Li, and J. Ye, "Electrochemical behavior of metal-organic framework MIL-101 modified carbon paste electrode: An excellent candidate for electroanalysis", *J. Electroanal. Chem. (Lausanne),* vol. 709, pp. 65-69, 2013.
[http://dx.doi.org/10.1016/j.jelechem.2013.09.017]

[99]   M. Zhong, L. Yang, H. Yang, C. Cheng, W. Deng, Y. Tan, Q. Xie, and S. Yao, An electrochemical immunobiosensor for ultrasensitive detection of *Escherichia coli* O157:H7 using CdS quantum dots-encapsulated metal-organic frameworks as signal-amplifying tags., *Biosens. Bioelectron.,* vol. 126, pp. 493-500, 2019.
[http://dx.doi.org/10.1016/j.bios.2018.11.001] [PMID: 30476880]

[100]   P. Dong, L. Zhu, J. Huang, J. Ren, and J. Lei, "Electrocatalysis of cerium metal-organic frameworks for ratiometric electrochemical detection of telomerase activity", *Biosens. Bioelectron.,* vol. 138, p. 111313, 2019.
[http://dx.doi.org/10.1016/j.bios.2019.05.018] [PMID: 31108380]

[101]   Q. Qiu, H. Chen, Y. Wang, and Y. Ying, "Recent advances in the rational synthesis and sensing applications of metal-organic framework biocomposites", *Coord. Chem. Rev.,* vol. 387, pp. 60-78, 2019.
[http://dx.doi.org/10.1016/j.ccr.2019.02.009]

[102]   S. Motoc, F. Manea, A. Iacob, A. Martinez-Joaristi, J. Gascon, A. Pop, and J. Schoonman, "Electrochemical selective and simultaneous detection of diclofenac and ibuprofen in aqueous solution using HKUST-1 metal-organic framework-carbon nanofiber composite electrode", *Sensors (Basel),* vol. 16, no. 10, p. 1719, 2016.
[http://dx.doi.org/10.3390/s16101719] [PMID: 27763509]

[103]   G. Luo, Y. Deng, X. Zhang, R. Zou, W. Sun, B. Li, B. Sun, Y. Wang, and G.A. Li, "ZIF-8 derived nitrogen-doped porous carbon and nitrogen-doped graphene nanocomposite modified electrode for simultaneous determination of ascorbic acid, dopamine and uric acid", *New J. Chem.,* vol. 43, pp. 16819-16828, 2019.
[http://dx.doi.org/10.1039/C9NJ04095A]

[104]   T.H. Le, Y. Kim, and H. Yoon, "Electrical and electrochemical properties of conducting polymers", *Polymers (Basel),* vol. 9, no. 4, pp. 150-182, 2017.
[http://dx.doi.org/10.3390/polym9040150] [PMID: 30970829]

[105]   N.F. Atta, G.G. Abdo, A. Elzatahry, A. Galal, and S.H. Hassan, "Designed electrochemical sensor based on metallocene modified conducting polymer composite for effective determination of tramadol in real samples", *Can. J. Chem.,* vol. 99, pp. 437-446, 2021.
[http://dx.doi.org/10.1139/cjc-2020-0199]

[106]   W. Su, and S. Cheng, "Electrochemical oxidation and sensitive determination of acetaminophen in pharmaceuticals at poly(3,4-ethylenedioxythiophene)-modified screen-printed electrodes", *Electroanal.,* vol. 22, pp. 707-714, 2010.
[http://dx.doi.org/10.1002/elan.200900455]

[107]   A. Kannan, and R. Sevvel, "A highly selective and simultaneous determination of paracetamol and dopamine using poly-4-amino-6-hydroxy-2-mercaptopyrimidine (Poly-AHMP) film modified glassy carbon electrode", *J. Electroanal. Chem. (Lausanne),* vol. 791, pp. 8-16, 2017.
[http://dx.doi.org/10.1016/j.jelechem.2017.03.002]

[108]   G.E. Uwaya, and O.E. Fayemi, "Electrochemical detection of serotonin in banana at green mediated

PPy/Fe3O4 NPs nanocomposites modified electrodes", *Sens. Biosensing Res.,* vol. 28, p. 100338, 2020.
[http://dx.doi.org/10.1016/j.sbsr.2020.100338]

[109] Z. Yu, H. Li, X. Zhang, N. Liu, W. Tan, X. Zhang, and L. Zhang, "Facile synthesis of NiCo2O4@Polyaniline core-shell nanocomposite for sensitive determination of glucose", *Biosens. Bioelectron.,* vol. 75, pp. 161-165, 2016.
[http://dx.doi.org/10.1016/j.bios.2015.08.024] [PMID: 26318785]

[110] Y. Fan, J.H. Liu, C.P. Yang, M. Yu, and P. Liu, "Graphene–polyaniline composite film modified electrode for voltammetric determination of 4-aminophenol", *Sens. Actuators B Chem.,* vol. 157, pp. 669-674, 2011.
[http://dx.doi.org/10.1016/j.snb.2011.05.053]

[111] K. Radhapyari, P. Kotoky, M.R. Das, and R. Khan, "Graphene-polyaniline nanocomposite based biosensor for detection of antimalarial drug artesunate in pharmaceutical formulation and biological fluids", *Talanta,* vol. 111, pp. 47-53, 2013.
[http://dx.doi.org/10.1016/j.talanta.2013.03.020] [PMID: 23622524]

[112] R. Jain, D.C. Tiwari, and S. Shrivastava, "A sensitive voltammetric sensor based on synergistic effect of polyaniline and zirconia nanocomposite film for quantification of proton pump inhibitor esomeprazole", *J. Electrochem. Soc.,* vol. 161, pp. B39-B44, 2014.
[http://dx.doi.org/10.1149/2.018404jes]

[113] X. Ma, T. Miao, W. Zhu, X. Gao, C. Wang, C. Zhaoa, and H. Ma, "Electrochemical detection of nitrite based on glassy carbon electrode modified with gold–polyaniline–graphene nanocomposites", *RSC Advances,* vol. 4, pp. 57842-57849, 2014.
[http://dx.doi.org/10.1039/C4RA08543D]

[114] R. Jain, and S. Shrivastava, "A graphene-polyaniline-Bi2O3hybrid film sensor for voltammetric quantification of anti-inflammatory drug Etodolac", *J. Electrochem. Soc.,* vol. 161, pp. H189-H194, 2014.
[http://dx.doi.org/10.1149/2.043404jes]

[115] R. Devi, M. Thakur, and C.S. Pundir, "Construction and application of an amperometric xanthine biosensor based on zinc oxide nanoparticles-polypyrrole composite film", *Biosens. Bioelectron.,* vol. 26, no. 8, pp. 3420-3426, 2011.
[http://dx.doi.org/10.1016/j.bios.2011.01.014] [PMID: 21324666]

[116] L. Xing, Q. Rong, and Z. Ma, "Non-enzymatic electrochemical sensing of hydrogen peroxide based on polypyrrole/platinum nanocomposites", *Sens. Actuators B Chem.,* vol. 221, pp. 242-247, 2015.
[http://dx.doi.org/10.1016/j.snb.2015.06.078]

[117] Y-S. Gao, J.K. Xu, L-M. Lu, L.P. Wu, K.X. Zhang, T. Nie, X.F. Zhu, and Y. Wu, "Overoxidized polypyrrole/graphene nanocomposite with good electrochemical performance as novel electrode material for the detection of adenine and guanine", *Biosens. Bioelectron.,* vol. 62, pp. 261-267, 2014.
[http://dx.doi.org/10.1016/j.bios.2014.06.044] [PMID: 25022509]

[118] H. Mao, J. Liang, H. Zhang, Q. Pei, D. Liu, S. Wu, Y. Zhang, and X-M. Song, "Poly(ionic liquids) functionalized polypyrrole/graphene oxide nanosheets for electrochemical sensor to detect dopamine in the presence of ascorbic acid", *Biosens. Bioelectron.,* vol. 70, pp. 289-298, 2015.
[http://dx.doi.org/10.1016/j.bios.2015.03.059] [PMID: 25840013]

[119] L. Lu, O. Zhang, J. Xu, X. Wen, and X. Duan, "Yu. H., Wu, L.A facile one-step redox route for the synthesis of graphene/poly (3,4-ethylenedioxythiophene) nanocomposite and their applications in biosensing", *Sens. Actuators B Chem.,* vol. 181, pp. 567-574, 2013.
[http://dx.doi.org/10.1016/j.snb.2013.02.024]

[120] V. Safran, I. Göktürk, A. Derazshamshir, F. Yılmaz, N. Sağlam, and A. Denizli, "Rapid sensing of Cu+2 in water and biological samples by sensitive molecularly imprinted based plasmonic biosensor", *Microchem. J.,* vol. 148, pp. 141-150, 2019.

[http://dx.doi.org/10.1016/j.microc.2019.04.069]

[121] Y. Liu, J. Liu, J. Liu, W. Gan, B-c. Ye, and Y. Li, "Highly sensitive and selective voltammetric determination of dopamine using a gold electrode modified with a molecularly imprinted polymeric film immobilized on flaked hollow nickel nanospheres", *Mikrochim. Acta,* vol. 184, pp. 1285-1294, 2017.
[http://dx.doi.org/10.1007/s00604-017-2124-4]

[122] J. Zhang, X-T. Guo, J-P. Zhou, G-Z. Liu, and S-Y. Zhang, "Electrochemical preparation of surface molecularly imprinted poly(3-aminophenylboronic acid)/MWCNTs nanocomposite for sensitive sensing of epinephrine", *Mater. Sci. Eng. C,* vol. 91, pp. 696-704, 2018.
[http://dx.doi.org/10.1016/j.msec.2018.06.011] [PMID: 30033304]

[123] M. Arvand, M. Farahpour, and M.S. Ardaki, "Electrochemical characterization of in situ functionalized gold organosulfur self-assembled monolayer with conducting polymer and carbon nanotubes for determination of rutin", *Talanta,* vol. 176, pp. 92-101, 2018.
[http://dx.doi.org/10.1016/j.talanta.2017.08.012] [PMID: 28917811]

[124] S. Jafari, M. Dehghani, N. Nasirizadeh, and M. Azimzadeh, "An azithromycin electrochemical sensor based on an anilineMIP film electropolymerized on a gold nano urchins/graphene oxide modified glassy carbon electrode", *J. Electroanal. Chem. (Lausanne),* vol. 829, pp. 27-34, 2018.
[http://dx.doi.org/10.1016/j.jelechem.2018.09.053]

[125] J. Cai, B. Sun, X. Gou, and Y. Gou, "LiW, Hu F. A novel way for analysis of calycosin via polyaniline functionalized grapheme quantum dots fabricated electrochemical sensor", *J. Electroanal. Chem. (Lausanne),* vol. 816, pp. 123-131, 2018.
[http://dx.doi.org/10.1016/j.jelechem.2018.03.035]

[126] F.M. Zahed, B. Hatamluyi, F. Lorestani, and Z. Es'haghi, "Silver nanoparticles decorated polyaniline nanocomposite based electrochemical sensor for the determination of anticancer drug 5-fluorouracil", *J. Pharm. Biomed. Anal.,* vol. 161, pp. 12-19, 2018.
[http://dx.doi.org/10.1016/j.jpba.2018.08.004] [PMID: 30142492]

[127] M. Mostafavi, M.R. Yaftian, F. Piri, and H. Shayani-Jam, "A new diclofenac molecularly imprinted electrochemical sensor based upon a polyaniline/reduced graphene oxide nano-composite", *Biosens. Bioelectron.,* vol. 122, pp. 160-167, 2018.
[http://dx.doi.org/10.1016/j.bios.2018.09.047] [PMID: 30265965]

[128] A.S. Rajpurohit, N.S. Punde, C.R. Rawool, and A.K. Srivastava, "Application of carbon paste electrode modified with carbon nanofibres/polyaniline/platinum nanoparticles as an electrochemical sensor for the determination of bezafibrate", *Electroanalysis,* vol. 30, pp. 571-582, 2018.
[http://dx.doi.org/10.1002/elan.201700781]

[129] S.K. Ponnaiah, and P. Periakaruppan, "A glassy carbon electrode modified with a copper tungstate and polyaniline nanocomposite for voltammetric determination of quercetin", *Mikrochim. Acta,* vol. 185, no. 11, p. 524, 2018.
[http://dx.doi.org/10.1007/s00604-018-3071-4] [PMID: 30374580]

[130] M. Dehghani, N. Nasirizadeh, and M.E. Yazdanshenas, "Determination of cefixime using a novel electrochemical sensor produced with gold nanowires/graphene oxide/electropolymerized molecular imprinted polymer", *Mater. Sci. Eng. C,* vol. 96, pp. 654-660, 2019.
[http://dx.doi.org/10.1016/j.msec.2018.12.002] [PMID: 30606577]

[131] A. Fatahi, R. Malakooti, and M. Shahlaei, "Electrocatalytic oxidation and determination of dexamethasone at a Fe3O4/PANI–Cu II microsphere modified carbon ionic liquid electrode", *RSC Advances,* vol. 7, pp. 11322-11330, 2017.
[http://dx.doi.org/10.1039/C6RA26125F]

[132] Q. Wang, R. Zhao, S. Wang, H. Guo, J. Li, H. Zhou, X. Wang, X. Wu, Y. Wang, and W. Chen, "A highly selective electrochemical sensor for nifedipine based on layer-by-layer assembly films from polyaniline and multiwalled carbon nanotube", *J. Appl. Polym. Sci.,* vol. •••, p. 133, 2016.

[http://dx.doi.org/10.1002/app.43452]

[133] J. Li, D. Kuang, Y. Feng, F. Zhang, Z. Xu, M. Liu, and D. Wang, "Green synthesis of silver nanoparticles-graphene oxide nanocomposite and its application in electrochemical sensing of tryptophan", *Biosens. Bioelectron.,* vol. 42, pp. 198-206, 2013.
[http://dx.doi.org/10.1016/j.bios.2012.10.029] [PMID: 23202352]

[134] L. Wen, Z. Lin, P. Gu, Z. Zhou, B. Yao, G. Chen, and J. Fu, "Extracellular biosynthesis of monodispersed gold nanoparticles by a SAM capping route", *J. Nanopart. Res.,* vol. 11, pp. 279-288, 2009.
[http://dx.doi.org/10.1007/s11051-008-9378-z]

[135] L. Du, H. Jiang, X. Liu, and E. Wang, "Biosynthesis of gold nanoparticles assisted by Escherichia coli DH5α and its application on direct electrochemistry of hemoglobin", *Electrochem. Commun.,* vol. 9, pp. 1165-1170, 2007.
[http://dx.doi.org/10.1016/j.elecom.2007.01.007]

[136] T. Wang, L. Yang, B. Zhang, and J. Liu, "Extracellular biosynthesis and transformation of selenium nanoparticles and application in H2O2 biosensor", *Colloids Surf. B Biointerfaces,* vol. 80, no. 1, pp. 94-102, 2010.
[http://dx.doi.org/10.1016/j.colsurfb.2010.05.041] [PMID: 20566271]

[137] S.S. Ahiwale, A.V. Bankar, S. Tagunde, and B.P. Kapadnis, "A bacteriophage mediated gold nanoparticles synthesis and their antibiofilm activity", *Indian J. Microbiol.,* vol. 57, no. 2, pp. 188-194, 2017.
[http://dx.doi.org/10.1007/s12088-017-0640-x] [PMID: 28611496]

[138] Y. Seo, S. Manivannan, I. Kang, S-W. Lee, and K. Kim, "Gold dendrites Co-deposited with M13 virus as a biosensor platform for nitrite ions", *Biosens. Bioelectron.,* vol. 94, pp. 87-93, 2017.
[http://dx.doi.org/10.1016/j.bios.2017.02.036] [PMID: 28262612]

[139] P-Y. Chen, X. Dang, M.T. Klug, N-M.D. Courchesne, J. Qi, M.N. Hyder, A.M. Belcher, and P.T. Hammond, "M13 virus-enabled synthesis of titanium dioxide nanowires for tunable mesoporous semiconducting networks", *Chem. Mater.,* vol. 27, pp. 1531-1540, 2015.
[http://dx.doi.org/10.1021/cm503803u]

[140] M. Gericke, and A. Pinches, "Microbial production of gold nanoparticles", *Gold Bull.,* vol. 39, pp. 22-28, 2006.
[http://dx.doi.org/10.1007/BF03215529]

[141] A. Ingle, M. Rai, A. Gade, and M. Bawaskar, "Fusarium solani: a novel biological agent for the extracellular synthesis of silver nanoparticles", *J. Nanopart. Res.,* vol. 11, pp. 2079-2085, 2009.
[http://dx.doi.org/10.1007/s11051-008-9573-y]

[142] R. Raliya, and J.C. Tarafdar, "Biosynthesis and characterization of zinc, magnesium and titanium nanoparticles: an eco-friendly approach", *Int. Nano Lett.,* vol. 4, pp. 93-103, 2014.
[http://dx.doi.org/10.1007/s40089-014-0093-8]

[143] S. Mondal, N. Roy, R.A. Laskar, I. Sk, S. Basu, D. Mandal, and N.A. Begum, "Biogenic synthesis of Ag, Au and bimetallic Au/Ag alloy nanoparticles using aqueous extract of mahogany (Swietenia mahogani JACQ.) leaves", *Colloids Surf. B Biointerfaces,* vol. 82, no. 2, pp. 497-504, 2011.
[http://dx.doi.org/10.1016/j.colsurfb.2010.10.007] [PMID: 21030220]

[144] D. Philip, and C. Unni, "Extracellular biosynthesis of gold and silver nanoparticles using Krishna tulsi (Ocimum sanctum) leaf", *Physica E,* vol. 43, pp. 1318-1322, 2011.
[http://dx.doi.org/10.1016/j.physe.2010.10.006]

[145] G.S. Ghodake, N.G. Deshpande, Y.P. Lee, and E.S. Jin, "Pear fruit extract-assisted room-temperature biosynthesis of gold nanoplates", *Colloids Surf. B Biointerfaces,* vol. 75, no. 2, pp. 584-589, 2010.
[http://dx.doi.org/10.1016/j.colsurfb.2009.09.040] [PMID: 19879738]

[146] S.P. Dubey, M. Lahtinen, and M. Sillanpää, "Tansy fruit mediated greener synthesis of silver and gold

nanoparticles", *Process Biochem.,* vol. 45, pp. 1065-1071, 2010.
[http://dx.doi.org/10.1016/j.procbio.2010.03.024]

[147] M.A. Khalilzadeh, and M. Borzoo, "Green synthesis of silver nanoparticles using onion extract and their application for the preparation of a modified electrode for determination of ascorbic acid", *J. Food Drug Anal.,* vol. 24, no. 4, pp. 796-803, 2016.
[http://dx.doi.org/10.1016/j.jfda.2016.05.004] [PMID: 28911618]

[148] K. Ramachandran, D. Kalpana, Y. Sathishkumar, Y.S. Lee, and K. Ravichandran, "Gnana kumar, G.; A facile green synthesis of silver nanoparticles using Piper betle 3 biomass and its catalytic activity toward sensitive and selective nitrite detection", *J. Ind. Eng. Chem.,* vol. 35, pp. 29-35, 2016.
[http://dx.doi.org/10.1016/j.jiec.2015.10.033]

[149] A.A. Mohamed, A. Fouda, M.A. Abdel-Rahman, S.E-D. Hassan, S. Gamal, S.S. Salem, and T.I. Shaheen, "Fungal strain impacts the shape, bioactivity and multifunctional properties of green synthesized zinc oxide nanoparticles", *Biocatal. Agric. Biotechnol.,* vol. 19, pp. 101103-101111, 2019.
[http://dx.doi.org/10.1016/j.bcab.2019.101103]

[150] S.E.L.D. Hassan, S.S. Salem, A. Fouda, M.A. Awad, M.S. El-Gamal, and A.M. Abdo, "New approach for antimicrobial activity and bio-control of various pathogens by biosynthesized copper nanoparticles using endophytic actinomycetes", *J. Radiat. Res. Appl. Sci.,* vol. 11, pp. 262-270, 2018.
[http://dx.doi.org/10.1016/j.jrras.2018.05.003]

[151] M. Nasrollahzadeh, R. Akbari, Z. Issaabadi, and S.M. Sajadi, "Biosynthesis and characterization of Ag/MgO nanocomposite and its catalytic performance in the rapid treatment of environmental contaminants (/paper Redirect/5863973)", *Ceram. Int.,* vol. 46, pp. 2093-2101, 2020.
[http://dx.doi.org/10.1016/j.ceramint.2019.09.191]

[152] Z. Vaseghi, O. Tavakoli, and A.N. Nematollahzadeh, "New insights into mechanistic aspects and structure of polycrystalline Cu/Cr/Ni metal oxide nanoclusters synthesized using Eryngium campestre and Froriepia subpinnata", *Korean Inst. Chem. Eng.,* vol. 36, pp. 489-499, 2019.
[http://dx.doi.org/10.1007/s11814-018-0216-4]

[153] P. Malik, R. Shankar, V. Malik, N. Sharma, and T.K. Mukherjee, "Green chemistry based benign routes for nanoparticle synthesis", *J. Nanopart.,* vol. 3, pp. 1-14, 2014.
[http://dx.doi.org/10.1155/2014/302429]

[154] J. Singh, T. Dutta, K-H. Kim, M. Rawat, P. Samddar, and P. Kumar, "'Green' synthesis of metals and their oxide nanoparticles: applications for environmental remediation", *J. Nanobiotech.,* vol. 16, no. 84, pp. 1-24, 2018.
[http://dx.doi.org/10.1186/s12951-018-0408-4]

[155] P. Li, Y. Ding, A. Wang, L. Zhou, S. Wei, Y. Zhou, Y. Tang, Y. Chen, C. Cai, and T. Lu, "Self-assembly of tetrakis (3-trifluoromethylphenoxy) phthalocyaninato cobalt(II) on multiwalled carbon nanotubes and their amperometric sensing application for nitrite", *ACS Appl. Mater. Interfaces,* vol. 5, no. 6, pp. 2255-2260, 2013.
[http://dx.doi.org/10.1021/am400152k] [PMID: 23452401]

[156] E.M. Egorova, and A.A. Revina, "Synthesis of metallic nanoparticles in reverse micelles in the presence of quercetin", *Colloids Surf. A Physicochem. Eng. Asp.,* vol. 168, pp. 87-96, 2000.
[http://dx.doi.org/10.1016/S0927-7757(99)00513-0]

[157] D.V. Ravi Kumar, S.R. Kumavat, V.N. Chamundeswari, P.P. Patra, A. Kulkarni, and B.L.V. Prasad, "Surfactant-free synthesis of anisotropic gold nanostructures: Can dicarboxylic acids alone act as shape directing agents?", *RSC Advances,* vol. 3, pp. 21641-21647, 2013.
[http://dx.doi.org/10.1039/c3ra43974g]

[158] A. González, E. Pérez, A. Almendarez, A. Villegas, and J. Vallejo-Montesinos, Calcium pimelate supported on TiO$_2$ nanoparticles as isotactic polypropylene prodegradant., *Poly. Bull,* 2016, pp. 39-51.

CHAPTER 2

Functionalized Nanomaterial-Based Electrochemical Sensors for Point-of-Care Devices

İsmail Mert ALKAÇ[1,2], Ramazan BAYAT[1,2], Muhammed BEKMEZCİ[1,2] and Fatih ŞEN[1,*]

[1] *Sen Research Group, Department of Biochemistry, University of Dumlupinar, 43000 Kütahya, Turkey*

[2] *Department of Materials Science & Engineering, Faculty of Engineering, Dumlupınar University, Evliya Çelebi Campus, 43100 Kutahya, Turkey*

Abstract: Electrochemical sensors used for the detection of specific biomolecules have attracted great attention in recent years due to their high sensitivity, selectivity, simple preparation, and fast response. In recent years, the integration of biological components into analytical tools, especially in biomedical research, has become a prerequisite for the early diagnosis of many diseases. In addition, enabling early diagnosis and treatment of diseases provides improved health outcomes and rapid medical decision-making. Here, we present the classification of biosensors and general concept, materials used in electrochemical sensors, their advantages, disadvantages, and main strategies in electrochemical biosensor technology such as electrodes and supporting substrates, materials for the bioreceptor materials for improved sensitivity and selectivity. This book chapter is focused on the functionalization of nanomaterials developed in recent years aimed at electrochemical biosensors and their functionalization for point of care devices. The various point of care device examples have been tried to form an idea about the usage areas.

Keywords: Diagnosis, Electrochemical Sensors, Nanomaterial, Point-of-Care Devices.

INTRODUCTION

Stable, cost-effective, and reliable detection of specific biomolecules or chemical products is of great importance in medicine, the environment, and industry [1]. Electrochemical sensors are frequently studied and used for their fast response, high sensitivity, specificity, and accuracy. They are also used in the production of miniaturized portable systems for point of care [1].

---

* **Corresponding author Fatih ŞEN:** Department of Materials Science & Engineering, Faculty of Engineering, Dumlupınar University, Evliya Çelebi Campus, 43100 Kutahya, Turkey; E-mail: fatihsen1980@gmail.com

J.G. Manjunatha (Ed.)

The production of electrochemical sensors is of great interest due to their high selectivity and sensitivity. It is increasingly used in industrial process monitoring, clinical diagnosis, analytical chemistry, environmental analysis, and food industries [2].

Biosensors are of interest to biomedical and pharmaceuticals due to their sensitivity, high selectivity, high cost/benefit ratios, specificity, simple use, fast data collection, and analysis [2].

The main advantages of biosensors over traditional analytical methods, which will lead to even more pronounced use in the biomedical field soon, are: analyte detection is often performed without prior separation; flexibility and simplicity of preparation; easy to use allows in-field or point-of-care (POC) measurements.; short possible response times of biological real-time monitoring and production processes; miniaturization and automation possibility; possibility of low production costs and mass production. The use of biosensors as components in medical devices has improved reliability in POC analysis and accessibility, portability, and real-time diagnosis [2].

Point of care testing facilitates disease diagnosis by providing rapid detection, monitoring, and management of analytes. Early diagnosis of the disease allows patients to start treatment early, allowing rapid medical decisions. Recently, several potential POC devices have been developed, leading to a new generation of POC tests [3].

Biosensors are directly responsible for bioanalytical performance and are critical components for POC devices because of these properties. They have therefore been investigated for forward-looking POC applications necessary for personalized healthcare management, as they predict and hence detect levels of disease-causing chemical reactions or biological markers, such as body fluids, by generating signals often associated with the concentration of an analyte. Their high sensitivity and selectivity allowed treatment of targeted diseases and early diagnosis. Therefore, in combination with timely treatment decisions, nanotechnology can improve the diagnosis and assessment of disease progression and facilitate treatment planning for many diseases [3].

Recently, various electrochemical biosensing methods have been miniaturized devices for *in situ* analysis and developed for simplicity. Besides, the high repeatability and sensitivity of the biosensor remain a challenge. The use of functional nanomaterials for a matrix to amplify signal amplification and high-performance electrochemical analysis draw attention [4]. Nanomaterials (NMs) have a wide surface area, resulting in a synergistic effect for signal amplification by supporting increased loading capacity and bulk transport of reactants [5].

Taking advantage of exceptional features such as being easy to use, economical, sensitive, portable and simple to make, great importance has been focused on the integration of recognition elements with electronic elements to develop electrochemical biosensors. Besides, the high repeatability and sensitivity of the biosensor remain a challenge. The use of functional nanomaterials for a matrix to amplify signal amplification and high-performance electrochemical analysis gained much attention.

The creation of new transmission networks of NMs, biosensors (sub-class electrochemical sensors) provide superior catalytic activity and strong paramagnetic properties and signal enhancement capabilities.

NANOMATERIALS

In the last decade, there has been a great development in nanotechnology due to the improvement in nanomaterials (NMs), including metallic and magnetic nanoparticles and carbon nanomaterials [6]. NMs commonly defined as materials with dimensions less than 100 nm have a large application area due to their small size properties (improved chemical reactivity, electrical conductivity, and tensile strength) achieved with increased surface area per unit weight [2].

Nanoparticles (NPs) are high sensitivity (thousands of atoms can be released from a nanoparticle), very stable (compared to enzyme labels), have large varieties of NPs, and are available on the market. NMs are mainly used as biomolecule tracer or electrode making. Nanoparticles are used today as electrochemical tags, as tools with hundreds or thousands of electroactive labels, reducing the limits of detection to several hundred biomolecules [7].

Carbon Nanotubes

As effective electrode materials, carbon nanotubes (CNT) have been used extensively for electrochemical sensing platforms. CNTs have a high electro-catalytic operation due to excellent electron transfer capabilities, ideal structural characteristics, wide surface areas, mechanical, electrical properties, and high sensitivity. NTs exhibit a high degree of immobilization of bio-recognition elements such as enzymes, aptamers, antibodies, imprinted polymers [8].

There are two types of CNTs; the first one is multi-walled carbon nanotubes (MWCNTs) and the other one is single-walled carbon nanotubes (SWCNT). Graphene, a 2-D hexagonal carbon atom model, is also used as an electrode due to its higher specific surface area than CNTs [9]. Graphene oxide (GOx) resolves this problem by increasing the removal of GO oxygen and hydrophilicity of the graphene layer, providing excellent ease of surface modification and electrical conductivity for the immobilization of biomolecules [10]. However, graphene has hydrophobicity and a low yield, which limits its use in biosensor applications [11].

Carbon nanomaterials (Fig. **1**) offer a wide range of building materials starting with zero-dimensional (0-D) structures, continuing with one-dimensional (1-D), two-dimensional (2-D), and three-dimensional structures (3-D) [2].

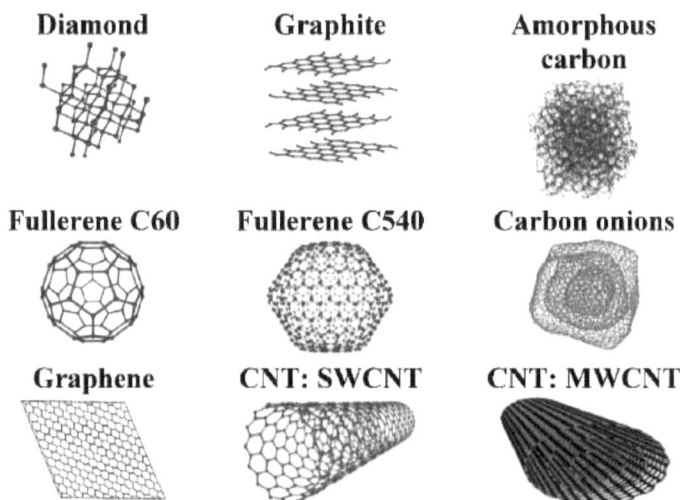

Fig. (1). Carbon entities at nanoscale grade [2].

1-D nanostructures such as semiconductor or conductive polymer nanowires and carbon nanotubes are particularly attractive materials for the operation of electrodes used in biosensors. Due to the new electron transport properties, the high surface-to-volume ratio of these nanostructures, electrical conductivity of 1-D nanostructures is greatly affected by small surface imperfections. Such 1-D materials are therefore unnecessarily redundant in fast (real-time) and precise unlabeled bioelectronic sensing and nano sensor arrays. These nanomaterials allow a large number of sensing elements to be packaged over a small area of an array device. Another remarkable feature of the CNT is the ability to develop many excellent electrochemical detection devices, from amperometric enzyme electrodes to unlabeled DNA hybridization biosensors [12]. Redox centers of the bulk of oxidoreductases are electrically segregated by protein shells. A major challenge with amperometric enzyme electrodes is the establishment of satisfactory electrical communication between the enzyme's active site and the electrode surface. Direct electron transfer between electrode surfaces and enzymes can pave the way for superior biosensing devices, as it eliminates the need for mediators or substrates and allows efficient transduction of the biosensing event. Self-assembled "trees" of aligned CNTs in the nanoforest will function as the molecular wire that enables electrical interaction between redox and redox proteins to lower electrodes that are covalently bonded to the ends of SWCNT (Fig. **2**) [13, 14].

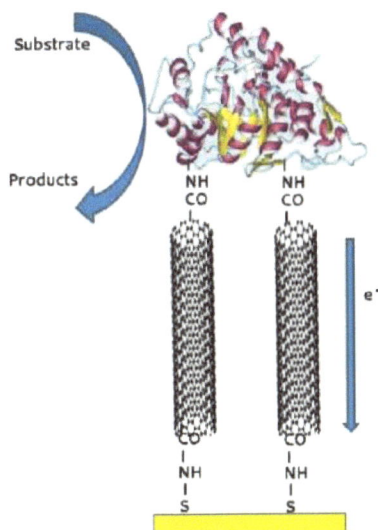

Fig. (2). Carbon nanotube-based enzyme biosensor [54].

A study has been done showing that the aligned reconstructed GOx at the edge of SWCNT can be attached to the surface of the electrode [15]. Wire arrays or nanoscopic gold tube arrays were prepared by electronically depositing metal into the pores of membranes with traces of polycarbonate particles [16]. This type of enzyme reconstruction at the tip of the CNT is a highly efficient approach to "plug" an electrode into the GOx. A detergent containing glassy carbon electrodes and natural receptor hemoglobin or CNT-modified carbon paste exhibit quite high acrylamide sensitivity observed in food samples [17, 18].

Liposomes are microscopic, fluid-filled sacs that consist of layers of phospholipids and have a cell membrane structure [19]. Wire arrays or nanoscopic gold tubes have been prepared by electronically depositing metal inside membrane pores with traces of polycarbonate particles [16]. İmmunoliposomes encapsulated with an electroactive marker are used as amplified electrochemical immunoassays signal (Fig. **3**) [20, 21].

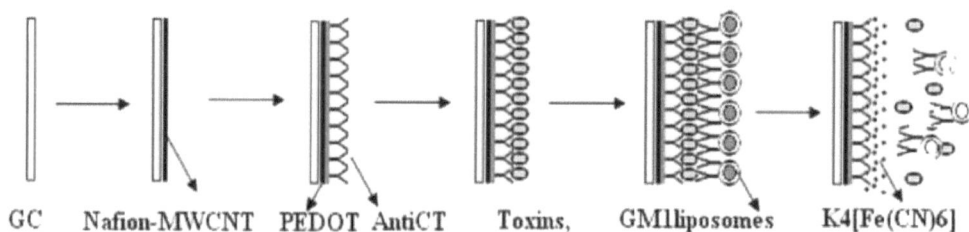

GC   Nafion-MWCNT   PEDOT AntiCT   Toxins,   GM1liposomes   K4[Fe(CN)6]

Fig. (3). Electrochemical immunosensor regimen using potassium ferrocyanide encapsulated in liposomes receptor Ganglioside in drinking water samples for the identification of cholera toxin: GM1 ganglioside, Anti CT Cholera toxin antibody, PEDOT polyethylene dioxythiophene, Toxins Cholera toxin, MWCNT multiwall carbon nanotube, GC glass carbon electrode [54].

Graphene Oxide and Graphene

Graphene is a nanocarbon of great interest by technologists and scientists over the past decade. Graphene has many important properties, such as mechanical strength, high surface area, high conductivity, and catalytic structure, leading to many wonderful electrochemical and physical properties. It is also used as an electrode for high-quality electrochemical sensors. In addition, graphene displays a rapid electron transfer rate, enhanced electrochemical behavior, and reduced load transfer resistance [8].

Graphene is used as the optimal type of support for a biosensor due to its excellent mechanical and electrical conductivity [9]. Chemically modified graphene contains a large number of defects/voids and functional groups, so graphene is used as the optimal source of support for a biosensor, owing to its exceptional mechanical and electrical conductivity. An unlabeled biosensor manufactured with material supporting reduced graphene oxide (rGO)/AgNP composites has been launched. Small AgNPs are more effective than big AgNPs in enhancing the electrical properties of rGO [22]. Widely used reducing agents and a small molecule such as sodium citrate have been used to manufacture AgNP/rGO composites with increased electrical conductivity [23].

Graphene oxide (GO), rGO, and nanoparticle/nanocomposite metal/metal oxide are good transducer platforms for electrochemical detection [8]

Electrochemical sensors are cost-efficient and effective due to the easy use of analytical methods, sensitivity detection, ease of miniaturization, and reproducibility of findings. The mechanical strength and electrical conductivity of CNTs and graphene provide advantages over other NMs [8].

Non-carbon Nanomaterials

Silica nanoparticle offers various benefits such as high surface-to-volume ratio, homogeneous structure, adjustable pore structure and chemically exchangeable surface [24]. Metal nanoparticles have a broad surface area to maximize the immobilization ability of biomolecules. They also show electron transfer, catalytic activity, and very good biocompatibility ability [25, 26]. Being 1D, nanowires have an outstanding potential due to electrical properties and limited dimensions in biosensing approaches [27, 28]. Hybrid materials can provide several advantages in biochemical analysis thanks to their properties such as functionality, polymer solubility, and flexibility. Organic polymers, which can be printed on various solid substrates and easily processed, are applied to manufacture NMs that form signal probes and electrodes of the electrochemical biosensor [29].

Metallic Nanoparticles (Other Materials)

To date, the use of various metallic NPs in electrochemical sensors increases the interest in this field day by day. Metal oxide and metal NPs are the most widely used the sensors, such as metal oxides (Co_3O_4, ZnO, CuOx, NiO, and TiO_2) and metals (Cu, Au, Pt, Ni, Pd). They show high conductivity, electronic properties, and good mechanical, structural strength, catalytic activity, biocompatibility [8].

Metallic NPs have attracted considerable attention in the production of electrochemical sensors due to their inherent properties, such as high surface area and porosity [8].

Nanowire

Nanowire has a remarkable ability as an alternative sensing technique due to its high surface-to-volume ratios, small size, and optical, magnetic, and electrical properties [27]. The nanowire was considered thinner and more flexible than larger wires. However, the one-dimensional structure shows a high width-to-long ratio which results in some physical properties comparable to quantum phenomena [28]. The electrical conductivity of nanowires can be regulated by synthesizing chemical compounds and different components such as semiconductors (GaN, Si, InP), metals (Cu, Pt, Ni, Au, *etc.*, and metal oxides (Fe_2O_3, ZnO, SnO_2)) [30].

Platinum nanowires are biocompatible with biomolecules (nucleic acids and proteins). They also have very good catalytic properties for hydrogen peroxide reactions [31]. Silver-nanowires have high electrical features (*e.g.*, efficient electrocatalytic activity, rapid reaction, and repeatability) and can be used as an active carrier to produce a particular signal for several electrochemical measurements. A silver nanowire-based electrochemical immunosensor study has been performed [24]. Semiconductor nanowire (Silicon Nanowire Field-Effect Transistors (Si-NWFETs)) have been recognized as real-time and unmarked features, a flexible, high-precision electrical sensing mechanism [32].

Compared to polymeric nanomaterials, metallic nanoparticles, and CNTs, it has many advantages such as low-temperature synthesis, adjustable electrical conductivity, purification, and catalytic deposition. However, as biosensor elements, polymeric nanomaterials are less suitable because of their comparatively poor electrical conductivity compared to CNTs and their non-oriented morphology of nanofibers, which results in low analytical efficiency [32].

The major disadvantage to nanowire electrochemical experiments is that the electrostatic potential of the nanowire due to the load on the molecule of the analyte decreases the distance to zero exponentially [33]. To overcome the intrinsic problem, capture by cleavage with a single chain fragment or a proteolytic enzyme or a reduction in the size of its antibody solves the technical problem [34]. It may also be known that the density control of the capture molecules on the electrode reduces the inherent limitations of the nanowire.

Organic Polymer Conductive

The development of low-cost, miniaturized, compact sensors based on high-efficiency organic electronics is expected to offer major benefits for biological and chemical recognition applications. Gradual maturation of organic materials, a synthetic technique in the field of electronics, has contributed to the rapid production of new materials and a deeper understanding of the interactions between semiconductor analytes. Thanks to these developments, there are continual increases instability, sensitivity, and precision, in addition to the identification of a wide spectrum of chemical analytes [29]. These conductive polymers, which can be conveniently manufactured and printed on several sturdy supports, have been used for the manufacture of transistors and electrodes.

Chemical conductive polymers have been added to the structure of ''Organic Thin Film Transistor (OTFT)'' containing GOx. The diffusion of protons released by glucose oxidation with the organic polymer can be increased by evaluating the reaction times of these sensors [32].

Due to the combination of outstanding electrochemical and optical properties and increased catalytic stability, the combination of nanocomposites with metal nanoparticles and conductive polymer attracted great interest [35].

Nanoparticles with Magnetic Properties

Considerable attempts have been made in recent years to produce magnetic nanoparticles (MNPs) due to their low cost of processing and inherent advantages (nanoscale, magnetism). MNPs with a diameter of 10-20 nm perform best due to super magnetism, making them particularly suitable for a fast response. They have a large surface area and high mass transfer capacity due to their inherent dimensions [36].

Magnetic particles are surrounded by a membrane that is non-magnetic. Due to their higher stability, iron oxides like maghemite (γ-Fe_2O_3) and magnetite (Fe_3O_4) are mostly used instead of iron [37]. Magnetic beads based on silica can be prepared in various pores, sizes, and shapes. Magnetic particles typically consist of nano-sized super magnetic iron oxide spread in a silica matrix or an organic polymer for biological applications [38, 39]. Many magnetic particles of different types, such as epoxy, amino, carboxyl, and hydroxyl, are currently used for various applications [2].

MNPs can be dispersed in the sample and integrated into transducer materials, then their attraction to the active sensing surface of the (bio) sensor is traced with an external magnetic field. Bio-recognition molecules such as antibodies, oligonucleotides, or enzymes immobilized on magnetic particles can be easily captured by magnets and held close to or above an electrode surface [36, 40]. Because the properties of MNPs depend largely on their size, preparation and synthesis must be designed to obtain particles with physicochemical properties depending on the size. MNPs with adequate physical, chemical, and special surface properties have been synthesized under applied and accurate conditions for sensors, detection systems of biosensors [41, 42].

SENSORS

The area of nanotechnology investigates the modification of matter at the molecular and atomic level for the development and creation of biological, chemical, and physical structures at a range of 1-100 nanometres. These materials, commonly known as NMs or nanoparticles, have ushered in a new era in the scientific world due to their outstanding physical, chemical, and biological properties compared to their counterparts [3, 43]. The special physicochemical properties of nanoparticles have led to the manufacture of POC disease diagnostic biosensors. Its small size also improves the efficacy of other methods, such as enzymatic biosensors and electrochemical, by increasing electron transfer rates and decreasing enzyme-electrode distances [44]. Localized surface plasmon resonance (SPR) may also be increased by noble metal nanoparticles and optical biosensors established accordingly [45]. For example, changes in color resulting from inter-particle plasmon binding of these nanoparticles are commonly used in nanoparticle aggregation-based biosensors [3].

Biosensors

Biosensors are generally independent miniature portable devices that can quickly detect the analyte with the help of transducers that convert the biological response into measurable analytical signals [42]. A biosensor is known as a measuring device consisting of a transducer probe and a biological recognition element. As a result of the interaction of the analyte with the bioreceptor, the information (*e.g.*, an electrical signal) can be measured by the transducer [46, 47]. There are four primary types of transducers: optical, electrochemical, mass, and thermal (heat-sensitive sensors or thermistors) [48].

Electrochemical biosensors are the most widely used due to well-known bio-interaction and short detection time [48]. The biosensor device, developed by Springs Instruments (Yellow Springs, OH) in 1975 and first commercially introduced to the market, was used for rapid glucose determination in blood samples taken from diabetics. The use of NMs in the development of biosensors helps many new signal processing methods to be used in their manufacture. Nanoprobes, nanosensors, and other nano-systems revolutionize the fields of chemical and biological analysis due to their size [49]. Especially, the ability to adapt the scale and shape of NMs and their properties provide excellent opportunities to develop new detection systems and enhance the efficiency of bioanalytical assays [50].

A biosensor includes a bio-sensible layer that contains biosensors or is composed of bio-receptors covalent to the transducer. The coating of receptor molecules capable of selectively binding analyte molecules must be pre-immobilized for bio sensation on the transducer surface. The immobilization technique must be reproducible and stable, and the stability and function of the receptor must be maintained. The immobilization of sensor molecules on the sensor surface is a crucial factor influencing the output of the sensor [12, 51, 52]. Generally, materials for enhancing electro-analytical efficiency, electrode materials and substrate support, materials for immobilizing biological elements, and biological recognition elements are materials used for electrochemical sensors [53].

Biosensing surfaces can include enzymes, antigens, antibodies, tissues, mammalian cells, microorganisms, or receptors. The nature of the biosensing surface, *i.e.*, the long-term use of the sensor and the expected long-term storage and operational stability, are very important [54].

Based on superior detection efficiency compared to other analytical methods, electrochemical sensors can be developed using a variety of materials (MIPs, MOFs, metals, metal oxides, graphene, CNTs) for proteins, toxins, drugs, hormones, neurochemicals (NCs), *etc.* (currently have a wide range of applications). Wide studies have considered electrochemical sensors based on NM with important NC detection applications [8].

Electrochemical Biosensors

Electrochemical biosensors are analytical devices that convert biochemical events such as antigen-antibody interaction into electrical signals and enzyme-substrate reactions (*e.g.*, current, impedance, voltage, *etc.*) [32].

Electrode materials: Solid electrodes have replaced conventional mercury electrodes due to their toxicity. However, the option of electrode working material in electrochemical sensing is central to the accuracy of electrochemical measurements [55]. Demand for disposable and low-cost biosensor sticks or strips for fast marketing has been achieved with screen-printed electrodes, which require the material accumulation for electrodes, noble metals, and in particular, carbon on inert PVC or ceramic substrate [56, 57]. Inexpensive, easy-to-use, miniaturized, disposable chips are essential for electrochemical analysis, and many groups are working in this direction. In recent years, various forms of silver, platinum, copper, gold, nickel-doped or doped carbon, dimensionally stable anions, *etc.* It is widely used in biosensor applications [58].

Fig. (**4**) shows the practical NMs (non-carbon-based and carbon-based) used in different types of electrochemical biosensors to enhance analytical sensitivity efficiency. Nanomaterials used as electrodes or co-matrices must meet requirements such as electro-catalytic aid for signal enhancement, excellent biocompatibility with capture biomolecules, and outstanding electron mobility. Electrochemical experience nanomaterials can be used for both paper and microfluid biosensor applications, with practical measurement platforms for POC variants of biomolecular detection [32].

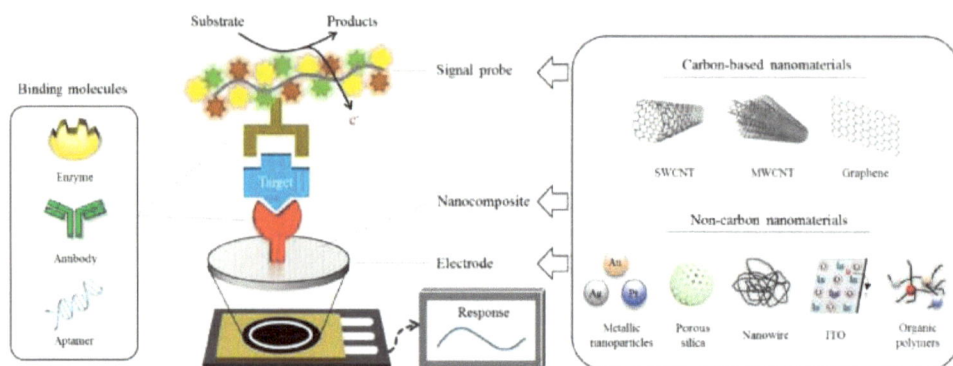

Fig. (4). Analytical principle of electrochemical biosensors composed of non-carbon nanomaterials and carbon [32].

Bioreceptors: Bioreceptors are molecules that have biological molecular types such as nucleic acids, proteins, enzymes, antibodies. Bioreceptors allow the specific analyte to be coupled for the sensor to provide measurement with minimal interference with other elements of mixtures. Recent research has focused on bio-receptors as they are important elements for the specificity of biosensor technologies in Electrochemical biosensors. Such a biosensor must use an enzyme that acts especially to transform the reactant molecule products. Cofactors are found in most enzymatic reactions. These cofactors are ions or molecules that may help to react. The cofactors must be chemically transformed during catalysis, and the subsequent physicochemical effects can be followed or identified by the enzymatic process. Another example is the case in the immune system, where antigens interfere with antibodies. The antigen is known as an international agent. By binding to it and trying to remove a particular antibody-antigen is produced acting against it. Antibodies can be grown *in vitro* for the detection of specific molecules. Antibodies are the cornerstone of the biosensor detection system in this manner. Even with only minor physicochemical changes, antigens can bind to appropriate antibodies. The lack of ability to detect low concentrations of analytes is a significant obstacle to the creation of unlabeled immunosensors. The value of immunosensors would be stronger if there were a feasible method for amplifying immunological associations resulting in more pronounced changes [59]. Enzyme immunoassays (EIAs) based on electrochemical identification have many possible benefits and have been used in biotechnological, biological, clinical, nutritional, and environmental research. The most popular enzyme markers used include glucose oxidase, alkaline phosphatase (ALP), and horseradish peroxidase (HRP) [60].

DNA is an important biomolecule as it has an important role in the replication of living organisms, determining hereditary characteristics and storage of necessary genetic information [61]. The basic idea of a DNA biosensor is to identify and transform molecular identification by DNA probe into an electrical signal using a converter. İdentification of DNA sequences is essential for the regulation, detection, and reading of molecular structures. In biosensor applications, aptamers act as biorecognition components. Aptamers are small and synthetic oligonucleotides that can directly classify and bind to almost any target, drugs, ions, low molecular weight ligands, cells, toxins, proteins, and peptides [62].

Classification of Electrochemical Biosensor

The principle of electrochemical sensors is based on the fact that the variation in electron flows leads to the formation of an electrochemical signal through oxidation or reduction of the electroactive analyte at the surface of the working electrode exposed to a constant or variable potential. A general scheme of the electrochemical sensor is shown in Fig. (5). Voltammetric and potentiometric biosensors are the two most important subclasses of electrochemical sensors [54].

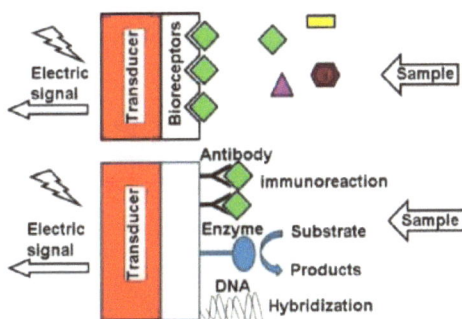

Fig. (5). General structure of a biosensor [54].

Voltammetric Sensors

Voltammetric sensors analyze the influence of concentration on the possible properties of the reduction or oxidation of a given reaction to being observed [63]. The operating concept of an amperometric sensor is based on the application of a constant potential to an electrochemical cell as a function of the current arising from a reduction reaction. The general structure of amperometric biosensors is shown in Fig. (6). Current is used to measure substances used in the reaction, and amperometric sensors are also a subset of voltammetric sensors [12, 64].

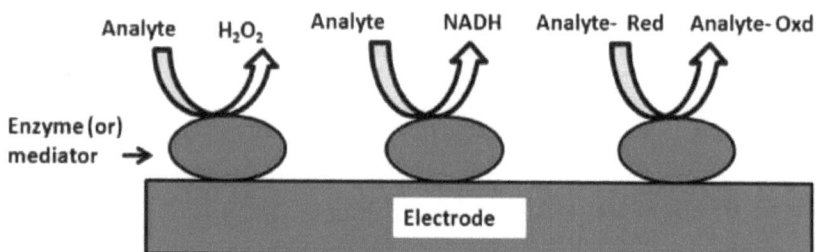

Fig. 6. Various types of amperometric biosensors [54].

The direct or indirect measurement capabilities of amperometric biosensors demonstrate their versatility. Although direct amperometry takes advantage of the close relationship between the measured current and redox reaction products, indirect amperometry uses traditional detectors to calculate metabolic substrate or interest analyte product [65]. Amperometric biosensors, usually glucose and lactate [66, 67], are used on a broad scale for analytics such as sialic acid [68, 69].

Potentiometric Biosensors

These biosensors analyze the potential difference between a reference electrode and an active electrode. They also track charge accumulation at zero current produced by selective bonding on electrode surface [63].

Potentiometric biosensors are based on ion-selective field-effect transistors (ISFET) and ion-selective electrodes (ISE). The potentiometric signal generation process is based on the isolation of loads between two phases as a result of the selective movement of analyte ions from the aqueous phase to the organic phase [70]. For *e.g.*, glucose has a minimal pH influence in the working world. Glucose oxidase can be immobilized on a surface of the pH electrode. However, enzymatically formed gluconate contributes to acidification [54].

Electrical surface potential relies on ions reacting with the semiconductor surface. The ISFET transistor gate surface is bound by a selective membrane. This potential modulates the flow of current through the semiconductor. This can be made from substances such as aluminum (Al_2O_3), zirconium oxide (ZrO_2), tantalum oxide (Ta_2O_5), and silicon nitride for pH determination (Si_3N_4).

Impedimetric Biosensors:

Such devices track impedance (Z), capacitance (C), and component resistance (R). Inductance has only a minimal effect in a typical electrochemical setup. The expression of impedance is shown below:

$$Z^2 = R^2 + \frac{1}{(2fC)^2}$$

The inverse value of resistance is called conductivity, which is why some researchers call such devices conductometric. Impedance biosensors usually contain three alternating voltage electrodes; voltages ranging from a few mV to 100 mV are used. A significant increase in impedance may lead to enzymatically produced ions. Alternatively, through the production of conductive metabolites, impedance biosensors have been used effectively to track the growth of microorganisms [71, 72]. Impedimetric biosensors are used less frequently compared to potentiometric and amperometric biosensors. False-positive results from electrolytes in samples are the main drawback of impedance biosensors. However, there have been some promising answers to this. Hybridization of DNA fragments, previously amplified by a polymerase chain reaction, was regulated by impedance tests [73]. The impedance immunosensor model containing captured avidin and electrodeposited polypyrrole film attached to anti-human IgG *via* biotin has been able to detect antibodies as low as 10 pg/ml in the sample [74].

Electrochemical impedance spectroscopy has been successfully used for the study of associations between amyloid β peptide and possible drugs and for direct determination of antigen SPI in culture media in appropriate immunizes [51, 75].

NANOMATERIAL-BASED ELECTROCHEMICAL SENSORS

Nanomaterial-based electrochemical sensors for Hydrogen Peroxide (H$_2$O$_2$) Biosensor

H$_2$O$_2$ is a well-known bleach, oxidizer, and also one of the important by-products of various regulatory biological processes. It causes the progression of aging, kidney disease, atherosclerosis, and other diseases. The determination of the trace amount of H$_2$O$_2$ in biological samples is therefore very important [76].

H$_2$O$_2$, an effective intermediate and strong oxidizing agent, is readily miscible with water and is commonly used in the light industry, medicinal sciences, pharmaceutical engineering, medical hygiene, and electronics. Reliable, quick, and precise identification of H$_2$O$_2$ is therefore highly significant in areas of food safety, bioanalysis, and environmental conservation. Various H$_2$O$_2$ measurement and detection methods have been developed, such as fluorometry, spectrophotometry, electrochemical analysis, and cell imaging. Electrochemical sensing technology is considered among these methods to be the most effective method due to its excellent advantages, such as rapid reaction, high sensitivity, and comfortable activity [76].

Enzyme-based biosensor: Catalase (CAT) and HRP are two types of enzymes that are widely used for the identification of H_2O_2. Whereas the enzyme stabilization and immobilization procedure are complex, and it is difficult to perform direct electron transfer to the bare electrode. Therefore, carbon-based nanomaterials, polymer, and metal nanoparticles were used and prepared in the manufacture of enzyme-based H_2O_2 biosensors. Enzyme loading within the NM can increase the area, thus making contact with increasing the rate of electron transfer and more detectors. NMs and their hybrid nanostructure will effectively increase the efficiency of the biosensor (Table **1**) [76].

Table 1. Enzyme-based various H_2O_2 biosensors [76].

Biosensors	LOD (nm)	Linear Range	Sensitivity
GCE/f-MWCNT/ PLL/CAT	8	1n-3.6µM	$0.39 \mu A \cdot mM^{-1} \cdot cm$-2
Cys-AuNPs/ MWCNTs-COOH/ CAT/NF/	0.5	1nM-1µM	--
MoS_2-Au/CAT	100	0.5µ-0.2mM	$0.18 \mu A \cdot mM^{-1} \cdot cm$-2
PANI/ glutaraldehyde/CAT	2.18µM	5µ-0.1mM	--
HRP/Graphene	0.026	80n-0.66µM	--
HRP-Au-CPE	210	0.48-50µM	--
HRP-Au thin-film gold	16µM	20-500µM	$12 \mu A \cdot mM^{-1} \cdot cm$-2

Non-enzymatic biosensor: Previous studies have shown that non-enzymatic electrochemical biosensor provides an effective means of hydrogen peroxide detection, avoiding the disadvantages of poor reproducibility and instability. The non-enzymatic biosensorrelies on electrocatalytic reduction occurring at the electrode surface.

Electrochemical detection using traditional electrodes requires a comparatively high degree of intense potential [76]. Transition metal oxides such as MnO_2, Co_3O_4, and CuO show electrocatalytic performance against H_2O_2. This form of NM has the benefit of being simple to prepare, low cost, and is environmentally friendly. However, its sensitivity and stability are unsatisfactory. A noble metal-based H_2O_2 biosensor of non-enzymatic attracted further interest. In reaction to this scenario, various help NMs, such as amorphous carbon, graphene, carbon nanodots, and poly microparticles, are used for the immobilization of Ag nanomaterials which are an important technique for enhancing the detection properties of prepared biosensors. Also, as shown in Table **2**, various Ag nanomaterial-based H_2O_2 biosensors of non-enzymatic are summarized in previous reports [76].

Compared to an enzyme-based biosensor, H_2O_2 biosensor of non-enzymatic typically has a broader linear spectrum and increased sensitivity. It is more suited for environmental and industrial monitoring. Also, Ag nanomaterial-based biosensors of non-enzymatic are being studied further due to their perfect properties and wide application possibilities. The enzymatic biosensor also has the lowest detection limit acceptable for biomedical applications [76].

Table 2. Ag nanomaterial-based various H_2O_2 biosensors of non-enzymatic [76].

Biosensors	LOD(μm)	Linear Range (mM)	Sensitivity
C-CPE/AgNPs	0.3		$309.4 \ \mu A \cdot mM^{-1} \cdot cm^{-2}$
AgNPs/MWCNT/GCE	0.5	0.05-17	$1.42 \ \mu A \cdot mM^{-1}$
AgNPs-rGO-/ITO	5	0.1-100	--
PQ11-AgNPs/GCE	33.9	0.1-180	--
AgNPs/PVA/Pt	1	0.04-6	$128 \ \mu A \cdot mM^{-1}$
AgNPs-rGO/GCE	4.3	0.1-70	--
AgNPs-PMPD/GCE	4.7	0.1-30	--
AgNPs-NFs/GCE	62	0.1-80	--
AgNPs-Co3O4-rGO	0.035	$0.1\mu A$-7.5	$146.5 \ \mu A \cdot mM^{-1} \cdot cm^{-2}$
ErGO-Ag/GCE	1.6	0.1-100	--
AgNPs-MWCNT-rGO/GCE	0.9	0.1-100	$833 \ \mu A \cdot mM^{-1}$
ITO/N-graphene-Ag NDs	0.26	0.1-80	$88.4 \ \mu A \cdot mM^{-1} \cdot cm^{-2}$

Electrochemical Nanomaterial-based Sensors for the Identification of Neurochemicals in Biological Matrices

Neurochemicals such as adenosine, glutamate, dopamine, serotonin, and Gamma-Aminobutyric acid (GABA) are effective indicators for measuring the dynamics of brain disorders. Detection strategies for these neurochemicals, both in vivo and *in vitro*, are important in the treatment of a variety of human diseases. In addition to traditional instruments (capillary electrophoresis, microelectrodes, flow injection, biosensors, fluorescence, chromatography, Raman, spectrophotometry, and Fourier transform infrared), nanomaterial-based electrochemical sensors (carbon nanotubes, graphene, metallic polymers, and metal-organic frames) have also arisen as effective tools for quantitation [8].

Sensors have many advantages over traditional analytical instruments in terms of expense, ease of use, precision, and speed. In specific, NM-based electrochemical sensors may provide additional benefits by increasing selectivity for NCs. Nanostructures based on graphene, including doped graphene (oxides) and graphene oxides, are commonly used for high resolution, stability, and selectivity sensors. CNT-based sensors of electrochemical can exhibit low detection limit (LOD) with rapid response despite signal enhancement provided with desirable electrode properties such as high surface area, rapid response, and low overvoltage. CNTs also have chemical stability, good mechanical strength, and high thermal conductivity, which are desirable properties for electrochemical sensing applications. Graphene, GO, rGO and CNTs are commonly used in the manufacture of NC sensing electrodes [8].

A variety of nanoparticles have been used as electrochemical sensors. For example, a carbon electrochemical sensor modified with gold nanoparticles (AuNPs) was used to detect dopamine (DA) in human serums. Similarly, ZnO nanotubes have been used for electrochemical DA detection of human urine samples. TiO_2 nanotube photonic crystals have been used for sensitive and selective detection of DA released from mouse brains as a photoelectrochemical sensing medium [8].

ELECTROCHEMICAL SENSORS FOR POINT-OF-CARE

POC better health management is very important in providing better health care [77, 78]. Higher quality in healthcare management can also be accomplished by making prompt decisions based on quick diagnosis, evidence interpretation, and smart data analysis. Intelligent science, clinical and medical services to promote health are included [79, 80]. The follow-up evaluation and successful control of disease development are essential for awareness. The treatment of illness epidemics also relies on the optimization of therapy [79]. Therefore, it is imperative to develop intelligent diagnostic systems for personalized health care such as POC. It allows rapid medical decisions as diseases can be diagnosed at an early stage, leading to improved health outcomes allowing patients to start treatment early. It also allows for quick identification of analytes near to patients, enabling improved diagnosis, control, and treatment of illness [81]. In recent years, a large number of potential POC have been developed, paving the way for the next generation of bedside testing [82].

Biosensors are analytical instruments that transform the biological reaction to a detectable signal [83]. The measurable signal can be piezoelectric, electrochemical, thermal, or optical, as shown in Fig. (7). Biosensors of electrochemical have been prospectively investigated for POC applications for customized healthcare administration, particularly because they anticipate any chemical reactions or levels of biological markers by effectively producing signals consistent with the concentration of an analyte. Also it can thus identify disease markers from body fluids [79, 83 - 85]. Their high sensitivity and selectivity allowed early detection and targeted diseases. It also encourages prompt decision-making on care [79]. The combination of biosensors and nanotechnology will improve the estimation of disease development and onset [83].

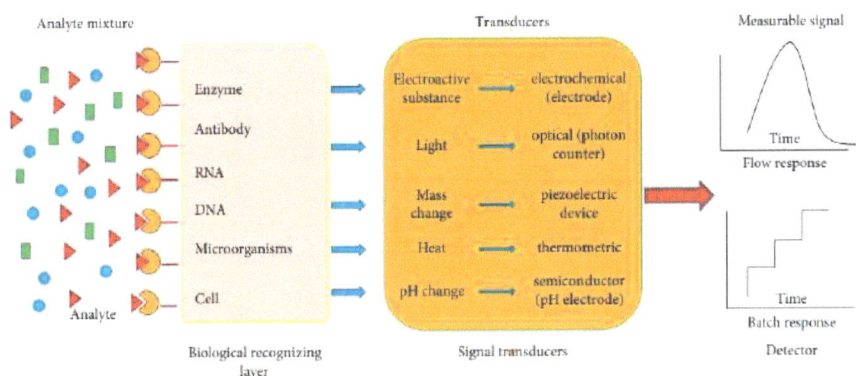

Fig. (7). A standard biosensor with all its components [3].

Materials with a scale of 1-100 nanometers, commonly known as NMs and nanoparticles, are transforming the scientific world due to outstanding biological, chemical and physical properties compared to their bulk counterparts [43]. Noble metal nanoparticles may increase localized surface resonance plasmon (SPR) and build optical biosensors accordingly [44]. For example, in biosensors based on the aggregation of nanoparticles, color variations due to inter-particle plasmon binding of these nanoparticles are widely used [3]. This chapter discusses the latest developments in these nanosensors in POCs.

Various Point-of-care Diagnostic Nanosensors

Nano-biosensors for Point-of-Care detection of Cancer

Cancer accounts for one in seven deaths in the world, not just in developing countries [86, 87]. There are more than 200 forms of cancer, but the more popular kinds involve breast cancer, liver cancer, lung cancer, skin cancer, bladder cancer, kidney cancer, ovarian cancer, prostate cancer, esophageal cancer, colorectal pancreatic cancer, lymphoma, and thyroid cancer [88]. Early screening and diagnosis are accepted procedures to minimize the risk of cancer mortality and regeneration, leading to a substantial decrease in cancer deaths [89].

Cancer diagnosis also requires recognizing signs and signs that suggest the existence of biomarker-related abnormalities [89, 90]. Nucleic acids, proteins, cytogenetic parameters, carbohydrates, entire cells, cytokinetic parameters are minor metabolites of body fluids. Blood contains a wide variety of protein biomarkers with potential applications for early cancer diagnosis [91].

An ideal bedside diagnostic device is a portable device that guarantees reliability. This requires research into the development of new devices that will enable drug efficacy, early diagnosis, effective drug delivery, and cost-effective real-time in vivo monitoring of cancer [92]. The use of nanotechnology for drug delivery, drug diagnosis, and cancer treatment has led to [93] uses of NMs for the extraction [94] and extracellular vesicles discarded by the tumor or detection of specific tumor biomarkers in circulating tumor cells.

Biosensors based on nanotechnology have transformed cancer diagnosis and prognostics. With the tremendous advancements in the field of biosensors, numerous systems have emerged as promising point-of-care options for cancer diagnosis. A number of them have been patented and marketed [95] or are in clinical trials. Combidex, a product of AMAG Pharmaceuticals Inc made up of ultra-small SPIONs, has been authorized by the FDA as a new MRI contrast agent for the identification of metastatic lymph nodes [96]. The OraMarkTM Test is a simple, non-invasive, and easy-to-use oral rinse treatment for the accurate diagnosis of oral malignancies. Similarly, Cxbladder provides a non-invasive, accurate urine-based laboratory test for the identification of bladder cancer. Commercially accessible lung and colorectal cancer detection methods include two blood-based liquid biopsy tests, Epi proLung® and Epi proColon®. Videssa® Breast, a product line based on Provista's patented ProteoMark® Technology, is a protein-based blood test for the identification of breast cancer that analyzes various tumor blood-borne protein biomarkers. NuView Life Sciences is actively developing NV-VPAC1 technology as an *in vitro* diagnostic kit that incorporates a fluorophore coupled to the NV-VPAC1 peptide for the detection of shed cancer cells in urine. Many on-site diagnostic procedures or tools are in the early stages of development and are projected to grow into widely utilized point-of-care techniques in the next years [95].

Nano-biosensors for Point-of-Care in Diabetes

Diabetes is a rapidly emerging epidemic that currently impacts millions of people around the world [97]. Lower limb amputations can lead to various complications such as diabetic kidney disease [98], cardiovascular diseases, and blindness [97]. Although there is no cure for diabetes, patients can reduce their complications by strictly monitoring and controlling their blood sugar levels [83, 97, 99]. To ensure optimal control of glucose levels, a small blood sample is usually taken from patients, typically through a finger hole, this is then placed on a sensor test strip, and the glucose concentration is then read by a handheld electronic reader reporting the blood [97, 99]. The glucose sensors used since the 1970s were based on screen-printed electrode enzymatic electrochemical measurements [99, 100] and laboratory analysis [97, 99, 100] provide accurate and fast measurements of blood glucose.

Study in nanotechnology involving nanosensors and NMs aims to continuously track diabetes diagnosis in POC devices by increasing sensor surface area, expanding the supply of nanoscale sensors and electrode catalytic properties [97]. A broad range of new biosensors with on-chip NMs in laboratory and nanosensor applications are currently being developed for in vivo and *in vitro* glucose detection [101]. Fig. (**8**) shows the progression of first generation glucose biosensors into nano glucose sensors by Chas and Clark [97].

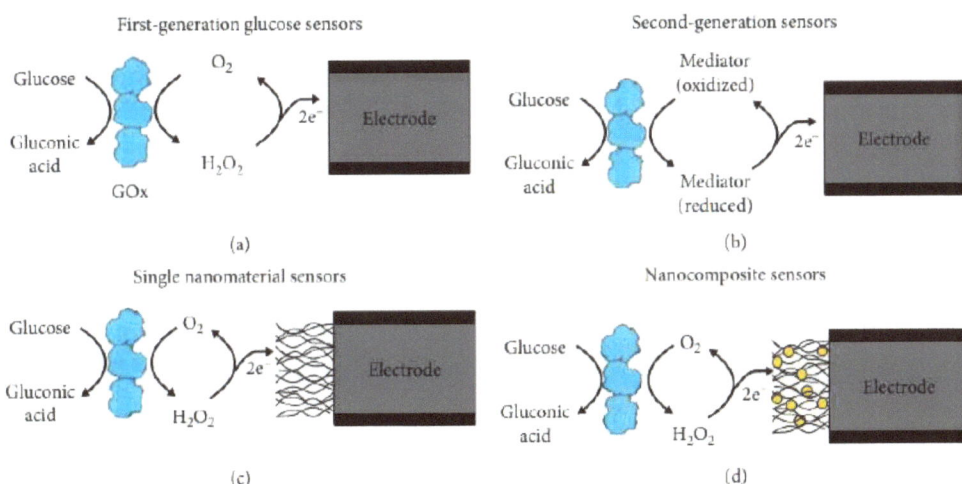

Fig. (8). Glucose sensors evolution from first generation to nanostructured materials used in glucose sensors. (a, b) Regular GOx biosensors of electrochemical. (c, d) shows the integration into the sensors of NMs such as nanocomposites composed or CNTs of several NMs in order to increase the surface area, change operating parameters, enhance catalytic action, improve the electron transfer from enzyme to electrode [97].

The most notable improvements have been made in the diabetes field, where many devices that are commercialized or in development focus on continuous glucose monitoring (CGM) in various bodily fluids. Commercial CGM systems consist of at least two components: a sensor placed or implanted beneath the skin to monitor glucose content in the interstitial fluid every 5–15 minutes, and a reader that receives and displays the wirelessly sent data [102]. Some CGMs are also linked to insulin pumps to construct open-loop or closed-loop systems for on-demand insulin administration. To detect glucose in interstitial fluid, most CGMs employ the glucose oxidase-FAD/FADH2– linked reaction (amperometric peroxide measurement) [103].

Because the devices employ first-generation glucometer technology for sensing, they are susceptible to interferences from acetaminophen and ascorbic acid.. To reduce interference from acetaminophen, the Dexcom G6® CGM has a permselective membrane coating [104]. Because of the different measurement methods of acetaminophen and ascorbic acid, they do not interact with the Eversense CGM; however, tetracycline and mannitol do [105]. There are no standards in place to check the quality of data. Most devices employ mean absolute relative deviation as a criterion for analytical accuracy (average stated ranges, 9 percent–12.3 percent) [106].

Aside from CGMs, many wearables (*e.g.*, patch, microneedle device, tattoo, electronic skin, contact lens [107]) and implantables are in the works, intending to make glucose measurement less invasive (*e.g.*, sweat or tear analysis), decreasing device size, or increasing sensor lifetime and user comfort level.

The introduction of painless blood-draw equipment is another technology development in the least invasive category. Seventh Sense Biosystems, for example, offers a technology called TAP that employs microneedles to collect up to 100 L of capillary whole blood with minimal discomfort. TAP is FDA-approved for blood collection for hemoglobin A1c analysis. Tasso, another prominent contender, offers a device called HemoLinkTM that employs microfluidics and suction to take capillary blood without the need for a puncture. A painless blood draw should make POCT more widely available [108].

Nano-biosensors for Point-of-Care in Infectious Diseases

Infections such as cholera, severe respiratory syndrome, bird flu, malaria, viral hepatitis, and dengue fever are often triggered by pathogenic microorganisms that have a profound impact on humanity due to their peculiar properties such as parasites, bacteria, fungi, and viruses. Their rapid reproduction and their unpredictability are what sets them apart from other diseases [3].

According to the World Health Organization (WHO), the optimal diagnostic system for infection should be precise, responsive, durable, user-friendly, reliable, and cost-effective [3]. Traditional diagnostic techniques for these diseases include culture and polymerase chain reaction (PCR) strategies, immunology, and microscopy [3]. While these techniques have contributed significantly to the detection and diagnosis of infectious diseases and have greatly promoted the treatment and prevention of various infectious diseases, they have demonstrated various limitations such as slowness, expensive, inaccuracy, require skilled expertise, and especially in developing countries [109]. Potential point-of-care testing is being designed to enable the development of new and updated diagnostic methods for high sensitivity, early identification, prevention, and treatment of these infectious diseases [109, 110].

Malaria rapid diagnostic tests (RDTs) have witnessed tremendous progress in the recent decade to enable fast and accurate testing in rural situations where clinical diagnostics resources are not frequently accessible [111 - 113]. A study's RDT lateral flow strips, for example, may identify proteins originating from malaria parasites in the blood, resulting in a sequence of plainly visible lines [114]. Rathod *et al.* effectively mimicked the capillary environment using microfluidic channels for more accurate in-field malaria testing [115].

According to WHO, around 10.4 million new cases of Tuberculosis (TB) were estimated in 2016, with 64 percent of cases detected [116], impeding prompt treatment treatments [117]. It would be difficult to meet the End TB strategy's aim of 90 percent reduction in incidence and 95 percent reduction in death by 2035 [118] without better TB diagnostic technologies to administer timely treatment treatments. QuantiFERON-TB, liquid culture, and smear microscopy are currently conventional diagnostic methods for tuberculosis [119], many of which involve expensive equipment, highly educated personnel, and enormous volumes of samples [120]. In recent years, there have been significant advances in the field of TB POC diagnostics. In TB-endemic countries, the WHO endorsed the POC Xpert® MTB/ RIF test developed by Alland *et al.* in December 2010 [121]. The Xpert® MTB/RIF assay is a cartridge-based integrated miniaturized PCR technology that requires little technical skill to acquire test results from unprocessed sputum samples in 90 minutes [122]. WHO-approved test tools, such as urine lateral flow lipoarabinomannan (LFLAM) and loop-mediated isothermal amplification (TB-LAMP), have also been created, eliminating the need for complex equipment (such as thermal cycle controlling systems) [116].

In general, nucleic acid and antibody (Ab) tests are the two types of quick POC testing that may identify COVID-19 infections [123]. The nucleic acid test is often conducted by looking for viruses in the patient's sputum (or saliva) or nasal secretions (snot) [124]. Such tests are effective in identifying viruses at an early stage of infection, even before symptoms occur. The antibody test strip (IgG/IgM test), on the other hand, is conducted by collecting patient blood samples containing antibodies against the virus [125]. In general, approximately 5 days after initial infection, the virus activates the immune response, stimulating the development of both IgM and IgG antibodies in the blood that fights the virus

[126]. These antibodies can be found in the plasma, serum, or whole blood of patients. When compared to existing POC biosensors, the gold standard for COVID-19, quantitative real-time polymerase chain reaction (qRT-PCR), has a clinical sensitivity and specificity of 79–96.7 and 100 percent, respectively [127]. Clinical sensitivity and specificity of a commercial POC biosensor (*i.e.*, IgG/IgM lateral flow test strip) are 86.43–93.75 and 90.63–100%, respectively [128].

Nano-biosensors for Point-of-Care in Human Immunodeficiency Virus (HIV)

HIV is a major health problem worldwide [129]. Established to include rheumatoid arthritis (RA) and cardiovascular diseases in a multisystem state (CVDs) [130]. In HIV patients, these cardiovascular diseases (CVDs) are the primary cause of mortality and morbidity. In most cases, early diagnosis of HIV infection is the only way to prevent the transmission of HIV infection and improve the potency of antiretroviral treatment, and HIV is thus diagnosed by viral load screening [131, 132]. As the gold standard technique, nucleic acid amplification tests (NAATs) are also used to detect low levels of viruses in the blood [132]. Conventional methods for determining HIV viral load include culturing polymerase chain reaction (PCR) and enzyme-linked immunosorbent assay (ELISA).

The detection of HIV p24 Ag and antibodies (followed by HIV-1/2 differentiation and rt-PCR confirmation) has drastically reduced the diagnostic window to within 2 weeks of transmission [133]. ARCHITECT HIV Ag/Ab EIA (Abbott Laboratories), GS HIV Combo Ag/Ab EIA (Bio-Rad Laboratories and Walter Reed Army Institute of Research), *Vitros* HIV combo assay (Ortho-Clinical Diagnostic), BioPlex 2200 HIV Ag-Ab assay (Bio-Rad Laboratories), and ADVIA Centaur HIV combo (Bio-Rad Laboratories) are among the FDA-approved fourth-generation HIV-Ag (Siemens Healthcare Diagnostics) [134], [135]. In the POC setting, commercially available HIV Rapid Diagnostic Tests (RDTs) such as the Multispot HIV-1/HIV-2 Rapid Test (Bio-Rad Laboratories), HIV 1/2/O rapid test device (ABON), Determine HIV 1/2 (Alere), OraQuick Rapid HIV-1/2 Antibody Test (OraSure Technologies), and DPP HIV 1/2 (Chembio) can detect and sometimes differentiate between antibodies to HIV-1/2 [136]. Since the number of CD4+ T-lymphocytes decreases with HIV infection and increases with successful ART [137], enumeration of CD4 + T-lymphocytes and quantification of HIV viral load can be used to monitor HIV infection [138, 139]. However, traditional techniques for quantifying CD4 + T-lymphocytes and HIV viral loads, such as flow cytometry EIAs and quantitative RT-PCR, are constrained by long turnaround times, the requirement for complex apparatus and well-trained operators, and the related high costs [140]. According to the World

Health Organization, there is an urgent need for point-of-care (POC) devices that can properly diagnose and monitor HIV/AIDS in resource-limited settings [141]. These point-of-care HIV detection devices should be accurate, affordable, simple to use, and disposable to identify HIV infection and quantify CD4+ T-lymphocytes and HIV viral load in resource-limited settings [142, 143]. To satisfy clinical demands, the devices' lower detection limit must be at least 200 CD4+ cells per μL and 400 copies of HIV per mL of whole blood [114].

Electrochemical Detection of Nucleic Acids by nanosensors

Detection of mismatched base pairs in DNA plays a very important role in the diagnosis of genetic-related diseases and conditions, especially for early-stage treatment. Advances in micro and nanotechnologies, new NMs, and particular processing processes have made it possible to create highly precise, highly sensitive sensors, making them desirable for the identification of minor sequence variations. Among the different biosensors used for DNA detection, EC sensors are very promising as they are capable of reliable DNA identification and effective signal transmission. In addition, the integration of sensors with sample preparation and fluid processes enable rapid, multiplexed detection of DNA necessary for the clinical diagnosis of POC [144].

Due to the complex nature of DNA, detection of a single or a small number of base mismatches requires high sensitivity and specificity [144]. Current detection methods are based on sample amplification combined with rigorous experimental stringency control [145]. For example, polymerase chain reaction (PCR) requires careful primer design and accurate temperature control to achieve sensitivities in the fM range with single-base mismatch specificity [144]. While these traditional technologies provide the gold standard for lab-based DNA diagnosis, they cannot meet the requirements of POC clinical diagnosis [146].

Microtechnology and nanotechnology have recently demonstrated new promise in the diagnosis of EC DNA. EC sensors provide the ideal interface for integrating these technologies, providing a wide range of innovative materials and production methods. It can be used in different areas of the detection system, including NMs, capture probes, molecular reporting, electrode production, and electrode coatings. Compared to conventional materials, these materials have greater biocompatibility, additional bonding areas, and higher signal intensities (through improved electrical properties) in EC sensors [144]. Nanofabrication makes the sensing process more efficient by allowing the sensor to be miniaturized, which increases sensitivity and reduces sample and reagent volumes. Nanofabrication is described here as two different categories, and recent trends combine these two elements in the design of new EC DNA diagnostic sensors (Fig. **9**) [144].

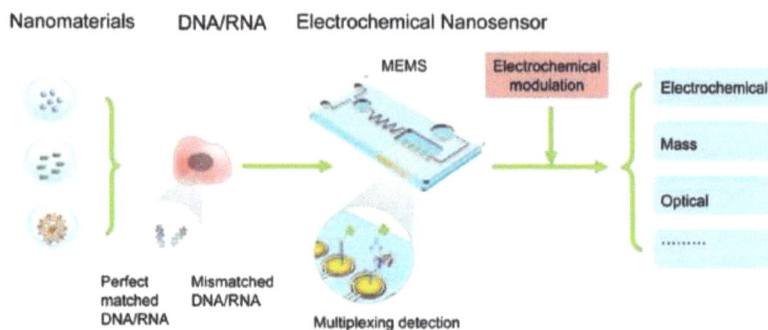

Fig. (9). Integration of NMs and processes of nano/microfabrication for EC DNA sensor development [144].

With the contribution of MEMS and microfluidics technology, EC sensors can be incorporated into compact platforms containing all necessary planning and fluidic processes contributing to the commercialization of clinical diagnostic instruments [147, 148]. The ultimate aim of creating an EC sensor is to create an overall efficient DNA biosensing analysis device that involves sample pretreatment, sample distribution, and detection [144].

A typical EC DNA sensor consists of a probe, a capture probe, and an electrode from a reporter. In order to respond to EC reactions, the reporter probe is a molecule that generates an EC signal. The capture probe is an organism used to identify and attach to the target DNA on a solid substratum such as the electrode surface and is usually immobilized. However, it is also possible to immobilize NMs or other biomolecules. Additional materials are now widely combined for improved sensor performance, such as intermolecular connectors and electrode coatings. Single-stranded oligonucleotides, DNA-related proteins, peptides, and aptamers include basic molecules used as probes (reporter and capture) [149]. The transmitter and capture probes have been integrated for advanced integration as a single device in some sensors. As seen in Fig. (**10**), probe, aim, and reporter molecules can all be substituted or related to properly incorporated NMs [144].

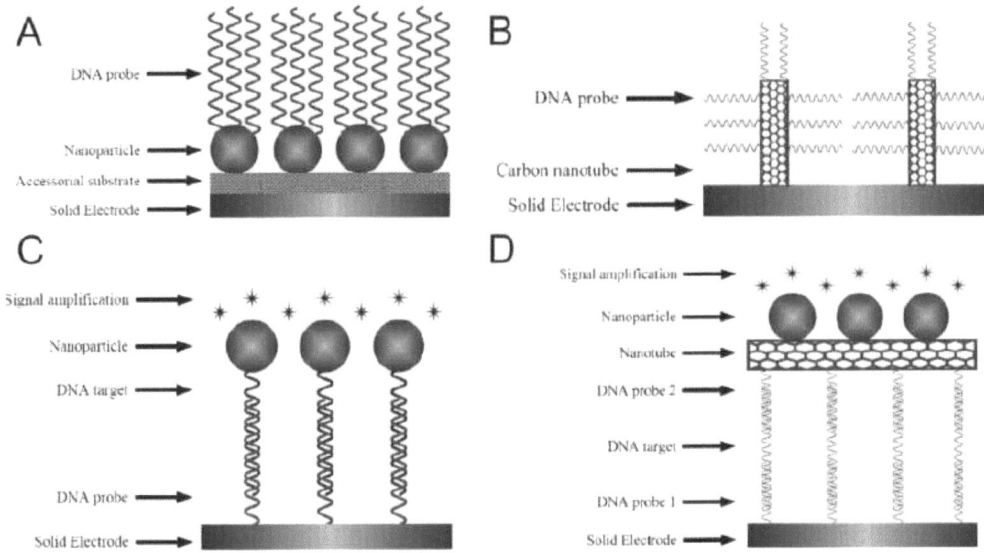

Fig. (10). Simple NMs used in EC RNA/DNA diagnostic biosensors; (a) NMs for electrode coatings, (b) NMs for probe labeling, (c) NMs for target labeling and (d) NMs for signal reporting [144].

A chip-based sandwich electrochemical genosensor for the quantitative measurement of RASSF1A DNA promoter methylation was created in 2016, employing Au/TMC/Fe$_3$O$_4$ nanocomposite as tracer tag to label DNA probe and polythiophene (PT) as electrode surface immobilization support. This technique detects DNA methylation in the nano- to the pico-molar range, with a detection limit of 2 1015 M [150]. Another electrochemical genosensor for early detection of circulating methylated DNA (E-cadherin) was recently described, which used an ssDNA probe conjugated to Fe$_3$O$_4$-citric acid nanocomposites and an antibody against 5-methylcytosine that was physically immobilized onto a reduced graphene oxide and polyvinylalcohol modified electrode. The designed biosensor has a dynamic range of 1 10 4 to 20 ng/mL and a sensitivity of 9 10 5 ng/mL for detecting circulating methylated DNA [151].

miRNAs are one of the newest biomarkers that may be used to diagnose cancer at an early stage. A sandwich-based electrochemical genosensor was developed by Daneshpour *et al.* for the ultrasensitive detection of microRNA (miR-106a) utilizing a double-specific probe method and Au/TMC/Fe$_3$O$_4$ nanocomposites as tracing tag. This system has a linear relationship with a range of 1×10^{-3} pM to 1×10^3 pM and a detection limit of roughly 3×10^{-4} pM [152]. By employing Au nanoparticles and CdSe@CdS quantum dots-contained magnetic nanocomposite as tracing tags, together with polythiophene/reduced graphene oxide-modified electrodes, this research hopes to produce an electrochemical genosensor that can identify two cancer-related miRNAs at the same time in 2018. The detection limits for let-7a and miR-106a were found to be at 0.02 fM and 0.06 fM, respectively, with the described POC system [153].

Synthesis and Applications to the Development of Electrochemical Sensors in the Determination of Drugs and Compounds of Clinical Interest

The analysis of pharmaceutical compounds and their relevance to clinical analysis is extremely important, especially in the healthcare field. Analysis of clinically interesting compounds makes it possible to make a more accurate and effective diagnosis. Both studies require sensitive, effective, and reliable analytical methods to ensure good drug quality and patient health [154].

Spectroscopic and chromatographic analytical methods are used for the quantification of drugs and compounds of clinically interesting interest. However, these techniques are relatively malignant, have some disadvantages regarding analysis time and pretreatment steps. In addition, electrochemical/electroanalytical sensors are promising to complement existing conventional techniques without sacrificing their sensitivity and selectivity due to their versatility, portability, and miniaturization possibilities of the system [154]. Electrochemical sensors use low-cost instruments with lower detection limits and less reagent usage compared to the instrumentation required for chromatographic and spectrophotometric analysis [155].

CONCLUSION

Nanoscale materials offer new applications in a wide variety of fields, and the variety of applications is constantly increasing. For this reason, many industries have embraced the improvements that nanoparticles can provide and developed industry-specific products.

It also allows quick medical decisions, especially as it leads to improved health outcomes, which allows diseases to be diagnosed at an early stage and patients can begin treatment early. In this study, a perspective was given to nanomaterial-based electrochemical sensors, which can be integrated into POC devices used in the health sector by giving examples of their basic usage areas.

CONSENT FOR PUBLICATION

Not applicable.

CONFLICT INTERESTS

The authors declare no conflict of interest, financial or otherwise.

ACKNOWLEDGMENTS

Dcelared none.

REFERENCES

[1]   J. Hovancová, I. Šišoláková, R. Oriňaková, and A. Oriňak, *J. Solid State Electrochem.,* vol. 21, pp. 2147-2166, 2017.
[http://dx.doi.org/10.1007/s10008-017-3544-0]

[2]   R. Sandulescu, M. Tertis, C. Cristea, and E. Bodoki, *Biosens.* Micro Nanoscale Appl., InTech, 2015.

[3]   N.M. Noah, and P.M. Ndangili, *J. Anal. Methods Chem.,* p. 2019, 2019.

[4]   J. Chen, B. Liu, X. Gao, and D. Xu, *RSC Advances,* vol. 8, pp. 28048-28085, 2018.
[http://dx.doi.org/10.1039/C8RA04205E]

[5]   S.A. Lim, and M.U. Ahmed, *RSC Advances,* vol. 6, pp. 24995-25014, 2016.
[http://dx.doi.org/10.1039/C6RA00333H]

[6]   T. Yamashita, K. Yamashita, H. Nabeshi, T. Yoshikawa, and Y. Yoshioka, "S. ichi Tsunoda, Y. Tsutsumi", *Materials (Basel),* vol. 5, pp. 350-363, 2012.
[http://dx.doi.org/10.3390/ma5020350] [PMID: 28817050]

[7]   M. Pumera, S. Sánchez, I. Ichinose, and J. Tang, *Sens. Actuators B Chem.,* vol. 123, pp. 1195-1205, 2007.
[http://dx.doi.org/10.1016/j.snb.2006.11.016]

[8]   A. Azzouz, K.Y. Goud, N. Raza, E. Ballesteros, S.E. Lee, J. Hong, A. Deep, and K.H. Kim, "TrAC -", *Trends Analyt. Chem.,* vol. 110, pp. 15-34, 2019.
[http://dx.doi.org/10.1016/j.trac.2018.08.002]

[9]   S.K. Krishnan, E. Singh, P. Singh, M. Meyyappan, and H.S. Nalwa, *RSC Advances,* vol. 9, pp. 8778-8781, 2019.
[http://dx.doi.org/10.1039/C8RA09577A]

[10] R. Justin, and B. Chen, "Strong and conductive chitosan-reduced graphene oxide nanocomposites for transdermal drug delivery", *J. Mater. Chem. B Mater. Biol. Med.,* vol. 2, no. 24, pp. 3759-3770, 2014.
[http://dx.doi.org/10.1039/c4tb00390j] [PMID: 32261722]

[11] J. Zheng, C.A. Di, Y. Liu, H. Liu, Y. Guo, C. Du, T. Wu, G. Yu, and D. Zhu, "High quality graphene with large flakes exfoliated by oleyl amine", *Chem. Commun. (Camb.),* vol. 46, no. 31, pp. 5728-5730, 2010.
[http://dx.doi.org/10.1039/c0cc00954g] [PMID: 20593087]

[12] J. Wang, *Analyst (Lond.),* vol. 130, pp. 421-426, 2005.
[http://dx.doi.org/10.1039/b414248a]

[13] J.J. Gooding, R. Wibowo, J. Liu, W. Yang, D. Losic, S. Orbons, F.J. Mearns, J.G. Shapter, and D.B. Hibbert, "Protein electrochemistry using aligned carbon nanotube arrays", *J. Am. Chem. Soc.,* vol. 125, no. 30, pp. 9006-9007, 2003.
[http://dx.doi.org/10.1021/ja035722f] [PMID: 15369344]

[14] X. Yu, D. Chattopadhyay, I. Galeska, F. Papadimitrakopoulos, and J.F. Rusling, *Electrochem. Commun.,* vol. 5, pp. 408-411, 2003.
[http://dx.doi.org/10.1016/S1388-2481(03)00076-6]

[15] F. Patolsky, Y. Weizmann, and I. Willner, *Angew. Chem. Int. Ed.*, vol. 43, pp. 2113-2117, 2004.
[http://dx.doi.org/10.1002/anie.200353275]

[16] M. Delvaux, and S. Demoustier-Champagne, "Immobilisation of glucose oxidase within metallic nanotubes arrays for application to enzyme biosensors", *Biosens. Bioelectron.*, vol. 18, no. 7, pp. 943-951, 2003.
[http://dx.doi.org/10.1016/S0956-5663(02)00209-9] [PMID: 12713918]

[17] A. Stobiecka, H. Radecka, and J. Radecki, "Novel voltammetric biosensor for determining acrylamide in food samples", *Biosens. Bioelectron.*, vol. 22, no. 9-10, pp. 2165-2170, 2007.
[http://dx.doi.org/10.1016/j.bios.2006.10.008] [PMID: 17097868]

[18] A. Krajewska, J. Radecki, and H. Radecka, "A Voltammetric Biosensor Based on Glassy Carbon Electrodes Modified with Single-Walled Carbon Nanotubes/Hemoglobin for Detection of Acrylamide in Water Extracts from Potato Crisps", *Sensors (Basel)*, vol. 8, no. 9, pp. 5832-5844, 2008.
[http://dx.doi.org/10.3390/s8095832] [PMID: 27873843]

[19] S. Kwakye, V.N. Goral, and A.J. Baeumner, "Electrochemical microfluidic biosensor for nucleic acid detection with integrated minipotentiostat", *Biosens. Bioelectron.*, vol. 21, no. 12, pp. 2217-2223, 2006.
[http://dx.doi.org/10.1016/j.bios.2005.11.017] [PMID: 16386889]

[20] A.J. Baeumner, R.N. Cohen, V. Miksic, and J. Min, "RNA biosensor for the rapid detection of viable Escherichia coli in drinking water", *Biosens. Bioelectron.*, vol. 18, no. 4, pp. 405-413, 2003.
[http://dx.doi.org/10.1016/S0956-5663(02)00162-8] [PMID: 12604258]

[21] S. Viswanathan, L.C. Wu, M.R. Huang, and J.A.A. Ho, "Electrochemical immunosensor for cholera toxin using liposomes and poly(3,4-ethylenedioxythiophene)-coated carbon nanotubes", *Anal. Chem.*, vol. 78, no. 4, pp. 1115-1121, 2006.
[http://dx.doi.org/10.1021/ac051435d] [PMID: 16478102]

[22] D.A. Dinh, K.S. Hui, K.N. Hui, Y.R. Cho, W. Zhou, X. Hong, and H.H. Chun, *Appl. Surf. Sci.*, vol. 298, pp. 62-67, 2014.
[http://dx.doi.org/10.1016/j.apsusc.2014.01.101]

[23] L. Han, C.M. Liu, S.L. Dong, C.X. Du, X.Y. Zhang, L.H. Li, and Y. Wei, "Enhanced conductivity of rGO/Ag NPs composites for electrochemical immunoassay of prostate-specific antigen", *Biosens. Bioelectron.*, vol. 87, pp. 466-472, 2017.
[http://dx.doi.org/10.1016/j.bios.2016.08.004] [PMID: 27591721]

[24] X. Cao, S. Liu, Q. Feng, and N. Wang, "Silver nanowire-based electrochemical immunoassay for sensing immunoglobulin G with signal amplification using strawberry-like ZnO nanostructures as labels", *Biosens. Bioelectron.*, vol. 49, pp. 256-262, 2013.
[http://dx.doi.org/10.1016/j.bios.2013.05.029] [PMID: 23774162]

[25] X. Cai, S. Weng, R. Guo, L. Lin, W. Chen, Z. Zheng, Z. Huang, and X. Lin, "Ratiometric electrochemical immunoassay based on internal reference value for reproducible and sensitive detection of tumor marker", *Biosens. Bioelectron.*, vol. 81, pp. 173-180, 2016.
[http://dx.doi.org/10.1016/j.bios.2016.02.066] [PMID: 26945184]

[26] H. Chang, H. Zhang, J. Lv, B. Zhang, W. Wei, and J. Guo, "Pt NPs and DNAzyme functionalized polymer nanospheres as triple signal amplification strategy for highly sensitive electrochemical immunosensor of tumour marker", *Biosens. Bioelectron.*, vol. 86, pp. 156-163, 2016.
[http://dx.doi.org/10.1016/j.bios.2016.06.048] [PMID: 27362254]

[27] B. He, T.J. Morrow, and C.D. Keating, "Nanowire sensors for multiplexed detection of biomolecules", *Curr. Opin. Chem. Biol.*, vol. 12, no. 5, pp. 522-528, 2008.
[http://dx.doi.org/10.1016/j.cbpa.2008.08.027] [PMID: 18804551]

[28] N.J. Ronkainen, and S.L. Okon, "Nanomaterial-Based Electrochemical Immunosensors for Clinically Significant Biomarkers", *Materials (Basel)*, vol. 7, no. 6, pp. 4669-4709, 2014.

[http://dx.doi.org/10.3390/ma7064669] [PMID: 28788700]

[29]   A.N. Sokolov, M.E. Roberts, and Z. Bao, *Mater. Today,* vol. 12, pp. 12-20, 2009.
       [http://dx.doi.org/10.1016/S1369-7021(09)70247-0]

[30]   I-H. Cho, J. Lee, J. Kim, M. Kang, J. Paik, S. Ku, H-M. Cho, J. Irudayaraj, and D-H. Kim, *Sensors (Basel),* vol. 18, p. 207, 2018.
       [http://dx.doi.org/10.3390/s18010207]

[31]   P. Wang, M. Li, F. Pei, Y. Li, Q. Liu, Y. Dong, Q. Chu, and H. Zhu, "An ultrasensitive sandwich-type electrochemical immunosensor based on the signal amplification system of double-deck gold film and thionine unite with platinum nanowire inlaid globular SBA-15 microsphere", *Biosens. Bioelectron.,* vol. 91, pp. 424-430, 2017.
       [http://dx.doi.org/10.1016/j.bios.2016.12.057] [PMID: 28064127]

[32]   I.H. Cho, D.H. Kim, and S. Park, *Biomater. Res.,* vol. 24, pp. 1-12, 2020.
       [http://dx.doi.org/10.1186/s40824-019-0181-y] [PMID: 31911841]

[33]   W. Huang, A.K. Diallo, J.L. Dailey, K. Besar, and H.E. Katz, "Electrochemical processes and mechanistic aspects of field-effect sensors for biomolecules", *J. Mater. Chem. C Mater. Opt. Electron. Devices,* vol. 3, no. 25, pp. 6445-6470, 2015.
       [http://dx.doi.org/10.1039/C5TC00755K] [PMID: 29238595]

[34]   R. Elnathan, M. Kwiat, A. Pevzner, Y. Engel, L. Burstein, A. Khatchtourints, A. Lichtenstein, R. Kantaev, and F. Patolsky, "Biorecognition layer engineering: overcoming screening limitations of nanowire-based FET devices", *Nano Lett.,* vol. 12, no. 10, pp. 5245-5254, 2012.
       [http://dx.doi.org/10.1021/nl302434w] [PMID: 22963381]

[35]   M. Gniadek, S. Modzelewska, M. Donten, and Z. Stojek, "Modification of electrode surfaces: deposition of thin layers of polypyrrole--Au nanoparticle materials using a combination of interphase synthesis and dip-in method", *Anal. Chem.,* vol. 82, no. 2, pp. 469-472, 2010.
       [http://dx.doi.org/10.1021/ac902426c] [PMID: 20038092]

[36]   T.A.P. Rocha-Santos, "TrAC -", *Trends Analyt. Chem.,* vol. 62, pp. 28-36, 2014.
       [http://dx.doi.org/10.1016/j.trac.2014.06.016]

[37]   A. Nomura, S. Shin, O.O. Mehdi, and J.M. Kauffmann, "Preparation, characterization, and application of an enzyme-immobilized magnetic microreactor for flow injection analysis", *Anal. Chem.,* vol. 76, no. 18, pp. 5498-5502, 2004.
       [http://dx.doi.org/10.1021/ac049489v] [PMID: 15362912]

[38]   X.F. Zhang, X.L. Dong, H. Huang, B. Lv, X.G. Zhu, J.P. Lei, S. Ma, W. Liu, and Z.D. Zhang, *Mater. Sci. Eng. A,* vol. 454–455, pp. 211-215, 2007.
       [http://dx.doi.org/10.1016/j.msea.2006.11.010]

[39]   C. Grüttner, S. Rudershausen, and J. Teller, *J. Magn. Magn. Mater.,* vol. 225, pp. 1-7, 2001.
       [http://dx.doi.org/10.1016/S0304-8853(00)01220-8]

[40]   J.F. Rusling, G. Sotzing, and F. Papadimitrakopoulosa, "Designing nanomaterial-enhanced electrochemical immunosensors for cancer biomarker proteins", *Bioelectrochemistry,* vol. 76, no. 1-2, pp. 189-194, 2009.
       [http://dx.doi.org/10.1016/j.bioelechem.2009.03.011] [PMID: 19403342]

[41]   K. Aguilar-Arteaga, J.A. Rodriguez, and E. Barrado, "Magnetic solids in analytical chemistry: a review", *Anal. Chim. Acta,* vol. 674, no. 2, pp. 157-165, 2010.
       [http://dx.doi.org/10.1016/j.aca.2010.06.043] [PMID: 20678625]

[42]   Y. Xu, and E. Wang, *Electrochim. Acta,* vol. 84, pp. 62-73, 2012.
       [http://dx.doi.org/10.1016/j.electacta.2012.03.147]

[43]   E. Gatebe, *J. Agric. Sci. Technol.,* vol. •••, p. 14, 2012.

[44]   X. Luo, A. Morrin, A.J. Killard, and M.R. Smyth, *Electroanalysis,* vol. 18, pp. 319-326, 2006.

[http://dx.doi.org/10.1002/elan.200503415]

[45] S. Eustis, and M.A. el-Sayed, "Why gold nanoparticles are more precious than pretty gold: noble metal surface plasmon resonance and its enhancement of the radiative and nonradiative properties of nanocrystals of different shapes", *Chem. Soc. Rev.,* vol. 35, no. 3, pp. 209-217, 2006.
[http://dx.doi.org/10.1039/B514191E] [PMID: 16505915]

[46] R. Brian, "Eggins", *Anal. Tech. Sci,* 2002.

[47] *G.G. Wagner G.* Food Biosensor Analysis: New York, 1994.

[48] S.P. Mohanty, and E. Koucianos, *IEEE Potentials,* vol. 25, pp. 35-40, 2006.
[http://dx.doi.org/10.1109/MP.2006.1649009]

[49] C. Jianrong, M. Yuqing, H. Nongyue, W. Xiaohua, and L. Sijiao, "Nanotechnology and biosensors", *Biotechnol. Adv.,* vol. 22, no. 7, pp. 505-518, 2004.
[http://dx.doi.org/10.1016/j.biotechadv.2004.03.004] [PMID: 15262314]

[50] "G. LI", *Anal. Sci.,* vol. 20, pp. 603-609, 2004.

[51] M. Wasowicz, S. Viswanathan, A. Dvornyk, K. Grzelak, B. Kłudkiewicz, and H. Radecka, "Comparison of electrochemical immunosensors based on gold nano materials and immunoblot techniques for detection of histidine-tagged proteins in culture medium", *Biosens. Bioelectron.,* vol. 24, no. 2, pp. 284-289, 2008.
[http://dx.doi.org/10.1016/j.bios.2008.04.002] [PMID: 18486465]

[52] C.M. Niemeyer, *Angew. Chem. Int. Ed.,* vol. 40, pp. 4128-4158, 2001.
[http://dx.doi.org/10.1002/1521-3773(20011119)40:22<4128::AID-ANIE4128>3.0.CO;2-S]

[53] M. Farré, L. Kantiani, S. Pérez, and D. Barceló, "TrAC -", *Trends Analyt. Chem.,* vol. 28, pp. 170-185, 2009.

[54] S. Viswanathan, H. Radecka, and J. Radecki, *Monatsh. Chem.,* vol. 140, pp. 891-899, 2009.
[http://dx.doi.org/10.1007/s00706-009-0143-5]

[55] Y. Liu, R. Yuan, Y. Chai, D. Tang, J. Dai, and X. Zhong, *Sens. Actuators B Chem.,* vol. 115, pp. 109-115, 2006.
[http://dx.doi.org/10.1016/j.snb.2005.08.048]

[56] J. Wang, M. Pedrero, H. Sakslund, O. Hammerich, and J. Pingarron, *Analyst (Lond.),* vol. 121, pp. 345-350, 1996.
[http://dx.doi.org/10.1039/an9962100345]

[57] P. Skládal, and T. Kaláb, *Anal. Chim. Acta,* vol. 316, pp. 73-78, 1995.
[http://dx.doi.org/10.1016/0003-2670(95)00342-W]

[58] A.P. Deng, and H. Yang, *Sens. Actuators B Chem.,* vol. 124, pp. 202-208, 2007.
[http://dx.doi.org/10.1016/j.snb.2006.12.023]

[59] I.H. El-Sayed, X. Huang, and M.A. El-Sayed, "Surface plasmon resonance scattering and absorption of anti-EGFR antibody conjugated gold nanoparticles in cancer diagnostics: applications in oral cancer", *Nano Lett.,* vol. 5, no. 5, pp. 829-834, 2005.
[http://dx.doi.org/10.1021/nl050074e] [PMID: 15884879]

[60] "M. ABDI", *Anal. Sci.,* vol. 20, pp. 1113-1126, 2004.

[61] D.R. Thévenot, K. Toth, R.A. Durst, and G.S. Wilson, "Electrochemical biosensors: recommended definitions and classification", *Biosens. Bioelectron.,* vol. 16, no. 1-2, pp. 121-131, 2001.
[PMID: 11261847]

[62] S. Tombelli, M. Minunni, and M. Mascini, "Aptamers-based assays for diagnostics, environmental and food analysis", *Biomol. Eng.,* vol. 24, no. 2, pp. 191-200, 2007.
[http://dx.doi.org/10.1016/j.bioeng.2007.03.003] [PMID: 17434340]

[63] P.D. Patel, "TrAC -", *Trends Analyt. Chem.,* vol. 21, pp. 96-115, 2002.

[http://dx.doi.org/10.1016/S0165-9936(01)00136-4]

[64] E. Bakker, "Electrochemical sensors", *Anal. Chem.,* vol. 76, no. 12, pp. 3285-3298, 2004.
[http://dx.doi.org/10.1021/ac049580z] [PMID: 15193109]

[65] B. Eggins, *Chemical Sensors and Biosensors.* Wiley: West Sussex, 2007.

[66] L.V. Shkotova, T.B. Goriushkina, C. Tran-Minh, J.M. Chovelon, A.P. Soldatkin, and S.V. Dzyadevych, *Mater. Sci. Eng. C,* vol. 28, pp. 943-948, 2008.
[http://dx.doi.org/10.1016/j.msec.2007.10.038]

[67] H. Ohnuki, T. Saiki, A. Kusakari, H. Endo, M. Ichihara, and M. Izumi, "Incorporation of glucose oxidase into Langmuir-Blodgett films based on Prussian blue applied to amperometric glucose biosensor", *Langmuir,* vol. 23, no. 8, pp. 4675-4681, 2007.
[http://dx.doi.org/10.1021/la063175g] [PMID: 17367170]

[68] S.A.M. Marzouk, S.S. Ashraf, and K.A. Tayyari, "Prototype amperometric biosensor for sialic acid determination", *Anal. Chem.,* vol. 79, no. 4, pp. 1668-1674, 2007.
[http://dx.doi.org/10.1021/ac061886d] [PMID: 17297971]

[69] L.D. Mello, and L.T. Kubota, *Food Chem.,* vol. 77, pp. 237-256, 2002.
[http://dx.doi.org/10.1016/S0308-8146(02)00104-8]

[70] J.M.G.G.G. Kauffmann, and G.G. Guilbault, "Potentiometric enzyme electrodes", *Bioprocess Technol.,* vol. 15, pp. 63-82, 1991.
[PMID: 1367610]

[71] M. Yuqing, G. Jianguo, and C. Jianrong, "Ion sensitive field effect transducer-based biosensors", *Biotechnol. Adv.,* vol. 21, no. 6, pp. 527-534, 2003.
[http://dx.doi.org/10.1016/S0734-9750(03)00103-4] [PMID: 14499153]

[72] M. Pohanka, D. Jun, and K. Kuca, "Mycotoxin assays using biosensor technology: a review", *Drug Chem. Toxicol.,* vol. 30, no. 3, pp. 253-261, 2007.
[http://dx.doi.org/10.1080/01480540701375232] [PMID: 17613010]

[73] F. Davis, M.A. Hughes, A.R. Cossins, and S.P.J. Higson, "Single gene differentiation by DNA-modified carbon electrodes using an AC impedimetric approach", *Anal. Chem.,* vol. 79, no. 3, pp. 1153-1157, 2007.
[http://dx.doi.org/10.1021/ac061070c] [PMID: 17263348]

[74] O. Ouerghi, A. Touhami, N. Jaffrezic-Renault, C. Martelet, H. Ben Ouada, and S. Cosnier, *Bioelectrochemistry.* Elsevier, 2002, pp. 131-133.

[75] I. Szymańska, H. Radecka, J. Radecki, and R. Kaliszan, "Electrochemical impedance spectroscopy for study of amyloid beta-peptide interactions with (-) nicotine ditartrate and (-) cotinine", *Biosens. Bioelectron.,* vol. 22, no. 9-10, pp. 1955-1960, 2007.
[http://dx.doi.org/10.1016/j.bios.2006.08.025] [PMID: 17000100]

[76] J. Liu, *Int. J. Biosens. Bioelectron.,* vol. 2, pp. 100-102, 2017.
[http://dx.doi.org/10.15406/ijbsbe.2017.02.00027]

[77] V. Bhardwaj, and A. Kaushik, *Micromachines (Basel),* vol. 8, p. 298, 2017.
[http://dx.doi.org/10.3390/mi8100298]

[78] A. Kaushik, A. Yndart, S. Kumar, R.D. Jayant, A. Vashist, A.N. Brown, C.Z. Li, and M. Nair, *Sci. Rep.,* vol. 8, pp. 1-5, 2018.
[PMID: 29311619]

[79] A. Kaushik, and M. Mujawar, *Sensors (Basel),* vol. 18, p. 4303, 2018.
[http://dx.doi.org/10.3390/s18124303]

[80] J. Artigas, C. Jiménez, C. Domínguez, S. Mínguez, A. Gonzalo, and J. Alonso, *Sens. Actuators B Chem.,* vol. 89, pp. 199-204, 2003.
[http://dx.doi.org/10.1016/S0925-4005(02)00464-1]

[81]   S. Vashist, *Biosensors (Basel)*, vol. 7, p. 62, 2017.
        [http://dx.doi.org/10.3390/bios7040062]

[82]   S.K. Vashist, P.B. Luppa, L.Y. Yeo, A. Ozcan, and J.H.T. Luong, "Emerging Technologies for Next-Generation Point-of-Care Testing", *Trends Biotechnol.*, vol. 33, no. 11, pp. 692-705, 2015.
        [http://dx.doi.org/10.1016/j.tibtech.2015.09.001] [PMID: 26463722]

[83]   S.K. Metkar, and K. Girigoswami, *Biocatal. Agric. Biotechnol.*, vol. 17, pp. 271-283, 2019.
        [http://dx.doi.org/10.1016/j.bcab.2018.11.029]

[84]   J. Wang, *Biosens. Bioelectron.* Elsevier, 2006, pp. 1887-1892.

[85]   P. Mehrotra, "Biosensors and their applications - A review", *J. Oral Biol. Craniofac. Res.*, vol. 6, no. 2, pp. 153-159, 2016.
        [http://dx.doi.org/10.1016/j.jobcr.2015.12.002] [PMID: 27195214]

[86]   F. Huber, H.P. Lang, J. Zhang, D. Rimoldi, and C. Gerber, *Swiss Med. Wkly.*, vol. 145, p. 14092, 2015.

[87]   B. Hayes, C. Murphy, A. Crawley, and R. O'Kennedy, *Diagnostics (Basel)*, vol. 8, p. 39, 2018.
        [http://dx.doi.org/10.3390/diagnostics8020039]

[88]   P. Paul, A.K. Malakar, and S. Chakraborty, "The significance of gene mutations across eight major cancer types", *Mutat. Res. Rev. Mutat. Res.*, vol. 781, pp. 88-99, 2019.
        [http://dx.doi.org/10.1016/j.mrrev.2019.04.004] [PMID: 31416581]

[89]   L. Wu, and X. Qu, "Cancer biomarker detection: recent achievements and challenges", *Chem. Soc. Rev.*, vol. 44, no. 10, pp. 2963-2997, 2015.
        [http://dx.doi.org/10.1039/C4CS00370E] [PMID: 25739971]

[90]   A. Mishra, and M. Verma, "Cancer biomarkers: are we ready for the prime time?", *Cancers (Basel)*, vol. 2, no. 1, pp. 190-208, 2010.
        [http://dx.doi.org/10.3390/cancers2010190] [PMID: 24281040]

[91]   P.M. Kosaka, V. Pini, J.J. Ruz, R.A. da Silva, M.U. González, D. Ramos, M. Calleja, and J. Tamayo, "Detection of cancer biomarkers in serum using a hybrid mechanical and optoplasmonic nanosensor", *Nat. Nanotechnol.*, vol. 9, no. 12, pp. 1047-1053, 2014.
        [http://dx.doi.org/10.1038/nnano.2014.250] [PMID: 25362477]

[92]   C.G. Siontorou, G.P.D. Nikoleli, D.P. Nikolelis, S. Karapetis, N. Tzamtzis, and S. Bratakou, *Next Gener.* Point-of-Care Biomed. Sensors Technol. Cancer Diagnosis, Springer Singapore, 2017, pp. 115-132.

[93]   Y. Okaie, T. Nakano, T. Hara, and S. Nishio, *SpringerBriefs Comput. Sci.* Springer, 2016, pp. 29-51.

[94]   E. Salvati, F. Stellacci, and S. Krol, "Nanosensors for early cancer detection and for therapeutic drug monitoring", *Nanomedicine (Lond.)*, vol. 10, no. 23, pp. 3495-3512, 2015.
        [http://dx.doi.org/10.2217/nnm.15.180] [PMID: 26606949]

[95]   R. Shandilya, A. Bhargava, N. Bunkar, R. Tiwari, I.Y. Goryacheva, and P.K. Mishra, "Nanobiosensors: Point-of-care approaches for cancer diagnostics", *Biosens. Bioelectron.*, vol. 130, pp. 147-165, 2019.
        [http://dx.doi.org/10.1016/j.bios.2019.01.034] [PMID: 30735948]

[96]   M.E. Grigore, *J. Med. Res. Heal. Educ. 1*, 2017.

[97]   K.J. Cash, and H.A. Clark, "Nanosensors and nanomaterials for monitoring glucose in diabetes", *Trends Mol. Med.*, vol. 16, no. 12, pp. 584-593, 2010.
        [http://dx.doi.org/10.1016/j.molmed.2010.08.002] [PMID: 20869318]

[98]   V. Kumar, S. Hebbar, A. Bhat, S. Panwar, M. Vaishnav, K. Muniraj, V. Nath, R.B. Vijay, S. Manjunath, B. Thyagaraj, C. Siddalingappa, M. Chikkamoga Siddaiah, I. Dasgupta, U. Anandh, T. Kamala, S.S. Srikanta, P.R. Krishnaswamy, and N. Bhat, "Application of a Nanotechnology-Based, Point-of-Care Diagnostic Device in Diabetic Kidney Disease", *Kidney Int. Rep.*, vol. 3, no. 5, pp.

1110-1118, 2018.
[http://dx.doi.org/10.1016/j.ekir.2018.05.008] [PMID: 30197977]

[99]   J. Zhang, W. Hodge, C. Hutnick, and X. Wang, "Noninvasive diagnostic devices for diabetes through measuring tear glucose", *J. Diabetes Sci. Technol.,* vol. 5, no. 1, pp. 166-172, 2011.
[http://dx.doi.org/10.1177/193229681100500123] [PMID: 21303640]

[100] J. Wang, "Electrochemical glucose biosensors", *Chem. Rev.,* vol. 108, no. 2, pp. 814-825, 2008.
[http://dx.doi.org/10.1021/cr068123a] [PMID: 18154363]

[101] M. Taguchi, A. Ptitsyn, E.S. McLamore, and J.C. Claussen, "Nanomaterial-mediated Biosensors for Monitoring Glucose", *J. Diabetes Sci. Technol.,* vol. 8, no. 2, pp. 403-411, 2014.
[http://dx.doi.org/10.1177/1932296814522799] [PMID: 24876594]

[102] D.C. Klonoff, D. Ahn, and A. Drincic, "Continuous glucose monitoring: A review of the technology and clinical use", *Diabetes Res. Clin. Pract.,* vol. 133, pp. 178-192, 2017.
[http://dx.doi.org/10.1016/j.diabres.2017.08.005] [PMID: 28965029]

[103] D. Olczuk, and R. Priefer, "A history of continuous glucose monitors (CGMs) in self-monitoring of diabetes mellitus", *Diabetes Metab. Syndr.,* vol. 12, no. 2, pp. 181-187, 2018.
[http://dx.doi.org/10.1016/j.dsx.2017.09.005] [PMID: 28967612]

[104] P. Calhoun, T.K. Johnson, J. Hughes, D. Price, and A.K. Balo, "Resistance to Acetaminophen Interference in a Novel Continuous Glucose Monitoring System", *J. Diabetes Sci. Technol.,* vol. 12, no. 2, pp. 393-396, 2018.
[http://dx.doi.org/10.1177/1932296818755797] [PMID: 29334775]

[105] C. Lorenz, W. Sandoval, and M. Mortellaro, "Interference Assessment of Various Endogenous and Exogenous Substances on the Performance of the Eversense Long-Term Implantable Continuous Glucose Monitoring System", *Diabetes Technol. Ther.,* vol. 20, no. 5, pp. 344-352, 2018.
[http://dx.doi.org/10.1089/dia.2018.0028] [PMID: 29600877]

[106] K. Obermaier, G. Schmelzeisen-Redeker, M. Schoemaker, H.M. Klötzer, H. Kirchsteiger, H. Eikmeier, and L. Del Re, *J. Diabetes Sci. Technol.* SAGE Publications Inc., 2013, pp. 824-832.

[107] F.J. Ascaso, and V. Huerva, "Noninvasive Continuous Monitoring of Tear Glucose Using Glucose-Sensing Contact Lenses", *Optom. Vis. Sci.,* vol. 93, no. 4, pp. 426-434, 2016.
[http://dx.doi.org/10.1097/OPX.0000000000000698] [PMID: 26390345]

[108] P. Wang, and L.J. Kricka, "Current and Emerging Trends in Point-of-Care Technology and Strategies for Clinical Validation and Implementation", *Clin. Chem.,* vol. 64, no. 10, pp. 1439-1452, 2018.
[http://dx.doi.org/10.1373/clinchem.2018.287052] [PMID: 29884677]

[109] Y. Wang, L. Yu, X. Kong, and L. Sun, "Application of nanodiagnostics in point-of-care tests for infectious diseases", *Int. J. Nanomedicine,* vol. 12, pp. 4789-4803, 2017.
[http://dx.doi.org/10.2147/IJN.S137338] [PMID: 28740385]

[110] T. Laksanasopin, T.W. Guo, S. Nayak, A.A. Sridhara, S. Xie, O.O. Olowookere, P. Cadinu, F. Meng, N.H. Chee, J. Kim, C.D. Chin, E. Munyazesa, P. Mugwaneza, A.J. Rai, V. Mugisha, A.R. Castro, D. Steinmiller, V. Linder, J.E. Justman, S. Nsanzimana, and S.K. Sia, "A smartphone dongle for diagnosis of infectious diseases at the point of care", *Sci. Transl. Med.,* vol. 7, no. 273, p. 273re1, 2015.
[http://dx.doi.org/10.1126/scitranslmed.aaa0056] [PMID: 25653222]

[111] A. Moody, "Rapid diagnostic tests for malaria parasites", *Clin. Microbiol. Rev.,* vol. 15, no. 1, pp. 66-78, 2002.
[http://dx.doi.org/10.1128/CMR.15.1.66-78.2002] [PMID: 11781267]

[112] M.E. Rafael, T. Taylor, A. Magill, Y.W. Lim, F. Girosi, and R. Allan, "Reducing the burden of childhood malaria in Africa: the role of improved", *Nature,* vol. 444, suppl. Suppl. 1, pp. 39-48, 2006.
[http://dx.doi.org/10.1038/nature05445] [PMID: 17159893]

[113] D. Bell, C. Wongsrichanalai, and J.W. Barnwell, "Ensuring quality and access for malaria diagnosis:

how can it be achieved?", *Nat. Rev. Microbiol.,* vol. 4, no. 9, suppl. Suppl., pp. S7-S20, 2006.
[http://dx.doi.org/10.1038/nrmicro1525] [PMID: 17003770]

[114] W.G. Lee, Y.G. Kim, B.G. Chung, U. Demirci, and A. Khademhosseini, "Nano/Microfluidics for diagnosis of infectious diseases in developing countries", *Adv. Drug Deliv. Rev.,* vol. 62, no. 4-5, pp. 449-457, 2010.
[http://dx.doi.org/10.1016/j.addr.2009.11.016] [PMID: 19954755]

[115] M. Antia, T. Herricks, and P.K. Rathod, " PLoS Pathog. 3", .

[116] A.L. García-Basteiro, A. DiNardo, B. Saavedra, D.R. Silva, D. Palmero, M. Gegia, G.B. Migliori, R. Duarte, E. Mambuque, R. Centis, L.E. Cuevas, S. Izco, and G. Theron, *Rev. Port. Pneumol.* 24th ed. English, 2018, pp. 73-85.

[117] R. McNerney, and P. Daley, "Towards a point-of-care test for active tuberculosis: obstacles and opportunities", *Nat. Rev. Microbiol.,* vol. 9, no. 3, pp. 204-213, 2011.
[http://dx.doi.org/10.1038/nrmicro2521] [PMID: 21326275]

[118] M. Uplekar, D. Weil, K. Lonnroth, E. Jaramillo, C. Lienhardt, H.M. Dias, D. Falzon, K. Floyd, G. Gargioni, H. Getahun, C. Gilpin, P. Glaziou, M. Grzemska, F. Mirzayev, H. Nakatani, and M. Raviglione, "WHO's new end TB strategy", *Lancet,* vol. 385, no. 9979, pp. 1799-1801, 2015.
[http://dx.doi.org/10.1016/S0140-6736(15)60570-0] [PMID: 25814376]

[119] K. Dheda, M. Ruhwald, G. Theron, J. Peter, and W.C. Yam, "Point-of-care diagnosis of tuberculosis: past, present and future", *Respirology,* vol. 18, no. 2, pp. 217-232, 2013.
[http://dx.doi.org/10.1111/resp.12022] [PMID: 23190246]

[120] V. Mani, S. Wang, F. Inci, G. De Libero, A. Singhal, and U. Demirci, "Emerging technologies for monitoring drug-resistant tuberculosis at the point-of-care", *Adv. Drug Deliv. Rev.,* vol. 78, pp. 105-117, 2014.
[http://dx.doi.org/10.1016/j.addr.2014.05.015] [PMID: 24882226]

[121] S.D. Lawn, and M.P. Nicol, "Xpert® MTB/RIF assay: development, evaluation and implementation of a new rapid molecular diagnostic for tuberculosis and rifampicin resistance", *Future Microbiol.,* vol. 6, no. 9, pp. 1067-1082, 2011.
[http://dx.doi.org/10.2217/fmb.11.84] [PMID: 21958145]

[122] D. Helb, M. Jones, E. Story, C. Boehme, E. Wallace, K. Ho, J. Kop, M.R. Owens, R. Rodgers, P. Banada, H. Safi, R. Blakemore, N.T. Lan, E.C. Jones-López, M. Levi, M. Burday, I. Ayakaka, R.D. Mugerwa, B. McMillan, E. Winn-Deen, L. Christel, P. Dailey, M.D. Perkins, D.H. Persing, and D. Alland, "Rapid detection of Mycobacterium tuberculosis and rifampin resistance by use of on-demand, near-patient technology", *J. Clin. Microbiol.,* vol. 48, no. 1, pp. 229-237, 2010.
[http://dx.doi.org/10.1128/JCM.01463-09] [PMID: 19864480]

[123] C. Sheridan, "Fast, portable tests come online to curb coronavirus pandemic", *Nat. Biotechnol.,* vol. 38, no. 5, pp. 515-518, 2020.
[http://dx.doi.org/10.1038/d41587-020-00010-2] [PMID: 32203294]

[124] J. Zhifeng, A. Feng, and T. Li, "Consistency analysis of COVID-19 nucleic acid tests and the changes of lung CT", *J. Clin. Virol.,* vol. 127, p. 104359, 2020.
[http://dx.doi.org/10.1016/j.jcv.2020.104359] [PMID: 32302956]

[125] Z. Li, Y. Yi, X. Luo, N. Xiong, Y. Liu, S. Li, R. Sun, Y. Wang, B. Hu, W. Chen, Y. Zhang, J. Wang, B. Huang, Y. Lin, J. Yang, W. Cai, X. Wang, J. Cheng, Z. Chen, K. Sun, W. Pan, Z. Zhan, L. Chen, and F. Ye, "Development and clinical application of a rapid IgM-IgG combined antibody test for SARS-CoV-2 infection diagnosis", *J. Med. Virol.,* vol. 92, no. 9, pp. 1518-1524, 2020.
[http://dx.doi.org/10.1002/jmv.25727] [PMID: 32104917]

[126] I. Thevarajan, T.H.O. Nguyen, M. Koutsakos, J. Druce, L. Caly, C.E. van de Sandt, X. Jia, S. Nicholson, M. Catton, B. Cowie, S.Y.C. Tong, S.R. Lewin, and K. Kedzierska, "Breadth of concomitant immune responses prior to patient recovery: a case report of non-severe COVID-19", *Nat. Med.,* vol. 26, no. 4, pp. 453-455, 2020.

[http://dx.doi.org/10.1038/s41591-020-0819-2] [PMID: 32284614]

[127] J.L. He, L. Luo, Z.D. Luo, J.X. Lyu, M.Y. Ng, X.P. Shen, and Z. Wen, "Diagnostic performance between CT and initial real-time RT-PCR for clinically suspected 2019 coronavirus disease (COVID-19) patients outside Wuhan, China", *Respir. Med.*, vol. 168, p. 105980, 2020.
[http://dx.doi.org/10.1016/j.rmed.2020.105980] [PMID: 32364959]

[128] J.R. Choi, "Development of Point-of-Care Biosensors for COVID-19", *Front Chem.*, vol. 8, p. 517, 2020.
[http://dx.doi.org/10.3389/fchem.2020.00517] [PMID: 32574316]

[129] M.A. Lifson, M.O. Ozen, F. Inci, S. Wang, H. Inan, M. Baday, T.J. Henrich, and U. Demirci, "Advances in biosensing strategies for HIV-1 detection, diagnosis, and therapeutic monitoring", *Adv. Drug Deliv. Rev.*, vol. 103, pp. 90-104, 2016.
[http://dx.doi.org/10.1016/j.addr.2016.05.018] [PMID: 27262924]

[130] S. Islam, S. Shukla, V.K. Bajpai, Y.K. Han, Y.S. Huh, A. Kumar, A. Ghosh, and S. Gandhi, "A smart nanosensor for the detection of human immunodeficiency virus and associated cardiovascular and arthritis diseases using functionalized graphene-based transistors", *Biosens. Bioelectron.*, vol. 126, pp. 792-799, 2019.
[http://dx.doi.org/10.1016/j.bios.2018.11.041] [PMID: 30557838]

[131] F. Inci, O. Tokel, S. Wang, U.A. Gurkan, S. Tasoglu, D.R. Kuritzkes, and U. Demirci, "Nanoplasmonic quantitative detection of intact viruses from unprocessed whole blood", *ACS Nano*, vol. 7, no. 6, pp. 4733-4745, 2013.
[http://dx.doi.org/10.1021/nn3036232] [PMID: 23688050]

[132] P.M. Kosaka, V. Pini, M. Calleja, and J. Tamayo, "Ultrasensitive detection of HIV-1 p24 antigen by a hybrid nanomechanical-optoplasmonic platform with potential for detecting HIV-1 at first week after infection", *PLoS One*, vol. 12, no. 2, p. e0171899, 2017.
[http://dx.doi.org/10.1371/journal.pone.0171899] [PMID: 28199410]

[133] M. Stone, J. Bainbridge, A.M. Sanchez, S.M. Keating, A. Pappas, W. Rountree, C. Todd, S. Bakkour, M. Manak, S.A. Peel, R.W. Coombs, E.M. Ramos, M.K. Shriver, P. Contestable, S.V. Nair, D.H. Wilson, M. Stengelin, G. Murphy, I. Hewlett, T.N. Denny, and M.P. Busch, *J. Clin. Microbiol.*, vol. •••, p. 56, 2018.

[134] M. Stone, J. Bainbridge, A.M. Sanchez, S.M. Keating, A. Pappas, W. Rountree, C. Todd, S. Bakkour, M. Manak, S.A. Peel, R.W. Coombs, E.M. Ramos, M.K. Shriver, P. Contestable, S.V. Nair, D.H. Wilson, M. Stengelin, G. Murphy, I. Hewlett, T.N. Denny, and M.P. Busch, *J. Clin. Microbiol.*, vol. •••, p. 56, 2018.

[135] X. Qiu, L. Sokoll, P. Yip, D.J. Elliott, R. Dua, P. Mohr, X.Y. Wang, M. Spencer, P. Swanson, G.J. Dawson, and J. Hackett Jr, "Comparative evaluation of three FDA-approved HIV Ag/Ab combination tests using a genetically diverse HIV panel and diagnostic specimens", *J. Clin. Virol.*, vol. 92, pp. 62-68, 2017.
[http://dx.doi.org/10.1016/j.jcv.2017.05.005] [PMID: 28535437]

[136] H. Chen, K. Liu, Z. Li, and P. Wang, "Point of care testing for infectious diseases", *Clin. Chim. Acta*, vol. 493, pp. 138-147, 2019.
[http://dx.doi.org/10.1016/j.cca.2019.03.008] [PMID: 30853460]

[137] V. Simon, and D.D. Ho, "HIV-1 dynamics in vivo: implications for therapy", *Nat. Rev. Microbiol.*, vol. 1, no. 3, pp. 181-190, 2003.
[http://dx.doi.org/10.1038/nrmicro772] [PMID: 15035022]

[138] M.R.G. O'Gorman, R. Gelman, J. Folds, R. Gelman, C. Wilkening, J. Spritzler, D. Weng, A. Landay, F. Mandy, M. O'Gorman, M. Waxdal, C. Monical, H. Paxton, S. Blum, C. Lowell, T. Hartz, T. Denny, S. Douglas, R. Gelman, C. Wilkening, F. Mandy, J. Nicholson, M. O'Gorman, H. Paxton, S. Plaeger, B. Ward, C. Schnitzlein-Bick, J. Kagan, and D. Livnat, "Inter- and intrainstitutional evaluation of automated volumetric capillary cytometry for the quantitation of CD4- and CD8-positive T

lymphocytes in the peripheral blood of persons infected with human immunodeficiency virus. Site Investigators and the NIAID New CD4 Technologies Focus Group", *Clin. Diagn. Lab. Immunol.,* vol. 4, no. 2, pp. 173-179, 1997.
[http://dx.doi.org/10.1128/cdli.4.2.173-179.1997] [PMID: 9067651]

[139] D.D. Ho, A.U. Neumann, A.S. Perelson, W. Chen, J.M. Leonard, and M. Markowitz, "Rapid turnover of plasma virions and CD4 lymphocytes in HIV-1 infection", *Nature,* vol. 373, no. 6510, pp. 123-126, 1995.
[http://dx.doi.org/10.1038/373123a0] [PMID: 7816094]

[140] S.A. Fiscus, B. Cheng, S.M. Crowe, L. Demeter, C. Jennings, V. Miller, R. Respess, W. Stevens, P. Balakrishnan, B. Branson, D. Burgess, I. Cabruja, P. Clayden, K. Condliffe, G. Corrigan, M. De Baar, R. Downing, E. Ekong, L. Fox, G. Gonsalves, M. Guarinieri, M. Guillerm, N. Hellmann, J. Kaplan, H. Lee, R. Lloyd, C. Major, L. Margherio, J. Martinez-Picardo, D. Newman, T. Peter, M. Rayfield, R. Ridzon, T.R. De Wit, J. Safrit, L. Spacek, T. Spira, K. Tamura, M. Ussery, F. Valentine, O. Varnier, and R. Ziermann, *PLoS Med.,* vol. 3, pp. 1743-1750, 2006.

[141] C. Willyard, "Simpler tests for immune cells could transform AIDS care in Africa", *Nat. Med.,* vol. 13, no. 10, p. 1131, 2007.
[http://dx.doi.org/10.1038/nm1007-1131] [PMID: 17917650]

[142] J. Cohen, *Science (80-.).,* vol. 304, 2004.

[143] V. Linder, S.K. Sia, and G.M. Whitesides, "Reagent-loaded cartridges for valveless and automated fluid delivery in microfluidic devices", *Anal. Chem.,* vol. 77, no. 1, pp. 64-71, 2005.
[http://dx.doi.org/10.1021/ac049071x] [PMID: 15623279]

[144] F. Wei, P.B. Lillehoj, and C.M. Ho, "DNA diagnostics: nanotechnology-enhanced electrochemical detection of nucleic acids", *Pediatr. Res.,* vol. 67, no. 5, pp. 458-468, 2010.
[http://dx.doi.org/10.1203/PDR.0b013e3181d361c3] [PMID: 20075759]

[145] K.M. Murphy, and K.D. Berg, "Mutation and single nucleotide polymorphism detection using temperature gradient capillary electrophoresis", *Expert Rev. Mol. Diagn.,* vol. 3, no. 6, pp. 811-818, 2003.
[http://dx.doi.org/10.1586/14737159.3.6.811] [PMID: 14628908]

[146] J. Tost, and I.G. Gut, "Genotyping single nucleotide polymorphisms by MALDI mass spectrometry in clinical applications", *Clin. Biochem.,* vol. 38, no. 4, pp. 335-350, 2005.
[http://dx.doi.org/10.1016/j.clinbiochem.2004.12.005] [PMID: 15766735]

[147] O.P. Kallioniemi, "Biochip technologies in cancer research", *Ann. Med.,* vol. 33, no. 2, pp. 142-147, 2001.
[http://dx.doi.org/10.3109/07853890109002069] [PMID: 11327117]

[148] E. Pavlovic, R.Y. Lai, T.T. Wu, B.S. Ferguson, R. Sun, K.W. Plaxco, and H.T. Soh, "Microfluidic device architecture for electrochemical patterning and detection of multiple DNA sequences", *Langmuir,* vol. 24, no. 3, pp. 1102-1107, 2008.
[http://dx.doi.org/10.1021/la702681c] [PMID: 18181654]

[149] F. Lucarelli, S. Capponcelli, G. Marrazza, L. Sangiorgi, and M. Mascini, *Analyst (Lond.),* vol. 134, pp. 52-59, 2009.
[http://dx.doi.org/10.1039/B806514D]

[150] M. Daneshpour, "L.S. moradi, P. Izadi, K. Omidfar", *Biosens. Bioelectron.,* vol. 77, pp. 1095-1103, 2016.
[http://dx.doi.org/10.1016/j.bios.2015.11.007] [PMID: 26562330]

[151] R. Khodaei, A. Ahmady, S.M. Khoshfetrat, S. Kashanian, S.M. Tavangar, and K. Omidfar, "Voltammetric immunosensor for E-cadherin promoter DNA methylation using a Fe3O4-citric acid nanocomposite and a screen-printed carbon electrode modified with poly(vinyl alcohol) and reduced graphene oxide", *Mikrochim. Acta,* vol. 186, no. 3, p. 170, 2019.
[http://dx.doi.org/10.1007/s00604-019-3234-y] [PMID: 30741341]

[152] M. Daneshpour, K. Omidfar, and H. Ghanbarian, "A novel electrochemical nanobiosensor for the ultrasensitive and specific detection of femtomolar-level gastric cancer biomarker miRNA-106a", *Beilstein J. Nanotechnol.*, vol. 7, pp. 2023-2036, 2016.
[http://dx.doi.org/10.3762/bjnano.7.193] [PMID: 28144550]

[153] M. Daneshpour, B. Karimi, and K. Omidfar, "Simultaneous detection of gastric cancer-involved miR-106a and let-7a through a dual-signal-marked electrochemical nanobiosensor", *Biosens. Bioelectron.*, vol. 109, pp. 197-205, 2018.
[http://dx.doi.org/10.1016/j.bios.2018.03.022] [PMID: 29567564]

[154] L.S. Porto, D.N. Silva, A.E.F. de Oliveira, A.C. Pereira, and K.B. Borges, *Rev. Anal. Chem.*, vol. •••, p. 38, 2020.

[155] H. Pandey, P. Khare, S. Singh, and S.P. Singh, *Mater. Chem. Phys.*, vol. 239, p. 121966, 2020.
[http://dx.doi.org/10.1016/j.matchemphys.2019.121966]

CHAPTER 3

Voltammetric Sensors for Diverse Analysis

Agnes Chinecherem Nkele[1] and **Fabian I. Ezema**[1,2,3,*]

[1] *Department of Physics and Astronomy, University of Nigeria, Nsukka, Enugu, Nigeria*

[2] *Nanosciences African Network (NANOAFNET), iThemba LABS-National Research Foundation, 1 Old Faure road, Somerset West 7129, P.O. Box 722, Somerset West, Western Cape Province, South Africa*

[3] *UNESCO-UNISA Africa Chair in Nanosciences/Nanotechnology, College of Graduate Studies, University of South Africa (UNISA), Muckleneuk Ridge, P.O. Box 392, Pretoria, South Africa*

Abstract: Voltammetric techniques are widely used in detecting and diagnosing the electrochemical properties of electrode systems. These techniques adopt sensing devices that find relevance in analyzing water, studying biological systems, tracing poisonous chemicals, *etc*. These sensors have different compositions that aim at carrying out different functions in diverse systems. The sensors would be useful in biological, environmental, and chemical systems. In-depth knowledge of voltammetric sensors will be discussed. This chapter will also highlight different sensors, independent features, sensor classifications, voltammetric electrodes, and their areas of application.

Keywords: Electrochemical properties, Electrode, Sensors, Voltammetry.

GENERAL INTRODUCTION

Voltammetry is an electrochemical process that involves measuring current when an alternating voltage passes through the working electrode [1]. The potential applied depends on time because the current is measured with respect to the potential. It is a commonly used method in detecting and diagnosing the electrochemical characteristics of a system of electrodes. The use of biological and chemical sensors in determining contaminants within an environment has become an interesting study area. Such sensors have a quick response time, a small limit for detection, and on-site usage [2]. The conductivity of the working electrode is an important element that increases the electron mobility and sensiti-

---

[*] **Corresponding author F. I. Ezema:** Department of Physics and Astronomy, University of Nigeria, Nsukka, Enugu, Nigeria and Nanosciences African Network (NANOAFNET), iThemba LABS-National Research Foundation, 1 Old Faure road, Somerset West 7129, P.O. Box 722, Somerset West, Western Cape Province, South Africa; E-mail: fabian.ezema@unn.edu.ng

J.G. Manjunatha (Ed.)

vity of the biosensor. Voltammetry significantly minimizes the number of harmful chemicals obtainable from the use and disposal of materials. Voltammetry incorporates materials like silver, nickel, platinum, copper, and gold because of its wide potential range and increased electron mobility. Platinum is commonly used as the auxiliary electrode while silver/silver chloride serves as the reference electrode. The shape of cyclic voltammograms is affected by anions present. Introducing surfactants reduces the amount of faradaic current available. Stripping is an important class of voltammetry that is useful in electroanalytical chemistry, involves relatively cheap pieces of equipment, is capable of analyzing different elements, and is highly sensitive [3]. It has low detection limits due to the accumulation of the analyte on the working electrode. Metals can be analyzed by stripping method in the absence of contaminants obtainable from electrolytes. Anodic stripping voltammetry (ASV), cathodic stripping voltammetry (CSV), adsorptive stripping voltammetry (ADSV), and catalytic ADSV are advanced forms of stripping voltammetry that effectively and quantitatively analyze inorganic anions, transition metals, organic molecules, *etc* [2]. For Anodic SV; anodic currents are obtained by undertaking a preconcentration step at the cathode and taking a voltage scan towards the positive range [3]. The reverse occurs for the cathodic stripping voltammetry. Voltammetry of microparticles is a valuable method to study the corrosive nature of archaeological metals by understanding their source and authenticity [4]. Studying nanoparticles using voltammetry can also be useful for data collection and mapping. Electrochemical cells used in voltammetry can be produced in various sizes and give room for solutions to be analyzed. They include a working electrode, auxiliary electrode, reference electrode, optional bar for stirring, and a nitrogen purge line for oxygen removal, as shown in Fig. (**1**). At the working electrode, oxidation of the analyte occurs and minimizes the solvent by transporting the resultant electrons to the auxiliary electrode through the potentiostat. Reducing the quantity of analyte at the working electrode causes current (Faradaic current) to flow to the cathode from the auxiliary electrode. Diffusion, migration, and convection are important factors that affect the rate of mass transport. The current flowing through voltammetric cells is affected by the rate of mass transport and electron mobility.

Coulometry and amperometry are useful forms of voltammetry, with coulometry exhibiting better sensitive capabilities [5]. A steady voltage is applied to the working electrode in amperometry as the current is measured with time, as represented in Fig. (**2**). The kind of voltammetric technique to be adopted depends on properties of the material like its position and concentration, kind of precision and accuracy required.

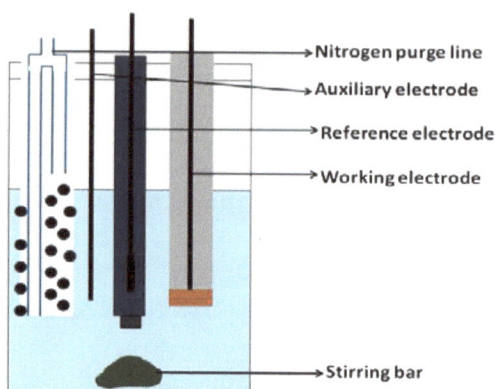

Fig. (1). Schematic diagram for a voltammetric electrochemical cell.

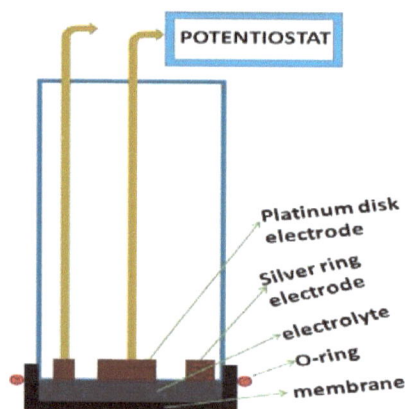

Fig. (2). Schematic representation of the amperometric sensor used to determine dissolved oxygen.

Cyclic voltammetry is a common form of voltammetry that records cyclic voltammograms through a scan in both directions and is carried out within a short time frame [6 - 11]. It estimates the activities of biocatalysts at the anode when subjected to various operating conditions [1]. Another measurement approach known as hydrodynamic voltammetry measures current as a differential pulse or linear scan, as mercury drops grow without oscillation. Voltammograms (current and applied potential plot) obtained from voltammetry provide meaningful information on the species undergoing reduction or oxidation [12, 13]. During the adsorption of redox species on an electrode, cyclic voltammograms, as seen in Fig. (**3**) [14], could be obtainable. The redox centers in Fig. (**3A**) get reduced at electrode potentials that are lower than E°. At higher potential values, oxidation occurs as the electrode and reduction potentials meet (Fig. **3B**). At increased potential, a measured positive current reduces to zero and oxidation occurs (Fig. **3C**). The manner of current measurement and inclusion of convection in mass transport affects the shape of the voltammogram.

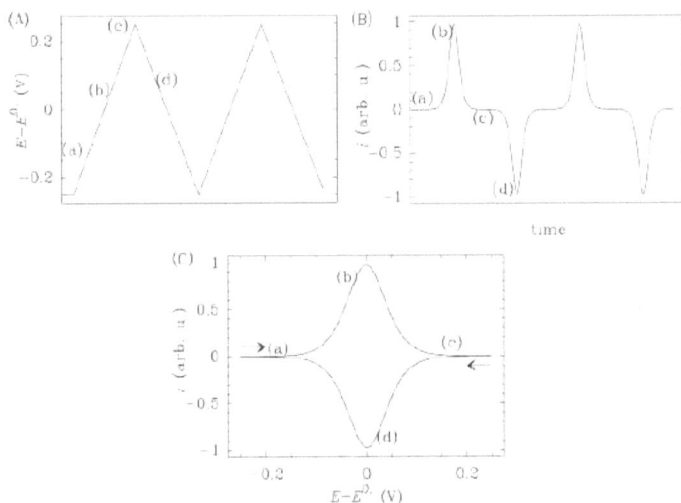

Fig. (3). Plots showing **A)** linear sweep of electrode potential in time **B)** observed current with time **C)** Cyclic voltammogram [14].

USEFUL VOLTAMMETRIC ELECTRODES

Electrodes are important tools that have shown improved electrochemical properties of voltammetric devices. Such electrodes are affected by the binding energy of the target material and the electrochemical potential [15]. They also encourage the production of miniature sensing devices.

Mercury electrodes can be incorporated because they have smooth surfaces, a wide operating negative voltage, and can be easily reproduced. The large surface-to-volume ratio accounts for the sensitive, precise, and stable nature of mercury electrode films. Incorporating liquid amalgam aids the removal of gas bubbles and impurities from the surface of the electrodes. Although mercury easily oxidizes, it serves as a good working electrode because of its high potential for hydrogen production, quick recovery of electrode surfaces, and ease of metal dissolution in mercury. Mercury electrodes have been successfully substituted with solid electrodes [2]. Mercury electrodes come in forms like hanging mercury drop electrode (HMDE), dropping mercury electrode (DME), and static mercury drop electrode (SMDE), as shown in Fig. (**4**). A hanging mercury drop electrode entails dropping mercury through a capillary tube due to the rotation of a micrometer screw. Dropping mercury electrode drops mercury under gravitational influence from a capillary tube end. A static mercury drop electrode regulates mercury flow by using a plunger driven by a solenoid.

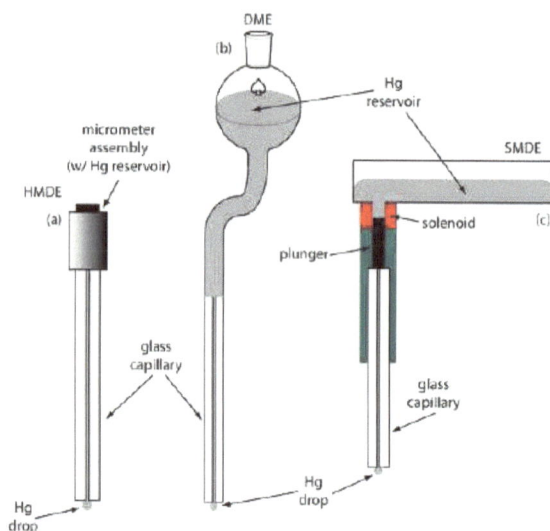

Fig. (4). Schematic representations of **a)** HMDE **b)** DME **c)** SMDE.

Solid electrodes suffer instability due to the interference of surfactants, polarization of electrodes, and deposition of reaction products. This instability can be curbed by cleaning the surface before measurement, although tedious. Solid electrodes made from silver, platinum, carbon, or gold have a wide range of potentials. For analyses involving negative potentials, mercury electrodes can be substituted with solid electrodes. The surface of a solid electrode is affected when oxide layers are formed and material species get adsorbed. Solid electrodes are usually encapsulated into disks and sealed at the end of an unreactive polymer supported with an electrical lead, as shown in Fig. (**5**). Solid amalgam electrodes can be electrochemically regenerated, easily prepared, mechanically stable, and durable. Although they have little positive range of potentials because of mercury; they still have a wide range of polarization potential.

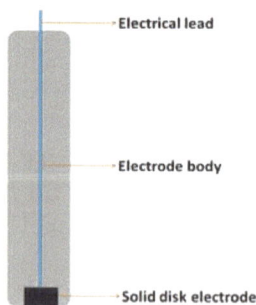

Fig. (5). Schematic diagram of a solid electrode.

Amalgam electrodes exhibit great reproducibility, minimal poisonous effect, are miniaturizable, and increased overpotential for hydrogen evolution reactions [2]. Such electrodes are very important in the electroanalysis of reducible analytes. Amalgam electrodes are also mechanically stable and simple to handle [16]. Amalgam electrodes in their diluted form serve as a useful reactant in voltammetric operations [17].

Bulk electrodes are easily synthesized, mechanically stable, and have easy usage. Electrodes with dimensions smaller than 25 μm are referred to as microelectrodes. Their small dimensions allow for improved transport features, easy attainment of steady-state situation, and incorporation into organic solvents [2].

Composite electrodes have several components with each component introducing individual characteristics. They are of great research interest and incorporate antibodies, bimetals, and zeolites. Composite electrodes made from graphene-polyaniline material are highly sensitive and exhibit increased analytical output [18].

Bismuth electrodes have high stability and solubility in water because of the bismuth ions which hydrolyze in basic and neutral solvents. The reaction efficacy depends on the pH of the solution.

Titanium electrodes oxidize in air and other oxidation systems. This oxidation effect can be reduced by coating the electrode with polycrystalline deposits of platinum, although it is unstable and poorly adheres between metals. Enhanced stability and resistivity to corrosion can be obtained by incorporating titanium carbide or titanium nitride [2].

Tubular electrodes comprise a single or several mM-sized rings positioned in an internal insulating tube wall in a concentric position. To allow swift passage of gas bubbles, the electrode is vertically positioned so that the bubbles of gas would not impede the electrode performance. Turbulent effects can be avoided by creating a smooth interaction between the insulator and the surface of the electrode. To reduce the impact of the flow, the diameters of the electrode and capillary should be equal. Coating the electrode surface with mercury film yields better hydrogen potential; thereby increasing the voltage limit of the cathode [2].

Annular band electrodes have two insulators separated by a conductor. Optimization of this electrode is attainable when the area of the electrode around the edges is enlarged. Metallic electrodes are influenced by electrolyte components because anions have adsorbing power on conducting surfaces; thereby blocking the active sites.

Carbon electrodes are chemically unreactive, applicable in detecting a positive range of potential values, and obtainable from single and multiple materials. Carbon electrodes can exist as glassy, paste, or screen-printed carbon materials [19 - 21]. Carbon electrodes made from graphite have improved electrochemical sensing properties [22]. Carbon paste electrodes help in detecting phloroglucinol [23] and can be modified with helianthin dye for increased voltammetric performance [24]. Glassy carbon electrodes are not affected by the oxygen contained in the voltammograms [2]. The sequential production of glassy carbon electrodes in Fig. (**6**) exhibited better electrocatalytic performance [25].

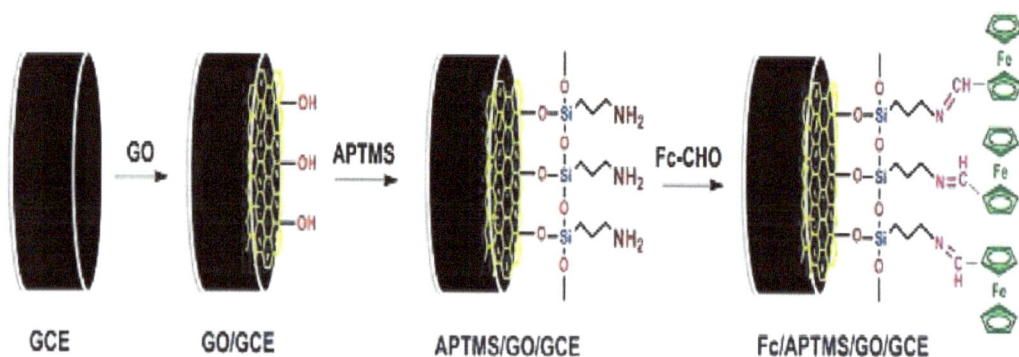

Fig. (6). Step-by-step fabrication process for a glassy carbon electrode [25].

CLASSIFICATIONS OF VOLTAMMETRIC SENSORS

During electroanalysis, the electrode performance can be improved by introducing working electrode sensors that are laterally positioned. Such sensors can be easily prepared, are miniaturizable, have no gas interference, and have amplified mass transport [2].

Electrochemical sensors adopt electroanalytical procedures to detect metal ions and little quantities of heavy metals, relatively affordable, highly sensitive, and robust with quick response [25]. Electrochemical methods entail relatively affordable means of detecting microorganisms. Ferrocene is a good component in electrochemical sensors as it maintains good stability in water and air, thereby leading to the production of different organometals [25, 26]. Ferrocene has a unique electronic and chemical buildup which contributes to its numerous electrochemical features, fast electron mobility. Ferrocene has been used to detect copper (ii) ions, lead ions, hydrogen peroxide, and analyze food [25]. Captopril can be electrochemically determined using p-Aminophenol multiwall carbon nanotube paste as an electrode, with p-Aminophenol acting as the mediator [27]. Carbon nanotubes exhibit features that are mechanically, structurally, chemically, and electronically unique with large electrode surfaces and increased carrier mobility. Upon investigation, Ensafi *et al*. obtained homogenously compact films, properly-defined peaks with a behavioral pattern that is controlled by the diffusion rate, and improved electrode characteristics that are displayed in Fig. (**7**) [27].

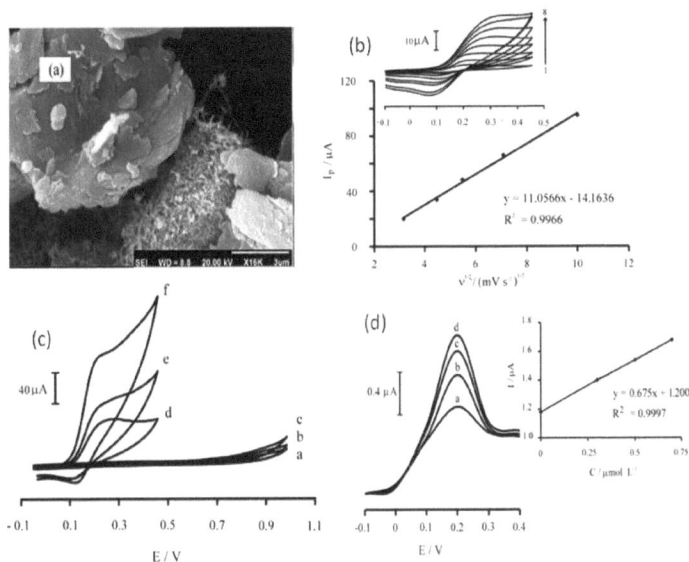

Fig. (7). Plots showing (**a**) Surface morphology (**b**) anodic peak *versus* square roots of the scan rate (**c**) cyclic voltammograms (**d**) square wave voltammograms of the synthesized materials.

Cylindrical sensors like Ag/AgCl reference electrodes have the surfaces of electrodes situated externally. This pattern of design makes it useful for usage in environs having high pressure and temperature as such sensors are positioned in cylindrical electrodes [2]. Sensors in cylindrical electrodes create room for understanding reversible voltammetric reactions using a suitable one-dimensional model [28]. Qian, Zhu, and Li fabricated cylindrical nanopore electrodes with exceptional size features, increased limiting current, and functionalized surface [29].

Aptasensors can be incorporated into voltammetric processes to alter electrode proximity to the redox indicator so that the electron mobility between them can be modified [15]. The concentration of the target determines how high or low the current flowing would be.

Enzyme-based sensors are dependent on how conductive the working electrode is. These sensors are achievable with indium-doped tin oxide, screen-printed, and fluorine-doped tin oxide electrodes [30]. The detective processes taken by enzyme-based sensors are represented in Fig. (**8**). Differential pulse and square-wave voltammetric techniques are involved in analyzing enzyme-based sensors.

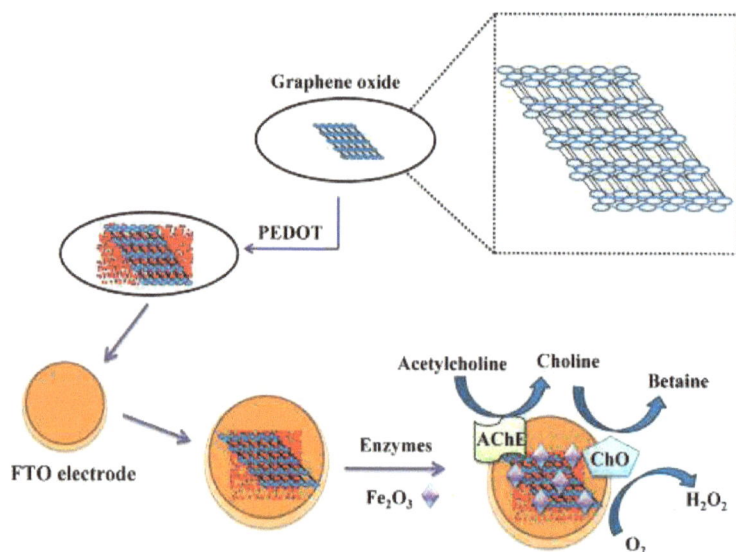

Fig. (8). Schematic representation of enzyme-based biosensors in detecting acetylcholine.

A taste sensor or electronic tongue has a data processor and an information collector that is useful in aqueous solutions. The sample composition can be detected from the pattern displayed and bulk potential data upon subjecting it to a heterogeneous compound. Helmholtz double layer is created when charges migrate to the surface of the electrode and create a pulse of current which then decays as the Faradaic current diffuses. The taste sensor involves lipid systems that are sensitive to ions and denotes basic tongue tastes like sweet, salty, bitter, or sour while the electronic tongue differs from this by classifying the quality of the measured material and involving potentiometry. Electrochemical techniques like voltammetry and potentiometry have been used in their description. Voltammetry is simple, unaffected by electrical interferences, sensitive, and entails measuring current on electrode surface upon applying a potential to the working electrode, while potentiometry measures the potential moving through a working electrode under balanced conditions. Pulse and staircase voltammetry are useful techniques in studying electronic tongues. The quantity and diffusion coefficients of the charged compounds can be determined from the size and shape of the transient response.

APPLICATIONS OF VOLTAMMETRIC MATERIALS

Voltammetry is employed as a detective mechanism in highly performing liquid chromatographs and also in pharmaceutical companies to estimate the concentration of drugs [5]. It determines the equilibrium constant in chemical reactions, verifies the chemical nature of electrochemical reactions, and determines the number of electrons transferred during reactions. Voltammetry is useful for remote and field monitoring [17]. Estriol can be analyzed with surfactant-layered carbon sensors [31, 32]. Wang and Li fabricated ZrO_2/Au electrode for detecting organic pesticides made of phosphorus with no sophisticated equipment involved [33]. Electrochemical sensors can be utilized in preparing pharmaceutical products and determining captopril in samples [27]. A cylindrical electrode made of carbon fiber is useful in determining the number of phenolic antioxidants in food substances [34] and characterizing silver nanoparticles [35]. Incorporating anionic surfactant into carbon electrodes helps in determining riboflavin [36]. Carbon nanotubes allow paracetamol to be detected by voltammetric means [37]. Amperometry finds application in chemical and glucose sensors. Amperometry can also be used in analyte detection, *in-vivo* analysis of blood, field analyses, and analyzing analytes. Amperometric sensors are useful for the environmental analysis of samples and clinical usage. Voltammetric sensors can be utilized in enzyme-based biosensing devices *via* cyclic voltammetric technique [30]. Amalgam electrodes are used in analyzing soil and water pollutants. Voltammetric sensors can also be used in detecting caffeic acid in food supplements [38]. Stripping voltammetry determines how concentrated trace metals are in materials like urine, tissue, and blood. Cyclic voltammetry is applicable in enzyme-based biosensors and in verifying protein response upon immobilization on an electrode [30]. It can also determine the concentration of analytes. Cathodic stripping voltammetry is usefully applied in analyzing natural water [3]. Electrochemical sensors modified with polymer can be developed to determine alizarin carmine [39, 40] while those immobilized with surfactants detect Indigotine [41]. Tyrosine can be investigated as a drug using electrochemical sensing devices made from carbon nanotubes [42].

CONCLUDING REMARKS

Voltammetry is an electrochemical process that involves measuring current as alternating voltage passes through the working electrode. The kind of voltammetric technique to be adopted depends on properties of the material like its position and concentration, kind of precision and accuracy. Electrochemical sensors adopt electroanalytical procedures to detect metal ions and little quantities of heavy metals, relatively affordable, highly sensitive, and robust with quick response. This chapter focuses on explaining voltammetric techniques, voltammetric sensors, and electrodes, as well as their application areas. Mercury, solid, amalgam, bulk, composite, bismuth, titanium, tubular, annular band, metallic, and carbon electrodes have been highlighted. Voltammetric sensors like electrochemical, cylindrical, enzyme-based, and electronic tongue sensors have been discussed. Relevant areas of applying voltammetric sensors have also been listed.

CONSENT FOR PUBLICATION

Not applicable.

CONFLICT OF INTEREST

The authors declare no conflict of interest, financial or otherwise.

ACKNOWLEDGEMENT

Declared none.

REFERENCES

[1]   S. Roy, and S. Pandit, "Microbial Electrochemical System: Principles and Application", In: *Microbial Electrochemical Technology* Elsevier, 2019, pp. 19-48.
        [http://dx.doi.org/10.1016/B978-0-444-64052-9.00002-9]

[2]   R. Porada, K. Jedlińska, J. Lipińska, and B. Baś, "Review—Voltammetric Sensors with Laterally Placed Working Electrodes: A Review", *J. Electrochem. Soc.,* vol. 167, no. 3, 2020.037536.
        [http://dx.doi.org/10.1149/1945-7111/ab64d6]

[3]   E.P. Achterberg, M. Gledhill, and K. Zhu, *Voltammetry—Cathodic Stripping* Elsevier, 2018.
        [http://dx.doi.org/10.1016/B978-0-12-409547-2.00553-9]

[4]   A. Doménech-Carbó, Electrochemical Analysis of Metallic Heritage Artefacts: Voltammetry of Microparticles (VMP).*Corrosion and Conservation of Cultural Heritage Metallic Artefacts* Elsevier, 2013, pp. 165-189.
        [http://dx.doi.org/10.1533/9781782421573.2.165]

[5]   K. Robards, P.R. Haddad, and P.E. Jackson, *High-Performance Liquid Chromatography-Instrumentation and Techniques* Elsevier Ltd.: London, 2004, pp. 227-303.

[6]   A.C. Nkele, A.C. Nwanya, U. Nwankwo, A.B.C. Ekwealor, R.U. Osuji, R. Bucher, M. Maaza, and F.I. Ezema, "Structural, Optical and Electrochemical Properties of SILAR-Deposited Zirconium-Doped Cadmium Oxide Thin Films", *Mater. Res. Express,* 2019.
        [http://dx.doi.org/10.1088/2053-1591/ab31f5]

[7]   A.C. Nwanya, C. Awada, D. Obi, K. Raju, K.I. Ozoemena, R.U. Osuji, A. Ruediger, M. Maaza, F. Rosei, and F.I. Ezema, "Nanoporous Copper-Cobalt Mixed Oxide Nanorod Bundles as High Performance Pseudocapacitive Electrodes", *J. Electroanal. Chem. (Lausanne),* vol. 787, pp. 24-35, 2017.
        [http://dx.doi.org/10.1016/j.jelechem.2017.01.031]

[8]   A.C. Nwanya, D. Obi, R.U. Osuji, R. Bucher, M. Maaza, and F.I. Ezema, "Simple Chemical Route for Nanorod-like Cobalt Oxide Films for Electrochemical Energy Storage Applications", *J. Solid State Electrochem.,* vol. 21, no. 9, pp. 2567-2576, 2017.
        [http://dx.doi.org/10.1007/s10008-017-3520-8]

[9]   A.C. Nwanya, S.U. Offiah, I.C. Amaechi, S. Agbo, S.C. Ezugwu, B.T. Sone, R.U. Osuji, M. Maaza, and F.I. Ezema, "Electrochromic and Electrochemical Supercapacitive Properties of Room Temperature PVP Capped Ni(OH)2/NiO Thin Films", *Electrochim. Acta,* vol. 171, pp. 128-141, 2015.
        [http://dx.doi.org/10.1016/j.electacta.2015.05.005]

[10]   U.K. Chime, A.C. Nkele, S. Ezugwu, A.C. Nwanya, N.M. Shinde, M. Kebede, P.M. Ejikeme, M. Maaza, and F.I. Ezema, "Recent Progress in Nickel Oxide-Based Electrodes for High-Performance Supercapacitors", *Curr. Opin. Electrochem.,* vol. 21, pp. 175-181, 2020.
        [http://dx.doi.org/10.1016/j.coelec.2020.02.004]

[11]   B.N. Ezealigo, A.C. Nwanya, A. Simo, R.U. Osuji, R. Bucher, M. Maaza, and F.I. Ezema, "Optical and Electrochemical Capacitive Properties of Copper (I) Iodide Thin Film Deposited by SILAR Method", *Arab. J. Chem.,* 2017.

[12]   A.C. Nkele, A.C. Nwanya, N.U. Nwankwo, R.U. Osuji, A.B.C. Ekwealor, P.M. Ejikeme, M. Maaza, and F.I. Ezema, "Investigating the Properties of Nano Nest-like Nickel Oxide and the NiO/Perovskite for Potential Application as a Hole Transport Material", *Adv. Nat. Sci: Nanosci. Nanotechnol.,* vol. 10, no. 4, 2019.045009.
[http://dx.doi.org/10.1088/2043-6254/ab5102]

[13]   A.C. Nkele, U.K. Chime, L. Asogwa, A.C. Nwanya, U. Nwankwo, K. Ukoba, T.C. Jen, M. Maaza, and F.I. Ezema, "A Study on Titanium Dioxide Nanoparticles Synthesized from Titanium Isopropoxide under SILAR-Induced Gel Method: Transition from Anatase to Rutile Structure", *Inorg. Chem. Commun.,* vol. 112, 2020.107705
[http://dx.doi.org/10.1016/j.inoche.2019.107705]

[14]   V. Fourmond, and C. Léger, An Introduction to Electrochemical Methods for the Functional Analysis of Metalloproteins.*Practical Approaches to Biological Inorganic Chemistry* Elsevier, 2020, pp. 325-373.
[http://dx.doi.org/10.1016/B978-0-444-64225-7.00009-2]

[15]   S. Tom, H-E. Jin, and S-W. Lee, Aptamers as Functional Bionanomaterials for Sensor Applications.*Engineering of Nanobiomaterials* Elsevier, 2016, pp. 181-226.
[http://dx.doi.org/10.1016/B978-0-323-41532-3.00006-3]

[16]   B. Yosypchuk, and L. Novotný, "Nontoxic Electrodes of Solid Amalgams", *Crit. Rev. Anal. Chem.,* vol. 32, no. 2, pp. 141-151, 2002.
[http://dx.doi.org/10.1080/10408340290765498]

[17]   Ø. Mikkelsen, and K.H. Schrøder, "Amalgam Electrodes for Electroanalysis", *Electroanalysis,* vol. 15, no. 8, pp. 679-687, 2003.
[http://dx.doi.org/10.1002/elan.200390085]

[18]   Y. Fan, J-H. Liu, C-P. Yang, M. Yu, and P. Liu, "Graphene–Polyaniline Composite Film Modified Electrode for Voltammetric Determination of 4-Aminophenol", *Sens. Actuators B Chem.,* vol. 157, no. 2, pp. 669-674, 2011.
[http://dx.doi.org/10.1016/j.snb.2011.05.053]

[19]   N. S. Prinith, J. G. Manjunatha, and C. Raril, "Electrocatalytic analysis of dopamine, uric acid and ascorbic acid at poly (adenine) modified carbon nanotube paste electrode: A cyclic voltammetric study", *Anal. Bioanal. Electrochem,* vol. 11, pp. 742-756, 2020.

[20]   N. Hareesha, J.G. Manjunatha, C. Raril, and G. Tigari, "Sensitive and Selective Electrochemical Resolution of Tyrosine with Ascorbic Acid through the Development of Electropolymerized Alizarin Sodium Sulfonate Modified Carbon Nanotube Paste Electrodes", *ChemistrySelect,* vol. 4, no. 15, pp. 4559-4567, 2019.
[http://dx.doi.org/10.1002/slct.201900794]

[21]   P.A. Pushpanjali, J.G. Manjunatha, and M.T. Shreenivas, "The Electrochemical Resolution of Ciprofloxacin, Riboflavin and Estriol Using Anionic Surfactant and Polymer☐Modified Carbon Paste Electrode", *ChemistrySelect,* vol. 4, no. 46, pp. 13427-13433, 2019.
[http://dx.doi.org/10.1002/slct.201903897]

[22]   N. Hareesha, and J.G. Manjunatha, "Fast and Enhanced Electrochemical Sensing of Dopamine at Cost-Effective Poly (DL-Phenylalanine) Based Graphite Electrode", *J. Electroanal. Chem. (Lausanne),* vol. 878, 2020.
[http://dx.doi.org/10.1016/j.jelechem.2020.114533]

[23]   G. Tigari, and J.G. Manjunatha, "Optimized Voltammetric Experiment for the Determination of Phloroglucinol at Surfactant Modified Carbon Nanotube Paste Electrode", *Instrum. Exp. Tech.,* vol. 63, no. 5, pp. 750-757, 2020.

[http://dx.doi.org/10.1134/S0020441220050139]

[24] N. Hareesha, and J. G. Manjunatha, "Elevated and Rapid Voltammetric Sensing of Riboflavin at Poly (Heleianthin Dye) Blended Carbon Paste Electrode with Heterogeneous Rate Constant Elucidation", *Journal of the Iranian Chemical Society,* pp. 1-13, 2020.

[25] H. Beitollahi, M.A. Khalilzadeh, S. Tajik, M. Safaei, K. Zhang, H.W. Jang, and M. Shokouhimehr, "Recent Advances in Applications of Voltammetric Sensors Modified with Ferrocene and Its Derivatives", *ACS Omega,* vol. 5, no. 5, pp. 2049-2059, 2020.
[http://dx.doi.org/10.1021/acsomega.9b03788] [PMID: 32064365]

[26] C. Raril, J.G. Manjunatha, D.K. Ravishankar, S. Fattepur, G. Siddaraju, and L. Nanjundaswamy, "Validated Electrochemical Method for Simultaneous Resolution of Tyrosine, Uric Acid, and Ascorbic Acid at Polymer Modified Nano-Composite Paste Electrode", *Surg. Eng. Appl. Electrochem.,* vol. 56, no. 4, pp. 415-426, 2020.
[http://dx.doi.org/10.3103/S1068375520040134]

[27] A.A. Ensafi, B. Rezaei, Z. Mirahmadi-Zare, and H. Karimi-Maleh, "Highly Selective and Sensitive Voltammetric Sensor for Captopril Determination Based on Modified Multiwall Carbon Nanotubes Paste Electrode", *J. Braz. Chem. Soc.,* vol. 22, no. 7, pp. 1315-1322, 2011.
[http://dx.doi.org/10.1590/S0103-50532011000700017]

[28] H. Le, E. Kätelhön, and R.G. Compton, "Reversible Voltammetry at Cylindrical Electrodes: Validity of a One-Dimensional Model", *J. Electroanal. Chem. (Lausanne),* vol. 859, 2020.113865
[http://dx.doi.org/10.1016/j.jelechem.2020.113865]

[29] Y. Qian, J. Zhu, and Y. Li, "Single Cylindrical Nanopore Electrodes: Surface Functionalization, Unusual Voltammetry, and Size-Exclusion Properties", *ChemElectroChem,* vol. 5, no. 2, pp. 292-299, 2018.
[http://dx.doi.org/10.1002/celc.201701096]

[30] Y-C. Zhu, L-P. Mei, Y-F. Ruan, N. Zhang, W-W. Zhao, J-J. Xu, and H-Y. Chen, Enzyme-Based Biosensors and Their Applications.*Advances in Enzyme Technology* Elsevier, 2019, pp. 201-223.
[http://dx.doi.org/10.1016/B978-0-444-64114-4.00008-X]

[31] N. Hareesha, and J.G. Manjunatha, "Surfactant and Polymer Layered Carbon Composite Electrochemical Sensor for the Analysis of Estriol with Ciprofloxacin", *Mater. Res. Innov.,* vol. 24, no. 6, pp. 349-362, 2020.
[http://dx.doi.org/10.1080/14328917.2019.1684657]

[32] M.M. Charithra, J.G.G. Manjunatha, and C. Raril, "Surfactant Modified Graphite Paste Electrode as an Electrochemical Sensor for the Enhanced Voltammetric Detection of Estriol with Dopamine and Uric acid", *Adv. Pharm. Bull.,* vol. 10, no. 2, pp. 247-253, 2020.
[http://dx.doi.org/10.34172/apb.2020.029] [PMID: 32373493]

[33] M. Wang, and Z. Li, "Nano-Composite ZrO2/Au Film Electrode for Voltammetric Detection of Parathion", *Sens. Actuators B Chem.,* vol. 133, no. 2, pp. 607-612, 2008.
[http://dx.doi.org/10.1016/j.snb.2008.03.023]

[34] M.L. Agui, A.J. Reviejo, P. Yanez-Sedeno, and J.M. Pingarron, "Analytical Applications of Cylindrical Carbon Fiber Microelectrodes. Simultaneous Voltammetric Determination of Phenolic Antioxidants in Food", *Anal. Chem.,* vol. 67, no. 13, pp. 2195-2200, 1995.
[http://dx.doi.org/10.1021/ac00109a044]

[35] J. Ellison, C. Batchelor-McAuley, K. Tschulik, and R.G. Compton, "The Use of Cylindrical Micro-Wire Electrodes for Nano-Impact Experiments; Facilitating the Sub-Picomolar Detection of Single Nanoparticles", *Sens. Actuators B Chem.,* vol. 200, pp. 47-52, 2014.
[http://dx.doi.org/10.1016/j.snb.2014.03.085]

[36] G. Tigari, J.G. Manjunatha, C. Raril, and N. Hareesha, "Determination of Riboflavin at Carbon Nanotube Paste Electrodes Modified with an Anionic Surfactant", *ChemistrySelect,* vol. 4, no. 7, pp. 2168-2173, 2019.

[http://dx.doi.org/10.1002/slct.201803191]

[37]   M.M. Charithra, and J.G. Manjunatha, "Enhanced Voltammetric Detection of Paracetamol by Using Carbon Nanotube Modified Electrode as an Electrochemical Sensor", *Journal of Electrochemical Science and Engineering,* vol. 10, no. 1, pp. 29-40, 2020.
[http://dx.doi.org/10.5599/jese.717]

[38]   A.V. Bounegru, and C. Apetrei, "Voltammetric Sensors Based on Nanomaterials for Detection of Caffeic Acid in Food Supplements", *Chemosensors (Basel),* vol. 8, no. 2, p. 41, 2020.
[http://dx.doi.org/10.3390/chemosensors8020041]

[39]   P.A. Pushpanjali, and J.G. Manjunatha, "Development of Polymer Modified Electrochemical Sensor for the Determination of Alizarin Carmine in the Presence of Tartrazine", *Electroanalysis,* vol. 32, no. 11, pp. 2474-2480, 2020.
[http://dx.doi.org/10.1002/elan.202060181]

[40]   A. Balliamada Monnappa, J.G. Manjunatha, A.S. Bhatt, R. Chenthattil, and P. Pemmatte Ananda, "Electrochemical Sensor for the Determination of Alizarin Red-S at Non-Ionic Surfactant Modified Carbon Nanotube Paste Electrode", *Physical Chemistry Research,* vol. 7, no. 3, pp. 523-533, 2019.

[41]   C. Raril, J.G. Manjunatha, L. Nanjundaswamy, G. Siddaraju, D.K. Ravishankar, S. Fattepur, and E. Niranjan, "Surfactant Immobilized Electrochemical Sensor for the Detection of Indigotine", *ANALYTICAL & BIOANALYTICAL ELECTROCHEMISTRY,* vol. 10, no. 11, pp. 1479-1490, 2018.

[42]   N. Hareesha, J.G.G. Manjunatha, C. Raril, and G. Tigari, "Design of novel Surfactant Modified Carbon Nanotube Paste Electrochemical Sensor for the Sensitive Investigation of Tyrosine as a Pharmaceutical Drug", *Adv. Pharm. Bull.,* vol. 9, no. 1, pp. 132-137, 2019.
[http://dx.doi.org/10.15171/apb.2019.016] [PMID: 31011567]

CHAPTER 4

Methyl Orange Electropolymerized Composite Carbon Paste Electrode as a Sensitive and Selective Sensor for the Electrochemical Determination of Riboflavin

Amrutha B. Monnappa[1, 2], J. G. Manjunatha[1,*], Aarti S. Bhatt[2], Akshatha Nemumoolya[1], Sharmila B. Medappa[1] and Geethanjali N. Karthammaiah[1]

[1] *Department of Chemistry, FMKMC College, Madikeri, Constituent College of Mangalore University, Karnataka, India*

[2] *Department of Chemistry, N.M.A.M. Institute of Technology, Visvesvaraya Technological University, Belgavi, Nitte, 574110, Udupi District, Karnataka, India*

Abstract: Riboflavin (RB) is a vital B group water soluble vitamin. RB is essential for all animals and humans. This essential constituent of flavoenzyme, which plays an important role in biochemical reactions, is investigated by Cyclic voltammetry (CV) and Differential pulse voltammetry (DPV) using Poly methyl orange modified composite carbon paste electrode (PMOMCCPE). RB exhibits an oxidation peak with the highest anodic current of 54.82 µA at a potential of − 0.4360 V in 0.2 M phosphate buffer solution (PBS) with a pH value of 7.5. The surface topography studies of fabricated PMOMCCPE were carried out by CV, Field emission scanning electron microscopy (FE-SEM) and Electronic impedance spectroscopy (EIS). Optimization of pH and polymer cycle was carried out, and the effect of scan rate was also studied. The anodic peak current obtained is linearly related to the variations in the concentration of RB in the range 4 to 50 µM. Other features include a detection limit of 1.67×10^{-7} M, and good reproducibility, repeatability and stability. Simultaneous determination of RB with dopamine (DA) was done and interferences by some potentially interfering compounds were tested. PMOMCCPE was successfully used for determining RB in pharmaceutical formulations which gave recoveries in the range from 96.50 to 99.53%.

Keywords: Anodic peak current, Composite carbon paste electrode, Cyclic voltammetry, Differential pulse voltammetry, Dopamine, Riboflavin.

* **Corresponding author J. G. Manjunatha:** Department of Chemistry, FMKMC College, Madikeri, Constituent College of Mangalore University, Karnataka, India; E-mail: manju1853@gmail.com

INTRODUCTION

RB is found in a variety of foods and dissolves in water. In 1879, RB was first isolated from milk as an impure form and it was named lactochrome. It crystallises as orange-yellow crystals and has a low water solubility in its pure form [1]. RB or vitamin B_2 is a B group vitamin and an important constituent of flavoenzymes, which plays a significant role in the biochemical reactions in the human body. It's the major component of flavin adenine dinucleotide (FAD) and flavin mononucleotide (FMN), and it's required for a range of flavoprotein enzyme activities, including vitamin activation. Its deficiency is connected with eye lesions and skin disorders. RB cannot be formed in the human body, so it must be obtained from food like liver, cheese, fruits, vegetables and pharmaceutical products [2]. RB is commonly used in combination with other members of the vitamin B group in B-complex products. In organisms, RB serves as a hydrogen carrier during redox reactions that affect carbohydrate, protein, fat and haemoglobin synthesis, as well as enables the preservation of visual function [3].

Several researchers have worked hard to analyse RB using various analytical processes, with the electroanalytical method proving to be the most advantageous due to its simple, quick, and cost-effective instrumentation. Chemically modified carbon-based electrodes are an excellent method for improving sensor performance. Carbon nanotubes (CNT), also known as buckytubes, are composed of a concentric arrangement of several cylinders, which have unique properties with promising applications in the field of nanotechnology. CNTs have unique electrical and optical features that distinguish them from other carbonaceous materials and nanoparticles of other sorts due to their outstanding one-dimensional nanostructures. CNTs are widely used in electronic and optoelectronic, biomedical, pharmaceutical, energy, catalytic, analytical, and material fields [4 - 6]. Carbon paste electrodes (CPE) have been widely utilised for many years due to their non-hazardous behaviour, low cost, and ease of availability. It is used for the fabrication of various biosensors and electrochemical sensors. CPEs are the most preferred electrode material due to their advantageous qualities such as low mass density, heat resistance, chemical tolerance, low residual current throughout the entire potential range, and ease of electrode surface modification [7 - 9]. Hence, in the present work CNTs and carbon powder in the composite form are homogenised in silicone oil as the sensitive sensing material. These composite electrodes, due to their beneficial properties like enlarged surface area and ability to boost the rate of electron transfer, improvise the electrocatalytic nature of the sensor.

Surface modification of these composite electrodes by azo dyes is effectually done using electro polymerisation technique, where the polymer forms a consistent deposition of polymer film on the composite carbon paste electrode [10, 11]. Azo dyes, which are commonly used in science, medicine, food, and the textile industry, can be used to modify electrode surfaces in a variety of ways. Methyl orange is an organic dye which has both hydrophilic as well as hydrophobic groups, and a molecular cavity. It's an acidic, aromatic synthetic dye that's widely used in textiles, leather, petroleum, the food industry, and pharmaceuticals as a colouring agent [12, 13]. In this chapter, methyl orange is electropolymerized on a composite carbon paste electrode surface for the enhanced voltammetry determination of RB.

MATERIALS AND METHODS

Instrumentation

Electrochemical work station CHI-6038E model (CH Instrument-6038 electrochemical workstation-USA) was used to conduct electrochemical experiments at lab temperature. The electrochemical cell was set up with an ideal tri electrode system consisting of BCCPE and PMOMCCPE as working electrodes, KCl saturated calomel and platinum counter electrode. The instrument working at 5.00 kV from DST-PURSE Laboratory, Mangalore University, was used to perform FE-SEM.

Chemicals and Solutions

Carbon nanotube (30-50 nm and a length of 10-30) was bought from Sisco research laboratories, Mumbai. Carbon powder, Silicone oil and Potassium chloride were procured from Nice Chemicals (Cochin) India. Mono sodium dihydrogen phosphate, disodium hydrogen phosphate and l-leucine were obtained from Himedia chemicals (India), RB and DA were obtained from Molychem laboratory, India. 25×10^{-4} M RB was prepared by dissolving the required quantity in sodium hydroxide and DA stock solution was prepared by dissolving the essential amount of solute in double distilled water. 25 mM stock solution of MO was prepared by suspending the required amount of solute in double distilled water. Appropriate quantity of 0.2 M Na_2HPO_4 and 0.2 M NaH_2PO_4 were intermixed to prepare 0.2 M Phosphate buffer solution (PBS) of different pH values. Real sample solution was prepared by weighing the appropriate quantity of RB samples commercially available and was dissolved in an amber coloured volumetric flask with double distilled water to get a standard solution of 25×10^{-4} M strength.

Preparation of BCCPE and PMOMCCPE

Carbon powder, carbon nanotube and silicone oil were mixed in the ratio 35:35:30 (w/w) respectively to get BCCPE. The resulting mixture was methodically homogenised in the agate mortar using a pestle. The paste was then neatly filled into the Teflon tube of 3mm internal diameter without any airgap. The copper wire, which was implanted at the centre of the paste packed Teflon tube made contact with the exterior circuit. The bare surface of the electrode was then impulsively scrubbed on the soft tissue paper to obtain a smooth and shiny, reproducible working surface. Polymerisation of MO onto the surface of the bare electrode was carried out by running 10 potential cycles on BCCPE between -0.6 to 1.4 V in an aqueous PBS containing 1×10^{-3} MO.

RESULTS AND DISCUSSIONS

Surface Topography Studies by FESEM and EIS

The surface topography studies of BCCPE and PMOMCCPE were done by FESEM. Fig. (1A) shows the topography of BCCPE and 1B shows the topography image of PMOMCCPE. Fig. (1A) displays a densely packed image of composite carbon paste with certain tube-like structures of carbon nanotubes dispersed throughout the paste. Fig. (1B) shows the polymer film that is adsorbed on the composite carbon paste electrode after modification.

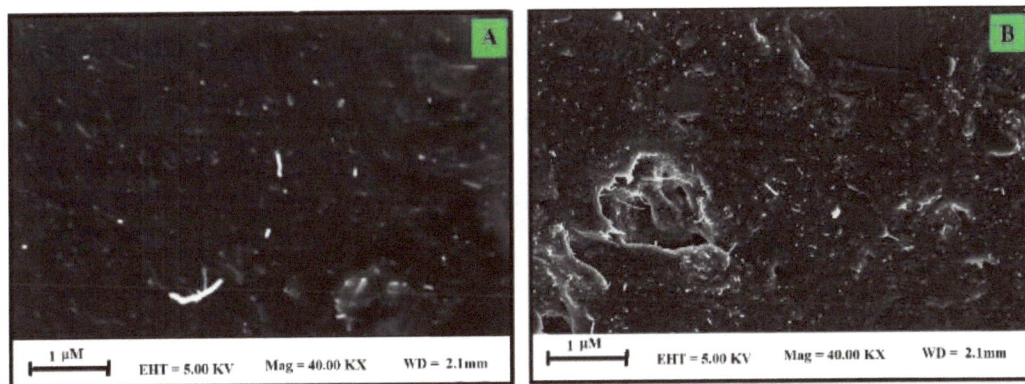

Fig. (1). A) FESEM depiction of BCCPE. **B)** FESEM depiction of PMOMCCPE.

The superior electrocatalytic performance of PMOMCCPE was analysed by electrochemical impedance spectroscopy (EIS) method. The electrical conductivity and surface-modification properties are investigated using EIS. The Randles circuit was used to fit the impedance output of BCCPE and PMOMCCPE in 0.1 M KCl containing 1 ml of 0. 2 mM $K_3[Fe(CN)_6]$. Fig. (**2**) displays the EIS spectra of BCCPE (curve a) and PMOMCCPE (curve b). From Fig. (**2**), it is apparent that the EIS curve obtained for BCCPE has a higher charge transfer resistance value (R_{ct}) compared to that of PMOMCCPE. The R_{ct} value for BCCPE and PMOMCCPE are 165.7 ohms and 0.6039 ohm, respectively. R_{ct} value obtained specifies that the improved conductivity of PMOMCCPE is because of the enhanced rate in charge transfer by the MO polymer film [14 - 16].

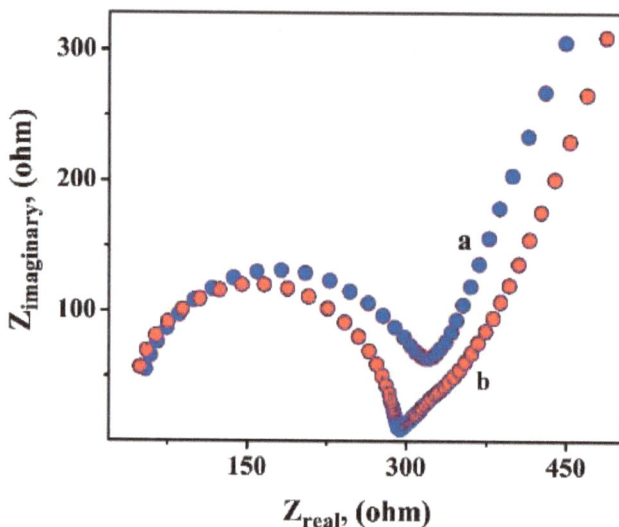

Fig. (2). Nyquist diagrams of EIS of BCCPE (curve a) and MOMCCPE (curve b).

Electrochemical Active Surface Area

The electrochemical surface area of BCCPE and PMOMCCPE was investigated by utilising CV method in 0.2 mM $[K_4 Fe(CN)_6]$ with 0.1 M KCl under optimum conditions (Fig. **3**). Randles-Sevcik equation (Equation 1) was applied for determining electrochemical active surface area of BCCPE and PMOMCCPE [17, 18].

$$I_{pa} = 2.69 \times 10^5 \, n^{3/2} \, A \, C_o \, D^{1/2} v^{1/2} \tag{1}$$

Where I_{pa} is the peak current, n is the number of transferred electrons, A is the active surface area, D is the diffusion coefficient, C is the concentration in mol/cm^3 and v is the scan rate. The calculated value of the active surface area of BCCPE was 0.006 cm^2 and PMOMCCPE was 0.01225. The increased surface area of the modified electrode is responsible for the PMOMCCPE's increased electrocatalytic activity.

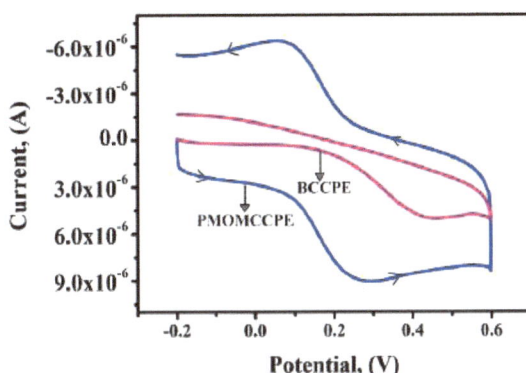

Fig. (3). Cyclic voltammograms of 0.1 mM [$K_4Fe(CN)_6$] with 0.1 M KCl at BCCPE and PMOMCCPE.

Electrochemical Polymerisation of MO

The PMOMCCPE was fabricated by running ten CV cycles for the electro polymerization of 1×10^{-3} M MO in 0.2 M PBS of pH 7.0, in a potential range of −0.6 to 1.4 V with a scan rate of 0.1 V/s. The voltammogram of MO is shown in Fig. (**4A**), with steadily growing redox curves with each potential cycle, indicating the deposition of MO polymer layer [19, 20]. The thicknesses of the polymer film have a considerable effect on the electrocatalytic activity of the BCCPE. By varying the number of scan cycles from 5 to 20, the MO film's coating on BCCPE was investigated. Fig. (**4B**) displays that the peak current of RB was boosted at the 10th cycle, but after the 10th cycle the peak current got diminished. Hence, the number of optimum scan cycles for electro polymerization of MO was scrutinised to be ten.

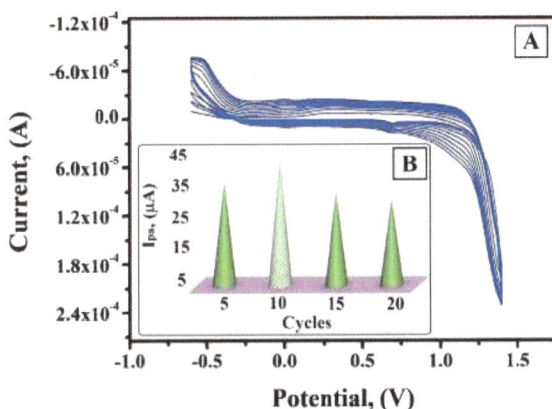

Fig. (4). A) Cyclic voltammograms of electrochemical polymerization of 1×10^{-3} M MO on the surface of BCCPE in 0.2 M PBS of pH 7.0 at a scan rate of 0.1 V/s. **B)** Plot of the anodic peak current of 1×10^{-3} M MO in 0.2 M PBS (pH 7.0) *versus* the number of polymerization cycles of MO.

Influence of Solution pH

The electrochemical response of the analyte depends on the pH of the solution, and hence it is important to optimize the pH to get a sharper response with higher sensitivity. To study the influence of pH of the solutions on the response of RB at PMOMCCPE, CV is recorded and displayed in Fig. (**5A**). It can be observed from the plot of E_{pa} *versus* pH (Fig. **5B**) that when pH increased, the oxidation peak potential migrated slightly to the negative side, indicating that protons are actively involved in the electrode reaction process [21, 22]. The linear relationship between pH and Epa is depicted by Epa's plot, with the linear regression equation; E_{pa} (V) = 0.0387 pH − 0.14207 and correlation coefficient; R^2 = 0.9942. In addition, the plot of Ipa *versus* pH (Fig. **5C**) indicates that the highest anodic peak current for PMOMCCPE is obtained at pH 7.5, hence pH 7.5 was chosen as the optimal.

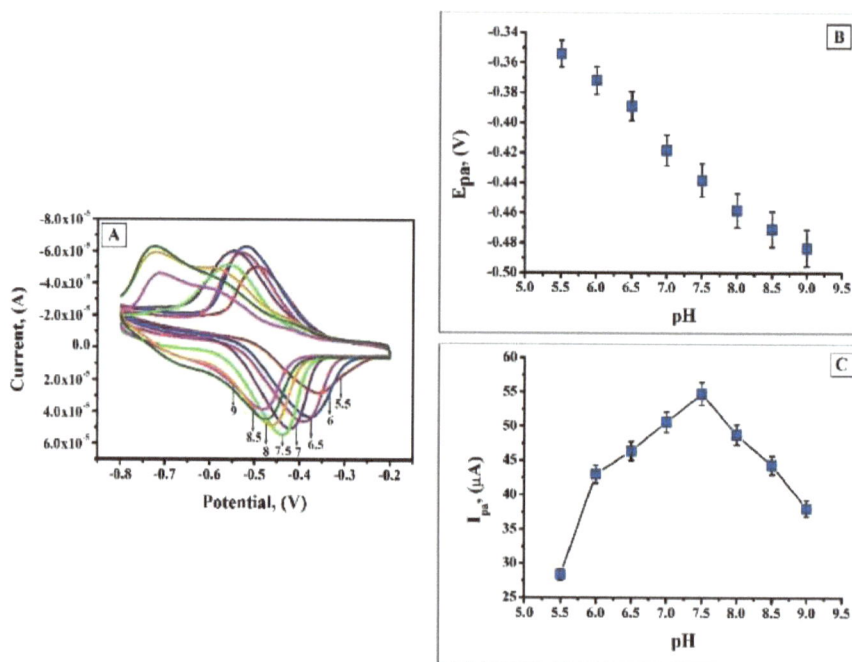

Fig. (5). A) CV response of 1 × 10⁻⁴ M RB at PMOMCCPE in 0.2 M PBS of different pH in the range 5.5 to 9.0 at the scan rate of 0.1 V/s. **B)** The plot of the anodic peak potential (E_{pa}) *versus* pH. **C)** The plot of oxidation peak current (I_{pa}) *versus* pH of the solution.

Electrochemical Behaviour of RB at Various Carbon Paste Electrodes

Fig. (**6**) shows the DPV of RB (1×10⁻⁴ M) at bare carbon paste electrode (BCPE), bare carbon nanotube paste electrode (BCNTPE), BCCPE and PMOMCCPE in 0.2 M PBS of 7.5 pH at the scan rate of 0.1V/s. BCPE displays a peak current of 8.66 µA, BCNTPE portrays a peak current of 13.27 µA, BCCPE shows a current of 22.91 µA on the contrary, in the identical condition PMOMCCPE displays an enhanced current sensitivity with an I_{pa} of 116.7 µA. The enhanced current sensitivity displayed at PMOMCCPE shows that BCCPE is effectively modified by a polymer film. This improves the catalytic activity towards the RB by increasing the surface area and conductivity of the modified electrode [23, 24].

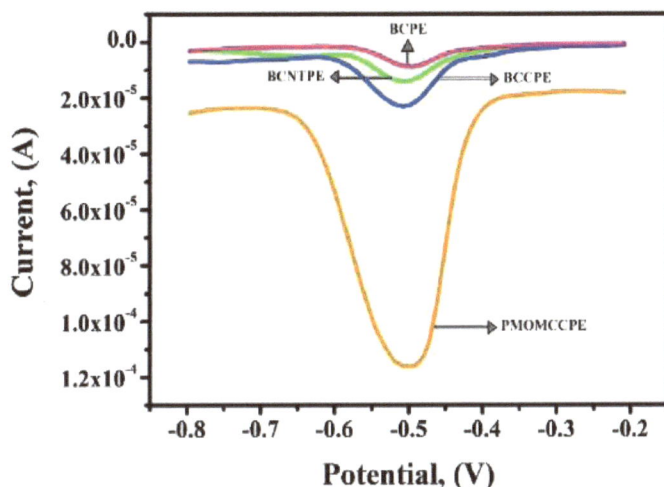

Fig. (6). DPV response of 1 × 10⁻⁴ M RB at BCPE, BCNTPE, BCCPE and PMOMCCPE under optimum condition.

Effect of Scan Rate

Using CV approach, scan rate optimization is essential to analyse various aspects like kinetic behaviour, reaction pathways, number of electrons transferred during the redox process at the surface of the PMOMCCPE electrode. Scan rate was optimised from 0.05 to 0.2 V/s, by taking 1 ×10⁻⁴ M RB in 0.2 M PBS of pH 7.5. The increase in the adsorption rate of the electroactive species on the electrode surface is related to the scan rate. According to Fig. (7A), the scan rate has an effect on the oxidation peak current values, with the peak current increasing as the scan rate increases. The plot between the scan rate and the anodic peak current value (Fig. 7B) shows linearity with a linear regression equation, I_{pa} (µA) = -26.92 + 774.054 v (V/s); R^2 = 0.9988. Linearity observed in the plot indicates that the process is dominated by an adsorption-controlled mechanism. In the RB analysis process, a potential shift in the E_{pa} and E_{pc} values was noted as the scan rate increased. If the magnitude of the peak potential separation increases, electron transfer kinetics will be slow, and this slower rate will become more pronounced as the scan rate increases. I_{pa}/I_{pc} values and ΔEp play a vital role in finding the optimum scan rate [25 - 27]. According to the experimental outcome, I_{pa}/I_{pc} value was less than 1 and ΔEp was greater than 59 mV, hence the process was considered to be a quasi-reversible reaction. The scan rate of 0.1 V/s is the best outcome of RB analysis since it has I_{pa}/I_{pc} 0.96 A, which is close to 1, and ΔEp is 0.112 V. The plot between E_{pa} and log v (Fig. **7C**) shows linearity with a linear regression equation, E_{pa} = -0.3256 + 0.1282 log v; R^2 = 0.9974.

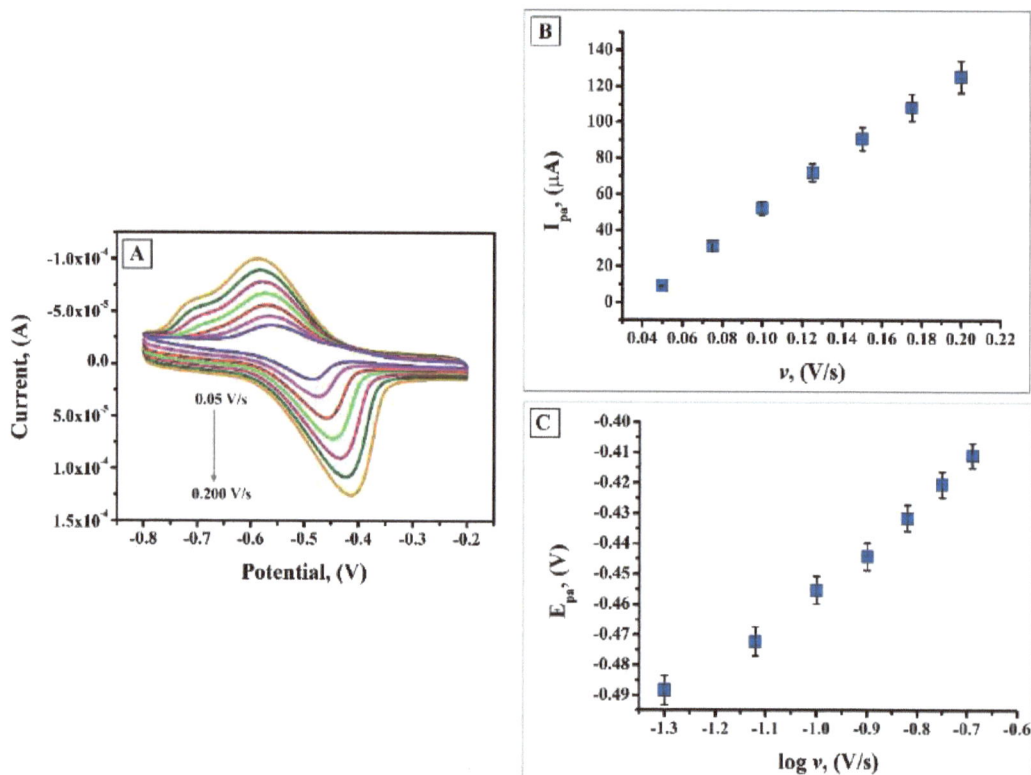

Fig. (7). (A) CV response of 1×10^{-4} M RB at the PMOMCCPE in pH 7.5 at different scan rates from (0.05 to 0.200 V/s) **(B)** Plot of I_{pa} *versus* v **(C)** Plot of E_{pa} *versus* log v.

Effect of Concentration on Peak Current

Analytical curves (Peak current *vs.* RB concentration) acquired under optimum experimental circumstances were used to assess the usability of the PMOMCCPE in DPV technique for the quantitative determination of RB. Fig. **(8A)** shows typical DP voltammograms recorded after addition of different concentrations of RB. The calibration plot between the anodic peak current and concentration of RB

(Fig. **8B**) shows a linear correlation in the concentration range of 4 mM to 50 mM. The graph shows two linear ranges for concentrations from 4 mM to 20 µM, with a corresponding linear equation: I_{pa} (µA) = 2.6222× 10^{-5} + 5.1592 [RB] (M); (R^2 = 0.9988). The second linear range is from 25 mM to 50 mM. The linear equation is stated as I_{pa} (µA) = 9.1146 × 10^{-5} + 1.7340 [RB] (M); (R^2 = 0.9930). The detection limit (LOD) and limit of quantification (LOQ) estimated by considering the first linear range, was 1.65 × 10^{-7} M and 5.52 ×10^{-7} M. (LOD = 3S/N, LOQ = 10 S/N where 'S' is the standard deviation of the peak current of blank solution and 'N' represents the slope of the calibration plot). The obtained values of LOD are in close approximate with the available values in the literature and it is depicted in Table **1**.

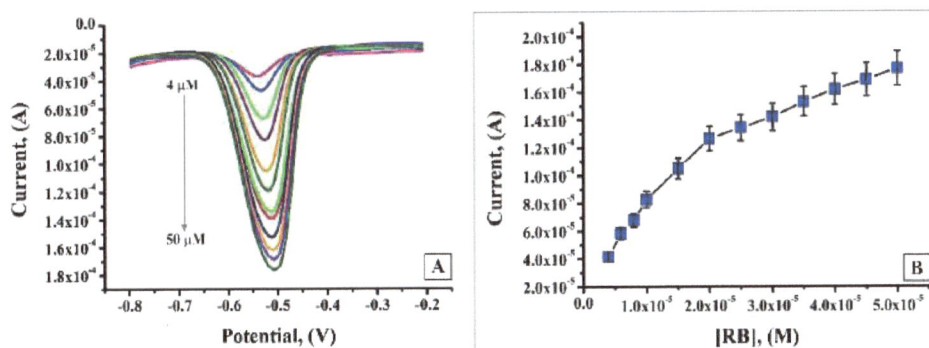

Fig. (8). A) DPV for RB at PMOMCCPE with different concentrations of RB in 0.2 M PBS of pH 7.5 at a scan rate of 0.1 V/s, B) I_{pa}Vs [RB].

Table 1. Comparison of the LOD values of the PMOMCCPE with the available values in the literature for voltammetric determination of RB.

Working Electrode	Method	Linear Range (mM)	LOD (M)	References
Zeolite modified carbon paste electrode	CV	1.7 – 34	0.71×10^{-6}	[28]
MnO$_2$ modified glassy carbon electrode	DPV	2.0 -110	2.7 ×10^{-8}	[29]
Poly leucine modified carbon paste electrode	DPV	2.0 – 50	3.4 ×10^{-8}	[30]
SnO$_2$/RGO nano composite modified glassy carbon electrode	SWV	0.1 - 150	34 × 10^{-9}	[31]
MnO$_2$ modified carbon paste electrode	DPV	0.02 - 9	15 × 10^{-9}	[32]
DNA modified pencil graphite electrode	DPV	1.0 – 186	0.9 × 10^{-6}	[33]
Methyl orange modified composite carbon paste electrode	DPV	4.0 – 20	1.65 ×10^{-7}	Present work

Selectivity of PMOMCCPE for Determination of RB

Research has shown that RB is one of the cost-effective neuroprotective drugs prescribed to patients with neurological disorders [34]. Parkinson's is a neurodegenerative disease which involves dopaminergic neurons. These disorders involve formation of reactive oxygen species, which will affect the dopaminergic system. Detection of DA level is essential to protect brain against these species. Hence PMOMCCPE is used for the simultaneous determination of RB and DA. Fig. (9) shows the DPV of RB (1×10^{-4} M) and DA (1×10^{-4} M) at BCCPE and PMOMCCPE in pH 7.5 at a scan rate of 0.1 V/s. DPV obtained for the mixture of RB and DA at BCCPE was less sensitive. But at PMOMCCPE, RB and DA shows a well-defined peak with an enhanced current response. An effective peak potential separation (0.4055 V) at PMOMCCPE was noted, which allows the selective resolution of RB and DA.

By adding 50 times highly concentrated sample solutions of metal ions and organic species found in the body, their interfering effect on the voltammetric response of the modified electrode was investigated under optimised circumstances. The results showed that concentrations of Mg^{2+}, Ca^{2+}, K^{+}, Na^{+}, Fe^{3+}, ascorbic acid, glucose, glycine and starch did not show significant influence on the height of the anodic peak currents [35 - 37]. The experimental outcome shows that PMOMCCPE can be used for selective determination of RB.

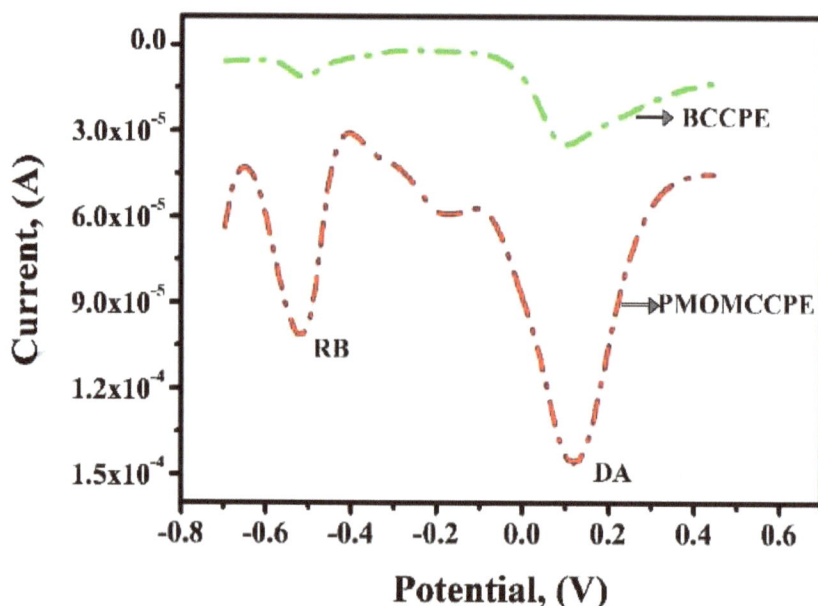

Fig. (9). DPV of RB (1×10^{-4} M) and DA (1×10^{-4} M) in 0.2 M PBS of pH 7.5 at BCCPE and PMOMCCPE.

Repeatability, Reproducibility, and Stability of PMOMCCPE

The reproducibility of the proposed sensor was assessed by fabricating PMOMCCPE five times using the same method and CV response of 1×10^{-4} M RB was recorded each time. Appreciable result was observed with a relative standard deviation 3.68% (n =4). The repeatability of the proposed sensor was examined by recording the CV response of RB under optimised conditions. From this study, obtained RSD value was 2.87% (n =3) which suggests convincing repeatability of PMOMCCPE. The stability of PMOMCCPE was evaluated by incessantly scanning the electrode for 30 cycles under optimal conditions. It was observed that the PMOMCCPE retains 95.08% of its initial response. This outcome was evaluated using the equation [38 - 40] % Degradation = I_{pn} / I_{pa} ×100, where I_{pn} is the peak current at n^{th} cycle and I_{pa} is the peak current at first cycle.

Pharmaceutical Sample Analysis

The application of the fabricated electrode for the detection of RB was demonstrated by the detection of RB in pharmaceutical samples [41]. The precision of the technique was calculated for multivitamin tablets. Real sample solution was prepared by weighing the required amount of commercially available multivitamin samples. The weighed sample was dissolved in sodium hydroxide in brown coloured bottles and double distilled water was added to get 25×10^{-4} M solution. Fig. (**10**) shows the DPV curves for the real sample analysis. Standard addition method of the tablet sample solution was done under optimised conditions and the results are tabulated in Table **2**. PMOMCCPE's significant recovery range suggests that the fabricated sensor with strong sensing capability is suitable for real sample analysis.

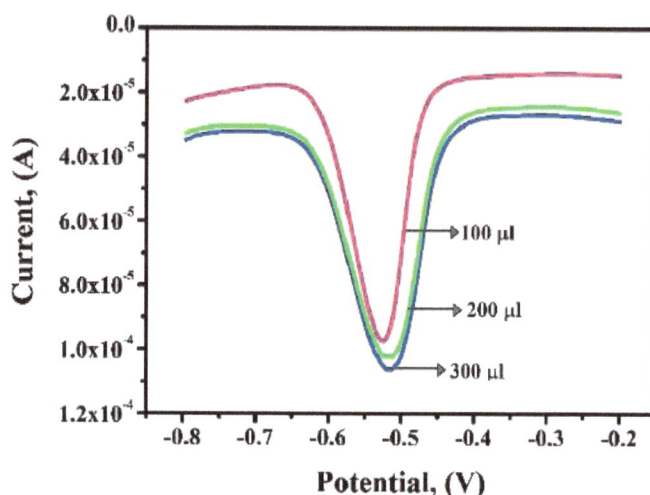

Fig. (10). DPV curves for analysis of RB in multivitamin tablets.

Table 2. Estimation of RB in multivitamin tablet sample.

Sample	Added (10^{-5} M)	Found (10^{-5} M)	Recovery (%)
Multivitamin tablet	10	9.65	96.50
	20	19.47	97.35
	30	29.86	99.53

CONCLUSION

The experimental outcome of this study reveals the suitability of PMOMCCPE for the voltammetry determination of RB in the presence of DA and other potential interferents. Remarkable improvement in the current sensitivity, high selectivity, low detection limit, easy fabrication procedure with good selectivity makes PMOMCPE a good sensor for RB analysis. The proposed sensor was effectively employed for determining RB content in multivitamin tablets. Scan rate studies reveal that the process was adsorption-controlled and the performance and efficacy of the PMOMCCPE were found to have acceptable repeatability and reproducibility.

CONSENT FOR PUBLICATION

Not applicable.

CONFLICT INTEREST

The authors declare no conflict of interest, financial or otherwise.

ACKNOWLEDGMENT

Dcelared none.

REFERENCES

[1]   H.J. Powers, "Riboflavin (vitamin B-2) and health", *Am. J. Clin. Nutr.,* vol. 77, no. 6, pp. 1352-1360, 2003.
[http://dx.doi.org/10.1093/ajcn/77.6.1352] [PMID: 12791609]

[2]   N.C. Catalina, "Electrochemical Sensors Used in the Determination of Riboflavin", *J. Electrochem. Soc.,* vol. 167, p. 037558, 2020.
[http://dx.doi.org/10.1149/1945-7111/ab6e5e]

[3]   N. Hareesha, and J. G. Manjunatha, "Fast and Enhanced Electrochemical Sensing of Dopamine at Cost-effective Poly (DL-phenylalanine) Based Graphite Electrode", *J. Electroanal. Chem,* vol. 878, .
[http://dx.doi.org/10.1016/j.jelechem.2020.114533]

[4]   N.E. Alireza, and P. Pouya, "Voltammetric Determination of Riboflavin based on Electrocatalytic Oxidation at Zeolite-Modified Carbon Paste Electrodes", *J. Ind. Eng. Chem.,* vol. 20, pp. 2146-2152, 2014.
[http://dx.doi.org/10.1016/j.jiec.2013.09.044]

[5]   S.D. Edwin, J.G. Manjunatha, C. Raril, T. Girish, and P.A. Pushpanjali, "Polymer Modified Carbon Paste Electrode as a Sensitive Sensor for the Electrochemical Determination of Riboflavin and Its Application in Pharmaceutical and Biological Samples", *Anal. Bioanal. Chem.,* vol. 7, pp. 461-472, 2020.
[http://dx.doi.org/10.22036/ABCR.2020.214882.1445]

[6]   H. Chengguo, and H. Shengshui, "Carbon Nanotube-Based Electrochemical Sensors: Principles and Applications in Biomedical Systems. ", *J. Sens,* 2009.
[http://dx.doi.org/10.1155/2009/187615]

[7]   J.G. Manjunatha, B.E. Kumara Swamy, G.P. Mamatha, C. Umesh, E. Niranjana, and B.S. Sherigara, "Cyclic Voltammetric Studies of Dopamine at Lamotrigine and TX-100 Modified Carbon Paste Electrode", *Int. J. Electrochem. Sci.,* vol. 4, pp. 187-196, 2009.

[8]   A.B. Monnappa, J.G. Manjunatha, and A.S. Bhatt, "Design of a Sensitive and Selective Voltammetric Sensor Based on a Cationic Surfactant-Modified Carbon Paste Electrode for the Determination of Alloxan", *ACS Omega,* vol. 5, no. 36, pp. 23481-23490, 2020.
[http://dx.doi.org/10.1021/acsomega.0c03517] [PMID: 32954201]

[9]   "Pushpanjali; P. A.; Manjunatha; J. G.; Amrutha; B. M.; Hareesha; N. Development of Carbon Nanotube-based Polymer-Modified Electrochemical Sensor for the Voltammetric Study of Curcumin", *Mater. Res. Innov.,* pp. 1-9, 2020.
[http://dx.doi.org/10.1080/14328917.2020.184258]

[10]   C. Raril, and J.G. Manjunatha, "Electropolymerization of Glycine at Carbon Paste Electrode and its Application for the Determination of Methyl Orange", *J. Sci.,* vol. 30, pp. 233-240, 2019.
[http://dx.doi.org/10.22059/JSCIENCES.2019.267743.1007328]

[11]   B.M. Amrutha, "Manjunatha. J. G.; Aarti, S. B.; Raril, C.; Pushpanjali, P. A. Electrochemical Sensor for the Determination of Alizarin Red-S at Non-ionic Surfactant Modified Carbon Nanotube Paste Electrode", *Phys. Chem. Res.,* vol. 7, pp. 523-533, 2019.
[http://dx.doi.org/10.22036/PCR.2019.185875.1636]

[12]　S. Vera, and K. Miloslav, "Chemically Modified Carbon Paste and Carbon Composite Electrodes", *Electroanal.,* vol. 1, pp. 251-256, 1989.
[http://dx.doi.org/10.1002/elan.1140010310]

[13]　T. Thomas, R.J. Mascarenhas, O.J. D'Souza, P. Martis, J. Dalhalle, and B.E. Swamy, "Multi-walled carbon nanotube modified carbon paste electrode as a sensor for the amperometric detection of L-tryptophan in biological samples", *J. Colloid Interface Sci.,* vol. 402, pp. 223-229, 2013.
[http://dx.doi.org/10.1016/j.jcis.2013.03.059] [PMID: 23628203]

[14]　N. Achargu, W. Chkairi, W. Chafiki, and M. Elbasri, "Electrochemical Behaviour of Carbon Paste Electrode Modified with Carbon Nanofibers: Application to detection of Bisphenol A", *Int. j. eng,* vol. 6, pp. 23-29, 2016.

[15]　M. Khodari, E.M. Rabie, and A.A. Shamroukh, "Carbon Paste Electrode Modified by Multiwalled Carbon Nanotube "for Electrochemical Determination" of Vitamin C", *International Journal of Biochemistry and Biophysics,* vol. 6, pp. 58-69, 2018.
[http://dx.doi.org/10.13189/ijbb.2018.060204]

[16]　N. Festinger, K. Morawska, V. Ivanovski, M. Ziąbka, K. Jedlińska, W. Ciesielski, and S. Smarzewska, "Comparative Electroanalytical Studies of Graphite Flake and Multilayer Graphene Paste Electrodes", *Sensors (Basel),* vol. 20, no. 6, p. 1684, 2020.
[http://dx.doi.org/10.3390/s20061684] [PMID: 32197336]

[17]　N. Hareesha, and J.G. Manjunatha, "Elevated and Rapid Voltammetric Sensing of Riboflavin at Poly (helianthin dye) Blended Carbon Paste Electrode with Heterogeneous Rate Constant Elucidation", *J Iran Chem Soc.,* vol. 17, pp. 1507-1519, 2020.
[http://dx.doi.org/10.1007/s13738-020-01876-4]

[18]　K. Giribabu, Y. Haldorai, M. Rethinasabapathy, S-C. Jang, R. Suresh, W-S. Cho, and Y-K. Han, "Changhyun Roh d, Yun Suk Huh, Vengidusamy Narayanan, Glassy Carbon Electrode Modified with Poly(methyl orange) as an Electrochemical Platform for the Determination of 4-Nitrophenol at Nanomolar Levels", *Curr. Appl. Phys.,* vol. 17, pp. 1114-1119, 2017.
[http://dx.doi.org/10.1016/j.cap.2017.04.016]

[19]　J.G. Manjunatha, "Poly (Adenine) Modified Graphene-Based Voltammetric Sensor for the Electrochemical Determination of Catechol, Hydroquinone and Resorcinol", *Open Chem. Eng. J.,* vol. 14, pp. 52-62, 2020.
[http://dx.doi.org/10.2174/1874123102014010052]

[20]　F. Karimian, G.H. Rounaghi, and M. Mohadeszadeh, "Electrochemical Determination of Riboflavin Using a Synthesized Ethyl[(methythio)carbonothioyl] Glycinate Monolayer Modified Gold Electrode", *J. Anal. Chem.,* vol. 71, pp. 1057-1062, 2016.
[http://dx.doi.org/10.1134/S1061934816080062]

[21]　P.A. Pushpanjali, J.G. Manjunatha, and M.T. Shreenivas, "The Electrochemical Resolution of Ciprofloxacin, Riboflavin and Estriol Using Anionic Surfactant and Polymer Modified Carbon Paste Electrode", *ChemistrySelect,* vol. 4, pp. 13427-13433, 2019.
[http://dx.doi.org/10.1002/slct.201903897]

[22]　S. Majid, M. Rhazi, A. Amine, A. Curulli, and G. Palleschi, "Carbon Paste Electrode Bulk-Modified with the Conducting Polymer Poly(1,8-Diaminonaphthalene): Application to Lead Determination", *Mikrochim. Acta,* vol. 143, pp. 195-204, 2003.
[http://dx.doi.org/10.1007/s00604-003-0058-5]

[23]　O. Gilbert, and C. Umesh, "Kumara Swamy, B. E.;Panduranga, C. M.; Nagaraj, C.; Sherigara, B. S, Poly (Alanine) Modified Carbon Paste Electrode for Simultaneous Detection of Dopamine and Ascorbic Acid", *Int. J. Electrochem. Sci.,* vol. 3, pp. 1186-1195, 2008.

[24]　N. Hareesha, and J.G. Manjunatha, "A Simple and Low-Cost Poly (dl-phenylalanine) Modified Carbon Sensor for the Improved Electrochemical Analysis of Riboflavin", *J Sci-Adv Mater Dev.,* vol. 5, pp. 502-511, 2020.

[http://dx.doi.org/10.1016/j.jsamd.2020.08.005]

[25]   G. Vijayaprasath, I. Habibulla, V. Dharuman, S. Balasubramanian, and R. Ganesan, "Fabrication of Gd2O3 Nanosheet-Modified Glassy Carbon Electrode for Nonenzymatic Highly Selective Electrochemical Detection of Vitamin B2", *ACS Omega,* vol. 5, no. 29, pp. 17892-17899, 2020.
[http://dx.doi.org/10.1021/acsomega.9b04284] [PMID: 32743160]

[26]   M. Abdul Jabbar, S. Salahuddin, A.J. Mahmood, and R.J. Mannan, "Voltammetric Evidences for the Interaction of Riboflavin with Cadmium in Aqueous Media", *J. Saudi Chem. Soc.,* vol. 20, pp. 158-164, 2016.
[http://dx.doi.org/10.1016/j.jscs.2012.06.005]

[27]   M.M. Charithra, and J.G. Manjunatha, "Fabrication of Poly (Evans Blue) Modified Graphite Paste Electrode as an Electrochemical sensor for Sensitive and Instant Riboflavin Detection, Mor", *J. Chem.,* vol. 9, 2020.
[http://dx.doi.org/10.48317/IMIST.PRSM/morjchem-v9i1.18239]

[28]   A. Nezamzadeh-Ejhieh, and P. Pouladsaz, "Voltammetric Determination of Riboflavin Based on Electrocatalytic Oxidation at Zeolite-Modified Carbon Paste Electrodes", *J. Ind. Eng. Chem.,* vol. 20, pp. 2146-2152, 2013.
[http://dx.doi.org/10.1016/j.jiec.2013.09.044]

[29]   D.Q. Huang, H. Wu, C. Song, Q. Zhu, H. Zhang, L.Q. Sheng, H.J. Xu, and Z.D. Liu, "(Vitamin B2) Using Manganese Dioxide Modified Glassy Carbon Electrode by Differential Pulse Voltammetry", *Int. J. Electrochem. Sci.,* vol. 13, pp. 8303-8312, 2018.
[http://dx.doi.org/10.20964/2018.09.02]

[30]   B.M. Amrutha, J.G. Manjunatha, S.B. Aarti, and P.A. Pushpanjali, "Fabrication of a Sensitive and Selective Electrochemical Sensing Platform Based on Polyleucine Modified Sensor for Enhanced Voltammetric Determination of Riboflavin", *Food Measure.,* vol. 14, pp. 3633-3643, 2020.
[http://dx.doi.org/10.1007/s11694-020-00608-9]

[31]   R. Sriramprabha, M. Divagar, N. Ponpandian, and C. Viswanathan, "Tin Oxide/Reduced Graphene Oxide Nanocomposite-Modified Electrode for Selective and Sensitive Detection of Riboflavin", *J. Electrochem. Soc.,* vol. 165, pp. 498-507, 2018.
[http://dx.doi.org/10.1149/2.0761811jes]

[32]   E. Mehmeti, D.M. Stanković, S. Chaiyo, Ľ. Švorc, and K. Kalcher, "Manganese dioxide-modified carbon paste electrode for voltammetric determination of riboflavin", *Mikrochim. Acta,* vol. 183, pp. 1619-1624, 2016.
[http://dx.doi.org/10.1007/s00604-016-1789-4] [PMID: 27217592]

[33]   A.A. Ensafi, E. Heydari-Bafrooei, and M. Amini, "DNA-functionalized biosensor for riboflavin based electrochemical interaction on pretreated pencil graphite electrode", *Biosens. Bioelectron.,* vol. 31, no. 1, pp. 376-381, 2012.
[http://dx.doi.org/10.1016/j.bios.2011.10.050] [PMID: 22099958]

[34]   A.V. Peraza, D.C. Guzmán, N.O. Brizuela, M.O. Herrera, H.J. Olguín, M.L. Silva, B.J. Tapia, and G.B. Mejía, "Riboflavin and pyridoxine restore dopamine levels and reduce oxidative stress in brain of rats", *BMC Neurosci.,* vol. 19, no. 1, p. 71, 2018.
[http://dx.doi.org/10.1186/s12868-018-0474-4] [PMID: 30413185]

[35]   J.G. Manjunatha, "Surfactant Modified Carbon Nanotube Paste Electrode for the Sensitive Determination of Mitoxantrone Anticancer drug", *Jese.,* vol. 7, pp. 39-49, 2017.
[http://dx.doi.org/10.5599/jese.368]

[36]   N.C. Catalina, "Electrochemical Sensors Used in the Determination of Riboflavin", *J. Electrochem. Soc.,* vol. 167, p. 037558, 2020.
[http://dx.doi.org/10.1149/1945-7111/ab6e5e]

[37]   T. Girish, and J.G. Manjunatha, "Optimized Voltammetric Experiment for the Determination of Phloroglucinol at Surfactant Modified Carbon Nanotube Paste Electrode", *Instrum. Exp. Tech.,* vol.

63, pp. 750-757, 2020.
[http://dx.doi.org/10.1134/S0020441220050139]

[38]   S.S. Khaloo, S. Mozaffari, P. Alimohammadi, H. Kargar, and J. Ordookhanian, "Sensitive and Selective Determination of Riboflavin in Food and Pharmaceutical Samples Using Manganese (III) Tetraphenylporphyrin Modified Carbon Paste Electrode", *Int. J. Food Prop.,* vol. 19, pp. 2272-2283, 2016.
[http://dx.doi.org/10.1080/10942912.2015.1130054]

[39]   A.B. Monnappa, J.G. Manjunatha, A.S. Bhatt, and K. Malini, "Sodium Dodecyl Sulfate Modified Carbon Nano Tube Paste Electrode for Sensitive Cyclic Voltammetry Determination of Isatin", *Adv. Pharm. Bull.,* vol. 11, no. 1, pp. 111-119, 2021.
[http://dx.doi.org/10.34172/apb.2021.012] [PMID: 33747858]

[40]   L.S. Anisimova, E.V. Mikheeva, and V.F. Slipchenko, "Voltammetric Determination of Riboflavin in Vitaminized Supplements and Feeds", *J. Anal. Chem.,* vol. 56, pp. 658-662, 2001.
[http://dx.doi.org/10.1023/A:1016748423494]

[41]   J.G. Manjunatha, M. Deraman, N.H. Basri, and I.A. Talib, "Fabrication of Poly (Solid Red A) Modified Carbon Nano Tube Paste Electrode and its Application for Simultaneous Determination of Epinephrine, Uric acid and Ascorbic acid", *Arab. J. Chem.,* vol. 11, pp. 149-158, 2018.
[http://dx.doi.org/10.1016/j.arabjc.2014.10.009]

CHAPTER 5

A Modified Nanostructured Gd-WO₃, Sensing Interface Morphology, their Voltammetric Determination and Applications in Advanced Energy Storage Devices

Vinayak Adimule[1,*], **Basappa C. Yallur**[2], **Debdas Bhowmik**[3] and **Santosh S. Nandi**[4]

[1] *Department of Chemistry, Angadi Institute of Technology and Management (AITM), Savagaon Road, Belagavi-5800321, Karnataka, India*

[2] *Department of Chemistry, M. S. Ramaiah Institute of Technology, Bangalore, Karnataka, India*

[3] *High Energy Materials Research Laboratory, Defence Research and Development Organization, Ministry of Defence, Government of India, Sutarwadi, Pune, India*

[4] *Chemistry Section, Department of Engineering Science and Humanities, KLE Dr. M.S. Sheshgiri College of Engineering & Technology, Udhyambagh, Belagavi-590008, Karnataka, India*

Abstract: Perovskite electrodes can be a pertinent method to enhance electrochemical voltammetric sensing performances. In the present research, Gd_xWO_3 (x = 2, 5, 10 wt. %) nanostructures (NS) are synthesized by a facile co-precipitation method, calcinated at 750-800 °C and subsequently fabricated using electron beam deposition method over glass substrate used for electrochemical voltammetric determination. Scanning electron microscopy and X-ray diffraction were employed to examine the crystal phase and morphology of the gadolinium doped nanostructures thus prepared. Surface chemical composition, chemical bond analysis, dispersion of Gd into WO_3 were confirmed by XPS (X-ray photoelectron spectral) studies. SEM (scanning electron microscopy) micrographs showed flake-like surface morphology can act as sensing interface for facile transfer of electrons over NS of Gd_xWO_3, and SEM cross section micrographs revealed agglomerated densely packed rod like structure. $Gd_x WO_3$ nanostructure showed ~ 89 nm grain, ~ 110 nm grain boundary distances, UV-visible absorptivity maxima observed between 280 nm to 340 nm for $Gd_x WO_3$ NS. Electrochemical cyclic voltammetry performance was tested in three electrodes assembled for high performance energy storage applications, involving cyclic voltammetric charge discharge cycles and sensing interface between fabricated NS, and cyclic voltammetry carried out in 6M KOH solution. Gd_xWO_3 demonstrates a specific capacitance of 450 F/g at a current density of 0.1 A/g. Gd-WO₃ NS has a capacitance retention rate of 88.5% after 5,000 cycles (cyclic stability). Voltammetric characteristics indicate that

* **Corresponding author Dr. Vinayak Adimule:** Department of Chemistry, Angadi Institute of Technology and Management (AITM), Savagaon Road, Belagavi-5800321, Karnataka, India; E-mail: adimulevinayak@yahoo.in

Gd$_x$ WO$_3$ NS is a promising electrode material for energy storage devices and can be used as high-performance super capacitor applications.

Keywords: Agglomeration, Charge-Discharge, Co precipitation, Cyclic stability, Cyclic Voltammetry, Electrochemical voltammetry, Energy storage devices, Gd-WO$_3$, Inorganic Nanostructures, Morphology, Nanoparticles, Nanostructures, Sensing, SEM, Specific Capacitance, Super capacitor, Synthesis, UV-Visible, XRD, XPS.

INTRODUCTION

In recent years, damage has been done to the environment and their subsequent causes such as gas greenhouse gas emission, air pollution, this necessitates better eco-friendly, cleaner energy sources. In this perspective, batteries and fuel cells were investigated and made use of as a viable options. However, disadvantages (lower power efficiency, high cost, low-rate capabilities, *etc.*) in the energy sources of batteries and fuel cells are envisaged for the development of new energy storage devices. Increasing demand for high energy devices in consumer electronics, portable electronics, electric vehicles, *etc.*, to fulfil the void and unleash energy storage super capacitors are widely attempted [1 - 3]. In general, super capacitor segmented into two main categories one is (EDLC) and another is pseudo capacitor [4]. EDLCs commonly use carbon containing materials such as graphene, carbon nanotubes (CNTs), reduced graphene oxide (rGO), having a large surface area, higher charge discharge cycles and high-power density. Structural morphology is maintained for a quite long period of time, which is due to adsorption and desorption of ions over the surface of the electrode. On the other hand, pseudo capacitors consist of transition metal oxides (ZnO, Gd$_2$O$_3$, Y$_2$O$_3$, CeO$_2$, *etc.*) and charge discharge cycles appear due to redox reactions occurring in the system. Electrochemical approaches offer practical benefits in addition to satisfactory sensitivity, wide linear concentration range, operation simplicity, the possibility of miniaturization, low expense of instruments, suitability for real-time detection and less sensitivity to matrix effects in comparison with spectral and separation methods. Recently, rare earth elements used as dopants have shown many potentials in enhanced microwave absorption, photoluminescence and photocatalytic activity.

Tungsten oxide (WO$_3$) is one of the interesting materials because it is highly stable, low-cost, eco-friendly, less toxic, and widely used as a pseudo capacitor material which has grasped much attention of the scientific community due to unique characteristic features such as being used in electrochromic devices, energy harvesting devices, enhanced I-V/C-V characteristics, photovoltaic, *etc.* [5 - 9]. Tungsten oxide (WO$_3$) as high-performance energy storage materials and their characteristic features published in recent years [10]. More recently, numerous reports were published for synthesis, characterization of WO_3 nanostructure, which includes chemical co-precipitation, vapour deposition, spray pyrolysis, *etc* [11 - 14]. Due to the small radius and desirable crystal structure of the hexagonal WO_3 NS, a prominent influence on the desired electrochemical performance was observed. The structural and morphological changes in the WO_3 NS were successively studied for photochromic properties by Huang *et al.* [15]. Similarly, Nagy *et al.* studied the effect of structural morphology variations and characterized them with respect to dependency in pH of the solution. Many researchers studied the photo catalytic efficiency of the WO_3 nanoparticles [16, 17]. Kalhori *et al.* synthesized flower shaped two different phases of WO_3 NS, one is orthorhombic and another is monoclinic by varying hydrothermal temperature. The precursor used was ammonium oxalate and WO_3 NS was studied for gas chromic responses [18]. WO_3 NS in crystalline nature possesses greater stability due to slower dissolution rate and denser accumulation of nanoparticles. As for WO_3, doping elements reported were mainly focused on transition metals such as Mo, Ni, Ti, V, Nb, Fe, Ru, *etc.*, by which the improvements both in ionic conductivity and high energy storage devices have been obtained mainly because of the introduction of oxygen vacancies. More recently, we have shown that gadolinium (Gd) doping can effectively improve the high energy storage device of the sputtered amorphous WO_3. Luminescent properties of Ta doped WO_3 NS having smooth crystallinity, moderate band gap and blue line emission in photoluminescence spectra. All the films showed 80% optical transference in the wavelength of 400 nm to 900 nm with increased optical reflectance [19]. A novel set of Ag doped WO_3 NS synthesized by hydrothermal method, heterostructure showed increased gas response towards 100 ppm of formalin (HCHO) almost 3.5 times higher response and recovery time was 3s/3s respectively [20]. $PtCl_2$ coated WO_3 NS deposited by hydrogen reduction of metal salts finds gas chromic application at 90 °C and becomes insignificant at temperatures below 90 °C [21]. Photocatalysis of nitrite ions under visible light performed for the Zn doped WO_3 nanostructure Enhanced Photo electrocatalysis illustrated highest degradation of nitrite ions for 2% Zn-WO_3 NS annealed at 400 °C [22]. A novel class of Tb-WO_3 NS fabricated using Radio Frequency (RF) magnetron sputtering technique shows porous nature of Tb^{+3} ions heavily doped exhibit strong blue, green and red emissions under UV excitation which suggest the possibility of using these luminescent films in white light emitting devices [23]. Enhanced photo electro-chemical properties of Bi doped WO_3 NS synthesized through hydrothermal method, displayed remarkable enhancement in photocurrent values from 0.401 mA cm^2 for undoped [24] WO_3 to 1.511 mA cm^2. Electrochemical sensor properties of PANI/Cellulose/ WO_3 NS used in acetone gas detection at room temperature [25]. In the present work, the author envisaged that doping appropriate amount of Gd (x = 2, 5, 10 wt. %) into WO_3 NS would enhance the electrochemical voltammetric performance of the NS. Gd-WO_3 NS synthesized by co-precipitation method and characterized by XRD, SEM, UV visible, XPS and studied their effects on the CV. The 10 wt. % of Gd-WO_3 NS showed enhanced CV [25 - 26], Electrochemical cyclic voltammetry performance tested in three electrodes assembly which involves galvanostatic charge discharge cycles [27 - 28], and cyclic voltammetry carried out in 6M KOH solution. The 10 wt. % of Gd doped WO_3 NS demonstrates specific capacitance of 450 F/g at a current density of 1 A/g. Gd doped tungsten oxide has a capacitance retention rate of 88.5% after 5000 cycles (cyclic stability). These characteristics indicate that Gd_xWO_3 nanomaterial showed promising electrode material for voltammetric determination for sensor applications their device fabrication and can be used as high-performance super capacitors.

EXPERIMENTATION

Materials and Methods

The required chemicals and reagents were purchased from Sigma Aldrich, Alfa Assar, S D Fine, *etc.* Crystallinity, doping of the Gd to WO $_3$ has been identified by X-ray diffractometer instrument made from Rigaku with 600 W power, 40 kV X-ray tube operating in 50 Hz/60Hz frequency using Cu Kα radiation ($\lambda = 1.5406$ A°) at wide-angle range (2θ value 5–80 °). Chemical compositional analysis of the bimetallic oxides measured by using X-ray photoelectron spectroscopy (XPS), AES instrument having auger electron spectroscopy with PHI 5000 versa probe II, Argon ion as well as C60 sputter Guns. UV-visible spectroscopy was performed using an instrument from specord 210 plus analytic jean with variable spectral resolution and cooled double beam detection. Morphology and grain size, distribution of the size of the nanoparticles was investigated by using an instrument from Nova Nano-SEM (5.0mm AWD with 35 ° take off angle) equipped with 5X to 1,000,000 X HR/UHR mode, high sensitivity, low kV directional backscattered detector (DBS), fitted with quadrant semiconductor diode. Galvanostatic charge and discharge cycles, cyclic stability, energy storage super capacity of the nanoparticles were investigated using cyclic voltammetry.

Synthesis of WO_3 Nanostructures (Step 1)

The WO_3 nanostructures were synthesized by using co-precipitation method. To begin with, sodium tungstate (Na_2WO_4. $2H_2O$) of 0.456 g was dissolved into distilled water in a round bottom flask, 0.100 g of sodium chloride, 0.234 g of urea added. The solution was stirred continuously for 1 h at room temperature. Further, 10 ml of concentrated hydrochloric acid was added and heated at 80 °C for 4 h. Reaction mixture was cooled, basified with ammonium hydroxide solution drop wise. The resulting precipitates were dried at 100-200 °C in an oven followed by calcination at 800 °C for 48 h to obtain fine WO_3 nanostructures.

REACTIONS

$$Na_2WO_4 + 2HCl \rightarrow H_2WO_4 + 2NaCl$$

$$H_2WO_4 \rightarrow WO_3 \text{ (crystal nucleus)} + H_2O$$

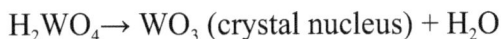

Gd Doping to WO_3 Nanostructures (Step 2)

Gadolinium (III) oxide (Gd_2O_3) with 1 wt. % to 10 wt. % taken in a 100 ml round bottom flask fitted with a reflux condenser. To this, add 10 mL of conc. hydrochloric acid, 0.150 g of cetrimonium bromide, 2 mL of triethanol amine were added, and the solution was heated at 80 °C, for 3 hours. The contents were slowly poured into the step 1 product stirring at room temperature in 25 mL of deionised water. The entire reaction mixture was heated at 80-90 °C for 3 hours. The precipitate was obtained by adding dropwise solution of ammonium hydroxide undercooling. The product was filtered, washed with an excess cold water, ethanol and was dried at 100-200 °C in an oven further calcinated at 700-800 °C to yield the Gd doped WO_3 NS. Similarly, 2.0, 6.0 and 10.0 wt. % Gd doped WO_3 nanostructures were prepared by adding an appropriate amount of gadolinium (III) oxide solution and following the above-mentioned procedure. Fig. (**1**) shows the flow chart of the synthetic pathway of Gd-WO_3 nanostructure.

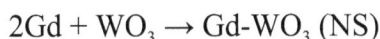

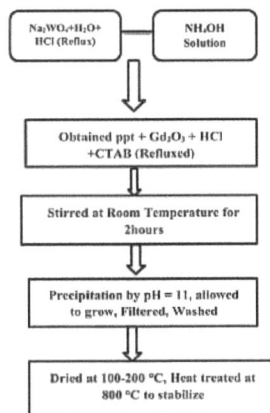

$$Gd_2O_3 + 2HCl \rightarrow 2Gd + H_2O + O_2 + Cl_2$$

$$2Gd + WO_3 \rightarrow Gd\text{-}WO_3 \text{ (NS)}$$

Fig. (1). Schematic flow chart of the synthetic pathways of Gd-WO_3 NS.

RESULTS AND DISCUSSIONS

X-ray Diffraction Pattern Studies

Crystallinity of the Gd-WO$_3$ NS was examined by XRD analysis with radiation from Cu-K alpha at λ =1.5406 Å. X-ray diffraction patterns were recorded over a range of 2θ from 20-80°. Diffraction peaks at planes corresponding to Gd$_2$O$_3$ with lattice constants a = 0.564 and b = 0.985, However NPs of Gd$_2$O$_3$ with orthorhombic crystal structure reported with the dramatic change in the decomposition behaviour under the reaction conditions. XRD lattice patterns confirm (Fig. **2**) the retention of size and shape of the NPs under applied conditions. By making use of the Scherer equation D = 0.9 λ/ β Cos θ where λ is the wavelength of the X-ray radiation (1.5418 Å) and β is the full width at half maximum (FWHM) of the peak at diffracted angle θ. A sharp, intense peak of WO$_3$ NPs orthorhombic crystal structure (agglomerated) corresponding to (101), (111), (301) and (311) at 2θ 27.6 °, 28.9 °, 42.6 °, 45.8 ° and the intense peak for Gd (201), (211) at 2θ 32.9 °, 33.8 ° confirms the configuration of the NPs which is well matched with JCPDS-72-0480 data. Comparison of the XRD patterns, the crystal structure of the Gd-WO$_3$ NPs is summarized in Table **1**. The crystal size of 48.2 nm is found for the Gd-WO$_3$ nanostructure using the Scherer method, and the sample showed flake-like crystalline phase. On the other hand, the incorporation of Gd decreases interplanar distance and the diffraction peaks shift towards a higher angle, enhancing the gadolinium concentration, as shown in Table **1**. Thus, a decrease in crystallite size is primarily due to the enhancement of Gd^{3+} in the pure WO$_3$ lattice. WO$_3$ NPs synthesized by microwave irradiation method showed improved crystallinity phases (monoclinic and orthorhombic) and all the reflection peaks could be assigned to the orthorhombic β-WO$_3$ (ICDD card no. 894477) [29] average crystalline size of γ-WO$_3$ and β-WO$_3$ 33 nm and 38 nm, respectively. NS of Gd doped WO$_3$ diffraction patterns confirms the formation of monoclinic phase with JCPDS file No 83-0951 data, and intensity of this triplet gradually decreases with an increase in Gd doping amount, which is attributed to successful replacement of W^{6+} ions by Gd $^{3+}$ ions [30]. Similar results have been observed in the case of Gd doped WO$_3$/TiO$_2$ NS confirms the monoclinic phase diffraction peaks were well matched with standard JCPDS card No. 21–1272 [31]. XRD patterns of Gd doped WO$_3$ diffraction peaks are highly crystalline in nature and crystal structure matching hexagonal wurtzite structure is according to JCPDS No.75–2187 [32]. Diffraction peaks shifted towards a higher angle with an increase in the Gd doping; the intensity of the diffraction peaks decreases. Silver doped WO$_3$ microstructure diffraction peaks match well with the JCPDS file No. 83-0950. Both pure and Ag doped WO$_3$ can be well indexed to the standard pattern of monoclinic WO$_3$ with the lattice constants a = 7.300, b = 7.538, c = 7.689 and b = 90.892 [33]. XRD pattern indicates the dispersion of Gd into WO$_3$

layered solid successfully with few extra peaks. Decrease in the interplanar distance due to decrease in the unit cell volume as a result of Gd impurity introduced into the WO_3 NS. Unit cell volume, Gd-WO_3 bond distance, length, bond energy, u parameters, crystal size (D) calculated by Debye-Scherer's formula $D = K\lambda/\beta\cos\theta$, where K is dimensionless shape factor, λ is X-ray wave length, β is line broadening at half maximum intensity (FWHM), θ is Bragg's angle.

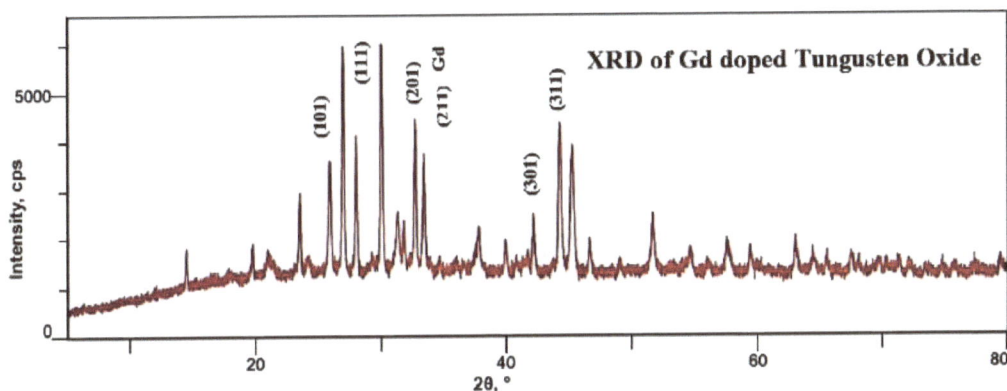

Fig. (2). XRD patterns of $Gd_x WO_3$ (x = 2, 5, 10 wt. %) NS.

Table 1. Crystalline phase, intense peaks and crystal size of the Gd-WO_3 NS.

Nanostructures	Crystalline Phase	XRD Peak Intensities	Prominent Peak Intensity Relative to (101) (201) (211)	Size of the Crystal (Scherer method)
Gd-WO₃ NS	Flake like	(101) (111)	(201) (211) (301) (311)	
		741 620	1104 44.2 49.2 98.5	48.2

Elemental Compositional Analysis of Gd-WO₃ NS

The purity of NPs and chemical composition are the two important parameters to be investigated before NPs deployed for experimentation. Elemental composit- ional analysis is one of the key experiments to understand the chemical constituents of metal or non-metal. NPs of Gd-WO_3 were obtained by co- precipitation method and subsequent reduction during the conversion from bulk to nanoparticles (NPs) solvent and solution react to form MOS NPs (metal oxide semiconductor nanoparticles). Gd-WO_3 NPs populated in the crystal structure and their composition can be understood with the elemental compositional analysis.

Elemental compositional analysis is one of the key experiments to understand the chemical constituents. Table **2** represents the total elemental percentage composition of various metals, non-metals and metal oxide for the different samples of pure and doped Gd-WO$_3$ chemical constituents. Bare Gd$_2$O$_3$ NPs clearly show only Gd$_2$O$_3$ with 96.85% purity; pure WO$_3$ exhibited the purity of 95.7% whereas Gd doped WO$_3$ portrayed 97.16% purity out of which the WO$_3$ of 45.94 and Gd$_2$O$_3$ of 51.22 with trace quantity of Na, O and K.

Table 2. Elemental compositional analysis of bare Gd$_2$O$_3$, WO$_3$ and Gd-WO$_3$ NS.

Samples	WO$_3$	Gd$_2$O$_3$	Na	N	O	Co	K	Si	C	P	Purity
Bare Gd$_2$O$_3$ NPs	0.05	96.8	1.12	-	1.7	-	0.2	-	-	-	96.85
Bare WO$_3$ NPs	95.7	0.08	1.11	-	1.57	-	0.11	-	-	-	95.7
Gd-WO$_3$ NS	45.94	51.22	0.48	-	1.64	-	0.21	-	-	-	97.16

SEM and Cross-sectional SEM Morphology

Surface morphology (act as a sensing membrane) of the synthesized Gd-WO$_3$ NPs has been investigated by SEM analysis. SEM images of Gd-WO$_3$ NS with different magnifications are as shown in Fig. (**3A, 3B, 3C, 3D**). As a result of well dispersion of the nanoparticle flakes like morphology which are grown vertically agglomerated structure with proper orientation, distribution of Gd$_2$O$_3$ and WO$_3$ NPs has been observed. The high density of NPs of Gd-WO$_3$ have a size of ~ 77 nm, grain diameter of ~ 89 nm, a distance of ~ 97 nm and ~ 34 μm in length. Morphology and distribution of the Gd-WO$_3$ NPs depend on different preparation approaches such as co precipitation, hydrothermal, microwave irradiation, *etc.*; in most cases, NS does not form a recognisable structure with agglomeration. Crystal morphology of the 2 wt. % Gd doped WO$_3$ showed shows the flower like and hierarchical design structures varying in their shape with some agglomeration and size of nanostructured material. The particle size for it lies in the range of ~ 60 nm which is a good covenant with the present investigation results [34]. In the case of ZnO decorated with WO$_3$ NPs exhibited long agglomerated structure with an average particle size of 89 nm [35]. Similar reports have been demonstrated the triclinic structure of the NPs of WO$_3$ deployed for acetone gas sensing characteristics; their SEM and TEM characterizations revealed highly porous and crystalline with a triclinic structure of WO$_3$ nano fibres [36]. It is evident from the CdO-WO$_3$ NS rod like structure. The size of the rod decreasing with the increase in concentration of CdO in WO$_3$ NPs up to specific limit nanorods gave less band gap due to the small size of the road, size of the nanorod in CdO: WO$_3$ NS varies from 25 to 55 nm [37]. Overall, from the top view of the SEM images, it can be concluded that orthorhombic Gd-WO$_3$ NS forms with more agglomeration of the NPs in the nanomatrix.

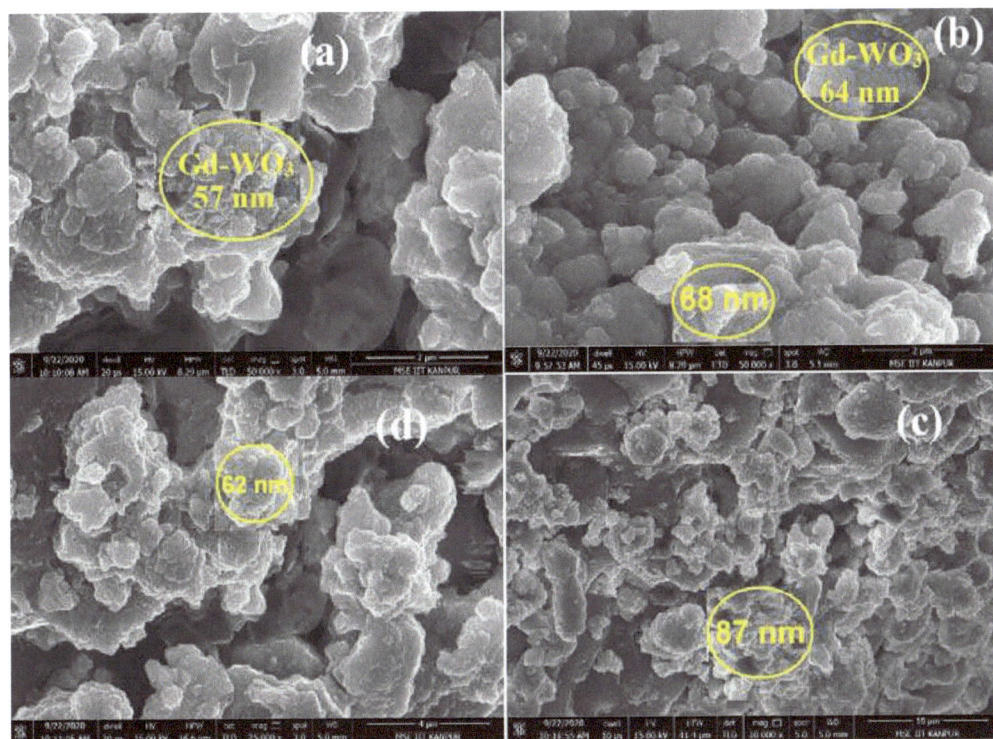

Fig. (3). SEM morphology of (a) 2 wt. % Gd-WO$_3$; (b) 5 wt. % Gd-WO$_3$; (c) 8 wt. % Gd-WO$_3$ and (d) 10 wt. % Gd-WO$_3$ NS.

UV-Visible Spectral Analysis

The Gd-WO$_3$ nanostructures were recorded for UV-Visible spectra (optical absorption) in powdered form with the aim of examining nanoparticles optical behaviour such as absorptivity, band gap and nanoparticles interaction with photons. Gd-WO$_3$ NPs showed near UV absorptivity when subjected to irradiation from the radiation source of 100 nm to 1000 nm. UV-Visible spectrum of Gd-WO$_3$NPs showed absorptivity maxima from 280 nm to 340 nm as shown in Fig. (**4**). Increase in the UV absorptivity is due to the transfer of Gd^{+3} ions into the active layer of WO$_3$. WO$_3$ NPs synthesized by pulsed laser ablation in water showed 335 to 387 nm near UV absorptivity due to factors like low zeta potential, agglomerations and aging of the nanoparticles. Reduction in the band gap of the WO$_3$ NPs increases the absorption coefficients causing more generation of the electron-hole pair [38]. In reports of pristine and cerium doped WO$_3$ NPs, Ce doping decreases band gap and creates localized sites. Absorption maxima were observed in between 340 nm to 380 nm upon an increase in the Ce concentration from 3% to 10% [39]. Absorptive wavelengths, band gap of Gd-WO$_3$ NPs, crystallinity are summarized in Table **3** and calculated using the formula hγ = A (hγ - E$_g$)^2 where, A absorptivity constant, γ frequency of the source radiation and E$_g$ is the optical gap of nanomaterials.

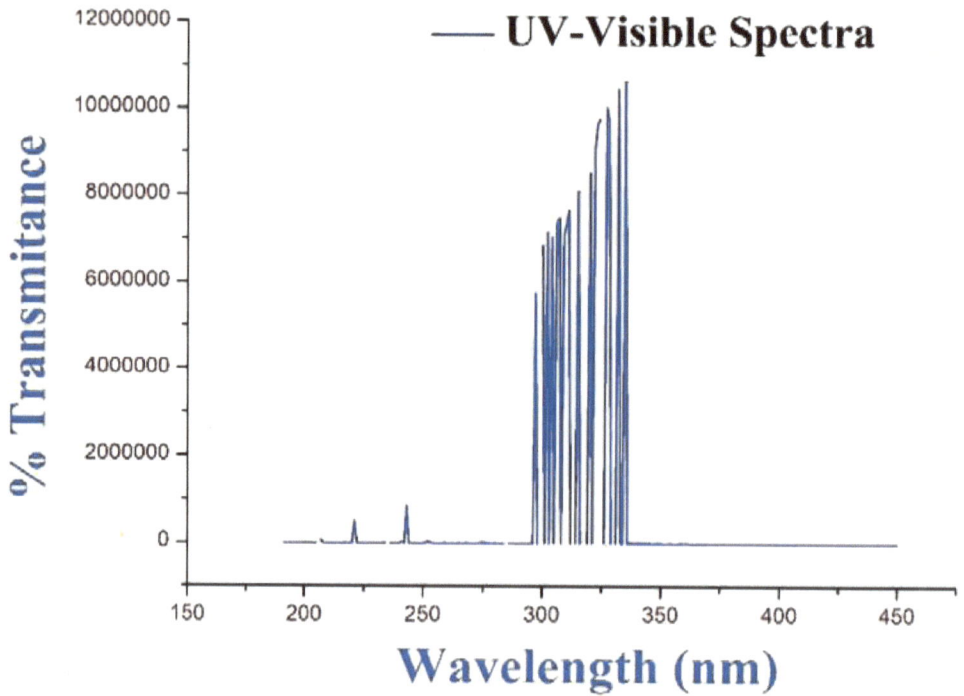

Fig. (4). UV-Visible spectrum of 10 wt. % Gd-WO$_3$ NS.

Table 3. Comparative UV-Visible spectrophotometer data of the bare Gd$_2$O$_3$, WO$_3$ and Gd-WO$_3$ NS.

Nano Structures	$\lambda_{Max 1}$ (nm)	$\lambda_{Max 2}$ (nm)	λ_{Min} (nm)	λ_{Dip} (nm)	Absorptivity Region
Bare Gd$_2$O$_3$ NPs	340	-	240	-	Ultraviolet
Bare WO$_3$ NPs	310	-	230	-	Ultraviolet
Gd-WO$_3$ NS	280	340	220	-	Ultraviolet

X-ray Photoelectron Spectroscopic (XPS) Analysis

Elemental composition, chemical states of synthesized Gd-WO$_3$ NS has been studied using XPS analysis and the overall results of the XPS are depicted in Fig. (**5**). The respective binding energies (B.E) of Gd-WO$_3$ NPs summarized in Table

4. Hybrid nanoparticles intensity of Gd 3d orbital peak increases with the upsurge of doping concentration and 3d orbital split into Gd 3d $_{3/2}$ and Gd 3d $_{5/2}$ with binding energies of 1297 eV and 1187 eV, respectively [40]. However, XPS investigation of vacuum annealed WO_3 NS showed W4f binding energies of 110 eV and O1s shift from 530.4 eV to 538.7 eV as reported in the literature [41]. XPS spectrum of Gd doped NPs, especially Gd 3d and Gd 4d orbital exposed binding energies of 1185 eV and 143.5 eV with singlet O1s occurred at 530 eV [42]. XPS survey scan of the ZnS: Gd (3 at %) sample Gd peaks were located at 141.05 for 4d $_{3/2}$, showing that the valance state of the Gd ion in the NPs is +3 and that Gd had been successfully doped into the ZnS NPs without any impurity phases [43] indicating that Gd dopant are at their highest oxidation states (Gd3þ), corresponding to Gd_2O_3 in the films.

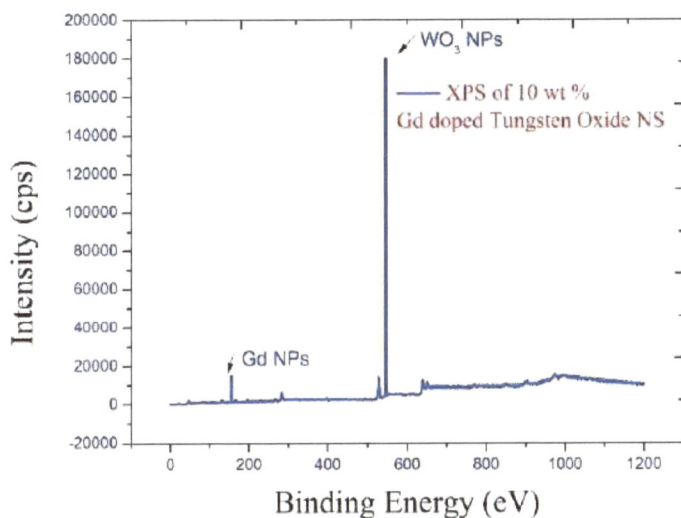

Fig. (5). XPS survey spectrum of Gd-WO_3 (Gd 4d $_{3/2}$, Gd 4d $_{5/2}$ and W 4f orbitals splitting) NS.

Table 4. Binding energies of Gd 3d $_{5/2}$, Gd 3d $_{3/2}$, W 4f, O1s together with doublet separation.

Peak Assignment	Binding Energy (eV)	Separation of Doublet (eV)
C1s	238.7	-
Gd 3d $_{3/2}$	1297	-
Gd 3d $_{5/2}$	1187	10.7
W 4f	110	2.5
O1s	530.4	-

CYCLIC VOLTAMMETRY EXPERIMENTATION FOR ADVANCED ENERGY STORAGE APPLICATIONS

Sample Preparation for Gd-WO$_3$ NS (x = 2, 5, 10 wt. %)

Cyclic voltammetry involves the application of potentials across the electrodes in both directions and find the total current gain. Redox processes involve the liberation of cation and anions into the solvent and their quantitative voltammetric determination can be done using three electrode systems. Samples or nano-structures need to be tested for their solubility in organic solvents like ethanol, methanol, acetonitrile, *etc.*, in order to obtain a homogeneous solution. High solubility, conductivity can be achieved by dissolving ammonium salts in water as electrolytes; KOH or NaOH solutions of different molarity can also be used as electrolytes. Electrolytes must have greater stability towards redox reactions taking place in the voltammetric system. Some of the commonly used solvents like tetra butyl ammonium (0.1 to 0.3 M) dissolved DMF, DCM, acetonitrile, Lithium salts like LiClO$_4$, TBAB in DMF widely used. In the present study, 6M KOH solution was used as an electrolyte. Ag/AgCl electrode as a reference, Pt (2mm thickness) and 2 mm glassy carbon electrode as working electrodes. Voltammetric experiments were performed under inert atmosphere (Ar) and electron migration happens from the electrode surface to the solvent to compensate for the charge-discharge processes. Different solvents were used, and cyclic voltammetric results were compared and the best suited solvent was used for the analysis of the experiments. In the present investigation, 6M KOH find better solvents as compared with other solvent systems.

Galvanostatic Voltammetric Charging and Discharging Studies (GCD)

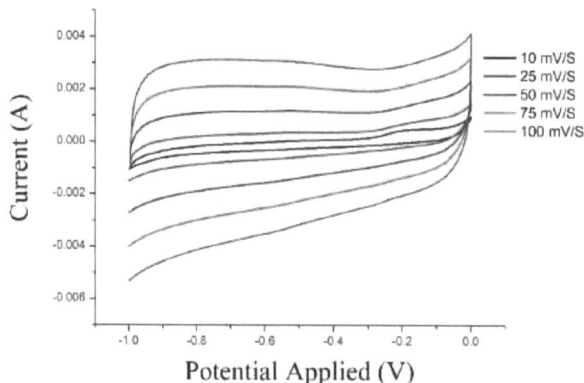

Fig. (6). Cyclic voltammogram of Gd-WO$_3$ NS in three electrode system at the scan rate of 10, 25, 50, 75, 100 mV/s in aqueous electrolyte system of 6M KOH.

Electrochemical Voltammetric Determination of Specific Capacitance

To explore the electrochemical voltammetric behaviour of Gd-WO$_3$ NS, three electrode systems were used. Three electrode systems consist of platinum as the counter electrode, Ag/AgCl (3M NaCl) as reference electrode and core-shell Gd-WO$_3$ NS coated on a glass substrate (uniform surface area of 1 cm^2 approximately 1.5 mg of the sample). We used 6M KOH solution as an electrolyte. CV (Fig. **6**) and galvanostatic charge-discharge (GCD) cycles performed with voltage range of 0-2 V *Vs.* Ag/AgCl with a scan rate of 10 mV/s to 100 mV/s of the sample were under investigation. GCD cycles (Fig. **7**) were performed under voltage 0-1 V with a constant current of 0.1 A/g to 1 A/g. Electrochemical voltammetric impedance analysis was carried out using electrochemical impedance analyser (EIS); the voltammetric investigation was carried out with voltage 2V. Supercapacitor cell was built by using C-coated Al foil electrodes with filter paper in between containing ~ 70 wt. % of Gd-WO$_3$ NPs, 10 wt. % of PVDF (polyvinylidene fluoride), 20 wt. % of acetylene black and 6M Aqueous KOH solution used as an electrolyte. Current densities were calculated using a mass of active materials.

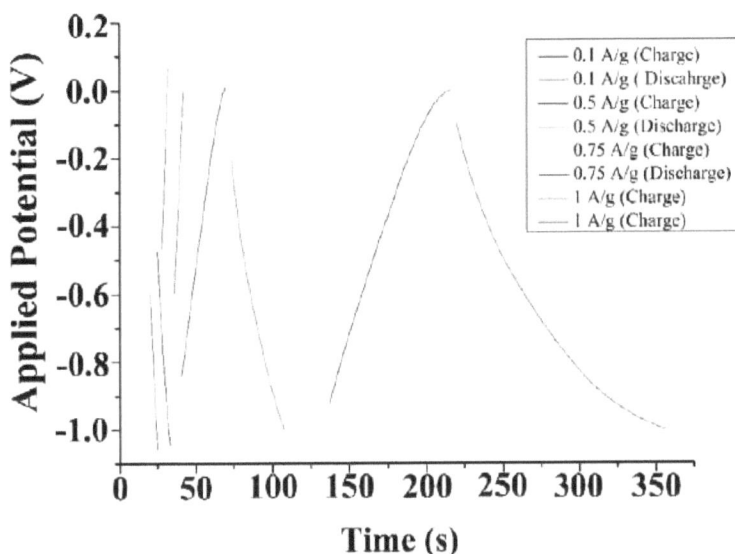

Fig. (7). Cyclic voltammetric determination of Galvanostatic charge discharge (GCD) cycles of Gd-WO$_3$ NS recorded in three electrode systems with different current densities. (0.1, 0.5, 0.75, 1 A/g).

High energy storage supercapacitor properties of Gd-WO$_3$ NS evaluated using 6M KOH solution and CV curves recorded at a scan rate of 10, 25, 50 75 and 100 mV/s and rectangular CV curves of Gd-WO$_3$ NS acting as sensing interface between solvent and solute which indicates good electrochemical nature as shown in Fig. (**6**). Strong redox peak attributed to flake like morphology of Gd-WO$_3$ covered with a layer of WO$_2$/WO$_3$. Upon insertion of alkali cations or protons, current density and number of active species (sensing cations and anions) on WO$_3$ NS increases [44 - 47]. High current density of the Gd-WO$_3$ NS corresponds to GCD and obtained voltammetric curves compared with bare Gd$_2$O$_3$ and WO$_3$ NS [48, 49]. Applied potential varied in between 0 to 1V and current densities from 0.1 A/g to 1 A/g. Table **5** lists the specific capacitance and constant phase element (CPE) values of 10 wt. % Gd-WO$_3$. At 0.1 A/g in 2 M KOH electrolyte, the calculated specific capacitance of 450 F/g for the 10 wt. % of Gd-WO$_3$ NPs and calculated using the formula of C= (iΔt)/ (ΔVm) where, C-is the specific capacitance obtained from the voltage-time response, i is the constant current, Δt is the discharge time, ΔV is the potential range, and m is the mass of the sample.

Table 5. Specific Capacitance, Rs, Rp, CPE values of Gd-WO$_3$ NS.

Material	Specific Capacitance (F/g) at 0.1A/g	R$_s$ (Ω)	R$_p$ (Ω)	CPE (mF)	C(mF)	W(m Ω)
Gd-WO$_3$	450.0	0.948	1.65	8.02	36.3	0.514

Thus, the synergistic effects in the crystal morphology of the Gd-WO$_3$ NS acting as a sensing interface structure resulted in high energy storage material used in supercapacitor applications. Specific capacitance 450.0 F/g at 0.1 A/g for Gd-WO$_3$ NS obtained from the cyclic voltammetric investigation was compared with previous results of Gd doped NS (Table **6**). Ohmic voltage drops in the discharge curves for bare WO$_3$ NS stabilized due to the migration of Gd^{+3} ions into WO$_3$ NS. Cycle voltammetric performance, stability, retention Gd-WO$_3$ NS evaluated based on observed capacitance over charge-discharge cycles up to 5000 as shown in Fig. (**7**). Retention was 88.5% after 5000 cycles; on the other hand, WO$_3$ electrode exhibits fast decay and less stability [50 - 52]. Based on charge balance theory, CV curves of most of the NS exhibit specific capacitance with an almost rectangular shape (Voltammetric curves) without an apparent redox peak even at the increased operating voltage up to 2.0 V. In the present investigation cyclic voltammetric (CV) experiments were performed with a scan rate of 10, 25, 50, 75 and 100 mV/s and CV curve maintains rectangular shape even at scan rate of 100 mV/s exhibiting good charge-discharge properties and rate capability. The relationship between the phase angles (θ) with the time elapsed during the charge discharge cycles of Gd-WO$_3$ NS is shown in Fig. (**9**). CV curve possessing enlarged voltage window which improves energy and power density. The energy density (E) and power density (P), calculated using the following equation E = ½ (CV)^2 and P = E/t, where C is the specific capacitance, V working voltage of the supercapacitor, and t is the elapsed times to discharge.

Table 6. Comparison of literature reports of electrode materials, their preparation methods, cycle stability and specific capacitance values.

Electrode Material	Preparation Method	Cyclic Stability	Capacitance	References
Cu-WO$_3$ NPs	Coprecipitation	-	1.04 V	[54]
Ce-MnO$_2$	Hydrothermal	-	147.25 F/g	[55]
Ni-WO$_3$/NiWO$_4$	Microwave Irradiation	-	171.2 F/g	[56]
Co-WO$_3$/CoWO$_4$	Microwave Irradiation	-	45 F/g	[57]
MWCNTs-WO$_3$	Hydrothermal	10000	429.6 F/g	[58]
Graphene-WO$_3$	Solvothermal	4000	465F/g	[59]
Gd-WO$_3$	Co precipitation	5000	450.0	This study

Relation Between Phase Angle and Time in Voltammetric Properties of Gd-WO$_3$ NS

Electrochemical impedance spectroscopy (EIS) (Fig. **8**) was performed to understand the electrochemical voltammetric determination and their performance of Gd-WO$_3$ NS. Gd-WO$_3$ NS presents Nyquist plots with smaller semicircle which is correlated to the charge transfer of the Gd^{+3} ions and O^{2-} on the surface of the WO$_3$ NS acting as sensing interface, which intern confirms the dominant electrochemical reactions. The phase angle initially increases and after the elapse of 150 seconds it decreases considerably, indicating Gd^{+3} ions from the surface of the WO$_3$ transferred effectively. The plot of phase angle *versus* time elapsed is shown in Fig. (**9**). Primarily, the phase angle varies slowly and as the electro-chemical reaction time increases phase angle also increases, indicating the reaction of O^{2-} ions from the surface of the WO$_3$ NS to the Gd [53] and electrons transferred from the surface. The magnitude of Z'-intercept correlates to the equivalent series resistance Rs (Fig. **10**), which includes the electrical resistance of electrodes. Observed resistance values are 1.578 Ω for Gd$_2$O$_3$ NPs, 1.0527 Ω for WO$_3$ NPs and 10 wt. % of Gd-WO$_3$ NPs has 0.948 Ω, respectively. Low charge transfer and electrical resistance of the nanostructure enable the formation of large capacitance with a good rate of discharge and recharge cycles.

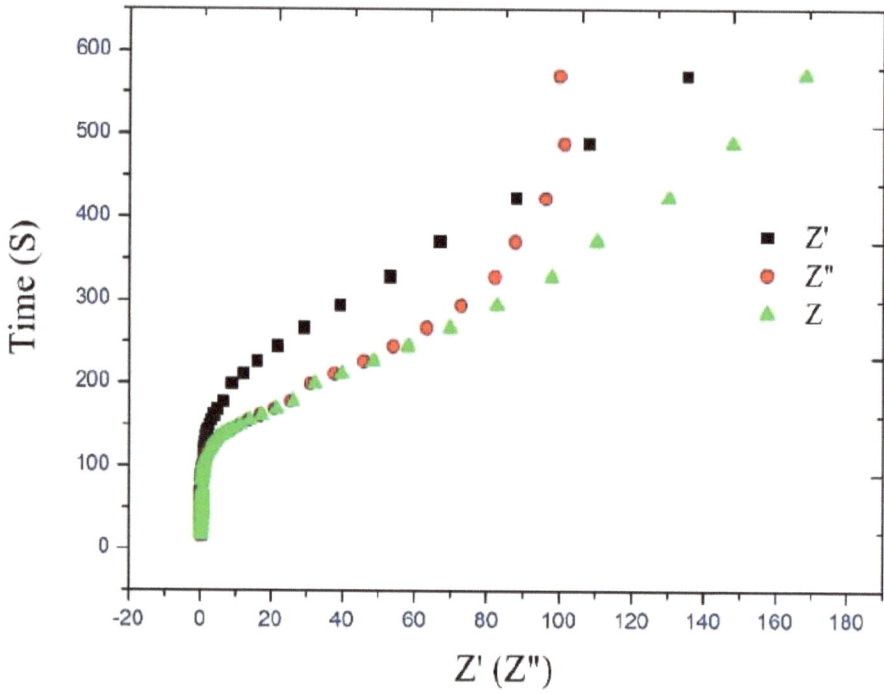

Fig. (8). Nequist analysis plots of all three electrodes of Gd-WO$_3$ NS.

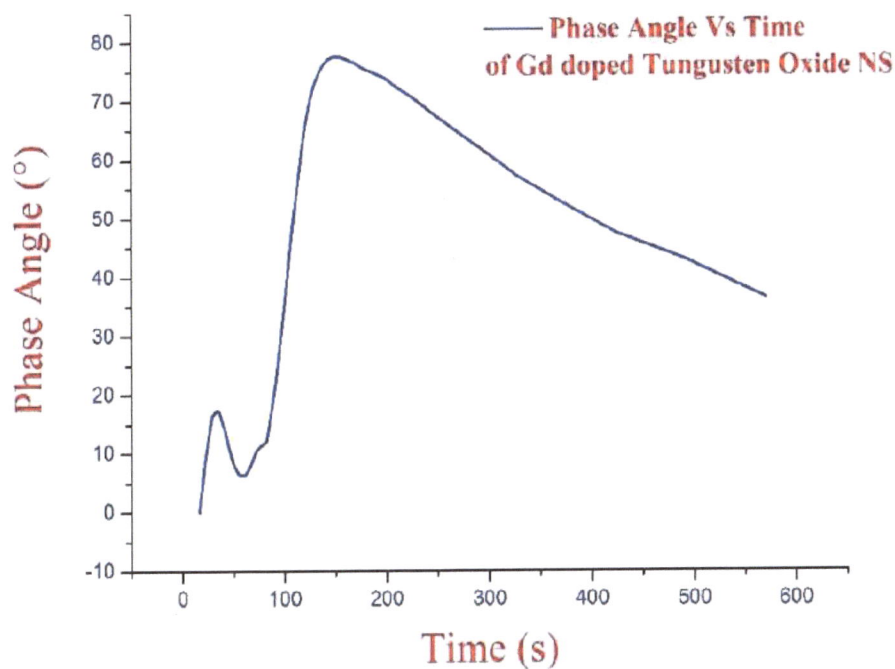

Fig. (9). Plot of phase angle *versus* time (s) of Gd-WO₃ NS.

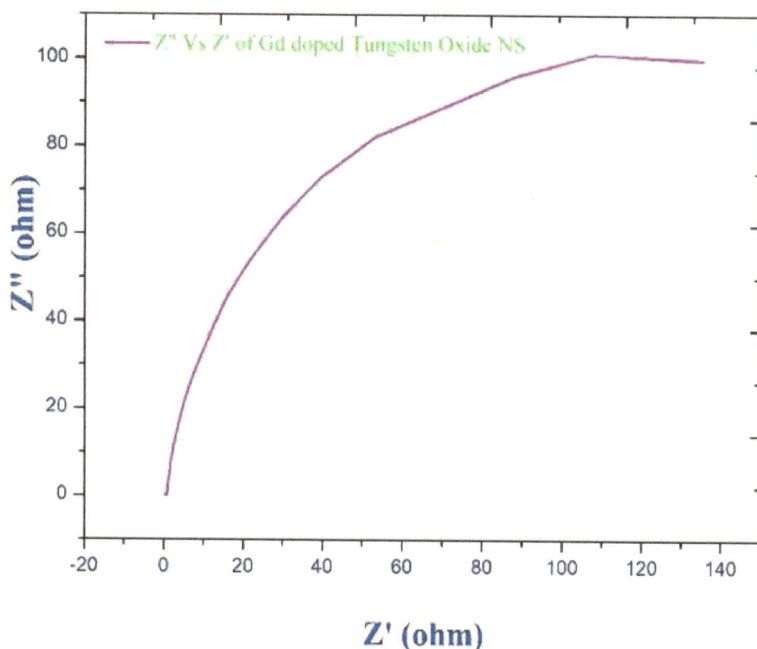

Fig. (10). Relation between observed resistivity (Z") *Vs.* Z' of Gd-WO$_3$ NS.

High energy storage supercapacitor properties were investigated by using a three-electrode system at room temperature. Due to the changes in the structure of nanomaterial after doping (act as sensing interface), energy storage properties, resistivity, surface conductivity in the presence of 6M KOH solution varies. Sensing mechanisms of Gd-WO$_3$ NS studied by alkaline electrolyte KOH (6M) and variation in the resistivity of nanostructures compared with the help of Nyquist plot as shown in Fig. (**8**). The relation between observed resistivity (Z") *Vs.* Z' of Gd-WO$_3$ NS observed during 5000 cycles of voltammetric performance is shown in Fig. (**10**). N-type of the 10 wt. % of the Gd doped WO$_3$ semiconductor electrode (having highest oxidation potential) reveals a larger semicircle with a total resistance of ~ 35 Ω in comparison with pure WO$_3$ NS of ~ 22 Ω, the total resistance was calculated by fitting with equivalent circuit analysis of the impedance data, which intern demonstrate the high rate of transfer of carrier charges on the surface of the electrodes. This phenomenon is due to the formation of heterojunction for the doped and undoped surfaces and the total resistance of the material decreases to 0.948 Ω while the pure electrode was lowered ~ 0.225 (Z") in the case of an isolated orthorhombic form of the WO$_3$ NS. This can be attributed to the resistance of Gd-doped WO$_3$ NS, which is approximately 4 times higher than the pure WO$_3$ NS. The EIS analysis demonstrated a better sensing response to the Gd doped WO_3 NS for high-energy storage supercapacitor applications.

CONCLUSION

The economical, low-cost and controlled synthesis of Gd-WO_3 NS was experimentally investigated for the high energy storage supercapacitor applications. Nanostructures were characterized by XRD, SEM, UV-Visible and XPS spectroscopic techniques. XRD patterns indicate orthorhombic crystal structure with fewer impurity peaks and JCPDS file matching to 72-0480 data. Flake-like structural morphology of the Gd-WO_3 NS revealed the formation of agglomerations inside the doped surface acting as sensing structural unit with average nanoparticles size was ~ 89nm and grain boundary distance of ~ 110 nm. Optical absorptivity exhibited maximum absorptivity observed between 280 nm to 340 nm (UV-region) and formation of UV-shoulders experiential at 310 nm and 330 nm. Chemical composition, formation of the oxides of nanostructures were investigated by using binding energies of Gd $3d_{3/2}$ 1297 eV, Gd $3d_{5/2}$ at 1187 eV for Gd_2O_3 NS and WO 4f at 110 eV. CV performed in between 0.1 V to 2 V and current 0.1 A/g to 2A/g achieved the highest capacitance of 450.0 F/g at 0.1 A/g, long term cyclic stability of the Gd-WO_3 NS determined by charge-discharge voltammetric experiments. After 5,000 cycles nanostructures retain 88.5%, thereby demonstrating excellent cyclic stability and reversibility. Nanostructure of Gd-WO_3 experiments with conductivity, resistivity, EIS, Rs and lower Rp values accomplish the requirement for a high-performance supercapacitor. Thus, doping rare earth metals is an effective strategy to enhance the energy storage and super capacitive properties of the hybrid NS. The results of energy storage supercapacitor measurements indicate that WO_3 modified by doping with Gd exhibits excellent features, like long-time stability, high reproducibility and low detection limit. Thus, we can conclude that the combination of doping as well as the passivation layer leads to the improvement in the energy storage performance of WO_3 vertical array films.

AUTHORS CONTRIBUTIONS

Dr. Vinayak Adimule contributed to synthesis, manuscript preparation, Dr. Yallur BC was involved in the characterization of the samples, Dr. D bowmik contributed to sample characterization and partly by preparation of the manuscript. Mr. Santosh Nandi was involved in the preparation of the manuscript.

CONSENT FOR PUBLICATION

Not applicable.

CONFLICT OF INTEREST

The authors declare no conflict of interest, financial or otherwise.

ACKNOWLEDGEMENTS

Authors are thankful to IIT Kanpur for SEM and XPS analysis, Rani-chennamma University for XRD and Elemental compositional investigations, UV-Visible and Elemental compositional experiments carried out in MSRIT, Bangalore and BMS R&D for CV analysis.

REFERENCES

[1]   D. Chen, Q. Wang, W.R. Rongming, and S.G. Guozhen, "Ternary oxide nanostructured materials for supercapacitors: a review", *J. Mater. Chem. A Mater. Energy Sustain.,* vol. 3, p. 10158, 2015.
[http://dx.doi.org/10.1039/C4TA06923D]

[2]   A. Burke, "Ultracapacitors: Why, How, and Where Is the Technology", *J. Power Sources,* vol. 91, p. 37, 2000.
[http://dx.doi.org/10.1016/S0378-7753(00)00485-7]

[3]   N. Hareesha, and J.G. Manjunatha, "A simple and low-cost poly (dl-phenylalanine) modified carbon sensor for the improved electrochemical analysis of Riboflavin. J. Science", *Advanced Materials and Devices.,* vol. 5, pp. 502-511, 2020.
[http://dx.doi.org/10.1016/j.jsamd.2020.08.005]

[4]   V.C. Lokhande, A.C. Lokhande, C.D. Lokhande, J.H. Kim, and T.S. Ji, "Supercapacitive composite metal oxide electrodes formed with carbon, metal oxides and conducting polymers", *J. Alloys Compd.,* vol. 682, pp. 381-403, 2016.
[http://dx.doi.org/10.1016/j.jallcom.2016.04.242]

[5]   C. Santato, M. Odziemkowski, M. Ulmann, and J. Augustynski, "Crystallographically oriented mesoporous WO3 films: synthesis, characterization, and applications", *J. Am. Chem. Soc.,* vol. 123, no. 43, pp. 10639-10649, 2001.
[http://dx.doi.org/10.1021/ja011315x] [PMID: 11673995]

[6]   S.H. Baeck, K.S. Choi, T.F. Jaramillo, G.D. Stucky, and E.W. McFarland, "Enhancement of photocatalytic and electrochromic properties of electrochemically fabricated mesoporous WO3 thin films", *Adv. Mater.,* vol. 15, p. 1269, 2003.
[http://dx.doi.org/10.1002/adma.200304669]

[7]   A. Srinivasan, and M. Miyauchi, "Chemically stable WO3 based thin-film for visible-light induced oxidation and superhydrophilicity", *J. Phys. Chem. C,* vol. 116, pp. 15421-15426, 2012.
[http://dx.doi.org/10.1021/jp303472p]

[8]   P.A. Pushpanjali, and J.G. Manjunatha, "Development of Polymer Modified Electrochemical Sensor for the Determination of Alizarin Carmine in the Presence of Tartrazine", *Electroanalysis,* vol. 32, p. 2474, 2020.
[http://dx.doi.org/10.1002/elan.202060181]

[9]   S.P. Nambudumada, G.M. Jamballi, and C. Raril, "Electrocatalytic Analysis of Dopamine, Uric Acid and Ascorbic Acid at Poly (Adenine) Modified Carbon Nanotube Paste Electrode: A Cyclic Voltametric Study", *Anal. Bioanal. Electrochem.,* vol. 6, pp. 742-756, 2019.

[10]   Q. Mi, A. Zhanaidarova, B.S. Brunschwig, H.B. Gray, and N.S. Lewis, "A quantitative assessment of the competition between water and anion oxidation at WO3 photoanodes in acidic aqueous electrolytes", *Energy Environ. Sci.,* vol. 5, p. 5694, 2012.
[http://dx.doi.org/10.1039/c2ee02929d]

[11] N. Hareesha, and J.G. Manjunatha, "Fast and enhanced electrochemical sensing of dopamine at cost-effective poly(DL-phenylalanine) based graphite electrode", *J. Electroanal. Chem. (Lausanne),* vol. 878, p. 114533, 2020.
[http://dx.doi.org/10.1016/j.jelechem.2020.114533]

[12] C. Raril, J.G. Manjunatha, K.R. Doddarasinakere, S. Fattepur, S. Gurumallappa, and L. Nanjundaswamy, "Validated Electrochemical Method for Simultaneous Resolution of Tyrosine, Uric Acid, and Ascorbic Acid at Polymer Modified Nano-Composite Paste Electrode", *Surg. Eng. Appl. Electrochem.,* vol. 56, pp. 415-426, 2020.
[http://dx.doi.org/10.3103/S1068375520040134]

[13] P. Yang, P. Sun, L. Du, Z. Liang, W. Xie, and X. Cai, "Quantitative analysis of charge storage process of tungsten oxide that combines pseudocapacitive and electrochromic properties", *J. Phys. Chem. C,* vol. 119, p. 16483, 2015.
[http://dx.doi.org/10.1021/acs.jpcc.5b04707]

[14] M.G. Hutchins, O. Abu-Alkhair, M.M. El-Nahass, and K. Abd El-Hady, "Structural and optical characterisation of thermally evaporated tungsten trioxide (WO3) thin films", *Mater. Chem. Phys.,* vol. 98, p. 401, 2006.
[http://dx.doi.org/10.1016/j.matchemphys.2005.09.052]

[15] M. Deepa, A.K. Srivastava, R. Sharma, S.N. Govind, and S.M. Shivaprasad, "Microstructural and electrochromic properties of tungsten oxide thin films produced by surfactant mediated electrodeposition", *Appl. Surf. Sci.,* vol. 254, p. 2342, 2008.
[http://dx.doi.org/10.1016/j.apsusc.2007.09.035]

[16] R. Huang, Y. Shen, L. Zhao, and M. Yan, "Effect of hydrothermal temperature on structure and photochromic properties of WO3 powder", *Adv. Powder Technol.,* vol. 23, pp. 211-214, 2012.
[http://dx.doi.org/10.1016/j.apt.2011.02.009]

[17] D. Nagy, I.M. Szilagyi, and X. Fan, "Effect of the morphology and phases of WO3 nanocrystals on their photocatalytic efficiency", *RSC Advances,* vol. 6, pp. 33743-33754, 2016.
[http://dx.doi.org/10.1039/C5RA26582G]

[18] H. Kalhori, M. Ranjbar, H. Salamati, and J.M.D. Coey, "Flower-like nanostructures of WO3: Fabrication and characterization of their in-liquid gasochromic effect", *Sensors Actuat. B.,* vol. 225, pp. 535-543, 2016.
[http://dx.doi.org/10.1016/j.snb.2015.11.044]

[19] V.S. Kavitha, S. Suresh, S.R. Chalana, and V.P. Mahadevan Pillai, "Luminescent Ta doped WO3 thin films as a probable candidate for excitonic solar cell applications", *Appl. Surf. Sci.,* vol. 466, pp. 289-300, 2019.
[http://dx.doi.org/10.1016/j.apsusc.2018.10.007]

[20] H. Yu, J. Li, Y. Tian, and Z. Li, "Gas Sensing and Electrochemical Behaviours of Ag-doped 3D Spherical WO3 Assembled by Nanostrips to Formaldehyde", *Int. J. Electrochem. Sci.,* vol. 13, pp. 9281-9291, 2018.
[http://dx.doi.org/10.20964/2018.10.52]

[21] N.T. Garavand, S.M. Mahdavi, and A.I. Zad, "Pt and Pd as catalyst deposited by hydrogen reduction of metal salts on WO3 films for gasochromic application", *Appl. Surf. Sci.,* vol. 273, pp. 261-267, 2013.
[http://dx.doi.org/10.1016/j.apsusc.2013.02.027]

[22] X.F. Cheng, W.H. Leng, D.P. Liu, J.Q. Zhang, and C.N. Cao, "Enhanced photoelectrocatalytic performance of Zn-doped WO(3) photocatalysts for nitrite ions degradation under visible light", *Chemosphere,* vol. 68, no. 10, pp. 1976-1984, 2007.
[http://dx.doi.org/10.1016/j.chemosphere.2007.02.010] [PMID: 17482660]

[23] V.S. Kavitha, R.R. Krishnan, R.S. Sreedharan, K. Suresh, C.K. Jayasankar, and V.P. Mahadevan Pillai, "Tb3þ-doped WO3 thin films: A potential candidate in white light emitting devices", *J. Alloys*

Compd., vol. 788, pp. 429-445, 2019.
[http://dx.doi.org/10.1016/j.jallcom.2019.02.222]

[24] S.S. Kalanur, I. Yoo, K. Eom, and H. Seo, "Enhancement of photoelectrochemical water splitting response of WO3 by Means of Bi doping", *J. Catal.,* vol. 357, pp. 127-137, 2018.
[http://dx.doi.org/10.1016/j.jcat.2017.11.012]

[25] E. Aparicio-Martinez, V. Osuna, R.B. Dominguez, A. Márquez-Lucero, A. Zaragoza-Contreras, and A. Vega-Rios, "Room Temperature Detection of Acetone by a PANI/Cellulose/WO3 Electrochemical Sensor", *J. Nanomater.,* vol. •••, p. 6519694, 2018.

[26] H. Nagarajappa, and J.G. Manjunatha, "Surfactant and polymer layered carbon composite electrochemical sensor for the analysis of estriol with ciprofloxacin", *Mater. Res. Innov.,* vol. 24, pp. 349-362, 2020.
[http://dx.doi.org/10.1080/14328917.2019.1684657]

[27] G. Tigari, J. G. Manjunatha, C. Raril, and N. Hareesha, "Determination of Riboflavin at Carbon Nanotube Paste Electrodes Modified with an Anionic Surfactant. ", *Chemistry selects,* vol. 4, pp. 2168-2173, 2019.
[http://dx.doi.org/10.1002/slct.201803191]

[28] C. Raril, and J.G. Manjunatha, "A simple approach for the electrochemical determination of vanillin at ionic surfactant modified graphene paste electrode", *Microchem. J.,* vol. 154, p. 104575, 2020.
[http://dx.doi.org/10.1016/j.microc.2019.104575]

[29] A.C. Anitha, N. Lavanya, K. Asokan, and C. Sekar, "WO3 nanoparticles based direct electrochemical dopamine sensor in the presence of ascorbic acid", *Electrochim. Acta,* vol. 167, pp. 294-302, 2015.
[http://dx.doi.org/10.1016/j.electacta.2015.03.160]

[30] J. Kaur, K. Anand, A. Kaur, and R.C. Singh, "Sensitive and selective acetone sensor based on Gd doped WO3/reduced graphene oxide nanocomposite", *Sens. Actuators B Chem.,* vol. 258, pp. 1022-1035, 2018.
[http://dx.doi.org/10.1016/j.snb.2017.11.159]

[31] G. Mathankumara, P. Bharathia, M.K. Mohana, S. Harisha, M. Navaneethana, J. Archana, P. Suresha, G.K. Manid, P. Dhivyad, S. Ponnusamya, and C. Muthamizhchelvan, "Synthesis and functional properties of nanostructured Gd-doped WO3/TiO2 composites for sensing applications", *Mater. Sci. Semicond. Process.,* vol. 105, p. 104732, 2020.
[http://dx.doi.org/10.1016/j.mssp.2019.104732]

[32] T. Govindaraj, C. Mahendran, R. Marnadu, M. Shkir, and V.S. Manikandan, "The remarkably enhanced visible-light-photocatalytic activity of hydrothermally synthesized WO3 nanorods: An effect of Gd doping", *Ceram. Int.,* 2020.

[33] S.M. Harshulkhan, K. Janaki, G. Velraj, R.S. Ganapthy, and M. Nagarajan, "Effect of Ag doping on structural, optical and photocatalytic activity of tungsten oxide (WO3) nanoparticles", *J. Mater. Sci. Mater. Electron.,* vol. 27, pp. 4744-4751, 2016.
[http://dx.doi.org/10.1007/s10854-016-4354-3]

[34] M.B. Tahira, and M. Sagir, "Carbon nanodots and rare metals (RM=La, Gd, Er) doped tungsten oxide nanostructures for photocatalytic dyes degradation and hydrogen production", *Separ. Purif. Tech.,* vol. 209, pp. 94-102, 2019.
[http://dx.doi.org/10.1016/j.seppur.2018.07.029]

[35] S.M. Lam, J.C. Sin, A.Z. Abdullah, and A.R. Mohamed, "ZnO nanorods surface-decorated by WO3 nanoparticles for photocatalytic degradation of endocrine disruptors under a compact fluorescent lamp", *Ceram. Int.,* vol. 39, pp. 2343-2352, 2013.
[http://dx.doi.org/10.1016/j.ceramint.2012.08.085]

[36] M. Imran, S.H. Rashid, Y. Sabri, N. Motta, T. Tesfamichael, P. Sonar, and M. Shafiei, "Template based sintering of WO3 nanoparticles into porous tungsten oxide nanofibers for acetone sensing applications", *J. Mater. Chem. C Mater. Opt. Electron. Devices,* vol. 7, pp. 2961-2970, 2019.

[http://dx.doi.org/10.1039/C8TC05982A]

[37]   M.B. Tahir, M. Sagir, and N. Abas, "Enhanced photocatalytic performance of CdO-WO3 composite for hydrogen production", *Int. J. Hydrogen Energy,* vol. 44, p. 2469, 2019.
[http://dx.doi.org/10.1016/j.ijhydene.2019.07.220]

[38]   M. Fakhari, M.J. Torkamany, S.N. Mirnia, and S.M. Elahi, "UV-visible light-induced antibacterial and photocatalytic activity of half harmonic generator WO3 nanoparticles synthesized by Pulsed Laser Ablation in water", *Opt. Mater.,* vol. 85, pp. 491-499, 2018.
[http://dx.doi.org/10.1016/j.optmat.2018.09.023]

[39]   M. Saleem, J. Iqbal, A. Nawaz, B. Islam, and I. Hussain, "Synthesis, characterization and performance evaluation of pristine and cerium-doped WO3 nanoparticles for photodegradation of methylene blue via solar irradiation", *Int. J. Appl. Ceram. Technol.,* vol. 17, pp. 1918-1929, 2020.
[http://dx.doi.org/10.1111/ijac.13496]

[40]   Y. Yin, C. Lan, S. Hu, and C. Li, "Effect of Gd-doping on electrochromic properties of sputter deposited WO3 films", *J. Alloys Compd.,* vol. 739, pp. 623-631, 2018.
[http://dx.doi.org/10.1016/j.jallcom.2017.12.290]

[41]   T.G.G. Maffeis, D. Yung, L. LePennec, M.W. Penny, R.J. Cobley, E. Comini, G. Sberveglieri, and S.P. Wilks, "STM and XPS characterisation of vacuum annealed nanocrystalline WO3 films", *Surf. Sci.,* vol. 601, pp. 4953-4957, 2007.
[http://dx.doi.org/10.1016/j.susc.2007.08.009]

[42]   H. Zhang, V. Malika, S. Mallapragada, and A. Mufit, "Synthesis and characterization of Gd-doped magnetite nanoparticles", *J. Magn. Magn.,* vol. 423, pp. 386-394, 2017.
[http://dx.doi.org/10.1016/j.jmmm.2016.10.005]

[43]   B. Poornaprakash, U. Chalapathi, M. Reddeppa, and S.H. Park, "Effect of Gd doping on the structural, luminescence and magnetic properties of ZnS nanoparticles synthesized by the hydrothermal method", *Superlattices Microstruct.,* vol. 97, pp. 104-109, 2016.
[http://dx.doi.org/10.1016/j.spmi.2016.06.013]

[44]   M.M. Charithra, J.G.G. Manjunatha, and C. Raril, "Surfactant Modified Graphite Paste Electrode as an Electrochemical Sensor for the Enhanced Voltammetric Detection of Estriol with Dopamine and Uric acid", *Adv. Pharm. Bull.,* vol. 10, no. 2, pp. 247-253, 2020.
[http://dx.doi.org/10.34172/apb.2020.029] [PMID: 32373493]

[45]   F. Winquist, "Voltammetric electronic tongues – basic principles and applications", *Mikrochim. Acta,* vol. 2008, no. 163, pp. 3-10, 2008.
[http://dx.doi.org/10.1007/s00604-007-0929-2]

[46]   G. Tigari, and J.G. Manjunatha, "Optimized Voltammetric Experiment for the Determination of Phloroglucinol at Surfactant Modified Carbon Nanotube Paste Electrode", *Instrum. Exp. Tech.,* vol. 63, pp. 750-757, 2020.
[http://dx.doi.org/10.1134/S0020441220050139]

[47]   J.G. Manjunatha, "Fabrication of Efficient and Selective Modified Graphene Paste Sensor for the Determination of Catechol and Hydroquinone", *Surfaces,* vol. 3, pp. 473-483, 2020.
[http://dx.doi.org/10.3390/surfaces3030034]

[48]   H.M. Shiri, and A. Ehsani, "Pulse electrosynthesis of novel wormlike gadolinium oxide nanostructure and its nanocomposite with conjugated electroactive polymer as a hybrid and high efficient electrode material for energy storage device", *J. Colloid Interface Sci.,* vol. 484, pp. 70-76, 2016.
[http://dx.doi.org/10.1016/j.jcis.2016.08.075] [PMID: 27592187]

[49]   C. Jo, I. Hwang, J. Lee, C.W. Lee, and S. Yoon, "Investigation of Pseudocapacitive Charge-Storage Behaviour in Highly Conductive Ordered Mesoporous Tungsten Oxide Electrodes", *J. Phys. Chem. C,* vol. 115, pp. 11880-11886, 2011.
[http://dx.doi.org/10.1021/jp2036982]

[50]   S. Yoon, E. Kang, J.K. Kim, C.W. Lee, and J. Lee, "Development of high-performance supercapacitor electrodes using novel ordered mesoporous tungsten oxide materials with high electrical conductivity", *Chem. Commun. (Camb.),* vol. 47, no. 3, pp. 1021-1023, 2011.
[http://dx.doi.org/10.1039/C0CC03594G] [PMID: 21069128]

[51]   H. Wei, X. Yan, S. Wu, Z. Luo, S. Wei, and Z. Guo, "Electropolymerized Polyaniline Stabilized Tungsten Oxide Nanocomposite Films: Electrochromic Behavior and Electrochemical Energy Storage", *J. Phys. Chem. C,* vol. 116, pp. 25052-25064, 2012.
[http://dx.doi.org/10.1021/jp3090777]

[52]   C.C. Huang, W. Xing, and S.P. Zhuo, "Capacitive performances of amorphous tungsten oxide prepared by microwave irradiation", *Scr. Mater.,* vol. 61, pp. 985-987, 2009.
[http://dx.doi.org/10.1016/j.scriptamat.2009.08.009]

[53]   J.G. Manjunatha, "Poly (Adenine) Modified Graphene-Based Voltammetric Sensor for the Electrochemical Determination of Catechol, Hydroquinone and Resorcinol", *Chem. Eng. J.,* vol. 14, pp. 52-62, 2020.
[http://dx.doi.org/10.2174/1874123102014010052]

[54]   B. Deepa, and V. Rajendran, "Pure and Cu metal doped WO3 prepared via co-precipitation method and studies on their structural, morphological, electrochemical and optical properties. Nano-Struct", *Nano-Objects.,* vol. 16, pp. 185-192, 2018.
[http://dx.doi.org/10.1016/j.nanoso.2018.06.005]

[55]   R. Rajagopal, and R. Kwang-Sun, "Temperature Controlled Synthesis of Ce–MnO2 Nanostructure: Promising Electrode Material for Supercapacitor Applications", *Sci. Adv. Mater.,* vol. 12, pp. 461-469, 2020.
[http://dx.doi.org/10.1166/sam.2020.3638]

[56]   R. Dhilip Kumar, Y. Andou, and S. Karuppuchamy, "Synthesis and characterization of nanostructured Ni-WO3 and NiWO4 for supercapacitor applications", *J. Alloys Compd.,* vol. 654, pp. 349-356, 2016.
[http://dx.doi.org/10.1016/j.jallcom.2015.09.106]

[57]   R. Dhilip Kumar, and S. Karuppuchamy, "Microwave mediated synthesis of nanostructured Co -WO3 and CoWO4 for supercapacitor applications", *J. Alloys Compd.,* vol. 674, pp. 384-391, 2016.
[http://dx.doi.org/10.1016/j.jallcom.2016.03.074]

[58]   P.A. Shinde, Y. Seo, C. Ray, and C. Jun, "S. Direct growth of WO3 nanostructures on multi-walled carbon nanotubes for high-performance flexible all-solid-state asymmetric supercapacitor", *Electrochim. Acta,* vol. 308, pp. 231-242, 2019.
[http://dx.doi.org/10.1016/j.electacta.2019.03.159]

[59]   A.K. Nayak, A.K. Das, and D. Pradhan, "High Performance Solid-State Asymmetric Supercapacitor using Green Synthesized Graphene–WO3 Nanowires Nanocomposite", *ACS Sustain. Chem.& Eng.,* vol. 5, pp. 10128-10138, 2017.
[http://dx.doi.org/10.1021/acssuschemeng.7b02135]

CHAPTER 6

Optimised Voltammetric Approaches for Clinical Sample Analysis

Gnanesh Rao[1], Raghu Ningegowda[2], B. P. Nandeshwarappa[3], Kiran Kumar Mudnakudu-Nagaraju[4], M. B Siddesh[5] and Sandeep Chandrashekharappa[6,7,*]

[1] *Department of Biochemistry, Bangalore University, Bangalore, Karnataka, India*

[2] *Jyoti Nivas College Autonomous, Department of Studies in Chemistry, Bangalore-560095, India*

[3] *Department of PG Studies and Research in Chemistry, Shivagangothri, Davangere University, Davanagere, Karnataka - 577 007, India*

[4] *Department of Biotechnology & Bioinformatics, Faculty of Life Sciences, JSS Academy of Higher Education and Research, Mysore 570015, Karnataka, India*

[5] *Department of Chemistry, KLE'S S. K. Arts College and H. S. K. Science Institute, Hubballi, Karnataka 580031, India*

[6] *Institute for Stem Cell Science and Regenerative Medicine, NCBS, TIFR, GKVK-Campus Bellary road, Bengaluru 560065, Karnataka, India*

[7] *Department of Medicinal Chemistry, National Institute of Pharmaceutical Education and Research (NIPER) Raebareli, Lucknow (UP)-226002, India*

Abstract: Extraordinary properties of nanomaterial lay a broad prospect for its application in various fields. Biosensors are a kind of special sensor which consists of biomolecule recognition element and transducer. Advances in biosensors with the use of micro/ nanomaterials are capable of detecting and analyzing living and chemical matter with high specificity, fastness, sensitivity, accuracy and low cost for the determination of proteins, hormone, enzyme, nucleic acid and other biological compounds in blood and urine samples. The basics of biosensors and optimized approaches for voltammetric clinical analysis are studied in this chapter. Exosomes are small extracellular vesicles (EV) involved in extracellular communication between cells, and are recognized as potential markers of human health and diseases. An extensive molecular characteristic comparison of the exosomal components of healthy and disordered cells or tissues reveals the differences.

• Electrochemical approaches for the analysis of clinical samples are developing at an incredible rate with advances in nanotubes and nanomaterials as biosensors.

---

* **Corresponding author Sandeep Chandrashekarappa:** Institute for Stem Cell Science and Regenerative Medicine, NCBS, TIFR, GKVK-Campus Bellary road, Bengaluru 560065, Karnataka, India And Department of Medicinal Chemistry, National Institute of Pharmaceutical Education and Research (NIPER) Raebareli, Lucknow (UP)-226002, India; E-mail: sandeep_m7@rediffmail.com, c.sandeep@niperraebareli.edu.in

• Exosomes play a crucial role in cell-to-cell communication in normal health and disease. They are potential biomarkers exploited for emerging non-invasive diagnosis.

• CNT sensors are capable of precise quantitative and qualitative detection of analytes.

Keywords: Affinity biosensor, Clinical analyte, CNT, CPE, Enzymatic biosensor, Exosome, Immunosensor, Voltammetric biosensor.

INTRODUCTION

The analysis of bodily fluids, especially urine, for the diagnosis of disease or disorder dates back to pre-3000 BC; writings and drawings serve as evidence. Chang Chung-Ching (AD 229), Charaka and Susruta (5th century) observed that individuals with sugar metabolism disorder presently termed as Diabetes mellitus are likely to be diagnosed by analyzing the urine of an individual for the presence of sugar. For this, dogs and ants served like sensors [1].

The chemical analysis of metabolite started with analyzing the urine for glucose and albumin in the 18th century. The knowledge of analysis has grown to an extent that the molecular characterization for diagnosis of a genetic disorder in a fetus is possible. The advances in sensors technology have resulted in the development of Point Of Care diagnosis, which is most convenient, faster and reliable [2 - 4].

The basics of a general biosensor are graphically represented in Fig. (1).

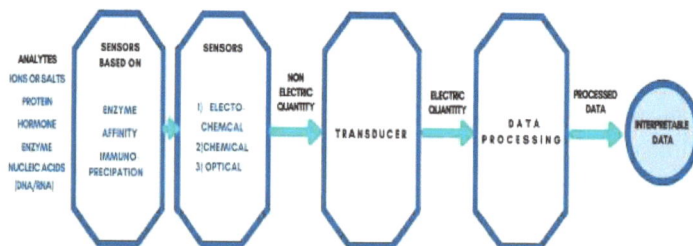

Fig. (1). General elements of a Biosensor.

CLINICAL ANALYTES

Cells are the product of millions of years of evolution; tissues are a step forward complex than cells with the capability of maintaining different chemical and physical conditions yet capable of performing a prefixed function.

The features of intracellular and intercellular signalling play a crucial role in homeostasis and disease. Cells communicate *via* a free or bound form of ions and biomolecules such as protein, hormone, enzyme, nucleic acid, *etc.*

Exosomes are 30-100nm vesicles that are released by most of the tissue systems in the body. The enclosed cargo of proteins, enzymes, and RNAs from the cell of origin renders exosomes the capability for intercellular signaling [5 - 10].

They are abundant in blood, which contributes to relay the function from proximal to target tissues. The ability to access exosomes from mostly any biological fluid offers a minimally invasive window for biochemical diagnosis [11 - 15].

Cell-to-cell communication is not only essential for repair and growth but also required for cell survival, cells failing to receive enough survival factor enter into the stage of apoptosis, possibly avoided by the use of many pharmaceutical drugs [16 - 25]. Signaling systems are initiated at the early stage of the differentiation of cells in the fertilized egg (Fig. **2**) [26, 27].

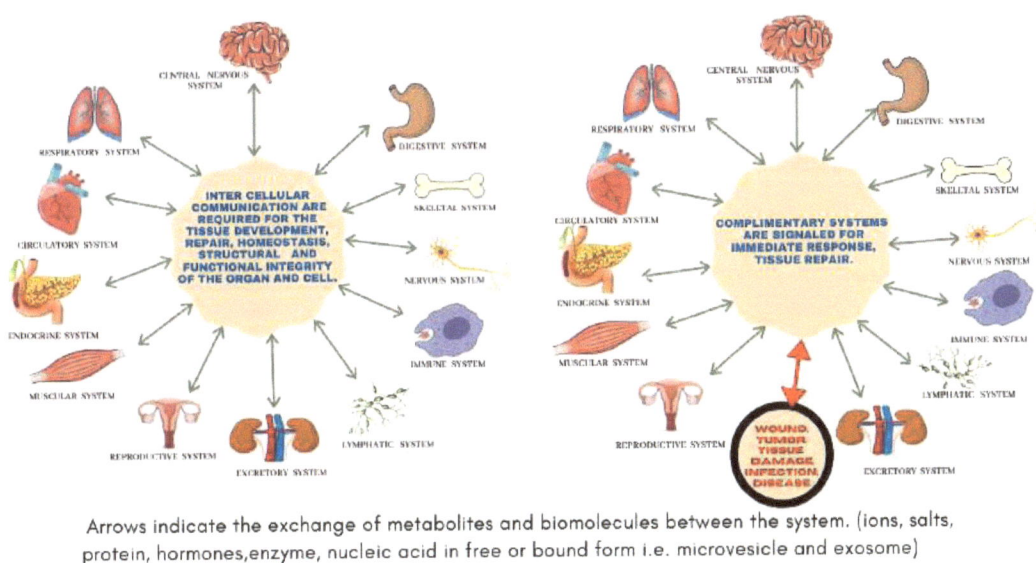

Arrows indicate the exchange of metabolites and biomolecules between the system. (ions, salts, protein, hormones,enzyme, nucleic acid in free or bound form i.e. microvesicle and exosome)

Fig. (2). Representative diagram of intercellular communication between organ systems.

Analysis of Clinical Analytes using Voltammetric Biosensor

Voltammetric biosensor is based on electrochemical techniques in which quantitative analysis of analyte is made by varying the potential and measuring the resulting current as an analyte reacts electrochemically with the surface of the working electrodes [28, 29]. Cyclic voltammetry (CV) and differential pulse voltammetry (DPV) are the most commonly applied techniques for the determination of redox potential and electrochemical reaction rates of analyte solutions (Fig. **3**) [30, 31].

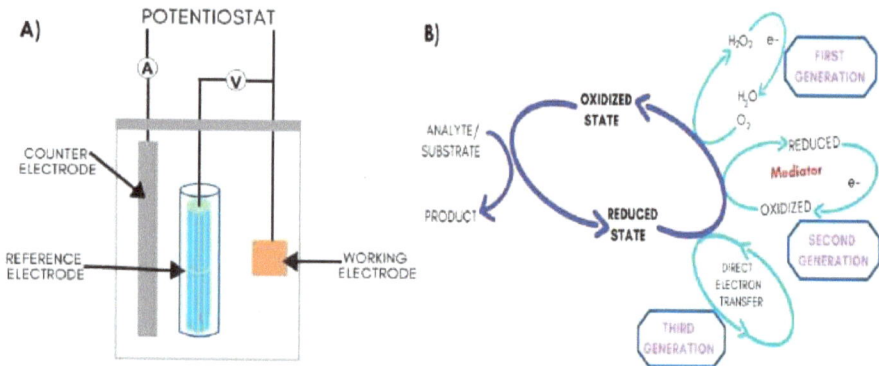

Fig. (3). A) General set up of voltammetric analysis; B) Types of an electrochemical biosensor.

CNT BIOSENSOR

The high electric conductivity of carbon and similar values of ionization potential and electron affinity contribute to its unique electric property to donate and accept electrons [32].

CNTs are a marvelous nanostructure having applications in the fields of agriculture, biofuel, chemical, pharmaceutical, cosmetics, electronics, pulping, and enzyme production [33, 34]. They are used in the field of diagnosis due to the increased surface area and the ability of the CNT to be covalently or non-covalently functionalized. Functionalized CNTs are suitable candidates for precise quantitative diagnosis [35, 36].

Terminal and sidewall functionalization of CNT by covalent or non-covalent modifications is used for the capture and quantification of a particular analyte [37]. The method of functionalization depends on the type of functionalization required. Analytes in free form can be analysed directly, and for those in a bound state, the isolation and purification have to be performed before analysis.

In 1996, CNT as an electrode material in the field of biosensors was explored by Britto and co-workers, showing the possibility of electrochemical oxidation of dopamine [38]. Functionalized CNTs are third-generation electrochemical biosensors capable of transferring the electron directly to the electrode [39, 40]. The transducer picks up the electric signals and forwards to the amplifier for amplification. Amplified signals are processed and then converted into interpretable values [41 - 43].

Electrochemical biosensor provides highly sensitive and specific measurements for a broad spectrum of biomolecules [44]. Biosensors have the potential to be easily transformed from a laboratory-based instrument to a commercializable POC device with a relatively fast detection time (Fig. **4**).

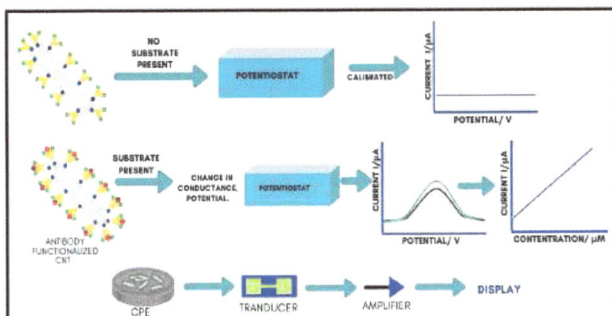

Fig. (4). Basic workflow of CNT-based electrochemical biosensor.

Types Based on Functionalization

Affinity sensor detects the binding of the analyte to sensing element. Metabolism sensor involves the measurement of a chemical change . Catalytic sensor is a biosensor that converts an auxiliary substrate.

1. *Enzymatic Biosensor*

In 1962, Leland C. Clark developed an enzyme-based electrochemical glucose biosensor [45]. The basic operation of glucose biosensor is based on the fact that the enzyme Glucose oxidase catalyzes the oxidation of glucose to gluconolactone, in turn reducing itself [46].

This change of state on the CNT surface is responsible for the change in potentiometric, impedometric, conductometric and amperometric readings of sensors in the presence of the substrate [47]. Oxidative enzymes such as galactose oxidase [48], lactate oxidase [49], and cholesterol oxidase [50] are used for the determination of galactose, lactose and cholesterol, respectively. Dehydrogenase which oxidizes substrate while reducing NAD^+ to NADH can be used in the detection of glutamate [51], alcohol [52], using glutamate dehydrogenase, alcohol dehydrogenase, *etc*. (Fig. **5**).

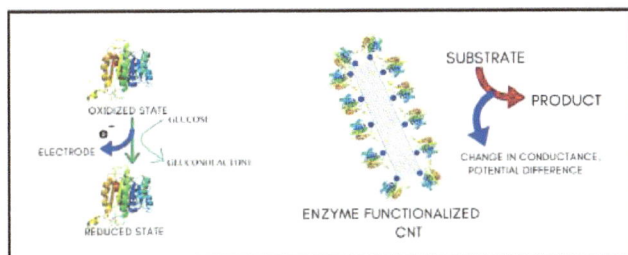

Fig. (5). Working of enzyme-based CNT voltammetric biosensor.

2. *Affinity Biosensors*

The ability of a nucleotide strand to bind to the synthetic complementary strand to form a hybrid DNA is exploited in the quantification of target strands. The hybridization of the strand on the CNT surface brings about changes in the voltammetric and amperometric reading of sensors in the presence of the target strand [53]. CNTs functioned with probes able to specifically bind with proteins, such as thrombin, TNF-α, TSH, and thyrotropin, are used to detect and quantify the same protein. The amount of change observed in the reading of sensors is proportional to the quantity of an analyte; an electric signal received is later processed and converted into interpretable reading [54 - 57].

Hybridization of DNA: The ssDNA-CNT-CPE electrode has to be immersed into a hybridization solution containing target DNA with the working potential and left for confined time for complete DNA hybridization followed by washing with 0.2% SDS to remove the unhybridized DNA. The same Hybridization reaction can also detect mismatched DNA and non-complementary DNA using suitable probe DNA [58].

Sensitivity of the hybridization detection can be increased by the use of external electroactive intercalators and groove binding indicators, such as daunomycin [59, 60], tris (2,2′-bipyridine) ruthenium (III) $[Ru(BPY)_3]^{2+}$ [61, 62], and tris (1,10-phenanthroline) cobalt (III) $[Co(phen)_3]^{3+}$ [63]. This can also be done by sandwich type hybridization using electroactive oligonucleotides (*e.g.*, ferrocene bound) (Fig. **6**) [64].

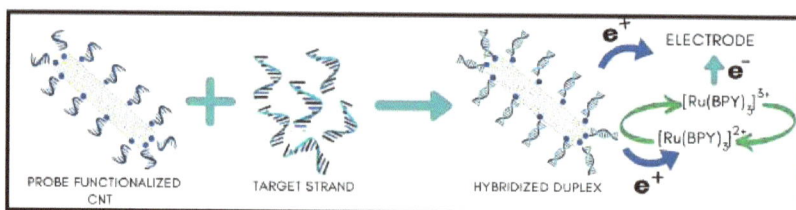

Fig. (**6**). Affinity-based CNT biosensor.

3. *Immunosensors*

Due to highly specific protein-protein interaction and acting as an antibody against an antigen, proteins are capable of binding specifically to form an immunoprecipitate; immunoprecipitation on the CNT surface results in changed voltammetric and amperometric reading in the presence of a specific antigen [65, 66].

Proteins such as streptavidin and avidin can be used for the detection of biotin [67]. Cancer and tumor markers, prostate-specific antigen [68], carcinoembryonic antigen [69], Interleukin-6 [70], and α-1-Fetoprotein [71] can be detected by a specific antibody. To increase the specificity and selectivity, a secondary antibody tagged with redox electroactive mediator like ferrocene, $[Ru(bpy)_3]^{2+}$, *etc.* can be used.

CNT functionalized with specific antibody immunoprecipitates with specific antigen fluorophore/chromophore conjugated secondary antibody produces fluorescence upon binding. The amount of fluorescence is proportional to the quantity of analyte; the optical signal is converted into an electric signal, later processed and converted into interpretable reading (Fig. **7**) [72].

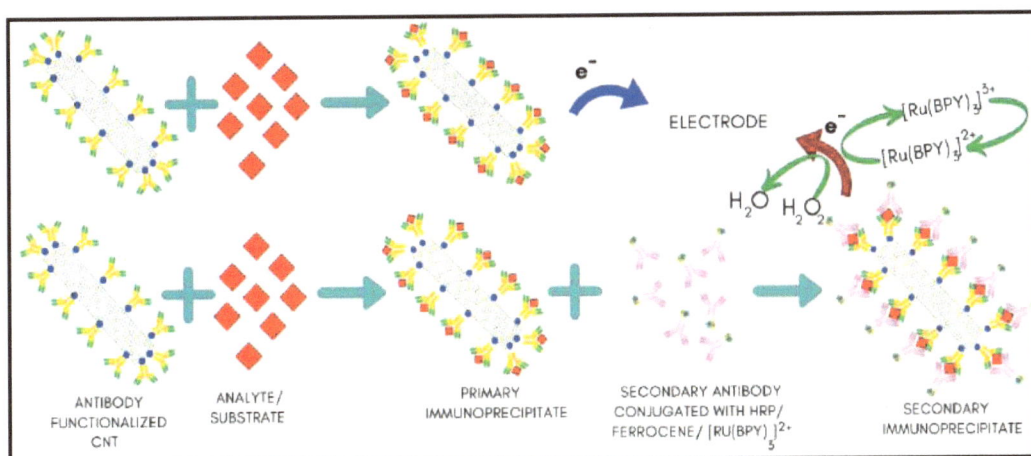

Fig. (7). Immunoprecipitation based CNT biosensor.

Preparation of Carbon Paste Electrode

Graphite powder and carbon nanotube finely ground in a mortar and pestle are mixed with silicone oil/paraffin oil/mineral oil as a binder to yield a homogenized mixture [73]. Binderless biocomposite based on mixing enzyme within CNT can also be prepared [74]. Obtained uniform paste has to be filled into the cavity of glass or teflon rod pre-inserted with conductive wire for electrical connection. The polishing of the surface of the electrode to obtain a uniform surface, followed by coating the bare electrode with a probe by electrodeposition/electrochemical polymerization rather than drop-casting, in order to avoid non-uniform coating [75, 76].

Future Perspective of Exosomal Analysis using Nanotube Biosensors in Personalised and Precision Diagnosis

Comparative data analysis can be used for the diagnosis of the disease before progression and also for evaluating the role of other factors responsible for disease progression. Exosomes have great potential as non-invasive biomarkers for liquid biopsy. Functionalized Carbon Nano Tube (CNT) sensors are highly precise but have lower level of detection. Thus, they are most suitable for the analysis of the molecular characterization of exosomes.

Research labs around the world are contributing to the process of understanding the molecular and chemical characteristics of exosomes. Simultaneously, the process of identifying the contents specified in case of disease and health is carried out. The process of diagnosis using exosomes involves the isolation of exosomes, followed by lysis, isolation of the desired group of molecules and analysis. These steps are involved in the identification of unique analyte or fold changes in case of a particular disease. The analyte can further be identified and quantitatively analysed using biosensor systems, serving as a diagnostic method (Fig. **8**).

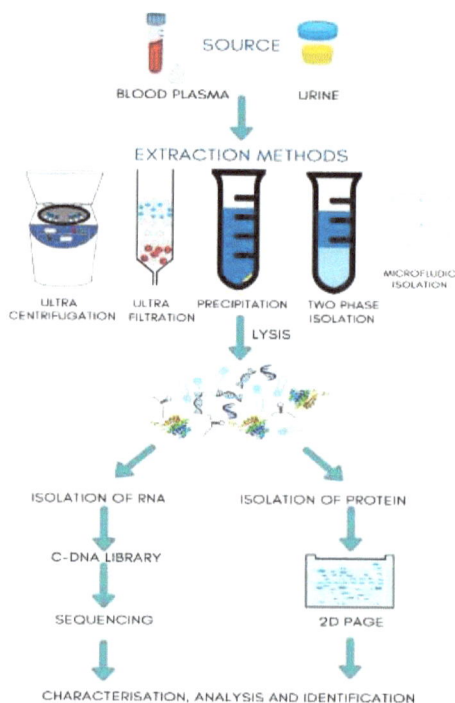

Fig. (8). Flow diagram of steps involved in the identification of markers.

ISOLATION METHODS

Different exosome isolation protocols are followed for minimally altered isolation, depending on the type of bodily fluids used for the analysis.

1. *Ultracentrifugation (UC)*: Centrifuge the sample at 1,600 g for 15 min at room temperature to remove cells. Transfer the supernatant to new tubes and centrifuge at 10,000 g for 30 min at room temperature to remove debris and large vesicles. Centrifuge the supernatant for 70 min at 100,000 g and 4° C to pellet the exosomes. Discard the supernatant and resuspend the pellet in the buffer for washing. Conduct a second UC for 70 min at 100,000g and 4°C to pelletize the purified exosomes [77].

2. *Ultrafiltration*: Centrifuge the sample at 1,600 g for 15 min at room temperature to remove cells. Sepharose CL-2B has to be stacked in a syringe and washed and equilibrated with PBS. Load the sample onto the column and collect fractions using elution buffer [78].

3. *Precipitation*: Centrifuge the sample at 1,600 g for 15 min at room temperature to remove the cells; then dilute the supernatant in cold 20 mM Tris-HCL, 20 mM EDTA buffer of pH 8.6, vortex for 90 s and subject it to centrifugation at 8000g for 15 min at 4°C, and finally, filter it with Whatman 1.2 μm cellulose acetate filters. Carry out ultracentrifugation at 27,500 rpm for 90 min at 4°C. Resuspend the pellet in 30 ml of cold PBS, and repeat the ultracentrifugation step [79].

4. *Two-phase Isolation*: Centrifuge the sample at 10,000 g for 10 minutes to remove larger contaminants. The supernatant is then passed through 0.22 μm filter to remove smaller contaminating particles and to reduce the concentrations of larger extracellular vesicles. Filtered supernatant is equally diluted in Polyethylene glycol and Dextran mixture at 7:3 (w/w) ratio (ATPS-EV isolation solution), and centrifuged at 1,000 g for 10 min for phase separation. Twice, from the upper layer, 80% of the volume is discarded, PEG-rich phase is replaced with washing solution, which is prepared by mixing ATPS-EV isolation solution with distilled water at a 1:1 volume ratio and centrifuging at 1,000g for 10 min [80].

5. *Microfluidic Devices*: Centrifuge the sample at 1,000 g for 10 minutes, and pass the supernatant through 0.22 μm filter. The buffer solution and prepared samples are delivered through the device by syringe pumps at a rate of 3.74ml h^{-1} and 0.15ml h^{-1}, respectively [81].

Exosome Lysis: Purified exosomes can be lysed by incubation in 1% SDS, 10 mM CHAPS, 5 mM MgCl$_2$ in 100 mM TEAB solution for 30 min under agitation at 4°C [82].

Protein Isolation: Carry out total protein isolation by TRIzol method.

Nucleic Acid Isolation: Carry out organic extraction of DNA and RNA [83, 84].

2D PAGE, RNA Sequencing and Analysis: Load the sample on an immobilized pH gradient strip for isoelectric focusing and equilibrate, and then treat it with SDS and apply on to a 12.5% polyacrylamide gel for electrophoresis, followed by silver staining for visualization [85]. For a comparative study of protein characterization by software like Delta 2D, the unique spots are identified and further analyzed. Prepare C-DNA library using the isolated RNA, sequence the strands and compare for analysis.

Identification of exosomal RNAs, DNAs, and proteins as novel biomarkers for early diagnosis of chronic kidney disease, central nervous system disorders, cardiovascular diseases, cancer, *etc.*, has been reported; however, there has been a rapid increase in interest in the use of exosomes as a diagnostic tool [86 - 90]. There is a need for the development of biosensors for precise exosomal diagnosis. CNTs for detection are being tested and advancing day by day. General clinical analytes can also be detected using CNT-based biosensor.

The advantages listed below are several key features that CNT-based electrochemical biosensors offer for clinical diagnostics.

Specificity and Sensitivity: Biosensors labeled with specific enzyme, antibody and complementary DNA are required to achieve highly specific detection of clinical analyte, DNA recognition with single-base mismatch specificity in clinical samples [91], with sensitivities down to several fM in case of many analytes, and aM for short oligonucleotides. In some cases, signal amplification seems necessary to detect particular analyte without the interference of heteroatoms used as pharmaceutical drugs [92 - 101]. Indeed, advances in nanomaterials have helped achieve the enhancement of signal amplification without any mediator and improve overall sensor performance and usage [102].

Detection Time: Using electrochemical biosensors, clinical analyte recognition can occur within minutes. Entire detection process can be speeded up by optimizing fast sample preparation and signal readout. Such rapid detection is ideal for Point of Care clinical diagnostics, especially for actual implementation of biomarkers whose long-term stability and selectivity in complex biological samples are not possible [103].

Multiplexing: A single biomarker is not sufficient for assessing the health condition using clinical samples. By simultaneous detection of several associated biomarkers [104], the diagnosis accuracy can be greatly improved [105].

Electrical fields generated by electrodes are highly localized because of which simultaneous detection within a miniaturized platform is possible, with optimized detection condition for each analyte and also the DNA [106, 107].

Repeatability, Stability and Reliability: The capability to produce repeatable and highly reliable measurements using biosensors is one of the major challenges, as a slight variation in electrode geometry, the probe used, uniformity of coating, *etc.*, can lead to inconsistent measurements. Storage stability of the sensors depends on the type of probe used for functionalization. Thus, optimization of the fabrication process, automation for consistent manufacturing and storage conditions can enhance the overall repeatability, stability and reliability of the sensors [108].

CONCLUSION

In this chapter, a brief ideology of CNT application in the field of diagnosis and optimized voltammetric approaches for clinical sample analysis have been presented. Exosomes are an emerging diagnostic biomarker; the load of cargo enclosed is very limited and precise for the signal relay. CNT limited by its lower limit of detection but havingmaximum precision can serve as a sensor to overcome the limitations and for providing early advances in exosomal diagnosis [109].

Periodic bodily fluid derived exosomal analysis of a person and comparison of the results with the identified biomarkers against the normal state of the individual can serve as the personalized diagnosis.

Advances in the field of biosensors using nanomaterials can markedly improve the sensitivity and specificity of biomolecule detection; these have great potential in applications of diagnosis. Nano diagnostics may be cost-effective in the mere future, allowing better clinical diagnosis before disease progression by applying to point-of-care diagnosis and lab-on-a-chip technologies and making them accessible to the general population.

CONSENT FOR PUBLICATION

Not applicable.

CONFLICT OF INTEREST

The authors declare no conflict of interest, financial or otherwise.

ACKNOWLEDGEMENT

Declared none.

REFERENCES

[1]   M. Kirchhof, N. Popat, and J. Malowany, "J. Diagnostic Review A Historical Perspective of the Diagnosis of Diabetes", *Uwomj,* vol. 78, no. 1, pp. 7-11, 2009.

[2]   S.R. Benedict, "A reagent for the detection of reducing sugars. 1908", *J. Biol. Chem.,* vol. 277, no. 16, 2002.e5
         [PMID: 11953443]

[3]   V. C. Medvei, *The History of Clinical Endocrinology : A Comprehensive Account of Endocrinology from Earliest Times to the Present Day,* 1993.

[4]   D.W. Guthrie, and S.S. Humphreys, "Diabetes urine testing: an historical perspective", *Diabetes Educ.,* vol. 14, no. 6, pp. 521-526, 1988.
         [http://dx.doi.org/10.1177/014572178801400615] [PMID: 3061764]

[5]   T.H. Lee, E. D'Asti, N. Magnus, K. Al-Nedawi, B. Meehan, and J. Rak, "Microvesicles as mediators of intercellular communication in cancer--the emerging science of cellular 'debris'", *Semin. Immunopathol.,* vol. 33, no. 5, pp. 455-467, 2011.
         [http://dx.doi.org/10.1007/s00281-011-0250-3] [PMID: 21318413]

[6]   X.B. Li, Z.R. Zhang, H.J. Schluesener, and S.Q. Xu, "Role of exosomes in immune regulation", *J. Cell. Mol. Med.,* vol. 10, no. 2, pp. 364-375, 2006.
         [http://dx.doi.org/10.1111/j.1582-4934.2006.tb00405.x] [PMID: 16796805]

[7]   A. Bobrie, M. Colombo, G. Raposo, and C. Théry, "Exosome secretion: molecular mechanisms and roles in immune responses", *Traffic,* vol. 12, no. 12, pp. 1659-1668, 2011.
         [http://dx.doi.org/10.1111/j.1600-0854.2011.01225.x] [PMID: 21645191]

[8]   R.M. Johnstone, M. Adam, J.R. Hammond, L. Orr, and C. Turbide, "Vesicle formation during reticulocyte maturation. Association of plasma membrane activities with released vesicles (exosomes)", *J. Biol. Chem.,* vol. 262, no. 19, pp. 9412-9420, 1987.
         [http://dx.doi.org/10.1016/S0021-9258(18)48095-7] [PMID: 3597417]

[9]   T. Pisitkun, R.F. Shen, and M.A. Knepper, "Identification and proteomic profiling of exosomes in human urine", *Proc. Natl. Acad. Sci. USA,* vol. 101, no. 36, pp. 13368-13373, 2004.
         [http://dx.doi.org/10.1073/pnas.0403453101] [PMID: 15326289]

[10]   http://onesearch.unifi.it/openurl/39UFI/39UFI_Services?&sid=EMBASE&issn=14789450&id=doi:10.
         1586%2Fepr.09.17&atitle=Exosomes%3A+Proteomic+insights+and+diagnostic+potential&stitle=Exp
         ert+Rev.+Proteomics&title=Expert+Review+of+Proteomics&volume=6&issue=3&spage=267&epage
         =283&aulast=Simpson&aufirst=Richard+J.&auinit=R.J.&aufull=Simpson+R.J.&coden=&isbn=&pag
         es=267-283&date=2009&auinit1=R&auinitm=J
         [http://dx.doi.org/10.1586/epr.09.17]

[11]   M. Colombo, G. Raposo, and C. Théry, "Biogenesis, secretion, and intercellular interactions of exosomes and other extracellular vesicles", *Annu. Rev. Cell Dev. Biol.,* vol. 30, pp. 255-289, 2014.
         [http://dx.doi.org/10.1146/annurev-cellbio-101512-122326] [PMID: 25288114]

[12]   H. Valadi, K. Ekström, A. Bossios, M. Sjöstrand, J.J. Lee, and J.O. Lötvall, "Exosome-mediated transfer of mRNAs and microRNAs is a novel mechanism of genetic exchange between cells", *Nat. Cell Biol.,* vol. 9, no. 6, pp. 654-659, 2007.
         [http://dx.doi.org/10.1038/ncb1596] [PMID: 17486113]

[13]   J. Skog, T. Würdinger, S. van Rijn, D.H. Meijer, L. Gainche, M. Sena-Esteves, W.T. Curry Jr, B.S. Carter, A.M. Krichevsky, and X.O. Breakefield, "Glioblastoma microvesicles transport RNA and proteins that promote tumour growth and provide diagnostic biomarkers", *Nat. Cell Biol.,* vol. 10, no. 12, pp. 1470-1476, 2008.
         [http://dx.doi.org/10.1038/ncb1800] [PMID: 19011622]

[14]   L. Musante, D.E. Tataruch, and H. Holthofer, "Use and isolation of urinary exosomes as biomarkers for diabetic nephropathy", *Front. Endocrinol. (Lausanne),* vol. 5, no. SEP, p. 149, 2014.

[http://dx.doi.org/10.3389/fendo.2014.00149] [PMID: 25309511]

[15]   X. Yang, Z. Weng, D.L. Mendrick, and Q. Shi, "Circulating extracellular vesicles as a potential source of new biomarkers of drug-induced liver injury", *Toxicol. Lett.,* vol. 225, no. 3, pp. 401-406, 2014. [http://dx.doi.org/10.1016/j.toxlet.2014.01.013] [PMID: 24462978]

[16]   B. Padmashali, B.N. Chidananda, G. Bhanuprakash, S.M. Basavaraj, S. Chandrashekharappa, and K.N. Venugopala, *J. Appl. Pharm. Sci.,* vol. 9, no. 05, 2019.

[17]   H.K. Nagesh, and B. Padmashali, "C, Sandeep, T. M. C. Yuvaraj, M. B. Siddesh, S. M. Mallikarjuna", *Int. J. Pharm. Sci. Rev. Res.,* vol. 28, pp. 6-10, 2014.

[18]   M.B. Siddesh, B. Padmashali, K.S. Thriveni, and C. Sandeep, *Heterocyclic Letters,* vol. 4, pp. 503-514, 2014.

[19]   M.B. Siddesh, B. Padmashali, K.S. Thriveni, C. Sandeep, and B.C. Goudarshivnnanava, *J of Appli Chem,* vol. 2, pp. 1281-1288, 2013.

[20]   S.M. Mallikarjuna, and C. Basavaraj Padmashali, "Sandeep", *Int. J. Pharm. Pharm. Sci.,* vol. 6, no. 7, pp. 423-427, 2014.

[21]   K.N. Venugopala, S. Chandrashekharappa, M. Pillay, S. Bhandary, M. Kandeel, F.M. Mahomoodally, M.A. Morsy, D. Chopra, B.E. Aldhubiab, M. Attimarad, O.I. Alwassil, S. Harsha, K. Mlisana, and B. Odhav, "Synthesis and Structural Elucidation of Novel Benzothiazole Derivatives as Anti-tubercular Agents: In-silico Screening for Possible Target Identification", *Med. Chem.,* vol. 15, no. 3, pp. 311-326, 2019. [http://dx.doi.org/10.2174/1573406414666180703121815] [PMID: 29968540]

[22]   K.N. Venugopala, M.A. Khedr, M. Pillay, S.K. Nayak, S. Chandrashekharappa, B.E. Aldhubiab, S. Harsha, M. Attimard, and B. Odhav, *J. Biomol. Struct. Dyn.,* pp. 1-13, 2018.

[23]   S.M. Mallikarjuna, C. Sandeep, and B. Padmashali, *Pharma Chem.,* vol. 8, pp. 262-268, 2016.

[24]   H.K. Nagesh, "B. Padmashali, C. Sandeep, T.E Musturappa, M.R Lokesh", *Pharma Chem.,* vol. 7, no. 12, pp. 129-136, 2015.

[25]   S.M. Mallikarjuna, and C. Sandeep, "Basavaraj Padmashali", *IJPSR,* vol. 8, no. 7, pp. 2879-2885, 2017.

[26]   C. Tetta, E. Ghigo, L. Silengo, M.C. Deregibus, and G. Camussi, "Extracellular vesicles as an emerging mechanism of cell-to-cell communication", *Endocrine,* vol. 44, no. 1, pp. 11-19, 2013. [http://dx.doi.org/10.1007/s12020-012-9839-0] [PMID: 23203002]

[27]   C.L. Chi, S. Martinez, W. Wurst, and G.R. Martin, "The isthmic organizer signal FGF8 is required for cell survival in the prospective midbrain and cerebellum", *Development,* vol. 130, no. 12, pp. 2633-2644, 2003. [http://dx.doi.org/10.1242/dev.00487] [PMID: 12736208]

[28]   J.G. Manjunatha, C. Raril, N. Hareesha, M.M. Charithra, P.A. Pushpanjali, G. Tigari, and D. Ravishankar, "Mallappaji, S. C.; Gowda, J. Electrochemical Fabrication of Poly (Niacin) Modified Graphite Paste Electrode and Its Application for the Detection of Riboflavin", *Open Chem. Eng. J.,* vol. 14, no. 1, pp. 90-98, 2020. [http://dx.doi.org/10.2174/1874123102014010090]

[29]   N. Hareesha, and J.G. Manjunatha, "Fast and Enhanced Electrochemical Sensing of Dopamine at Cost-Effective Poly(DL-Phenylalanine) Based Graphite Electrode", *J. Electroanal. Chem. (Lausanne),* vol. 878, 2020.114533 [http://dx.doi.org/10.1016/j.jelechem.2020.114533]

[30]   N. Hareesha, J.G.G. Manjunatha, C. Raril, and G. Tigari, "Design of novel Surfactant Modified Carbon Nanotube Paste Electrochemical Sensor for the Sensitive Investigation of Tyrosine as a Pharmaceutical Drug", *Adv. Pharm. Bull.,* vol. 9, no. 1, pp. 132-137, 2019. [http://dx.doi.org/10.15171/apb.2019.016] [PMID: 31011567]

[31] J.N. Tiwari, V. Vij, K.C. Kemp, and K.S. Kim, "Engineered Carbon-Nanomaterial-Based Electrochemical Sensors for Biomolecules", *ACS Nano,* vol. 10, no. 1, pp. 46-80, 2016. [http://dx.doi.org/10.1021/acsnano.5b05690] [PMID: 26579616]

[32] M. Miyake, Electrochemical Functions.*Carbon Alloys.* Novel Concepts to Develop Carbon Science and Technology, 2003. [http://dx.doi.org/10.1016/B978-008044163-4/50026-7]

[33] D.G. Panpatte, and Y.K. Jhala, "Nanotechnology for Agriculture: Advances for Sustainable Agriculture", *Nanotechnol. Agric. Adv. Sustain. Agric.,* no. November, pp. 1-305, 2019. [http://dx.doi.org/10.1007/978-981-32-9370-0]

[34] R. Farma, M. Deraman, A. Awitdrus, I. A. Talib, E. Taer, N. H Basri, and J.G. Manjunatha Hashmi, "Preparation of Highly Porous Binderless Activated Carbon Electrodes from Fibres of Oil Palm Empty Fruit Bunches for Application in Supercapacitors", *Bio.*

[35] N. Hareesha, and J.G. Manjunatha, "Elevated and Rapid Voltammetric Sensing of Riboflavin at Poly(Helianthin Dye) Blended Carbon Paste Electrode with Heterogeneous Rate Constant Elucidation", *J. Iran. Chem. Soc.,* vol. 17, no. 6, pp. 1507-1519, 2020. [http://dx.doi.org/10.1007/s13738-020-01876-4]

[36] G. Tigari, J.G. Manjunatha, C. Raril, and N. Hareesha, "Determination of Riboflavin at Carbon Nanotube Paste Electrodes Modified with an Anionic Surfactant", *ChemistrySelect,* vol. 4, no. 7, pp. 2168-2173, 2019. [http://dx.doi.org/10.1002/slct.201803191]

[37] V. Schroeder, S. Savagatrup, M. He, S. Lin, and T.M. Swager, "Carbon Nanotube Chemical Sensors", *Chem. Rev.,* vol. 119, no. 1, pp. 599-663, 2019. [http://dx.doi.org/10.1021/acs.chemrev.8b00340] [PMID: 30226055]

[38] P.J. Britto, K.S.V. Santhanam, and P.M. Ajayan, "Carbon Nanotube Electrode for Oxidation of Dopamine", *Bioelectrochem. Bioenerg.,* vol. 41, no. 1, pp. 121-125, 1996. [http://dx.doi.org/10.1016/0302-4598(96)05078-7]

[39] E. Mehmeti, D. M. Stanković, S. Chaiyo, J. Zavasnik, K. Žagar, and K. Kalcher, "Wiring of Glucose Oxidase with Graphene Nanoribbons: An Electrochemical Third Generation Glucose Biosensor", *Microchim. Acta,* 2017. [http://dx.doi.org/10.1007/s00604-017-2115-5]

[40] D. Ivnitski, B. Branch, P. Atanassov, and C. Apblett, "Glucose Oxidase Anode for Biofuel Cell Based on Direct Electron Transfer", *Electrochem. Commun.,* 2006. [http://dx.doi.org/10.1016/j.elecom.2006.05.024]

[41] J.G.G. Manjunatha, "A novel poly (glycine) biosensor towards the detection of indigo carmine: A voltammetric study", *J. Food Drug Anal.,* vol. 26, no. 1, pp. 292-299, 2018. [http://dx.doi.org/10.1016/j.jfda.2017.05.002] [PMID: 29389566]

[42] J.G. Manjunatha, and B.E. Kumara Swamy, "G.P.Mamatha2, Ongera Gilbert, M. T.Shreenivas AndB.S.Sherigara", *Int. J. Electrochem. Sci.,* vol. 4, p. 1706, 2009.

[43] M.M. Charithra, and J.G. Manjunatha, "Enhanced Voltammetric Detection of Paracetamol by Using Carbon Nanotube Modified Electrode as an Electrochemical Sensor", *J. Electrochem. Sci. Eng.,* vol. 10, no. 1, pp. 29-40, 2019. [http://dx.doi.org/10.5599/jese.717]

[44] J.G. Manjunatha, "Electroanalysis of Estriol Hormone Using Electrochemical Sensor", *Sens. Biosensing Res.,* vol. 16, pp. 79-84, 2017. [http://dx.doi.org/10.1016/j.sbsr.2017.11.006]

[45] L.C. Clark Jr, and C. Lyons, "Electrode systems for continuous monitoring in cardiovascular surgery", *Ann. N. Y. Acad. Sci.,* vol. 102, pp. 29-45, 1962. [http://dx.doi.org/10.1111/j.1749-6632.1962.tb13623.x] [PMID: 14021529]

[46] C. Cai, and J. Chen, "Direct electron transfer of glucose oxidase promoted by carbon nanotubes", *Anal. Biochem.,* vol. 332, no. 1, pp. 75-83, 2004.
[http://dx.doi.org/10.1016/j.ab.2004.05.057] [PMID: 15301951]

[47] K., V.; Suresh, S.; K., M.; Gupta, A.; Vijayaraghav. R. Carbon Nanotubes - A Potential Material for Affinity Biosensors. In Carbon Nanotubes - Growth and Applications, 2011.
[http://dx.doi.org/10.5772/16836]

[48] J. Tkac, J.W. Whittaker, and T. Ruzgas, "The use of single walled carbon nanotubes dispersed in a chitosan matrix for preparation of a galactose biosensor", *Biosens. Bioelectron.,* vol. 22, no. 8, pp. 1820-1824, 2007.
[http://dx.doi.org/10.1016/j.bios.2006.08.014] [PMID: 16973345]

[49] A.C. Pereira, M.R. Aguiar, A. Kisner, D.V. Macedo, and L.T. Kubota, "Amperometric Biosensor for Lactate Based on Lactate Dehydrogenase and Meldola Blue Coimmobilized on Multi-Wall Carbon-Nanotube", *Sens. Actuators B Chem.,* vol. 124, no. 1, pp. 269-276, 2007.
[http://dx.doi.org/10.1016/j.snb.2006.12.042]

[50] Y-C. Tsai, S-Y. Chen, and C-A. Lee, "Amperometric Cholesterol Biosensors Based on Carbon Nanotube–Chitosan–Platinum–Cholesterol Oxidase Nanobiocomposite", *Sens. Actuators B Chem.,* vol. 135, no. 1, pp. 96-101, 2008.
[http://dx.doi.org/10.1016/j.snb.2008.07.025]

[51] S. Chakraborty, and C. Retna Raj, "Amperometric Biosensing of Glutamate Using Carbon Nanotube Based Electrode", *Electrochem. Commun.,* vol. 9, no. 6, pp. 1323-1330, 2007.
[http://dx.doi.org/10.1016/j.elecom.2007.01.039]

[52] J. Wang, and M. Musameh, "A Reagentless Amperometric Alcohol Biosensor Based on Carbon-Nanotube/Teflon Composite Electrodes", *Anal. Lett.,* vol. 36, no. 9, pp. 2041-2048, 2003.
[http://dx.doi.org/10.1081/AL-120023628]

[53] G. Ijeomah, F. Obite, and O. Rahman, "Development of Carbon Nanotube-Based Biosensors", *International Journal of Nano and Biomaterials,* 2016.
[http://dx.doi.org/10.1504/IJNBM.2016.079682]

[54] B.M. Amrutha, J.G. Manjunatha, S. Aarti Bhatt, and N. Hareesha, "Electrochemical Analysis of Evans Blue by Surfactant Modified Carbon Nanotube Paste Electrode, J. Mater", *Environ. Sci,* vol. 10, p. 668e676, 2019.

[55] P.A. Pushpanjali, J.G. Manjunatha, and M.T. Shreenivas, "The Electrochemical Resolution of Ciprofloxacin, Riboflavin and Estriol Using Anionic Surfactant and Polymer☐Modified Carbon Paste Electrode", *ChemistrySelect,* vol. 4, no. 46, pp. 13427-13433, 2019.
[http://dx.doi.org/10.1002/slct.201903897]

[56] B.M. Amrutha, J.G. Manjunatha, A.S. Bhatt, C. Raril, and P.A. Pushpanjali, "Electrochemical Sensor for the Determination of Alizarin Red-S at Non-Ionic Surfactant Modified Carbon Nanotube Paste Electrode", *Phys. Chem. Res.,* vol. 7, no. 3, pp. 523-533, 2019.
[http://dx.doi.org/10.22036/pcr.2019.185875.1636]

[57] A. Pemmatte, "Pushpanjali, Jamballi G Manjunatha, Girish Tigari, S. F. Poly(Niacin) Based Carbon Nanotube Sensor for the Sensitive and Selective Voltammetric Detection of Vanillin with Caffeine", *Anal. Bioanal. Electrochem.,* vol. 12, no. 4, pp. 553-568, 2020.

[58] "N.; YANG, T.; JIANG, C.; DU, M.; JIAO, K. Highly Sensitive Electrochemical Impedance Spectroscopic Detection of DNA Hybridization Based on Aunano–CNT/PANnano Films", *Talanta,* vol. 77, no. 3, pp. 1021-1026, 2009.
[http://dx.doi.org/10.1016/j.talanta.2008.07.058] [PMID: 19064085]

[59] K. Hashimoto, K. Ito, and Y. Ishimori, "Novel DNA Sensor for Electrochemical Gene Detection", *Anal. Chim. Acta,* vol. 286, no. 2, pp. 219-224, 1994.
[http://dx.doi.org/10.1016/0003-2670(94)80163-0]

[60] H. Cai, X. Cao, Y. Jiang, P. He, and Y. Fang, "Carbon nanotube-enhanced electrochemical DNA biosensor for DNA hybridization detection", *Anal. Bioanal. Chem.*, vol. 375, no. 2, pp. 287-293, 2003.
[http://dx.doi.org/10.1007/s00216-002-1652-9] [PMID: 12560975]

[61] K.M. Millan, and S.R. Mikkelsen, "Sequence-selective biosensor for DNA based on electroactive hybridization indicators", *Anal. Chem.*, vol. 65, no. 17, pp. 2317-2323, 1993.
[http://dx.doi.org/10.1021/ac00065a025] [PMID: 8238927]

[62] J.I.A. Rashid, and N.A. Yusof, "The Strategies of DNA Immobilization and Hybridization Detection Mechanism in the Construction of Electrochemical DNA Sensor: A Review", *Sens. Biosensing Res.*, vol. 16, pp. 19-31, 2017.
[http://dx.doi.org/10.1016/j.sbsr.2017.09.001]

[63] J. Wang, X. Cai, G. Rivas, and H. Shiraishi, "Stripping Potentiometric Transduction of DNA Hybridization Processes", *Anal. Chim. Acta,* vol. 326, no. 1–3, pp. 141-147, 1996.
[http://dx.doi.org/10.1016/0003-2670(96)00042-6]

[64] T. Ihara, M. Nakayama, M. Murata, K. Nakano, and M. Maeda, "Gene Sensor Using Ferrocenyl Oligonucleotide", *Chem. Commun. (Camb.),* no. 17, pp. 1609-1610, 1997.
[http://dx.doi.org/10.1039/a703401f]

[65] C.B. Jacobs, M.J. Peairs, and B.J. Venton, "Analytica Chimica Acta Review", *Carbon Nanotube Based Electrochemical Sensors for Biomolecules.,* vol. 662, pp. 105-127, 2010.
[http://dx.doi.org/10.1016/j.aca.2010.01.009] [PMID: 20171310]

[66] B.V. Chikkaveeraiah, A. Bhirde, R. Malhotra, V. Patel, J.S. Gutkind, and J.F. Rusling, "Single-wall carbon nanotube forest arrays for immunoelectrochemical measurement of four protein biomarkers for prostate cancer", *Anal. Chem.*, vol. 81, no. 21, pp. 9129-9134, 2009.
[http://dx.doi.org/10.1021/ac9018022] [PMID: 19775154]

[67] L. Fang, Z. Lü, H. Wei, and E. Wang, "Quantitative electrochemiluminescence detection of proteins: Avidin-based sensor and tris(2,2′-bipyridine) ruthenium(II) label", *Biosens. Bioelectron.*, vol. 23, no. 11, pp. 1645-1651, 2008.
[http://dx.doi.org/10.1016/j.bios.2008.01.023] [PMID: 18337079]

[68] J.P. Kim, B.Y. Lee, J. Lee, S. Hong, and S.J. Sim, "Enhancement of sensitivity and specificity by surface modification of carbon nanotubes in diagnosis of prostate cancer based on carbon nanotube field effect transistors", *Biosens. Bioelectron.*, vol. 24, no. 11, pp. 3372-3378, 2009.
[http://dx.doi.org/10.1016/j.bios.2009.04.048] [PMID: 19481922]

[69] S. Viswanathan, C. Rani, A. Vijay Anand, and J.A. Ho, "Disposable electrochemical immunosensor for carcinoembryonic antigen using ferrocene liposomes and MWCNT screen-printed electrode", *Biosens. Bioelectron.*, vol. 24, no. 7, pp. 1984-1989, 2009.
[http://dx.doi.org/10.1016/j.bios.2008.10.006] [PMID: 19038538]

[70] B.S. Munge, C.E. Krause, R. Malhotra, V. Patel, J.S. Gutkind, and J.F. Rusling, "Electrochemical Immunosensors for Interleukin-6. Comparison of Carbon Nanotube Forest and Gold Nanoparticle platforms", *Electrochem. Commun.*, vol. 11, no. 5, pp. 1009-1012, 2009.
[http://dx.doi.org/10.1016/j.elecom.2009.02.044] [PMID: 20046945]

[71] J. Lin, C. He, L. Zhang, and S. Zhang, "Sensitive amperometric immunosensor for α-fetoprotein based on carbon nanotube/gold nanoparticle doped chitosan film", *Anal. Biochem.*, vol. 384, no. 1, pp. 130-135, 2009.
[http://dx.doi.org/10.1016/j.ab.2008.09.033] [PMID: 18848914]

[72] Y. Piao, Z. Jin, D. Lee, H.J. Lee, H.B. Na, T. Hyeon, M.K. Oh, J. Kim, and H.S. Kim, "Sensitive and high-fidelity electrochemical immunoassay using carbon nanotubes coated with enzymes and magnetic nanoparticles", *Biosens. Bioelectron.*, vol. 26, no. 7, pp. 3192-3199, 2011.
[http://dx.doi.org/10.1016/j.bios.2010.12.025] [PMID: 21242086]

[73] F. Valentini, A. Amine, S. Orlanducci, M.L. Terranova, and G. Palleschi, "Carbon nanotube

purification: preparation and characterization of carbon nanotube paste electrodes", *Anal. Chem.,* vol. 75, no. 20, pp. 5413-5421, 2003.
[http://dx.doi.org/10.1021/ac0300237] [PMID: 14710820]

[74]   J. Wang, and M. Musameh, "Enzyme-dispersed carbon-nanotube electrodes: a needle microsensor for monitoring glucose", *Analyst (Lond.),* vol. 128, no. 11, pp. 1382-1385, 2003.
[http://dx.doi.org/10.1039/b309928h] [PMID: 14700233]

[75]   M. Jg, "A NEW ELECTROCHEMICAL SENSOR BASED ON MODIFIED CARBON NANOTUBE-GRAPHITE MIXTURE PASTE ELECTRODE FOR VOLTAMMETRIC DETERMINATION OF RESORCINOL", *Asian J. Pharm. Clin. Res.,* vol. 10, no. 12, p. 295, 2017.
[http://dx.doi.org/10.22159/ajpcr.2017.v10i12.21028]

[76]   W. Boumya, N. Taoufik, M. Achak, and N. Barka, "Chemically modified carbon-based electrodes for the determination of paracetamol in drugs and biological samples", *J. Pharm. Anal.,* vol. 11, no. 2, pp. 138-154, 2021.
[http://dx.doi.org/10.1016/j.jpha.2020.11.003] [PMID: 34012690]

[77]   K. Takov, D.M. Yellon, and S.M. Davidson, "Comparison of small extracellular vesicles isolated from plasma by ultracentrifugation or size-exclusion chromatography: yield, purity and functional potential", *J. Extracell. Vesicles,* vol. 8, no. 1, 2018.1560809
[http://dx.doi.org/10.1080/20013078.2018.1560809] [PMID: 30651940]

[78]   A. Gámez-Valero, M. Monguió-Tortajada, L. Carreras-Planella, Ml. Franquesa, K. Beyer, and F.E. Borràs, "Size-Exclusion Chromatography-based isolation minimally alters Extracellular Vesicles' characteristics compared to precipitating agents", *Sci. Rep.,* vol. 6, no. September, p. 33641, 2016.
[http://dx.doi.org/10.1038/srep33641] [PMID: 27640641]

[79]   M. Puhka, M.E. Nordberg, S. Valkonen, A. Rannikko, O. Kallioniemi, P. Siljander, and T.M. Af Hällström, "KeepEX, a simple dilution protocol for improving extracellular vesicle yields from urine", *Eur. J. Pharm. Sci.,* vol. 98, pp. 30-39, 2017.
[http://dx.doi.org/10.1016/j.ejps.2016.10.021] [PMID: 27771514]

[80]   O.K. Kırbaş, B.T. Bozkurt, A.B. Asutay, B. Mat, B. Ozdemir, D. Öztürkoğlu, H. Ölmez, Z. İşlek, F. Şahin, and P.N. Taşlı, "Optimized Isolation of Extracellular Vesicles From Various Organic Sources Using Aqueous Two-Phase System", *Sci. Rep.,* vol. 9, no. 1, p. 19159, 2019.
[http://dx.doi.org/10.1038/s41598-019-55477-0] [PMID: 31844310]

[81]   S.M. Santana, M.A. Antonyak, R.A. Cerione, and B.J. Kirby, "Microfluidic isolation of cancer-cel--derived microvesicles from hetergeneous extracellular shed vesicle populations", *Biomed. Microdevices,* vol. 16, no. 6, pp. 869-877, 2014.
[http://dx.doi.org/10.1007/s10544-014-9891-z] [PMID: 25342569]

[82]   X. Wang, R. Wilkinson, K. Kildey, J. Potriquet, J. Mulvenna, R.J. Lobb, A. Möller, N. Cloonan, P. Mukhopadhyay, A.J. Kassianos, and H. Healy, "Unique molecular profile of exosomes derived from primary human proximal tubular epithelial cells under diseased conditions", *J. Extracell. Vesicles,* vol. 6, no. 1, 2017.1314073
[http://dx.doi.org/10.1080/20013078.2017.1314073] [PMID: 28473886]

[83]   J. Marmur, "A Procedure for the Isolation of Deoxyribonucleic Acid from Micro-Organisms", *J. Mol. Biol.,* 1961.
[http://dx.doi.org/10.1016/S0022-2836(61)80047-8]

[84]   S. N. Peirson, and J. N. Butler, "RNA Extraction From Mammalian Tissues",
[http://dx.doi.org/10.1007/978-1-59745-257-1_22]

[85]   H. Zhou, T. Pisitkun, A. Aponte, P.S.T. Yuen, J.D. Hoffert, H. Yasuda, X. Hu, L. Chawla, R.F. Shen, M.A. Knepper, and R.A. Star, "Exosomal Fetuin-A identified by proteomics: a novel urinary biomarker for detecting acute kidney injury", *Kidney Int.,* vol. 70, no. 10, pp. 1847-1857, 2006.
[http://dx.doi.org/10.1038/sj.ki.5001874] [PMID: 17021608]

[86]   R. Khurana, G. Ranches, S. Schafferer, M. Lukasser, M. Rudnicki, G. Mayer, and A. Hüttenhofer,

"Identification of urinary exosomal noncoding RNAs as novel biomarkers in chronic kidney disease", *RNA,* vol. 23, no. 2, pp. 142-152, 2017.
[http://dx.doi.org/10.1261/rna.058834.116] [PMID: 27872161]

[87]   S. Saman, W. Kim, M. Raya, Y. Visnick, S. Miro, S. Saman, B. Jackson, A.C. McKee, V.E. Alvarez, N.C.Y. Lee, and G.F. Hall, "Exosome-associated tau is secreted in tauopathy models and is selectively phosphorylated in cerebrospinal fluid in early Alzheimer disease", *J. Biol. Chem.,* vol. 287, no. 6, pp. 3842-3849, 2012.
[http://dx.doi.org/10.1074/jbc.M111.277061] [PMID: 22057275]

[88]   X. Loyer, I. Zlatanova, C. Devue, M. Yin, K.Y. Howangyin, P. Klaihmon, C.L. Guerin, M. Kheloufi, J. Vilar, K. Zannis, B.K. Fleischmann, D.W. Hwang, J. Park, H. Lee, P. Menasché, J.S. Silvestre, and C.M. Boulanger, "Intra-Cardiac Release of Extracellular Vesicles Shapes Inflammation Following Myocardial Infarction", *Circ. Res.,* vol. 123, no. 1, pp. 100-106, 2018.
[http://dx.doi.org/10.1161/CIRCRESAHA.117.311326] [PMID: 29592957]

[89]   B.N. Hannafon, Y.D. Trigoso, C.L. Calloway, Y.D. Zhao, D.H. Lum, A.L. Welm, Z.J. Zhao, K.E. Blick, W.C. Dooley, and W.Q. Ding, "Plasma exosome microRNAs are indicative of breast cancer", *Breast Cancer Res.,* vol. 18, no. 1, p. 90, 2016.
[http://dx.doi.org/10.1186/s13058-016-0753-x] [PMID: 27608715]

[90]   Y. Zhan, L. Du, L. Wang, X. Jiang, S. Zhang, J. Li, K. Yan, W. Duan, Y. Zhao, L. Wang, Y. Wang, and C. Wang, "Expression signatures of exosomal long non-coding RNAs in urine serve as novel non-invasive biomarkers for diagnosis and recurrence prediction of bladder cancer", *Mol. Cancer,* vol. 17, no. 1, p. 142, 2018.
[http://dx.doi.org/10.1186/s12943-018-0893-y] [PMID: 30268126]

[91]   F. Wei, P.B. Lillehoj, and C-M. Ho, "DNA diagnostics: nanotechnology-enhanced electrochemical detection of nucleic acids", *Pediatr. Res.,* vol. 67, no. 5, pp. 458-468, 2010.
[http://dx.doi.org/10.1203/PDR.0b013e3181d361c3] [PMID: 20075759]

[92]   S.K. Rashmi, T.H. Suresha Kumara, and H.B.V. Gopalpur Nagendrappa, "Sowmya, P.S. Sujan Ganapathy, C. Sandeep, Sunil S. More", *Int. J. Pharm. Pharm. Sci.,* vol. 7, no. 2, pp. 493-497, 2015.

[93]   C. Sandeep, B. Padmashali, S.K. Rashmi, S.M. Mallikarjuna, M.B. Siddesh, H.K. Nagesh, and K.S. Thriveni, *Heterocyclic Letters,* vol. 4, pp. 371-376, 2014.

[94]   K.N. Venugopala, S. Chandrashekharappa, S. Bhandary, D. Chopra, M.A. Khedr, B.E. Aldhubiab, M. Attimarad, and B. Odhav, *Curr. Org. Synth.,* vol. 15, pp. 400-407, 2018.
[http://dx.doi.org/10.2174/1570179414666171024155051]

[95]   O.I. Alwassil, S. Chandrashekharappa, S.K. Nayak, and K.N. Venugopala, "Design, synthesis, and structural elucidation of novel NmeNANAS inhibitors for the treatment of meningococcal infection", *PLoS One,* vol. 14, no. 10, 2019.e0223413
[http://dx.doi.org/10.1371/journal.pone.0223413] [PMID: 31618227]

[96]   K.N. Venugopala, O.H.A. Al-Attraqchi, C. Tratrat, S.K. Nayak, M.A. Morsy, B.E. Aldhubiab, M. Attimarad, A.B. Nair, N. Sreeharsha, R. Venugopala, M. Haroun, M.B. Girish, S. Chandrashekharappa, O.I. Alwassil, and B. Odhav, "Novel Series of Methyl 3-(Substituted Benzoyl)-7-Substituted-2-Phenylindolizine-1-Carboxylates as Promising Anti-Inflammatory Agents: Molecular Modeling Studies", *Biomolecules,* vol. 9, no. 11, p. 661, 2019.
[http://dx.doi.org/10.3390/biom9110661] [PMID: 31661893]

[97]   C. Sandeep, B. Padmashali, and R.S. Kulkarni, *J. Appl. Chem. (Lumami, India),* vol. 2, pp. 1049-1056, 2013.

[98]   S. Chandrashekharappa, B. Padmashali, R.S. Kulkarni, R. Venugopala, and B. Odhav, *Asian J. Chem.,* vol. 28, no. 5, pp. 1043-1048, 2016.
[http://dx.doi.org/10.14233/ajchem.2016.19582]

[99]   S. Chandrashekharappa, K.N. Venugopala, M.A. Khedr, B. Padmashali, R.S. Kulkarni, R. Venugopala, and B. Odhav, *Indian Journal of Pharmaceutical Education and Research,* vol. 51, pp.

452-460, 2017.
[http://dx.doi.org/10.5530/ijper.51.3.73]

[100] S. Chandrashekharappa, K.N. Venugopala, R. Venugopala, and B. Padmashali, *J. Appl. Pharm. Sci.,* vol. 9, no. 02, pp. 124-128, 2019.
[http://dx.doi.org/10.7324/JAPS.2019.90217]

[101] K. N. Venugopala, C. Sandeep, M. Pillay, H. Hassan Abdallah, F. M. Mahomoodally, S. Bhandary, D. Chopra, M. Attimarad, B.E Aldhubiab, A.B. Nair, N. Sreeharsha, M.A. Morsy, S. Pottathil, R. Venugopala, B. Odhav, and K. Mlisana, "Plosone",

[102] P. Rijiravanich, M. Somasundrum, and W. Surareungchai, "Femtomolar electrochemical detection of DNA hybridization using hollow polyelectrolyte shells bearing silver nanoparticles", *Anal. Chem.,* vol. 80, no. 10, pp. 3904-3909, 2008.
[http://dx.doi.org/10.1021/ac701867m] [PMID: 18407674]

[103] C.B. Jacobs, M.J. Peairs, and B.J. Venton, Https://Doi.Org/10.1016/j.Aca.2010.01.009.Sensor
[http://dx.doi.org/10.1016/j.aca.2010.01.009]

[104] J.G. Manjunatha, M. Deraman, N.H. Basri, and I.A. Talib, "Fabrication of Poly (Solid Red A) Modified Carbon Nano Tube Paste Electrode and Its Application for Simultaneous Determination of Epinephrine, Uric Acid and Ascorbic Acid", *Arab. J. Chem.,* vol. 11, no. 2, pp. 149-158, 2018.
[http://dx.doi.org/10.1016/j.arabjc.2014.10.009]

[105] J.G. Manjunatha, M. Deraman, and N.H. Basri, "Electrocatalytic Detection of Dopamine and Uric Acid at Poly (Basic Blue b) Modified Carbon Nanotube Paste Electrode", *Asian J. Pharm. Clin. Res.,* vol. 8, no. 5, pp. 48-53, 2015.

[106] J.G. Manjunatha, M. Deraman, N.H. Basri, N.S.M. Nor, I.A. Talib, and N. Ataollahi, "Sodium Dodecyl Sulfate Modified Carbon Nanotubes Paste Electrode as a Novel Sensor for the Simultaneous Determination of Dopamine, Ascorbic Acid, and Uric Acid", *C. R. Chim.,* vol. 17, no. 5, pp. 465-476, 2014.
[http://dx.doi.org/10.1016/j.crci.2013.09.016]

[107] E. Pavlovic, R.Y. Lai, T.T. Wu, B.S. Ferguson, R. Sun, K.W. Plaxco, and H.T. Soh, "Microfluidic device architecture for electrochemical patterning and detection of multiple DNA sequences", *Langmuir,* vol. 24, no. 3, pp. 1102-1107, 2008.
[http://dx.doi.org/10.1021/la702681c] [PMID: 18181654]

[108] F. Ahour, and M.K. Ahsani, "An electrochemical label-free and sensitive thrombin aptasensor based on graphene oxide modified pencil graphite electrode", *Biosens. Bioelectron.,* vol. 86, pp. 764-769, 2016.
[http://dx.doi.org/10.1016/j.bios.2016.07.053] [PMID: 27476058]

[109] A. Makler, and W. Asghar, "Exosomal biomarkers for cancer diagnosis and patient monitoring", *Expert Rev. Mol. Diagn.,* vol. 20, no. 4, pp. 387-400, 2020.
[http://dx.doi.org/10.1080/14737159.2020.1731308] [PMID: 32067543]

CHAPTER 7

Recent Advances on Electrochemical Sensors for Detection and Analysis of Heavy Metals

Monima Sarma[1,*]

[1] *Department of Chemistry, KL Deemed to be University (KLEF), Greenfields, Vaddeswaram, Andhra Pradesh 522502, India*

Abstract: Since the beginning of modern civilization, heavy metals have been used in various industrial, domestic, technological, medical applications, *etc*. Often, if not all the time, the uncontrolled release of sewage to the water resources and emission from the industrial plants in the environment raises alarms over their potential impacts on human health and the environment. These metallic elements, being potentially toxic, are known to cause multiple organ impairment, even at extremely low exposure limits. Thus, it is highly imperative to develop simple and sensitive methods for their detection. Electrochemical techniques are one of the most promising methods for heavy metal sensing because of their short analytical time, easy accessibility, and high sensitivity for *in-situ* measurements. In this chapter, we discuss heavy metals and their potential adverse effects on human health and the environment, strategies for the design of materials for sensing the heavy metals/ions and their mechanisms, and the recent developments of electrochemical sensors in heavy metal detection. The chapter puts more emphasis on materials than methods.

Keywords: Biochemical detection, Biomaterials, Electrochemistry, Electro-chemical detection, Electrochemical sensing, Enzyme, Fuel cells, Heavy metal ions, Inorganic nanomaterials, Microorganism, Organic nanomaterials, Photo-catalysis, Recognition, Redox reactions, Screen-printed electrochemical sensors, Self-powered electrochemical sensors, Sensors, Toxicity, Voltammetry, Wastewater analysis.

INTRODUCTION

What are electrochemical sensors? The word sensor originated from the Latin word *sentire,* meaning 'to feel.' Sensors have become progressively more important with the advancement of modern civilization. The invention of the Internet of Things (IoT) has taken the importance of sensors to an even greater level, and basically, human life these days is as dependent on man-made sensors

---

* **Corresponding author Monima Sarma:** Department of Chemistry, KL Deemed to be University (KLEF), Green fields, Vaddeswaram, Andhra Pradesh 522502, India; Email: monima.22@gmail.com

J.G. Manjunatha (Ed.)

as it is on the biological ones to sustain life. Sensors are mechanical devices that respond to a physical or chemical perturbation such as light, temperature, magnetism, pressure, electric field, movement, pH, chemical reactions, *etc.* which is understood by a computer after a converter converts these analog responses into digital electrical impulses and an amplifier amplifies them. In nature, each sensor is different and operates by different target-specific mechanisms. Chemosensors are the type of sensors that analyze and provide information on the qualitative and quantitative chemical composition of their environment [1]. The basic operating principle of a chemosensor is recognition of the analyte by the receptor fragment prompting a chemical change, followed by generation of the output signal by the transducer, which is then read by the built-in computer. These chemical sensors have widespread importance whereby the key parameters of such sensors are selectivity, sensitivity, detection limit, response time, and lastly, the size of the device. Ideally, the sensor should have great selectivity and sensitivity, a faster response time, and acceptable packaging size [2]. Since the boom in microelectronics and microengineering, devices have been miniaturized without compromising selectivity and sensitivity. Alike the physical sensors, the chemical sensors also have target-specific operating principles where the sensing mechanism could be one of the various possibilities such as absorption or emission of light, electrochemical responses, absorption or release of heat, *etc.*

Of the various types of known chemical sensors, electrochemical sensors have gained popularity due to some obvious advantages, as we shall discuss in this chapter. All these sensors contain electrodes that act as the transducer. These sensors have far-reaching applications in modern life, such as (a) gas sensors to detect the level of carbon monoxide, (b) water analysis for heavy-metal pollutants, (c) automobile exhaust analysis, (d) environmental monitoring, (e) food and health monitoring, (f) medical diagnostics (for example, blood glucose monitoring) and so on [3]. Perhaps the most common type of electrochemical sensor is a typical pH meter where an electrode assembly (the glass electrode) selectively responds to the change in acidity or alkalinity of water causing a fluctuation of the cell potential which is converted into the pH scale by a built-in computer.

The electrical response and introduction to the electroanalytical techniques:
The electrochemical sensors work mainly based on the variation of one of these properties in the presence of an analyte: potential difference (potentiometry), current (amperometry and voltammetry), resistance (conductometry), and capacitance. The potentiometric method involves the interface between an electrode surface and a solution and is an example of the interface method. On the contrary, the conductance of the solution is directly measured using two electrodes, and this is a non-interface method [3]. A common electroanalytical

method is potentiometry which involves a change in electrode potential in the presence of the analyte according to the following Nernst equation:

$$E_{cell} = E^0_{cell} - \frac{RT}{nF} \ln \frac{a_{ox}}{a_{red}} \approx E^0_{cell} - \frac{RT}{nF} \ln \frac{[ox]}{[red]} \approx E^0_{cell} - \frac{0.0591}{n} \log \frac{[ox]}{[red]}$$

where, E_{cell} is the cell potential (formal potential), E^0_{cell} is the standard cell potential, R is the universal gas constant (~ 8.314 J mol^{-1} K^{-1}), T is the temperature in Kelvin, n is the number of electrons involved, F is Faraday constant (~ 96485 C mol^{-1}), and those in bracket are the concentrations (more accurately, activity) of the oxidized and reduced species in solution, respectively. The concerned instrument is called a potentiostat which also uses a reference electrode, for example, saturated calomel electrode, silver-silver chloride electrode, *etc*. The electrode potential of the reference electrode is usually insensitive to the chemical composition of the solution and remains constant throughout the experiment. The traditional electrochemical cells are three-electrode assemblies and consist of a working electrode on the surface of which occur the redox reactions, the reference electrode, and a counter electrode (usually a platinum wire) to complete the electric circuit [4].

Besides potential difference, the current is another electrical response that is used as a sensory mechanism; the concerned technique is known as chronoamperometry which measures the current flowing through the working electrode as a function of time. The concentration of the redox-active species is then calculated by correlating with the current flow using the following Cottrell equation:

$$I_t = \frac{nFAc^0_0 D_0^{1/2}}{\pi^{1/2}t^{1/2}} = bt^{1/2}I_i$$

where I_t is current at time t (s), n is the number of electrons, F is Faraday constant (96485 C mol^{-1}), A is the area of the electrode (cm^2), c_o is the concentration of the oxidized species (mol L^{-1}), D_o is the diffusion coefficient of the oxidized species (cm^2 s^{-1}) [5]. This method allows processes to be studied in which the current is directly measured as a function of a constant potential applied to the electrochemical system.

The most used electroanalytical technique is based on voltammetry, which measures the flow of current (I) between two electrodes as the potential difference (E) between them is varied [6]. The two popular voltammetry-based techniques are cyclic voltammetry (CV) and differential pulse voltammetry (DPV). Cyclic voltammetry measures the current flowing through a redox-active analyte solution as a potential difference across the electrodes is scanned for a period. The relevant electrical response is oxidation or reduction of an electrochemically active species at the electrode surface, which occurs at certain potentials and is characteristic to the species involved. In this technique, a potential sweep is performed, positive and negative mode separately, by linearly increasing the applied voltage until a certain time, reversing the sweep direction, and so on. A faradaic current passes through the system which is recorded and plotted against the applied potential difference generating a histogram known as cyclic voltammogram. The performance of the sensor is analyzed based on the cyclic voltammogram thus obtained.

Differential pulse voltammetry (DPV), as the name suggests, is a differential technique like the first derivative of a linear voltammogram in which the formation of a peak is observed for a given redox process. In this technique, a base potential is applied at which no electrochemical reactions occur. Thereafter, the base potential is gradually increased between pulses, the current is measured at the start and the end of the pulse, and their difference is recorded [5]. Usually, the pulsed techniques, such as DPV, are more sensitive than the linear sweep methods like CV due to the reduction of capacitive current in the former. However, CV has remained the most popular electroanalytical technique due to several advantages, such as relatively inexpensive, fast analysis time, can be used for both organic and inorganic species, excellent sensitivity, ability to determine multiple analytes simultaneously, *etc.*

The analytes – heavy metal ions in water: Although the industrial revolution has diverse positive implications on mankind, it poses a serious threat to civilization – pollution. Uncontrolled release of sewage from industrial plants has been continuously polluting the water resources, and heavy metals are especially intimidating water pollutants. Besides the industrial runoffs, the other sources of heavy metal pollutants in the environment are human activities such as the burning of fossil fuels, emission from smelters and waste incinerators, waste from mining, among many others. The term 'heavy metal' alludes to metals with a density higher than $5 \ g \ cm^{-3}$ and atomic mass between 63.5 and 200.6 amu, which is usually considered toxic [7]. Usually, these include all metals but Al, Na, Ca, Mg, and K. Some of these metals are very toxic, such as Pb, As, Hg, Cd, Sb, Cr(VI), which are non-biodegradable, tend to accumulate in the environment and remain in the environment for decades or even centuries, transport through the food chain, and pose a serious threat to the environment and human health, even at low concentration [8]. The biological action of these metals depends on the nature of the metals (hardness, softness, *etc.*). For example, the soft metal ion

Hg(I) is extremely toxic as it alters the biochemical reactions by forming stable bonds through soft-soft interaction with the thiol proteins [9]. Even some essential elements like iron, cobalt, zinc, copper, manganese, *etc.*, fall under the category of heavy metals as they are toxic at a higher concentration than required. Once these metals make their way into the human body, they can cause nausea, vomiting, diarrhea, allergic reactions, or even long-term chronic diseases such as growth impairment and development, cancers, damages to organs or nervous system, and even death. The toxicity of the heavy metals is normally due to enzyme inhibition, oxidative stress, and impaired antioxidant metabolism. These mechanisms show unpleasant health effects through the free radical generation that might even cause DNA damage [7].

Analysis of heavy metals and the importance of electrochemical sensors: Detection and determination of heavy metals in the environment is an important research topic where the focus has been on the development of convenient and cost-effective, selective, and sensitive analytical techniques. For the trace detection of heavy metals, spectroscopic techniques such as atomic absorption spectroscopy (AAS), x-ray fluorescence, inductively coupled plasma mass spectrometry (ICP-MS), neutron activation analysis, *etc.*, are popular. These techniques come with some notable benefits, such as their versatility over a broad range of elements, high sensitivity, detection limit even in the femtomolar concentration range, *etc.* Be that as it may, these techniques have some serious limitations such as complex sample preparation, instrumentation, and data analysis which require qualified personnel and prevent on-site measurement and transient phenomena monitoring [10].

The shortcomings of the above techniques could be addressed by using electrochemical-based techniques and electrochemical sensors [11]. The primary advantage of electrochemical sensors and analytical techniques is that they are easy to use and analyze, inexpensive, easy sample preparation, and the instruments are relatively smaller compared to spectroscopy-based instruments. These reduce the possibility of contamination of the samples and drastically bring down the analysis time and cost. Even *in-situ* measurements are possible as the electrochemical sensors also allow high temporal resolution when associated with flow injection analysis. The combination of electrochemical techniques with microfluidics is another promising strategy for the detection of environmental contaminants such as heavy metal ions and has gained popularity [12].

Objectives of the chapter: The recent progress on electrochemical sensing of heavy metal ion environmental pollutants is systematically reviewed in this chapter. The electrochemical sensors have widespread applications in sensing organic materials such as explosives, drug molecules, nutrients, and so on [13 -

22]. However, the main objective of this chapter is to give the reader a general overview of the sensing of heavy metal ions which has an important impact on humankind. The chapter is divided into several sections for the sake of clarity, and each section discusses the modification of the electrodes for subsequent metal ion detection. On a broad note, it can be said that the goal of electrochemical sensing of heavy metal ions is to modify the bare electrodes with special materials to achieve fast electron transfer kinetics. These modifying materials range from metal nanoparticles, carbon-based nanomaterials to biomaterials such as enzymes, microorganisms, *etc*. As the effective surface area and the electron transfer rate of these modified electrodes are considerably higher than those of the bare electrodes, the research attention in this subject is centered on developing better-performing electrode modifiers. The readers will have a good idea of these varieties of wonder materials which will encourage them to think up even better next-generation materials. The performance of the sensors discussed in this chapter is summarized in Table **1**.

Table 1. Performance of the sensors discussed in this chapter.

Electrode	Method	Heavy Metal	Linear Range	Limit of Detection	Refs.
AgNPs/RGO/GCE	SWASV	Pb(II) Cd(II) Cu(II) Hg(II)	—	0.141 µM 0.254 µM 0.178 µM 0.285 µM	[24]
BiNPs/CPE	SWASV	Cd(II) Pb(II) Ni(II)	1-100 ppb 1-100 ppb 10-150 ppb	0.81 ppb 0.65 ppb 5.47 ppb	[25]
SnNPs/GO/GCS	SWASV	Cd(II) Pb(II) Cu(II)	—	0.63 nM 0.60 nM 0.52 nM	[26]
TiO$_2$/GCE	SWASV	Hg(II)	—	0.017 µM	[34]
DNA modified Fe$_3$O$_4$@Au/GCE	SWV	Ag(I) Hg(II)	10-150 nM 10-100 nM	3.4 nM 1.7 nM	[35]
BiNPs@Ti$_3$C$_2$T$_x$/GCE	SWASV	Pb(II) Cd(II)	—	2.24 ppb 1.39 ppb	[39]
trGO/Fc-NH$_2$-UiO-66/GCE NH$_2$-UiO-66 is a metal-organic framework (MOF)	DPASV	Cd(II) Pb(II) Cu(II)	0.01–2 µM 0.1–2 µM 0.1–2 µM	8.5 nM 0.6 nM 0.8 nM	[45]
Ti$_3$C$_2$T$_x$/MWCNTs/Au/PET	SWASV	Zn(II) Cu(II)	350–830 ppb 10–600 ppb	1.5 ppb 0.1 ppb	[49]
MWCNTs-Chitosan/GCE	SWASV	Cd(II)	1 – 50 µg/L	0.09 µg/L	[50]
AuNP/PANI/GR/GCE	DPASV	Pb(II)	0.5–10 nM	0.1 nM	[57]

(Table 1) cont.....

Electrode	Method	Heavy Metal	Linear Range	Limit of Detection	Refs.
GO/AP/GCE	SWASV	Cd(II) Cu(II)	10-500 ppb 10-500 ppb	3.3 ppb 3.3 ppb	[58]
AuND@GPL	DPASV	Pb(II) Cu(II) Hg(II)	1-50 ppb 1-50 ppb 1-50 ppb	0.12 ppb 0.19 ppb 0.18 ppb	[59]
AuNP/ERGO/SPCE	DPASV	Pb(II)	—	0.321 nM	[61]
Bi-SPCNTE	SIA-ASV	Pb(II) Cd(II) Zn(II)	— 	0.2 µg/L 0.8 µg/L 11 µg/L	[62]
Bi/MWCNT-IL/SPCE	SWASV	Cd(II) Pb(II)	1-60 µg/L 1-60 µg/L	0.5 µg/L 0.12 µg/L	[63]
SPCEs | GS-Nafion/Au	DPSV	Pb(II) Cd(II)	0.5–60 nM 0.8–50 nM	0.23 nM 0.35 nM	[65]
SPE/ERGO/BiF	SWASV	Cd(II) Pb (II)	—	0.5 µg /L 0.8 µg /L	[66]

AgNPs: silver nanoparticles, GR: graphene, GO: graphene oxide, RGO: reduced graphene oxide, trGO: thermally reduced graphene oxide, MWCNT: multi-walled carbon nanotube, AuNP: gold nanoparticle, PANI: polyaniline, AuND: Au nanodendrites, AP: p-aminophenyl, ERGO: electro-reduced graphene oxide, GS-Nafion/Au: graphene sheets-Nafion Gold Nanoparticles, IL: ionic liquid, Bi-SPCNTE: Bismuth-scree--printed carbon nanotubes electrode; GCE: glassy carbon electrode, CPE: carbon paste electrode, GCS: glassy carbon sheets, PET: polyethylene terephthalate, GPL: graphite pencil lead, SPCE: screen-printed carbon electrode; SWASV: square wave anodic stripping voltammetry, SWV: square wave voltammetry, DPASV: differential pulse anodic stripping voltammetry, DPSV: differential pulse stripping voltammetry, SIA-ASV: sequential injection analysis-anodic stripping voltammetry.

Metal-Based Nanomaterials as Electrochemical Sensors

The inorganic nanomaterials have intrinsic advantages such as periodic structure, chemical, thermal stability, large surface-to-volume ratio, high surface reaction activity due to strong adsorptivity, and catalytic efficiency, which make them popular in the electrochemical sensing of heavy metal ions [23]. The nanomaterials are used extensively as electrode modifiers in the sensitive and anti-interference detection and determination of heavy metal pollutants in the environment. In this section, we will discuss some representative examples of inorganic nanomaterial-based electrochemical sensors.

Metal nanoparticles in electrochemical sensing: The metal nanoparticles are known to display exciting electrical, optical, and catalytic properties. Further modification of the nanoparticles by attaching small functional groups or biomolecules on the surface leads to new materials with unique properties, and these materials are being tested in heavy metal sensing. The early electroanalytical research to sense heavy metals involved the incorporation of hanging mercury electrodes due to their many advantages, such as wide cathodic potential range, high sensitivity, and repeatability. However, the toxicity of mercury raised the concern of using it as the electrode material, and a call for developing mercury-free solid-state electrodes for heavy metal sensing was made [23].

Among the various metal nanoparticles, silver nanoparticles (AgNPs) are known to exhibit fascinating quantum physical characteristics such as large specific surface area, small particle diameter, fast electron transferability, *etc.*, which make them promising electrode materials. Sang *et al.* developed different mass ratios silver nanoparticles – reduced graphene oxide (AgNPs–RGO) nanocomposites as electrochemical sensors for heavy metals [24]. The graphene oxide (GO) and reduced graphene oxide has gained popularity due to their high sensitivity, low cost, and environmental friendliness. The AgNPs–RGO nanocomposites were synthesized by *in situ* synthetic methods by dispersing silver nitrate in hot RGO-DMF solution, adding ascorbic acid and beta-cyclodextrin, followed by centrifuging the solution. The SEM and TEM images of the nanocomposite reveals the homogeneous distribution of the Ag nanoparticles on the RGO sheets, indicating a strong interaction between them. The relevant electrochemical studies comprised the three-electrode assembly where the working electrode was a bare or modified magnetic glassy carbon electrode (MGCE, 3 mm in diameter), Ag/AgCl (3 M KCl) was the reference electrode, and the counter electrode was platinum. The AgNPs–RGO nanocomposites were found to have excellent electrochemical activity and sensitivity to detect heavy metals such as lead, cadmium, copper, and mercury, as revealed by square wave anodic stripping voltammetry (SWASV). The relevant electrochemical behavior is shown in Fig. (**1**). The sensitivities and limits of detection (LOD) were 48.69, 40.06, 15.66 and 43.18 $\mu A\ \mu M^{-1}$ and 0.141, 0.254, 0.178 and 0.285 μM respectively for Pb(II), Cd(II), Cu(II) and Hg(II). This study reveals that the nanocomposite was better than the bare RGO film as it exhibited considerably greater electrochemical activity toward the detection of the four metal ions. Also, the nanocomposite modified electrode exhibited good anti-interference properties and the enhanced stability of the nanocomposite indicates its potentiality as an electrode material for sensing heavy metal pollutants.

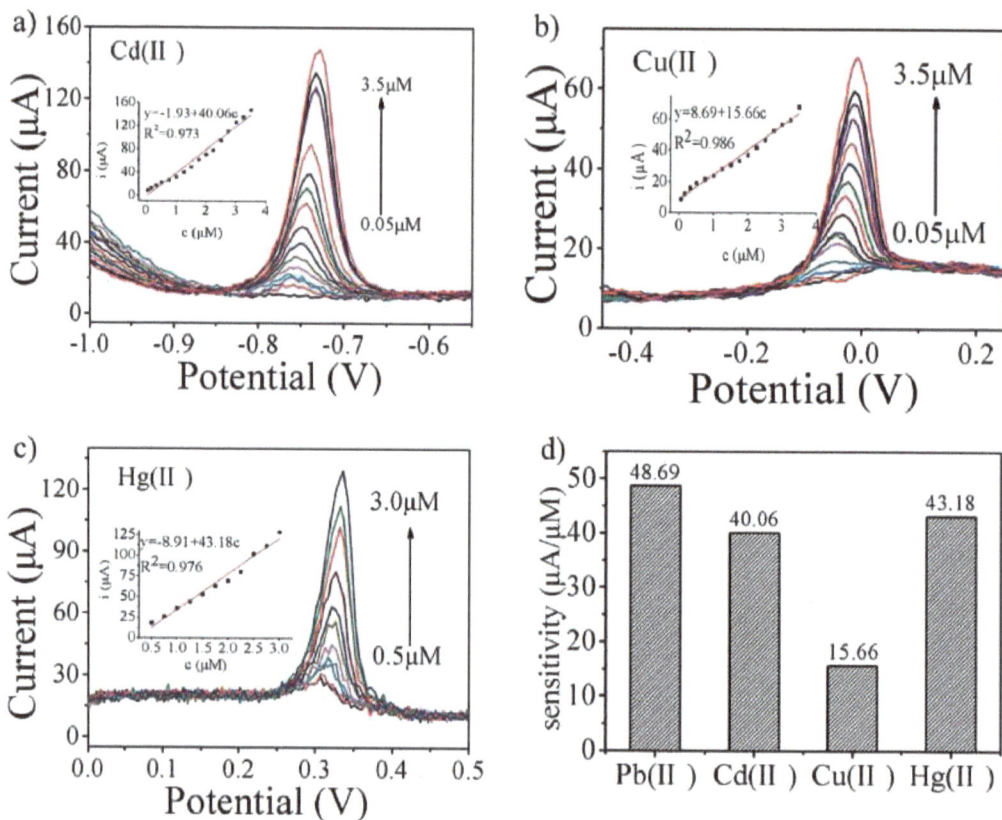

Fig. (1). (a-c) SWASV responses and the corresponding calibration plots of the AgNPs/RGO nanocomposite-modified MGCE for the individual determination of Cd(II), Cu(II), and Hg(II) after 150 s in 0.1 M HAc–NaAc solution (pH = 5.0). (d) Comparison between the sensitivity *versus* Pb(II), Cd(II), Cu(II), and Hg(II) of the AgNPs/RGO nanocomposite-modified electrode. Reproduced with permission from reference 24. Published by The Royal Society of Chemistry.

The bismuth nanoparticles are found to be useful too. Niu *et al.* reported detailed analytical activity of Bi nanoparticle – porous carbon paste based working electrode and electrochemical sensing of Cd(II), Pb(II), and Ni(II) in different types of water samples such as tap water, groundwater, wastewater, river water, *etc.* The concerned electrode comprised nicely dispersed metal nanoparticles in a highly porous carbon matrix which provided a large active area for electrochemical activity. Notably, detection limits up to 0.81, 0.65, and 5.47 ppb for Cd(II), Pb(II), and Ni(II) respectively were observed under four minutes overall analysis time. Also, the carbon-bismuth nanocomposite could be used as ink to produce printed electrodes for miniature devices [25].

Reduced graphene oxide – tin nanoparticles nanocomposites were also tested as the electrochemically active component in heavy metal ion sensing by Lee *et al.*

The ratio between tin and graphene oxide was found to be crucial during device fabrication, and a low value was maintained to prevent the aggregation of GO which might hinder the reduction process. The relevant electrodes exhibited high sensitivity for individual and simultaneous detection of Cd(II), Pb(II), and Cu(II) with good stability and limits of detection reaching up to 0.63 nM, 0.60 nM, and 0.52 nM, respectively. However, the stability of the simultaneous detection was found to be relatively low, perhaps due to the formation of some intermetallic compounds. The obtained LOD values were below those specified by the World Health Organization (WHO) [26].

Metal oxide-based nanomaterials in electrochemical sensing: Though the metal-based nanoparticles are promising electrode materials for heavy metal sensing, the metal oxide-based nanomaterials such as ZnO, Fe_3O_4, NiO, SnO_2, ZrO_4, TiO_2, MgO, MnO_2 *etc.* are proven to be better candidates due to their superior electrochemical performance, biocompatibility, non-toxicity, catalytic properties, *etc.* These nanomaterials display augmented electron transfer kinetics and strong adsorption ability. Besides, they have distinct surface properties such as surface defects and oxygen vacancies, definite exposed crystal face and crystal phase, *etc.*, which govern their catalytic and electroanalytical activities. Finding out the structure-property relationship of these materials is important to design efficient electrochemical sensors. The consequences of defects and oxygen vacancies in metal oxide nanoparticles have gained special attention, and extensive research in this direction has proved that the nano metal oxides with special surface properties boost their electroanalytical performance [23].

There are three general methods for the synthesis of the metal oxide nanoparticles – bottom-up strategy, top-down strategy, and hydrothermal synthesis; the crystal growth depends on both thermodynamic and kinetic conditions. In the bottom-up crystal growth strategy, the thermodynamic and kinetic factors are controlled by the adsorption of protecting agents such as organic surfactants, polymers, small molecules, *etc.* From the thermodynamic point of view, the adsorption of protectants causes reduction of the surface energy, while the kinetic view suggests preservation of the crystal facet as the protectants alter the growth rate and growth energy barrier [27]. The top-down crystal growth method, on the other hand, involves the generation of nanoparticles with regular crystal facets followed by selective *etc*hing of partial crystal faces. The *etc*hing speed of various crystal facets is found to be different even at the same condition due to the anisotropy of the crystal. The *etc*hing reaction is stopped when the exposed surface is saturated with oxygen atoms resulting in a pagoda-like morphology [28]. Hydrothermal method has remained as another popular synthetic method for preparing nanomaterials due to low cost, easy control over morphology and facet, high purity of the obtained materials, good dispersibility, *etc.* The raw materials, feed ratio, the extent of stirring, reaction temperature and time, *etc.*, normally influence the crystal growth. A variation of this method is a microwave-assisted hydrothermal synthesis which enhances the synthesis efficiency by shortening the reaction time and improving uniformity of the phase [29].

The active sites of the metal oxide nanomaterials are the defects and oxygen vacancies on the surface which enhances the adsorption ability, catalytic and electrochemical activity [30]. The defects are generally introduced by doping of heterogeneous atoms, absence of atoms or ions and dislocations, *etc.* The oxygen vacancies are usually found in d- and f-block metal oxides and alter the surface properties of these materials including electron transport, selectivity, catalytic activity, *etc.* Titanium dioxide (TiO_2) is a low-cost, highly abundant, and stable metal oxide which is found to be promising in the electrochemical sensing of heavy metals. Pure TiO_2 has low conductivity, poor reactivity, and a wide bandgap (3.2 eV) which makes it unsuitable for electrochemical detection [31]. However, modified TiO_2 such as $DNA/C/TiO_2$ and Ti/TiO_2 nanotube/Au composites, *etc.* exhibit good electrochemical performance and have gained popularity lately [32]. Defective TiO_2 crystals with specific exposed high-energy facets are known to be efficient electrocatalysts for oxygen reduction reactions [33].

Zhou *et al* demonstrated electrochemical sensing of heavy metal ions by a defective single-crystalline (001) titanium dioxide nanosheet which is modified by surface Ti^{3+} ion and oxygen vacancy [34]. Usage of defective TiO_2 nanosheet-modified electrodes eliminated the necessity of other types of modifications. The authors synthesized the TiO_2 nanosheets by a hydrothermal method employing 98% concentrated sulfuric acid as the solvent followed by heat treatment at different temperatures *viz.* 25 °C (T-1), 300 °C (T-2), 600 °C (T-3), and 800 °C (T-4). The first three samples had similar structures, whereas crystal transformation and morphology changes were observed in the fourth sample. Using the SWASV method, electrochemical detection of Hg^{2+} was then performed which revealed sensitivities of T-1, T-2, T-3, and T-4 to be 270.83, 159.52, 20.51, and 95.81 $\mu A\ \mu M^{-1}\ cm^{-2}$, respectively that were higher than bare glassy carbon electrode (13.75 $\mu A\ \mu M^{-1}\ cm^{-2}$) (Fig. **2**). As T-1, T-2, and T-3 had similar morphology, the difference in their sensitivity in Fig. (**2A**) alluded to the surface defects, and the authors predicted the enhanced sensitivity of T-3 as compared to T-4 to be due to an anatase-rutile heterostructure. Interestingly, the data of T-1- and T-2-based electrodes satisfied the requirement of the drinking water safety standard (0.03 μM) set by the World Health Organization (WHO) (Fig. **2B**).

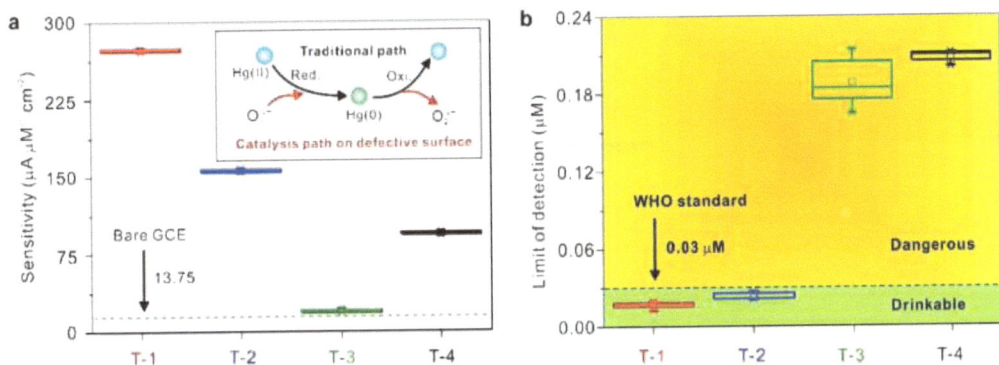

Fig. (2). The statistical distribution of electrochemical performance of TiO_2 electrodes toward Hg^{2+}. (a) Comparison of sensitivities of different heat-treated samples toward Hg^{2+}. The role of the oxygen vacancy is illustrated in the inset. b) Limits of detection (LOD) of sensors T-1 to T-4. Reproduced with permission from reference 34. Copyright (2017) American Chemical Society.

Miao *et al* proposed an electrochemical sensing strategy for highly sensitive and selective detection of Ag^+ and Hg^{2+} ions using DNA modified Fe_3O_4@Au nanoparticles and magnetic glassy carbon electrodes (MGCE) [35]. The magnetic nanomaterials have widespread technological applications where Fe_3O_4 nanoparticles with inherent high surface-to-volume ratio and exciting magnetic properties are one of the most widely used magnetic nanomaterials. The Fe_3O_4 nanoparticles are used in different forms such as $MnFe_2O_4$, Fe_3O_4@Au, Fe_3O_4@Pt, Fe_3O_4@carbon dot nanocomposites, *etc*. Metal ion coordinated DNA base pairs such as cytosine-cytosine (C-C) and thymine-thymine (T-T) mismatches in DNA duplexes on the surface of nanoparticles could be utilized to bind heavy metal ions. The synthesis and sensing mechanism of the DNA-modified Fe_3O_4@Au nanoparticles are shown in Fig. **(3)**, where the DNA probes labeled with two electrochemically independent species – ferrocene and methylene blue – are used for simultaneous detection of different types of heavy metal ions. The third DNA probe is thiolated to attach to the surface of Fe_3O_4@Au nanoparticles *via* sulfur-gold chemistry. The electrochemical sensing behavior of the obtained nanocomposites was analyzed by square wave voltammetry which revealed simultaneous detection of Ag^+ and Hg^{2+} with concentrations lower than the toxicity levels in drinking water defined by the U.S. Environmental Protection Agency (USEPA) (Ag^+: 460 nM; Hg^{2+}: 10 nM).

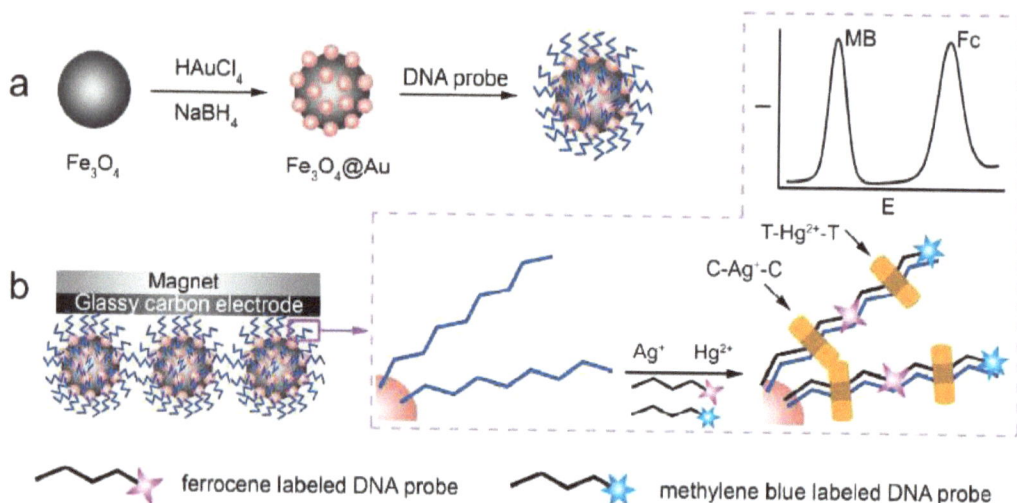

Fig. (3). (a) Synthesis scheme of DNA modified Fe3O4@Au nanoparticles. (b) Proposed sensing scheme for simultaneous detection of Ag^+ and Hg^{2+}. Reproduced with permission from reference 35. Copyright (2017) American Chemical Society.

Metal nanocomposites as electrochemical sensors: It has already been mentioned in a previous section that bismuth nanoparticle-based electrodes are being tried out as alternative electrodes to replace mercury-based electrodes. Like any other nanoparticles, the bismuth nanoparticles have a higher surface area and catalytic activity compared to the relevant films, and modification of the nanoparticles by some means is expected to alter their electrochemical behavior. A major setback of bismuth nanoparticles is their high surface free energy which stimulates aggregation of the particles thereby reducing their activity. This is sometimes avoided by tethering the nanoparticles in a support matrix such as carbon nanotubes, graphene, meta-organic framework, *etc* [36].

A special type of supporting matrix for bismuth nanoparticles is MXenes which are two-dimensional inorganic compounds consisting of few-atoms-thick layers of transition metal carbides, nitrides, or carbonitrides. These materials have potential applications in batteries, supercapacitors, optoelectronic devices, *etc.* due to their exciting properties such as good conductivity, stability, large hydrophilic surface area, *etc* [37]. Many scientists consider the MXenes as good anchorage to the bismuth nanoparticles. For example, Jiang *et al* developed an acetylcholinesterase biosensor modified with Ag@MXene nano-composite which was used to detect Malathion [38].

He *et al* demonstrated electrochemical sensors based on $BiNPs@Ti_3C_2T_x$ MXene nanosheets embedded in glassy carbon electrode (GCE) for detection of heavy-metal ions such as lead, cadmium [39]. The $Ti_3C_2T_x$ nanosheets were first prepared and the Bi^{3+} ions were then adsorbed to the surface of the nano-sheets which served as the nucleation site for the BiNPs. The active area of the BiNPs@$Ti_3C_2T_x$ nano-composite could be calculated using the equation:

$$Q(t) = \frac{2nFAcD^{1/2}t^{1/2}}{\pi^{1/2}} + Q_{dl} + Q_{ads}$$

where *Q(t)* is the absolute value of reduction charge at time *t*, *n* is the number of electrons, *F* is the Faraday constant, *A* is the area, *c* is the analyte concentration, *D* is the diffusion coefficient, Q_{dl} is the double-layer charge, and Q_{ads} is the charge consumed by the adsorbed species. The effective area of the BiNPs@$Ti_3C_2T_x$/GCE electrode was calculated to be 0.04857 cm² which was larger than bare GCE and $Ti_3C_2T_x$/GCE both. The electrode was then employed for simultaneous detection of Pb^{2+} and Cd^{2+}, and the corresponding SWASV signals were composed of well-defined peaks for both the ions. The better resolution of the composite electrode signals as compared to the bare GCE and $Ti_3C_2T_x$/GCE electrodes were presumably due to some sort of synergic effect between BiNPs and $Ti_3C_2T_x$ nanosheets. The detection limits were found to be 10.08 nM for Pb^{2+} and 12.4 nM for Cd^{2+}, which were lower than the WHO recommended concentration of the cations in drinking water. The electrochemical signals of lead and cadmium ions were found to remain almost unaffected in presence of a variety of ions such as Mg^{2+}, Fe^{3+}, Co^{2+}, Ni^{2+}, Mn^{2+}, Zn^{2+}, Al^{3+} *etc*. However, Cu^{2+} ion was found to interfere with the signals due to the formation of intermetallic compounds which could be avoided by adding ferrocyanide to the water sample [39].

Metal-organic frameworks: Even though significant progress has been made, the major problem associated with the electrochemical sensors is poor reproducibility. Ratiometric detection strategy which monitors changes in the ratio between two electrochemical signals (analyte and reference signals) is found to be useful in enhancing reproducibility and reliability of electrochemical sensing processes [40]. The functional materials modified electrodes are turned out to be better and one such material is the metal-organic frameworks (MOFs). The MOFs have engineerable porous structures, large surface area, and active metal sites which made them suitable in many applications such as gas storage, separation, sensing, drug delivery, catalysis, and so on. The structures of the MOFs could be easily manipulated and the adjustable pores in them assist diffusion of heavy metal ions through them which facilitate interaction between the guest molecules and the host matrix. The large surface area together with a variety of active metal centers and organic ligands are helpful for the recognition of heavy metal pollutants.

These factors give an impetus to the MOFs for their application in electrochemical sensing of heavy metal ions. Fluorescent MOFs are known to recognize heavy metals [41, 42], but their applications in electrochemical sensing are limited by poor electrical conductivity and sparse water solubility. It comes as no surprise that scientists have turned things around, and water-soluble, conductive MOFs are now known which has proved to be useful in electrochemical sensing of heavy metals [43, 44].

Wang and coworkers developed a ferrocene carboxylic acid-functionalized MOF (Fc-NH$_2$-UiO-66) for ratiometric electrochemical sensing of heavy metal analytes [45]. The redox-active ferrocene sites in the MOF serve well as the internal reference and participate in ratiometric sensing. Also, the ferrocene sites induce conductivity to the MOFs by a charge hopping mechanism. Thermally reduced graphene oxide (trGNO) was incorporated to further enhance the electrical conductivity of the material, and the prepared composite is designated as trGNO/Fc-NH2-UiO-66. This could be understood by enhanced electrochemical signal [Fe(CN)$_6$]$^{3-/4-}$ on the trGNO/GCE electrode compared to the bare GCE. The electrochemical signal is further augmented on the trGNO/Fc-NH$_2$-UiO-66/GCE electrode and the reason for this is the fast electron transfer due to excellent conductivity and large surface area of reduced graphene oxide; and a good degree of dispersion of Fc-NH$_2$-UiO-66 on trGNO nanosheets and the good contact between them. The electrode was further used in simultaneous ratiometric sensing of Cd^{2+}, Pb^{2+}, and Cu^{2+} which is indeed a challenge due to competitive adsorption and difference in affinity of the metal ions toward the MOF. Differential pulse anodic stripping voltammetry (DPASV) was used as the electroanalytical tool. The peak current of the three metal ions was gradually increased with increasing concentration while the ferrocene signal remained mostly unaffected. The weak signal between lead and copper might be due to the formation of some Pb-Cu intermetallic compounds. The limits of detection were calculated to be 8.5 nM for Cd^{2+}, 0.6 nM for Pb^{2+} and 0.8 nM for Cu^{2+}. The selectivity of the developed sensor was analyzed by recording data in presence of excess amounts of other ions such as Zn^{2+}, Al^{3+}, Ca^{2+}, Ni^{2+}, Mn^{7+}, Zr^{4+}, Na$^+$ and K$^+$. The DPASV response signals are little affected by the interfering ions demonstrating excellent selectivity of the trGNO/Fc-NH$_2$-UiO-66/GCE electrode toward simultaneous detection of Cd^{2+}, Pb^{2+}, and Cu^{2+}. Also, the sensor electrode was found to be reproducible and had quite good stability [45].

Carbon Nanomaterials in Heavy Metal Electrochemical Sensing

Carbon nanomaterials like any other nanomaterials are packed with interesting properties but are mostly composed of one element, carbon. Of the variety of carbon-based nanomaterials, the single and multi-walled carbon nanotubes

(SWNTs, MWNTs), fullerenes and graphene are the most common. The accessibility of various hybridized states (sp, sp^2, sp^3) of carbon due to the low energy difference between the 2s and the 2p orbitals opens diverse structural possibilities. The outcome is multiple nanostructures with a distinctive topology such as planar (graphene), spherical (fullerenes), tubular (nanotubes), and so on. The majority of these carbon-based nanomaterials have sp^2 hybridized carbon centers and most of these materials could be functionalized through appropriate chemical reactions to modify their physicochemical, optical, and electrical properties. The carbon nanotubes have enormous mechanical strength, can sustain extreme strain, and one of the most durable materials known. On balance, carbon nanomaterials are among the scientific marvels, and it comes as no surprise that they have a rich level of applications. Electrochemical sensing of heavy metals is one of those [46].

Voltammetric stripping analysis is a sensitive technique with a low signal-to-noise ratio that can detect multiple metal ions simultaneously. The three important steps of this method are (a) accumulation, preconcentration, sorption, or complexation of metal ions on the carbon nanomaterial modified electrode surface; (b) reduction of the accumulated metal ions to the corresponding neutral species under negative potential; (c) stripping of the reduced metal species back into the solution by applying a positive potential. The stripping current in the subsequent process is proportional to the number of accumulated metal ions. As the carbon-based nanomaterials offer fast electron transfer rates and large surface areas, stripping current with these modified electrodes is significantly higher than that of the bare electrodes [47]. Potentiometric stripping analysis of heavy metal ions is a relatively younger technique than voltammetric striping analysis of the same. As the name suggests, this analysis method scans the potential of the working electrode as a function of time, where the time needed for stripping the accumulated substrate is related to its concentration in the relevant solution. The potentiometric stripping analysis is less vulnerable to intermetallic or surfactant effects than voltammetric stripping analysis [47].

Carbon nanotube-based electrochemical sensors: The carbon-based nanomaterials display a slew of interesting properties that make them suitable as electrode materials for this chapter. According to the scientific evidence, the presence of reactive groups on the surface of the carbon nanotubes, or edge-plane type sites at the end of them and in defect areas are responsible for their electrocatalytic activities. Also, the carbon nanotubes have exciting electrical properties such as large electrical conductivity. In short, the carbon nanotubes are excellent materials for electrode modification due to the lower over-voltage and higher peak current they exhibit compared to the bare electrodes. The most popular methods to modify the electrode surface with carbon nanotubes are casting of colloidal dispersion and direct growth onto electrode surfaces by methods such as chemical vapor deposition [48].

Solubility is a major concern while developing modified electrodes, and there is a setback for the carbon nanotubes – they are sparsely soluble in most of the solvents. The insolubility stems from the tendency to form aggregates due to large dispersion forces between the tubes, and the direct dispersion of the carbon nanotubes in solvents is desired. Surfactants at high concentrations are found to be useful in dispersing carbon nanotubes in water. Another approach involves functionalizing the nanotubes with an appropriate functional group which improves dispersion. Carbon nanotubes are normally less reactive, and it is known that the ends of the tubes are more reactive than the cylindrical part. Oxidation of the tubes results in open tubes and the pentagonal defects at the ends and curvature make the open ends more reactive which can undergo a variety of chemical transformations. Due to the heterogeneous reactivity of the carbon nanotubes, the tips could be polarized which enhances the dispersibility of the aqueous medium. The downside of oxidation is that it must be done with great care to avoid disruption of the extended network of the carbon atoms which might change the electrical properties of the carbon nanotubes.

Chemical vapor deposition (CVD) is a great technique to grow carbon nanotubes directly on diverse surfaces. The obtained nanotubes are normally highly pure and subsequent purification is often not required. In this technique, the nanotubes are generated by the breakdown of a hydrocarbon or carbon monoxide gas, and the nanotube growth is seeded by some catalyst nanoparticles on the substrates. The growth kinetics is regulated by the hydrocarbon carrier gas, growth time, temperature, and composition of the catalyst [48].

As the whole food chain has a labyrinthine network and the contamination in water ultimately leads to the pollutants ending up in humans, detection of the heavy metals in human biofluids is as important as detecting them in water. A heavy metal electrochemical sensor for noninvasive detection of copper and zinc ions in human biofluids was developed by Hui *et al* which was based on a layer-by-layer assembly of $Ti_3C_2T_x$/MWNTs nanocomposites [49]. It has already been mentioned previously that MXenes have good electrical conductivity, outstanding hydrophilicity, large surface area and they are soft to the environment. Thus, a combination of carbon nanotubes and MXene as electrode material brings in many promises. The manufacturing and modification process of the said electrode is schematically represented in Fig. (**4**). Zn^{2+} and Cu^{2+} ions were simultaneously detected first in acetate buffer solution and then in human excretory samples. The stripping current of Cu^{2+} increased linearly as the concentration of copper was increased in presence of zinc ions where the stripping current of the latter was barely affected by the former. When the concentration of zinc ions was linearly increased and the level of copper remained fixed, the stripping current of zinc increased linearly and that of copper remained almost unaffected. The limit of detection for the $Ti_3C_2Tx/MWNTs/Au$ nanocomposite modified electrode was 1.5 ppb and 0.1 ppb for zinc and copper respectively using the SWASV technique and the detection the range for the said metal ions were 350–830 ppb and 10–600 ppb, respectively. The performance of this carbon nanotube nanocomposite was way ahead of many other contemporary reports. To check out the practicality of the fabricated electrode, it was applied to detect the two metal ions in raw urine and sweat samples where results similar to acetate buffer were obtained. Also, the sensor had good repeatability and flexibility for heavy metal ion detection.

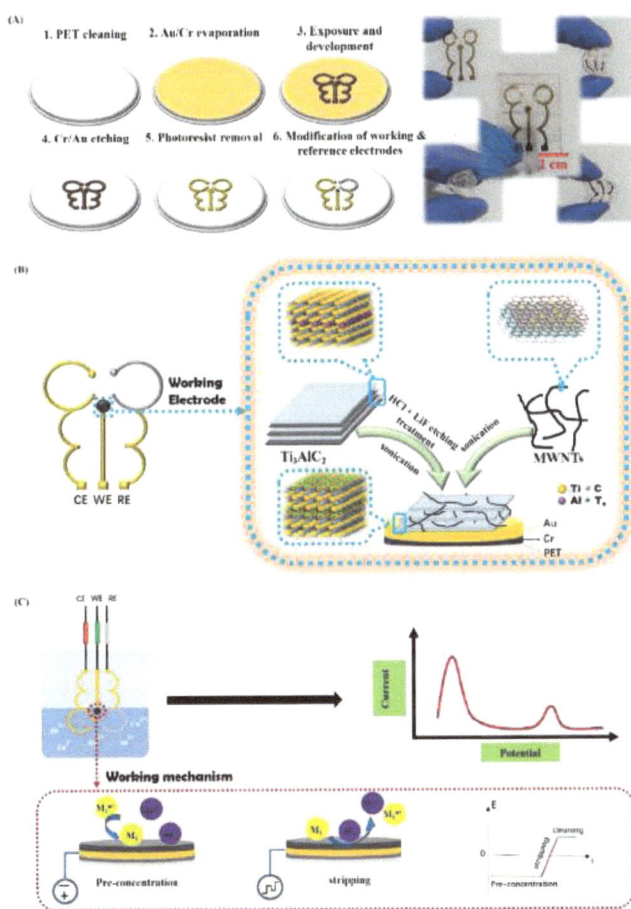

Fig. (4). (a) Schematic representation for the fabrication of $Ti_3C_2T_x/MWNTs/Au/PET$ electrode. (b) Modification of the working electrode. (c) The mechanism of heavy metal detection by the MXene-carbon nanotube nanocomposite modified electrode. Reproduced with permission from reference 49. Copyright (2020) American Chemical Society.

A MWCNTs-chitosan modified glassy carbon electrode for trace detection of cadmium (II) was developed by Galicia and coworkers [50]. The carbon nanotubes were first oxidized by nitric acid to form some carboxylic acid group which acted as the hydrogen bonding tethers to the chitosan molecules. Based on impedance and electrochemical experimentation it was found that the kinetics of the redox processes became faster after modification of the glassy carbon electrode with chitosan. This alluded to the better electron transfer ability and larger surface area of carbon nanotubes than GCE. The maximum phase angle (φ) could be selected as the parameter to measure the electron transfer kinetics using the equation:

$$\varphi = \tan^{-1}\frac{1}{1 + 2R_{ct}R_s}$$

where R_{ct} is charge transfer resistance and R_s is the resistance of the electrolyte solution. Thus, the phase angle decreases as the rate of charge transfer at the electrode interface increases. The effective working area of the electrode could be calculated using the de Randles–Sevcik equation analyzing the reversible redox process of $[Fe(CN)_6]^{3-}$:

$$i_p = 0.4463 \left(\frac{F^3}{RT}\right)^{1/2} n^{3/2} A C v^{1/2} D^{1/2}$$

Here, i_p is the peak current (μA), F is the Faraday constant, R and T are the gas constant and temperature, respectively, n is the number of electrons involved in the subsequent redox reaction, A is the electrode area (cm^2), C is analyte concentration (mol L^{-1}), v represents scan rate (V s^{-1}) and D is the diffusion coefficient (cm^2 s^{-1}). This equation illustrates that the mass transport mechanism at the interface is diffusion-limited. The cadmium ion sensing of the chitosan-CNT modified GCE electrode was validated using the SWASV technique which showed a well-resolved peak at -0.8 V *vs* SCE due to electrochemical oxidation of Cd (II), even at low concentration. In contrast, the bare GCE electrode signal contained mostly noise pointing out the higher efficiency of the CNT-chitosan modified electrode. The experiment was extended to several samples of tap water and drinking water where the electrode was found to offer excellent precision and accuracy. Also, the electrode was found to be useful in presence of several other toxic heavy metals such as Pb, Co, Hg, *etc.*

Graphene-based electrochemical sensors: Graphene is another carbon-based wonder nanomaterial that has a two-dimensional honeycomb network of sp^2

hybridized carbon atoms. The graphite structure is a stack of these carbon sheets held together by weak dispersion forces. The single-layer graphene sheets were first exfoliated by Novoselov and Geim who shared the physics Nobel Prize in 2010. Graphene is the building block for many other carbon-based nanomaterials, *i.e.,* it can be rolled into carbon nanotubes, stacked to obtain graphite, or it can be folded to form the fullerenes. Graphene has unique electronic, mechanical, physicochemical properties and is a subject of intensive modern research. It has the higher specific surface area of all materials which means it has a substantial fraction of surface atoms exposed to the analyte. The electron mobility of graphene is as high as 2×10^5 cm^2 V^{-1} s^{-1} even at room temperature. The large two-dimensional electrical conductivity of graphene together with a significant portion of electrochemically active edge carbon atoms per mass enhances electron transfer between substrate and electrode and this makes graphene an exciting material in electrochemistry [51].

Several methods to obtain single-layer graphene sheets or their analogs are now known. The most efficient method to obtain high purity and defect-free graphene sheets is the exfoliation method. Unfortunately, this method is not practical for the scaled-up synthesis of graphene as it is highly labor-intensive. Therefore, graphene is usually synthesized on an industrial scale by thermal decomposition of silicon carbide, chemical vapor deposition of hydrocarbons mediated by epitaxial growth of graphene on transition metals (Ni, Pd, Ru, Ir, Cu), or chemical reduction of graphite oxide which is an oxidized form of graphene. On a side note, the graphene obtained in the last method has a slightly different structure from pristine graphene, due to the presence of carbon vacancies, residual oxygen content, and polygonal carbon structures other than the normal hexagonal carbon system [52].

Before diving into the graphene-based electrochemical sensors for heavy metal ion detection, it is important to understand how the electron transfer from graphene-modified electrodes to the substrates occurs. As it is known that graphene has unusually high electrical conductivity and thereby little resistance $(1.00 \times 10^{-8}$ Ω cm), it might be tempting to state that graphene has a fast electron transfer rate between it and the analyte. Ironically, graphene has the intrinsic ability to impede electron transfer, making its implication as an electrode material more complicated than the concept based on conductivity alone [53]. Evidently, the electrochemical activity at the solid-electrolyte interface is governed by factors other than conductivity alone. It is found that the symmetry of orbitals plays an important role in electron transfer at the solid-electrolyte interface. The applied potential for a feasible electron transfer reaction to occur at the electrode surface must overcome the total energy required for the electron transfer and the subsequent reorganization of the oxidized or reduced species. The commonly used metals in electrodes (gold, platinum, *etc.*) have cubic closed packed structures with conductivity in multiple directions allowing electron flow in several directions. In contrast, due to planar atomic arrangement, conductivity in a graphene sheet is limited to only the plane in which the atoms are bonded together (conventionally, the xy plane), and conductivity along the perpendicular direction is puny. In other words, graphene has a high intramolecular conductivity but low intermolecular conductivity, setting up a paradox for its use as an efficient electrode material.

The conundrum of graphene electrochemistry is untangled by considering the orbital structure of graphene. Based on extensive simulation studies, it is realized that the orbitals active for electron transfer from graphene to other substrates are located at the edge of the sheets. If we assume that the graphene sheet is aligned parallel to the electrode surface, the electron that is transferred to graphene from the electrode is immediately available at the edge due to the high conductivity of the graphene sheet. This electron at the edge plane of graphene can be efficiently transferred to the electrochemically active species if the former is exposed to the solution. In summary, the electrical anisotropy, not the conductivity is the key to graphene's use as an electrode material. Though there are different views on this explanation, it is being consolidated by new evidence [54].

Graphene-modified electrodes have gained popularity for sensing heavy metal ions [52, 55, 56]. Nafion-graphene nanocomposite-based electrochemical sensors with high sensitivity are known for the detection of various toxic heavy metals such as Pb^{2+}, Cd^{2+}, Zn^{2+}, Cu^{2+} *etc.* The high sensitivity of this electrode system alluded to the combined effect of enhanced electron conduction of reduced graphene oxide and cation exchange capacity of Nafion. Most of these experiments used SWASV as the analytical tool, and the performance of the Naphion-graphene electrode was comparable to that of inductively coupled plasma mass spectrometry (ICP-MS). One problem of this Naphion-graphene nanocomposite electrode was the chance of aggregation of the graphene sheets after the solution dried up. This problem was minimized by incorporating nanoparticles into the graphene sheets [55]. A gold nanoparticle/polyaniline/graphene nanocomposite modified glassy carbon electrode was developed which offered good detection ability for Pb^{2+} (detection limit 0.1 nm) [57]. Detection at such low concentration is important as many metals are present only in trace amounts in the biological system.

As pointed out previously, the presence of Cu^{2+} often inhibits simultaneous detection of metal ions in water samples due to the formation of intermetallic complexes. A graphene oxide terminated p-aminophenyl modified glassy carbon electrode was developed which exhibited simultaneous detection ability of Cu^{2+}

and Cd^{2+} with detection limits of 3.3 x 10^{-12} M and 5.0 x 10^{-10} M, respectively without any interference [58]. Some other graphene-based electrodes offer simultaneous detection of Cd^{2+} and Pb^{2+} such as nitrogen-doped quantum dots-graphene oxide hybrid, activated carbon modified multilayer graphene paste electrode, graphene nanosheets with MgFe-layered double hydroxide on the surface, *etc* [56].

The graphite pencil lead (GPL) is a good conductor of electricity, inexpensive, abundant, consistent in shape and quality, and could be a cheaper alternative in electrochemical sensing. GPL covered with gold nanodendrites (AuNDs) was tried out as an electrochemical sensor for simultaneous detection of Pb^{2+} and Cu^{2+}. The AuNDs can provide a large electroactive surface area, and it incorporated onto the GPL surface by simple electrodeposition. The AuND@GPL electrode was used for concurrent detection of Pb^{2+}, $Cu^{2+,}$ and Hg^{2+} where detection limit as low as 0.2 ppb was achieved with good sensor-to-sensor reproducibility [59].

Screen-Printed Electrochemical Sensors

With technological advancement, the miniaturization of devices has become an obvious trend and electrochemical devices are no exception. Compared to the bulky three-electrode system, the screen-printed electrodes (SPEs) are the smaller versions, and they consist of a three-electrode assembly on an insulating substrate such as plastics and ceramics (Fig. **5**). The SPEs have gained widespread research attention due to their promise to obtain disposable miniature electrodes. These printed devices have some notable advantages such as (a) smaller sample size due to small electrode size, (b) enhanced signal-to-noise ratio, (c) improved mass transmission rate, (d) rapid detection ability, (e) omission of complicated pre-steps like electrode polishing, *etc*. The SPEs are unusually thin electrodes with thickness in the range of 100 nm to a few micrometers and are normally made of conductive inks (carbon ink, silver ink, gold ink, *etc*.). The nonconducting impurities in these inks impede the performance of the electrodes, calling for research attention on getting over this setback [60].

Integration of the nanomaterials such as metal and metal-oxide nanomaterials, carbon nanomaterials, nanocomposites, *etc*. in the conducting inks is proved to be effective in fabricating high-performance SPEs. We have already discussed the nanomaterial-modified electrodes in the previous sections and observed that those modified electrodes outdid the bare electrodes which are due to the unique properties of the nanomaterials. The same goes for the screen-printed electrodes too. However, the performance of the SPEs is found to depend on the modification methods. The nanomaterials could be incorporated into the conducting inks either by directly adding them to the ink or by surface modification of the electrodes (drop-casting, electrodeposition, electrospray, electrospun, *etc.*).

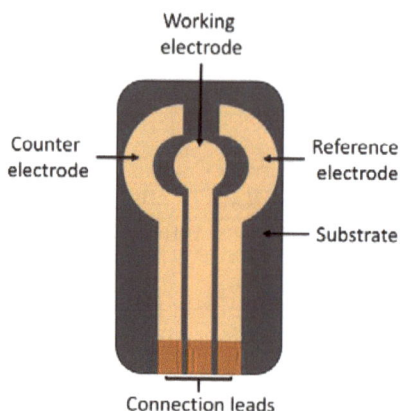

Fig. (5). Representation of a screen-printed electrode showing the three-electrode assembly on a substrate.

The carbon-based nanomaterials such as carbon nanoparticles (CNPs), single-walled and multi-walled carbon nanotubes (SWNTs, MWNTs), graphene, graphene oxide (GO), carbon nanofibers (CNFs), carbon nano horns (CNHs), *etc.* are tried out as the modifying material for SPEs, graphene, and CNTs being most common. Both the single-walled and multi-walled CNTs are found to be effective in the detection of toxic heavy metals such as Pb, Cd, Hg, Tl, *etc.*, the techniques used to be mostly SWASV and DPASV. Sometimes a metal nanoparticle is added to the CNT-modified electrode or a thin film of a metal compound is electrodeposited on it to increase the sensitivity of the subsequent SPEs. Interestingly, simultaneous detection of multiple heavy metal ions with great sensitivity in various samples such as lake, drinking, river, sea, tap water, wastewater, soil, biological samples, *etc.* are possible using the SPE technology which is the future of electrochemical sensing. As stated in the previous paragraph, the nanomaterials are either directly added to the ink or are used to modify the electrode surface.

An SWCNT-based disposable electrode was fabricated by drop-casting carbon nanotube solution onto the screen-printed carbon electrode, followed by electrodeposition of gold nanoparticles on it [61]. This electrode exhibited excellent electrochemical stripping detection of Pb^{2+} using DPASV as the technique, for example, a linear range of 4-38 nM and sensitivity of 72.128 nA nM^{-1}. Another typical example of a disposable electrode for simultaneous detection of Pb^{2+}, Cd^{2+}, and Zn^{2+} includes MWCNT modified electrode where the carbon nanotube was added directly to the ink [62]. A thin layer of BiF was electroplated on top of this electrode to increase the sensitivity of the electrode.

The optimized experimental condition realized wide liner ranges of 2.0-100.0 μg L^{-1} and detection limits as low as 0.2 μg L^{-1} (Pb^{2+}), 0.8 μg L^{-1} (Cd^{2+}), and 11.0 μg L^{-1} (Zn^{2+}). Sometimes ionic liquids are found to be useful for surface modification of SPEs by CNTs by drop-casting a solution of CNTs and ionic liquid on the SPE surface followed by drying it. The subsequent sensor was successfully employed for the determination of Pb^{2+} and Cd^{2+} in a soil sample [63]. Like the CNTs, graphene-based nanomaterials too are found to be effective in obtaining high-performance SPEs. For example, a disposable graphene-based SPE was developed by direct screen-printing with graphene paste which was more efficient than graphite-based or the glassy carbon electrode. Detection of Cd^{2+} ion at as low as 10^{-7}M limit of detection without any interference from added Na^+, K^+, Mg^{2+}, Ca^{2+} ions and successful recognition of Cd^{2+} ions in rice sample was possible using the electrode [64]. Graphene-based surface-modified SPEs are also known to simultaneously detect Cd^{2+} and Pb^{2+} in river water and pond water samples with excellent sensitivity. The electrode was prepared by reducing graphene oxide with hydrazine followed by dispersing the obtained graphene sheets in a Nafion solution and dropping onto the surface of a screen-printed carbon electrode and finally, electrodeposition of gold nanoparticles on top of it [65]. A problem of using graphene oxide as the starting material to prepare electrodes is the toxicity of the reducing agent which might make the graphene-based SPEs less attractive. An alternate greener method was proposed that called for one-step electrodeposition by directly placing the screen-printed carbon electrode in graphene oxide suspension followed by applying a potential difference for a longer time [66]. The major advantages of this electrode fabrication approach are the omission of the environmentally unfriendly reducing agents and direct deposition of graphene on the electrode surface without the necessity of an extra deposition step.

The metal and metal oxide nanomaterials have unique physical and chemical properties that make them attractive electrode materials. Applying them to the screen-printed electrodes offers exciting outcomes. The types of metal-based nanomaterials used in this area are mostly mercury, gold, silver, and bismuth nanoparticles; and metal oxide nanoparticles such as Fe_3O_4, Bi_2O_3, CuO, ZnO, Cr_2O_3, *etc.* As discussed earlier, the metal nanomaterial might have catalytic activity and facilitate electron transfer kinetics during the deposition and stripping processes. On the other hand, the metal oxide nanomaterials might adsorb the heavy metal ions speeding up the deposition amounts during the detection process. Like the carbon nanomaterial, the modification method of the metal nanomaterials too is crucial to obtain optimum performance of the sensors. Direct addition of the nanomaterials to the ink seems to be the most practical approach, but this method offers less sensitivity as a portion of the nanoparticles penetrates the bulk and are not exposed to the surface. The surface modification method, on the other hand, is more sensitive as all the nanoparticles are exposed to the analytes. It comes with a limitation though – it is less convenient than the direct addition. Nevertheless, a wide range of metal ions could be detected with great sensitivity in a variety of water samples using the screen-printed electrode technology [60].

Polymer-Based Electrochemical Sensors

As demand mounts to develop relatively inexpensive and easily operable analytical tools capable of rapid qualitative and quantitative data processing, it has become indispensable to search for new low-cost materials. Organic polymers are promising materials in this area due to their low cost, strong adsorption ability, compatibility, and so on. Efforts have been made to improve the adsorption of the polymers by grafting, branching, cross-linking, copolymerization, combining polymers with other substances, *etc*. The conducting polymers (CPs) have molecular structures that are flexible and allow the movement of electrons across the polymers. The electrical properties of these polymers could be altered through appropriate chemical modification, and they are popular materials in materials chemistry [67]. Some common examples of conducting polymers are polyanilines, polypyrroles, poly(3,4-ethylene dioxythiophene) (PEDOT), *etc*. The CPs are normally blended with inorganic nanomaterials to obtain heavy metal ion sensors. These polymer-based sensors are often biocompatible and have good electrical conductivity. The polypyrrole-based sensors were successfully used in the recognition of heavy metal ions such as Cd^{2+}, Hg^{2+}, Pb^{2+}, Cu^{2+} *etc*. with a detection limit as low as 0.004 nM employing SWASV, DPV, or DPSV as the analytical technique. Another important conducting polymer in this area is PEDOT which is electrochemically stable, highly conductive, optically transparent, and has outstanding electrocatalytic property. They, too, are found to be useful in heavy metal ion detection. For example, a PEDOT nanorod/GO nanocomposite modified GCE was developed which when combined with the DPSV technique, exhibited good linearity (10.0 nM–3.0 mM) and a low detection limit of 2.78 nM for Hg^{2+} ion sensing. The excellent synergy between the polyanilines and the heavy metal ions, superior physicochemical stability, environmental stability, and biocompatibility of the polyanilines made their way up to the electrochemical sensing of heavy metal ions. An illustrative example is a core-shell type ferrosoferric oxide (Fe_3O_4) @ polyaniline nanocomposite which was good at detecting Pb^{2+} ions in the range of $0.1 - 10.0$ nM with the detection limit of 0.3 nM [68].

The molecularly imprinted polymers (MIPs) are a special type of tailored polymers in which affinity toward specific analytes is imprinted by molecular imprinting technique. They are gradually gaining pace in the electrochemical sensing of heavy metal ions due to their obvious benefits such as high selectivity, easy availability, low synthesis cost, scalable synthesis, robust physical and chemical properties, reusability, stability, and so forth and so on. The application of MIPs in electrochemistry is facilitated by electropolymerization which incorporates polymerization of electroactive molecules under an electrochemical condition. The MIP technology goes back to the 1970s when it was mainly associated with separation and extraction. These days, MIP has been coupled with screen-printed electrode technology which together could take the miniaturization of electrochemical sensors to another level. The wonderful thing about the MIPs is their ability to act as the receptor in the sensor. Target-specific binding sites could be easily incorporated into their molecular structures. Though there are various synthetic methods to obtain the MIPs, the basic outline of the syntheses is (a) generation of a polymer containing the template of target molecules attached to a functional group of the host, and (b) removal of the template from the polymer matrix leaving behind a cavity complementary to the target. In other words, it can be said that the templated synthesis of the MIPs induces a molecular memory in the polymer which can now selectively recognize the specific analyte molecules. The mechanism of recognition is comparable to the lock-and-key mechanism of the enzymes and offers high specificity [69].

The electrochemistry of the MIP-based sensors is similar to those of the other electrodes – a fluctuation of the electric signal depending on the concentration of the analyte molecule. The most used techniques to detect the relevant electrochemical events are voltammetry (LSV, CV, DPV, SWV, *etc.*) and electrochemical impedance spectroscopy (EIS). In the case of the electroactive molecules, the current can be measured directly, where the current originates from the redox reaction of the target molecules or ions on binding the MIPs. However, if the analyte is non-electroactive, an electrical signal due to the redox process of a secondary redox system such as $[Fe(CN)_6]^{3-}/[Fe(CN)_6]^{4-}$ is used instead. The second mechanism is called the gate-controlled mechanism where the signal of the probe molecules has an inverse relationship with the analyte concentration. MIPs were successfully used in electrochemical recognition of pharmaceuticals, antibiotics, anthelmintics, hormones, pesticides, insecticides, herbicides, fungicides, biocides, heavy metal ions, and other environmental contaminants such as industrial explosives and byproducts, ingredients of cosmetics, brominated flame retards, *etc.* Surprisingly, only a handful number of research publications demonstrate the use of the MIPs in sensing heavy metals [70]. Only a few metal ions, such as Be^{2+}, Co^{2+}, $Cu^{2+,}$ and Zn^{2+} were recognized using these sensors. The imprinted sensor for beryllium ion was produced by complex formation between beryllium ion and 4-(2-pyridylazo)-resorcinol followed by copolymerization with o-phenylenediamine which was the functional monomer in this case. Interestingly, the electrochemical studies on the imprinted polymer revealed recognition of the beryllium complex template, not the individual metal ion or the ligand. An electrochemiluminescence sensor was designed for sensing Co^{2+} containing bovine serum albumin and the metal ion as the template. An interesting approach was however taken for sensing copper ion in water and soil samples – biosorption. It is known that some algae tend to biosorption of heavy metals. Therefore, a blue-green alga was used as the functional monomer as well as the crosslinker to obtain the imprinted polymer [70]. Existence of such a limited number of data nevertheless opens new possibilities for metal ion sensing using imprinted polymers.

Fuel Cell-Based Self-Powered Electrochemical Sensors

For any sensor to operate, it must be fed by electrical power as the sensing mechanism is coupled with an electric circuit that converts the sensory output into an electric signal. This external power usually comes from the battery attached to the device. Self-powered sensors do not require this auxiliary circuit and this type of sensors caught the attention of researchers in the early 2000s. The fuel cell and nanogenerator-based self-powered electrochemical devices are most popular among the self-powered sensors and significant research attention has been given to them. The fuel cells convert chemical energy into electrical energy through redox reactions in the presence of a metal catalyst. The fuel cells have some indigenous benefits such as, high efficiency, environment friendliness, non-limitation of its energy conversion by the Carnot cycle, *etc.* and they are being hailed as the next generation clean power sources. As the catalyst plays an important role in fuel cells, considerable research is focused on developing high-efficiency catalysts. In a biofuel cell, the conventional transition metal catalyst is replaced with an enzyme or other microorganism such as microbes, organelles, non-enzymatic proteins, enzyme mimics, *etc.* The type of fuel fed into the biofuel cells ranges from ethanol and glucose to various metabolic products, fermentation products, or even sewage. In contrast to the conventional fuel cells which normally require elevated working temperature, the biofuel cells operate at ordinary temperature and pressure, making them more attractive. As the biofuel cells utilize the physiologically produced glucose as fuel, they have excellent biocompatibility too. Another type of recently developed fuel cell is the photocatalytic fuel cell which converts solar energy to electricity where the energy conversion stems from photocatalytic oxidation of organic compounds in the presence of semiconducting photocatalysts. The importance of biofuel cells and self-powered electrochemical cells was first realized in 2001. The cell produced electricity *via* reduction of oxygen by cytochrome c oxidase immobilized at the cathode with simultaneous oxidation of glucose, by glucose oxidase immobilized at the anode. Self-powered sensors based on piezoelectric and triboelectric nanogenerators were also subsequently introduced. On a side note, the nanogenerators do not depend on biochemical reactions and they convert mechanical energy into electricity. The nanogenerator-based devices come with their promises and they have different types of applications that are beyond the scope of the present chapter [71].

Before we discuss the application of the self-powered electrochemical sensors on metal ion detection, we briefly describe the different types of biofuel cells and the chemistry within them. The fuel cells for self-powered electrochemical sensors are divided into three broad categories based on the fuel and the catalyst: enzymatic biofuel cells, microbial fuel cells, and photocatalytic fuel cells. The enzymatic biofuel cells make use of the enzyme-catalyzed oxidation of naturally occurring fuels such as alcohols, saccharides, and amines, *etc*. The microbial fuel cells take care of the microorganism-mediated conversion of chemical energy into electrical energy in an anaerobic environment. The photocatalytic fuel cells on the other hand produce electricity as they degrade the organic pollutants using semiconductors as the photocatalysts.

The self-powered electrochemical sensors have been tried out in the field of electrochemical sensing of heavy metal ions with promising results. An Enterobacter cloacae modified anode was developed for the detection of As^{5+} and As^{3+} using the reversible inhibition of the ions on laccase [72]. The enzymatic biofuel cells were found useful in the detection of Hg^{2+} at a detection limit as low as 56 pM and a wide linear range of 0.1 nM − 5 μM [73]. The inhibition of enzymatic effect could also be utilized to electrochemically recognize Hg^{2+} ions in tap, ground, lake water, *etc*. An enzymatic biofuel cell was configured in this regard which consisted of alcohol dehydrogenase (ADH) supported on carbon nano horn-based mediator system as the anode and bilirubin oxidase (BOD) as the cathodic biocatalyst. The enzymatic activity could be inhibited by the presence of Hg^{2+} ions as they can interact with the active sites of the enzymes, decreasing the open-circuit voltage. The relevant sensor was found to exhibit a linear range of 1-500 nM with a detection limit of 1nM [74]. The self-powered electrochemical sensors based on microbial fuel cells exhibit fluctuation of voltages based on the toxicity of the analytes and are good at assaying the toxicity of heavy metal ions in wastewater. The respiratory activity of the electrochemically active bacteria could thus be detected in a programmed manner where a toxicity sequence of Hg^{2+} > Cu^{2+} > Cd^{2+} > Zn^{2+} > Pb^{2+} > Cr^{3+} was determined based on a dual-chamber microbial fuel cell-based self-powered electrochemical cell [75].

Biomaterials in Electrochemical Sensing of Heavy Metal Ions

The role of biomaterials on the electrochemical sensing of heavy metal ions was introduced in the previous section. These materials have some binding sites which can specifically interact with the selective metal ions and this is used as a mechanism in the electrochemical sensing of heavy metal ions. The enzymes such as oxidase, peroxidase, urease, *etc.* are common in this area where the sensing mostly involves inhibition of enzymatic activity and the fluctuation of the electrical signal that follows thereafter. However, the enzyme-based sensors have some setbacks such as tedious device fabrication, poor stability, and reproducibility that impede their application in practical devices.

The most extensively used biomaterials in heavy metal sensing are nevertheless based on nucleic acids. The functional nucleic acids have gained significant research attention lately because of the ability of the metal ion to bind to specific nucleic bases. In contrast to the enzymes, the nucleic acid-based sensors are easier to fabricate, chemically and thermally more stable, justifying their popularity as the electrochemical sensing of heavy metal ions. For example, selective detection of Hg^{2+} was possible due to the tendency of the ion to form stable complexes with the thymine bases ($T–Hg^{2+}–T$). The stability of these mercury-thymine complexes is even higher than the Watson-Crick adenine-thymine pair. Sometimes, a conformational change of the DNA hairpins is induced by Hg^{2+} ions and this phenomenon too was explored to electrochemically sense the relevant heavy metal ion. The DNA molecule was tagged with redox-active species, ferrocene, in the middle of the loop. The conformational change caused opening the hairpins which increased the redox current. In another case, a methylene blue labeled thymine-rich probe was developed which was able to quantify Hg^{2+} ion at a low detection limit of 0.2 nM using amperometry as the technique. A combination of nanomaterials with thymine-rich probes was also tested for the recognition of mercury ions. A 'turn on' Hg^{2+} sensor was synthesized by functionalizing reporter DNA with silver nanoparticles which were able to detect the said ion at a low detection limit of 0.5 nM [23a].

Whereas the enzyme-based electrochemical sensors suffer from some setbacks due to the stability issues, the DNAzymes are highly thermally stable, can catalyze specific biochemical reactions, and are being tried out for recognizing toxic metal ions. An interesting Pb^{2+} sensor was constructed by labeling the DNAzyme with redox-active species methylene blue and then tethering the enzyme to a gold electrode using the gold-thiol interaction, the redox-active species being away from the electrode surface. The addition of lead ion sliced the substrate strand into two fragments, where the fragments being more flexible assisted the electron transfer between the electrode and the redox species. This change generated an electrochemical signal depending on the concentration of the analyte and a detection limit of 300 nM was achieved [75]. DNAzyme based sensors for Cu^{2+} were also developed. Unlike the Pb^{2+} sensor which works by breaking the strand, the copper sensor, on the other hand, was designed to function by reorganizing the enzyme with copper-catalyzed oxidation of ascorbic acid, causing adsorption of the oxidation products on the electrode surface leading to a fluctuation of the electrical response [76].

Other DNA-based materials for heavy metal ion sensing are DNA G-quadruplexes (G4) which have a highly dynamic and polymorphic four-stranded structure and can be obtained from guanine-rich DNA sequences. A conformational flipping from random-coil to metal-stabilized G-quadruplex could be stimulated by many metal ions such as sodium, potassium, lead ions, *etc*. One such sensor for Pb^{2+} ion used ferrocyanide/ferricyanide redox probe and electrical impedance spectroscopy as the technique where the metal ion caused switching to a G4 structure from G-rich DNA strands, resulting in decreasing of charge transfer resistance. The relevant sensor could detect the metal ion at a concentration as low as 0.5 nM [77]. Coupling of the conformational switching with intercalation of small molecules into DNA quadruplex was explored as another possibility for electrochemical sensing of heavy metal ions. In one such instance, a 'turn on' electrochemical sensor for Pb^{2+} was developed that made use of intercalation of crystal violet as the G4 binding indicator, the conformational change, and the electrical signal that followed. This sensor could detect Pb^{2+} ions in a linear range of 1.0 nM – 1.0 mM with a detection limit of 0.4 nM [78].

CONCLUSION AND PERSPECTIVE

On balance, the research on electrochemical sensors for heavy metal ions is mostly centered on modifying electrode surfaces where diverse materials ranging from metal nanoparticles, carbon-based nanomaterials to biomaterials are tested as electrode modifiers. With the advancement of modern civilization, our environment has been progressively polluted with a world of toxic substances where heavy metal ions are potential candidates. The primary sources of these metal ions are wastewater, waste gas, solid waste from industry, *etc*. Due to interconnection between the food chain, these toxic substances ultimately make their way into the human body, causing a range of diseases. Therefore, on-field monitoring of the toxic substances in water, soil samples, *etc*., is gaining paramount importance. Compared to the other sensing techniques such as fluorescence, mass spectrometry, atomic absorption spectroscopy, *etc*., the electrochemical techniques are relatively cheap, easy to operate, and offer the possibility of miniaturization. Therefore, it comes as no surprise that it has gained immense popularity among researchers to develop heavy metal sensors. The on-site determination of the toxic metal ions comes with its challenges. The water environment is usually complicated where each water sample has different properties such as pH, composition, turbidity, and coexistence of multiple interfering metal ions, *etc*. Enhanced sensitivity, low detection limit, and wide detection range are the primary criteria for developing good electrochemical sensors for heavy metal ions, including good stability, repeatability, and reproducibility. Simultaneous detection of multiple metal ions is a plus point.

As breaking the matters down to nanomaterials imparts unique physicochemical properties, the rate of electron transfer between the substrate and the electrodes is directly affected by the application of these modifiers. Consequently, the coupling of nanomaterials with electrochemical techniques boosted the performance of the electrodes, such as faster response time, high sensitivity, reproducibility, low detection limit due to the large surface area to volume ratio, high catalytic activity, and strong adsorption ability of nanoparticles. Therefore, the central aim of the research in this area is to discover better-performing electrode modifying materials. In this chapter, we attempted to briefly overview these electrode modifiers with some typical examples of the sensors.

The metal and metal oxide-based nanomaterials are found to be useful electrode modifiers, and substantial progress is made in this direction. However, there are some key issues in this regard. For instance, precise prediction of the electroanalytical ability of these metal-based nanoparticles from the existing physicochemical data often does not work out. To establish a structure-function relationship, different approaches such as simulation studies based on quantum mechanics and electrode kinetics are considered, only to end up with an ambiguous conclusion. The theoretical modeling is further complicated by the tendency of aggregate formation by the deposited and adsorbed metal ions on the electrode surface. The cluster formation makes the explanation of the experimental data difficult because the adsorption of heavy metal ions on electrodes is assumed to be a monolayer or multilayer model evenly distributed on the electrode surface. The interference of the metal ions on the operation of metal-based nanomaterial modified electrodes is another annoying problem. The interfering ions cause alteration of the effective concentration of the analyte metal ion, inducing error in analysis. Thus, apart from focusing on sensitivity, developing electrodes with high selectivity and tolerance toward interference is imperative.

Carbon-based nanomaterials such as carbon nanotube, graphene, *etc.*, are promising electrode modifiers to remove environmental pollutants. This is driven by their exciting properties such as charge transport, large surface area, huge mechanical strength, electrocatalytic properties, relative chemical non-reactivity, *etc.* The properties of these carbon-based nanomaterials are greatly explored, and there is still a plethora of possibilities to be explored with these materials. The electrochemical sensors based on molecularly imprinted polymers and screen-printed electrodes are quite interesting. These polymers are easily synthesizable, can be tailored, inexpensive, and compatible with water. The sensitivity and selectivity of the molecularly imprinted polymers are further enhanced by coupling them with nanomaterials. Both the metal nanoparticles and carbon nanomaterials are found to be useful in fabricating screen-printed electrodes, which still need further optimization of performance.

The development of self-powered fuel cells is a big leap toward the miniaturization of electrochemical devices. To date, various recognition elements such as enzymes, aptamers, antibodies, molecularly imprinted polymers, *etc.*, are coupled with fuel cells to develop self-powered electrochemical sensors for rapid and selective detection of various analytes. The allure of these devices is due to many interesting facts such as (a) the possibility of fabricating wearable devices to monitor health issues, (b) long-term online environmental monitoring of toxic pollutants, (c) possible application in food safety risk assignment is based on electrochemical response of the toxic substances present in food and so on. However, incorporation of the biological molecules in the sensory devices raises some concerns – performance degradation of the electrochemical devices as the majority of the biomolecules lack stability.

Considering all these points, the electrochemical sensing of heavy metal ions is a promising technique to monitor environmental pollutants. Nevertheless, there are some serious issues, repeatability and reproducibility, which limit practical applications. Analytical data in real water samples, soil, food, or other systems are more important, and more emphasis should be given. Thus, in a nutshell, the research in this area focusing on screen-printed technology and self-powered fuel cells could be very useful technologies for in-field detection of heavy metals, and there is a plethora of research scope to explore these recent findings towards the miniaturization of electrochemical devices.

CONSENT FOR PUBLICATION

Not applicable.

CONFLICT OF INTEREST

The author declares no conflict of interest, financial or otherwise.

ACKNOWLEDGEMENTS

The author would like to thank Science and Engineering Research Board (SERB), DST, India (Project No. SRG/2019/001365) for funding.

REFERENCES

[1]   V. Schroeder, S. Savagatrup, M. He, S. Lin, and T.M. Swager, "Carbon Nanotube Chemical Sensors", *Chem. Rev.,* vol. 119, no. 1, pp. 599-663, 2019.
[http://dx.doi.org/10.1021/acs.chemrev.8b00340] [PMID: 30226055]

[2]   P. Kassal, M.D. Steinberg, and I.M. Steinberg, "Wireless chemical sensors and biosensors: A review", *Sens. Actuators B Chem.,* vol. 266, pp. 228-245, 2018.
[http://dx.doi.org/10.1016/j.snb.2018.03.074]

[3]   F.R. Simões, and M.G. Xavier, *Nanoscience and its Applications.* 1st ed. Elsevier Inc., 2017.

[4]   G.D. Christian, P.K. Dasgupta, and K.A. Schug, *Analytical Chemistry.* 7th ed. Wiley & Sons, 2013.

[5]   A.M.O. Brett, and C.M. Brett, *Electrochemistry Principles, Methods, and Applications.* Oxford University Press: Coimbra, 1994.

[6]   F.J. Holler, D.A. Skoog, and S.R. Crouch, *Principles of Instrumental Analysis.* 7th ed. Cengage Learning: Boston, USA, 2018.

[7]   M.B. Gumpu, S. Sethuraman, U.M. Krishnan, and J.B.B. Rayappan, "A review on detection of heavy metal ions in water – An electrochemical approach", *Sens. Actuators B Chem.,* vol. 213, pp. 515-533, 2015.
[http://dx.doi.org/10.1016/j.snb.2015.02.122]

[8]   A. Odobašić, I. Šestan, and S. Begić, "Biosensors for Environmental Monitoring", In: *Intech Open Limited*London, 2019.
[http://dx.doi.org/10.5772/intechopen.84139]

[9]   D.L. Nelson, and M.M. Cox, *Lehninger Principles of Biochemistry.* 4th ed. W.H. Freeman: New York, 2008.

[10]   aL. Pujol, D. Evrard, K. Groenen-Serrano, M. Freyssinier, A. Ruffien-Cizsak, and P. Gros, "Electrochemical sensors and devices for heavy metals assay in water: the French groups' contribution", *Front Chem.,* vol. 2, no. 19, p. 19, 2014.
[http://dx.doi.org/10.3389/fchem.2014.00019] [PMID: 24818124] bL. Ming, and G. Honglei, "Israa; A.-O.; Nianqiang, W. Nanostructured Sensors for Detection of Heavy Metals: A Review", *ACS Sustain. Chem.& Eng.,* vol. 1, no. 7, pp. 713-723, 2013.
[http://dx.doi.org/10.1021/sc400019a]

[11]   B. Bansod, T. Kumar, R. Thakur, S. Rana, and I. Singh, "A review on various electrochemical techniques for heavy metal ions detection with different sensing platforms", *Biosens. Bioelectron.,* vol. 94, pp. 443-455, 2017.
[http://dx.doi.org/10.1016/j.bios.2017.03.031] [PMID: 28340464]

[12]   S. Li, C. Zhang, S. Wang, Q. Liu, H. Feng, X. Ma, and J. Guo, "Electrochemical microfluidics techniques for heavy metal ion detection", *Analyst (Lond.),* vol. 143, no. 18, pp. 4230-4246, 2018.
[http://dx.doi.org/10.1039/C8AN01067F] [PMID: 30095826]

[13]   P.S. Nambudumada, J.G. Manjunatha, and R. Chenthattil, "Electrocatalytic Analysis of Dopamine, Uric Acid and Ascorbic Acid at Poly(Adenine) Modified Carbon Nanotube Paste Electrode: A Cyclic Voltammetric Study", *Analytical and Bioanalytical Electrochemistry,* vol. 11, pp. 742-756, 2019.

[14]   N. Hareesha, and J.G. Manjunatha, "Fast and enhanced electrochemical sensing of dopamine at cost-effective poly(DL-phenylalanine) based graphite electrode", *J. Electroanal. Chem. (Lausanne),* vol. 878, p. 114533, 2020.
[http://dx.doi.org/10.1016/j.jelechem.2020.114533]

[15]   R. Chenthattil, J.G. Manjunatha, D.K. Ravishankar, F. Santosh, S. Gurumallappa, and N. Lingappa, "Validated Electrochemical Method for Simultaneous Resolution of Tyrosine, Uric Acid, and Ascorbic Acid at Polymer Modified Nano-Composite Paste Electrode", *Surg. Eng. Appl. Electrochem.,* vol. 56, pp. 415-426, 2020.
[http://dx.doi.org/10.3103/S1068375520040134]

[16] N. Hareesha, and J.G. Manjunatha, "Surfactant and Polymer Layered Carbon Composite Electrochemical Sensor for the Analysis of Estriol with Ciprofloxacin", *Mater. Res. Innov.*, vol. •••, pp. 1-14, 2019.

[17] M.M. Charithra, and J.G. Manjunatha, "Enhanced Voltammetric Detection of Paracetamol by Using Carbon Nanotube Modified Electrode as an Electrochemical Sensor", *Journal of Electrochemical Science and Engineering*, vol. 10, pp. 29-40, 2020.
[http://dx.doi.org/10.5599/jese.717]

[18] N.S. Prinith, and J.G. Manjunatha, "Polymethionine Modified Carbon Nanotube Sensor for Sensitive and Selective Determination of L-tryptophan", *Journal of Electrochemical Science and Engineering*, vol. 10, pp. 305-315, 2020.
[http://dx.doi.org/10.5599/jese.774]

[19] E.S. Dsouza, J.G. Manjunatha, C. Raril, G. Tigari, and P.A. Pushpanjali, "Polymer Modified Carbon Paste Electrode as a Sensitive Sensor for the Electrochemical Determination of Riboflavin and Its Application in Pharmaceutical and Biological Samples", *Analytical and Bioanalytical Chemistry Research*, vol. 7, pp. 461-472, 2020.

[20] J.G. Manjunatha, "Fabrication of Efficient and Selective Modified Graphene Paste Sensor for the Determination of Catechol and Hydroquinone", *Surfaces*, vol. 3, pp. 473-483, 2020.
[http://dx.doi.org/10.3390/surfaces3030034]

[21] J.G. Manjunatha, "A Promising Enhanced Polymer Modified Voltammetric Sensor for the Quantification of Catechol and Phloroglucinol", *Analytical and Bioanalytical Electrochemistry*, vol. 12, pp. 893-903, 2020.

[22] J.G. Manjunatha, "Poly (Adenine) Modified Graphene-Based Voltammetric Sensor for the Electrochemical Determination of Catechol, Hydroquinone and Resorcinol", *Open Chem. Eng. J.*, vol. 15, pp. 52-62, 2021.

[23] aL. Cui, J. Wu, and H. Ju, "Electrochemical sensing of heavy metal ions with inorganic, organic and bio-materials", *Biosens. Bioelectron.*, vol. 63, pp. 276-286, 2015.
[http://dx.doi.org/10.1016/j.bios.2014.07.052] [PMID: 25108108] bM. Yang, P.-H. Li, S.-H. Chen, X.-Y. Xiao, X.-H. Tang, C.-H. Lin, X.-J. Huang, and W.-Q. Liu, *Nanometal Oxides with Special Surface Physicochemical Properties to Promote Electrochemical Detection of Heavy Metal Ions.*, .
[http://dx.doi.org/10.1002/smll.202001035]

[24] S. Sang, D. Li, H. Zhang, Y. Sun, A. Jian, Q. Zhang, and W. Zhang, "Facile synthesis of AgNPs on reduced graphene oxide for highly sensitive simultaneous detection of heavy metal ions", *RSC Advances*, vol. 7, pp. 21618-21624, 2017.
[http://dx.doi.org/10.1039/C7RA02267K]

[25] P. Niu, C. Fernández-Sánchez, M. Gich, C. Ayora, and A. Roig, "Electroanalytical Assessment of Heavy Metals in Waters with Bismuth Nanoparticle-Porous Carbon Paste Electrodes", *Electrochim. Acta*, vol. 165, pp. 155-161, 2015.
[http://dx.doi.org/10.1016/j.electacta.2015.03.001]

[26] P.M. Lee, Z. Chen, L. Li, and E. Liu, "Reduced graphene oxide decorated with tin nanoparticles through electrodeposition for simultaneous determination of trace heavy metals", *Electrochim. Acta*, vol. 174, pp. 207-214, 2015.
[http://dx.doi.org/10.1016/j.electacta.2015.05.092]

[27] Z.G. Liu, Y.F. Sun, W.K. Chen, Y. Kong, Z. Jin, X. Chen, X. Zheng, J.H. Liu, X.J. Huang, and S.H. Yu, "Facet-dependent stripping behavior of Cu2O microcrystals toward lead ions: a rational design for the determination of lead ions", *Small*, vol. 11, no. 21, pp. 2493-2498, 2015.
[http://dx.doi.org/10.1002/smll.201402146] [PMID: 25630388]

[28] X.G. Han, Y.Q. Jiang, S.F. Xie, Q. Kuang, X. Zhou, D.P. Cai, Z.X. Xie, and L.S. Zheng, "Control of the Surface of ZnO Nanostructures by Selective Wet-Chemical Etching", *J. Phys. Chem. C*, vol. 114, pp. 10114-10118, 2010.

[http://dx.doi.org/10.1021/jp101284p]

[29] G.J. Wilson, G.D. Will, R.L. Frost, and S.A. Montgomery, "Efficient microwave hydrothermal preparation of nanocrystalline anatase TiO2 colloids", *J. Mater. Chem.,* vol. 12, pp. 1787-1791, 2002.
[http://dx.doi.org/10.1039/b200053a]

[30] H. Shang, M. Li, H. Li, S. Huang, C. Mao, Z. Ai, and L. Zhang, Oxygen Vacancies Promoted the Selective Photocatalytic Removal of NO with Blue TiO$_2$ *via* Simultaneous Molecular Oxygen Activation and Photogenerated Hole Annihilation., *Environ. Sci. Technol.,* vol. 53, no. 11, pp. 6444-6453, 2019.
[http://dx.doi.org/10.1021/acs.est.8b07322] [PMID: 31050293]

[31] K. Woan, G. Pyrgiotakis, and W. Sigmund, "Photocatalytic Carbon□Nanotube–TiO2 Composites", *Adv. Mater.,* vol. 21, pp. 2233-2239, 2009.
[http://dx.doi.org/10.1002/adma.200802738]

[32] aM. Liu, G. Zhao, Y. Tang, Z. Yu, Y. Lei, M. Li, Y. Zhang, and D. Li, "A simple, stable and picomole level lead sensor fabricated on DNA-based carbon hybridized TiO(2) nanotube arrays", *Environ. Sci. Technol.,* vol. 44, no. 11, pp. 4241-4246, 2010.
[http://dx.doi.org/10.1021/es1003507] [PMID: 20441178] bW. Jin, G. Wu, and A. Chen, "Sensitive and selective electrochemical detection of chromium(VI) based on gold nanoparticle-decorated titania nanotube arrays", *Analyst (Lond.),* vol. 139, no. 1, pp. 235-241, 2014.
[http://dx.doi.org/10.1039/C3AN01614E] [PMID: 24191278]

[33] D-N. Pei, L. Gong, A-Y. Zhang, X. Zhang, J-J. Chen, Y. Mu, and H-Q. Yu, "Defective titanium dioxide single crystals exposed by high-energy 001 facets for efficient oxygen reduction", *Nat. Commun.,* vol. 6, p. 8696, 2015.
[http://dx.doi.org/10.1038/ncomms9696] [PMID: 26493365]

[34] W-Y. Zhou, J-Y. Liu, J-Y. Song, J-J. Li, J-H. Liu, and X.J. Huang, "Surface-Electronic-State-Modulated, Single-Crystalline (001) TiO2 Nanosheets for Sensitive Electrochemical Sensing of Heavy-Metal Ions", *Anal. Chem.,* vol. 89, no. 6, pp. 3386-3394, 2017.
[http://dx.doi.org/10.1021/acs.analchem.6b04023] [PMID: 28221774]

[35] P. Miao, Y. Tang, and L. Wang, "DNA Modified Fe3O4@Au Magnetic Nanoparticles as Selective Probes for Simultaneous Detection of Heavy Metal Ions", *ACS Appl. Mater. Interfaces,* vol. 9, no. 4, pp. 3940-3947, 2017.
[http://dx.doi.org/10.1021/acsami.6b14247] [PMID: 28079364]

[36] L. Shi, Y. Li, X. Rong, Y. Wang, and S. Ding, "Facile fabrication of a novel 3D graphene framework/Bi nanoparticle film for ultrasensitive electrochemical assays of heavy metal ions", *Anal. Chim. Acta,* vol. 968, pp. 21-29, 2017.
[http://dx.doi.org/10.1016/j.aca.2017.03.013] [PMID: 28395771]

[37] https://en.wikipedia.org/wiki/MXenes

[38] Y.J. Jiang, X.N. Zhang, L.J. Pei, S. Yue, L. Ma, L.Y. Zhou, Z.H. Huang, Y. He, and J. Gao, "Silver nanoparticles modified two-dimensional transition metal carbides as nanocarriers to fabricate acetylcholinesterase-based electrochemical biosensor", *Chem. Eng. J.,* vol. 339, pp. 547-556, 2018.
[http://dx.doi.org/10.1016/j.cej.2018.01.111]

[39] Y. He, L. Ma, L. Zhou, G. Liu, Y. Jiang, and J. Gao, "Preparation and Application of Bismuth/MXene Nano-Composite as Electrochemical Sensor for Heavy Metal Ions Detection", *Nanomaterials (Basel),* vol. 10, no. 5, p. 866, 2020.
[http://dx.doi.org/10.3390/nano10050866] [PMID: 32365912]

[40] L. Zhang, and Y. Tian, "Designing Recognition Molecules and Tailoring Functional Surfaces for In Vivo Monitoring of Small Molecules in the Brain", *Acc. Chem. Res.,* vol. 51, no. 3, pp. 688-696, 2018.
[http://dx.doi.org/10.1021/acs.accounts.7b00543] [PMID: 29485847]

[41] N.D. Rudd, H. Wang, E.M.A. Fuentes-Fernandez, S.J. Teat, F. Chen, G. Hall, Y.J. Chabal, and J. Li, "Highly Efficient Luminescent Metal-Organic Framework for the Simultaneous Detection and

Removal of Heavy Metals from Water", *ACS Appl. Mater. Interfaces,* vol. 8, no. 44, pp. 30294-30303, 2016.
[http://dx.doi.org/10.1021/acsami.6b10890] [PMID: 27736058]

[42] W.P. Lustig, S. Mukherjee, N.D. Rudd, A.V. Desai, J. Li, and S.K. Ghosh, "Metal-organic frameworks: functional luminescent and photonic materials for sensing applications", *Chem. Soc. Rev.,* vol. 46, no. 11, pp. 3242-3285, 2017.
[http://dx.doi.org/10.1039/C6CS00930A] [PMID: 28462954]

[43] Y. Wang, L. Wang, W. Huang, T. Zhang, X.Y. Hu, J.A. Perman, and S. Ma, "A metal-organic framework and conducting polymer-based electrochemical sensor for high-performance cadmium ion detection", *J. Mater. Chem. A Mater. Energy Sustain.,* vol. 5, pp. 8385-8393, 2017.
[http://dx.doi.org/10.1039/C7TA01066D]

[44] M. Lu, Y. Deng, Y. Luo, J. Lv, T. Li, J. Xu, S.W. Chen, and J. Wang, "Graphene Aerogel-Meta--Organic Framework-Based Electrochemical Method for Simultaneous Detection of Multiple Heavy-Metal Ions", *Anal. Chem.,* vol. 91, no. 1, pp. 888-895, 2019.
[http://dx.doi.org/10.1021/acs.analchem.8b03764] [PMID: 30338985]

[45] X. Wang, Y. Qi, Y. Shen, Y. Yuan, L. Zhang, C. Zhang, and Y. Sun, "A ratiometric electrochemical sensor for simultaneous detection of multiple heavy metal ions based on ferrocene-functionalized metal-organic framework", *Sens. Actuators B Chem.,* vol. 310, p. 127756, 2020.
[http://dx.doi.org/10.1016/j.snb.2020.127756]

[46] A.G. Sharpe, and C.H. Housecroft, *Inorganic Chemistry.* 4th ed. Pearson Education Limited: England, 2012.D. Shriver, M. Weller, T. Overton, J. Rourke, and F. Armstrong, *Inorganic Chemistry.* 6th ed. W. H. Freeman and Company: New York, 2014.

[47] J. Wang, *Stripping Analysis: Principles, Instrumentation, and Applications.* Wiley-VCH Verlag GmbH, 1985.

[48] A.K. Wanekaya, "Applications of nanoscale carbon-based materials in heavy metal sensing and detection", *Analyst (Lond.),* vol. 136, no. 21, pp. 4383-4391, 2011.
[http://dx.doi.org/10.1039/c1an15574a] [PMID: 21894336]

[49] X. Hui, M. Sharifuzzaman, S. Sharma, X. Xuan, S. Zhang, S.G. Ko, S.H. Yoon, and J.Y. Park, "High-Performance Flexible Electrochemical Heavy Metal Sensor Based on Layer-by-Layer Assembly of Ti3C2Tx/MWNTs Nanocomposites for Noninvasive Detection of Copper and Zinc Ions in Human Biofluids", *ACS Appl. Mater. Interfaces,* vol. 12, no. 43, pp. 48928-48937, 2020.
[http://dx.doi.org/10.1021/acsami.0c12239] [PMID: 33074662]

[50] J.C. Solís, and M. Galicia, "High Performance of MWCNTs -Chitosan Modified Glassy Carbon Electrode for Voltammetric Trace Analysis of Cd (II)", *Int. J. Electrochem. Sci.,* vol. 15, pp. 6815-6828, 2020.
[http://dx.doi.org/10.20964/2020.07.56]

[51] A.K. Geim, and K.S. Novoselov, "The rise of graphene", *Nat. Mater.,* vol. 6, no. 3, pp. 183-191, 2007.
[http://dx.doi.org/10.1038/nmat1849] [PMID: 17330084]

[52] Y. Wang, and S. Hu, "Applications of Carbon Nanotubes and Graphene for Electrochemical Sensing of Environmental Pollutants", *J. Nanosci. Nanotechnol.,* vol. 16, no. 8, pp. 7852-7872, 2016.
[http://dx.doi.org/10.1166/jnn.2016.12762]

[53] D.A.C. Brownson, D.K. Kampouris, and C.E. Banks, "Graphene electrochemistry: fundamental concepts through to prominent applications", *Chem. Soc. Rev.,* vol. 41, no. 21, pp. 6944-6976, 2012.
[http://dx.doi.org/10.1039/c2cs35105f] [PMID: 22850696]

[54] E.P. Randviir, and C.E. Banks, *Carbon-Based Nanosensor Technology.,* C. Kranz, Ed., vol. Vol. 17. Springer, 2018, pp. 141-164.
[http://dx.doi.org/10.1007/5346_2018_25]

[55] J. Chang, G. Zhou, E.R. Christensen, R. Heideman, and J. Chen, "Graphene-based sensors for

detection of heavy metals in water: a review", *Anal. Bioanal. Chem.,* vol. 406, no. 16, pp. 3957-3975, 2014.
[http://dx.doi.org/10.1007/s00216-014-7804-x] [PMID: 24740529]

[56]   M. Coro,, S. Pruneanu, and R-I.S-v. Staden, "Review– Recent Progress in the Graphene-Based Electrochemical Sensors and Biosensors", *J. Electrochem. Soc.,* vol. 167, p. 037528, 2020.
[http://dx.doi.org/10.1149/2.0282003JES]

[57]   Y.P. Dong, Y. Zhou, Y. Ding, X.F. Chua, and C.M. Wang, "Sensitive detection of Pb(ii) at gold nanoparticle/polyaniline/graphene-modified electrode using differential pulse anodic stripping voltammetry", *Anal. Methods,* vol. 6, pp. 9367-9374, 2014.
[http://dx.doi.org/10.1039/C4AY01908C]

[58]   V.K. Gupta, M.L. Yola, N. Atar, Z. Ustundag, and A.O. Solak, "A novel sensitive Cu(II) and Cd(II) nanosensor platform: Graphene oxide terminated p-aminophenyl modified glassy carbon surface", *Electrochim. Acta,* vol. 112, pp. 541-548, 2013.
[http://dx.doi.org/10.1016/j.electacta.2013.09.011]

[59]   N.Q. Giao, V.H. Dang, P.T.H. Yen, P.H. Phong, V.T.T. Ha, P.K. Duy, and H. Chung, "Au nanodendrite incorporated graphite pencil lead as a sensitive and simple electrochemical sensor for simultaneous detection of Pb(II), Cu(II) and Hg(II)", *J. Appl. Electrochem.,* vol. 49, pp. 839-846, 2019.
[http://dx.doi.org/10.1007/s10800-019-01326-x]

[60]   X. Liu, Y. Yao, Y. Ying, and J. Ping, "Recent advances in nanomaterial-enabled screen-printed electrochemical sensors for heavy metal detection", *Trends Analyt. Chem.,* vol. 115, pp. 187-202, 2019.
[http://dx.doi.org/10.1016/j.trac.2019.03.021]

[61]   B. Molinero-Abad, D. Izquierdo, L. Pérez, I. Escudero, and M.J. Arcos-Martínez, "Comparison of backing materials of screen printed electrochemical sensors for direct determination of the sub-nanomolar concentration of lead in seawater", *Talanta,* vol. 182, pp. 549-557, 2018.
[http://dx.doi.org/10.1016/j.talanta.2018.02.005] [PMID: 29501191]

[62]   U. Injang, P. Noyrod, W. Siangproh, W. Dungchai, S. Motomizu, and O. Chailapakul, "Determination of trace heavy metals in herbs by sequential injection analysis-anodic stripping voltammetry using screen-printed carbon nanotubes electrodes", *Anal. Chim. Acta,* vol. 668, no. 1, pp. 54-60, 2010.
[http://dx.doi.org/10.1016/j.aca.2010.01.018] [PMID: 20457302]

[63]   H. Wang, G. Zhao, Z.H. Zhang, Y. Yi, Z.Q. Wang, and G. Liu, "A Portable Electrochemical workstation using Disposable Screen-Printed Carbon Electrode decorated with Multiwall Carbon Nanotube-Ionic Liquid and Bismuth Film for Cd(II) and Pb(II) Determination", *Int. J. Electrochem. Sci.,* vol. 12, pp. 4702-4713, 2017.
[http://dx.doi.org/10.20964/2017.06.73]

[64]   Y.J. Teng, Y.C. Zhang, K. Zhou, and Z.X. Yu, "Screen Graphene-printed Electrode for Trace Cadmium Detection in Rice Samples Combing with Portable Potentiostat", *Int. J. Electrochem. Sci.,* vol. 13, pp. 6347-6357, 2018.
[http://dx.doi.org/10.20964/2018.07.61]

[65]   F. Wu, H.W. Lin, X. Yang, and D.Z. Chen, "GS-Nafion-Au Nanocomposite Film Modified SPCEs for Simultaneous Determination of Trace Pb2+ and Cd2+ by DPSV", *Int. J. Electrochem. Sci.,* vol. 8, pp. 7702-7712, 2013.

[66]   J. Ping, Y. Wang, J. Wu, and Y. Ying, "Development of an electrochemically reduced graphene oxide modified disposable bismuth film electrode and its application for stripping analysis of heavy metals in milk", *Food Chem.,* vol. 151, pp. 65-71, 2014.
[http://dx.doi.org/10.1016/j.foodchem.2013.11.026] [PMID: 24423503]

[67]   M.A. Rahman, P. Kumar, D.S. Park, and Y.B. Shim, "Electrochemical Sensors Based on Organic Conjugated Polymers", *Sensors (Basel),* vol. 8, no. 1, pp. 118-141, 2008.

[http://dx.doi.org/10.3390/s8010118] [PMID: 27879698]

[68]   H. Wang, C. Xu, and B. Yuan, "Polymer-based Electrochemical Sensing Platform for Heavy Metal Ions Detection - A Critical Review", *Int. J. Electrochem. Sci.,* vol. 14, pp. 8760-8771, 2019.
[http://dx.doi.org/10.20964/2019.09.22]

[69]   J.J. BelBruno, "Molecularly Imprinted Polymers", *Chem. Rev.,* vol. 119, no. 1, pp. 94-119, 2019.
[http://dx.doi.org/10.1021/acs.chemrev.8b00171] [PMID: 30246529]

[70]   P. Rebelo, E. Costa-Rama, I. Seguro, J. G. Pacheco, H. P. A. Nouws, M. N. D.S. Cordeiro, and C. Delerue-Matos, "Molecularly imprinted polymer-based electrochemical sensors for environmental analysis. ", *Biosens. Bioelectron,* vol. 172, 15, p. 112719.
[http://dx.doi.org/10.1016/j.bios.2020.112719]

[71]   Y. Chen, W. Ji, K. Yan, J. Gao, and J. Zhang, "Fuel cell-based self-powered electrochemical sensors for biochemical detection", *Nano Energy,* vol. 61, pp. 173-193, 2019.
[http://dx.doi.org/10.1016/j.nanoen.2019.04.056]

[72]   M. Rasmussen, and S.D. Minteer, "Long-term arsenic monitoring with an Enterobacter cloacae microbial fuel cell", *Bioelectrochemistry,* vol. 106, no. Pt A, pp. 207-212, 2015.
[http://dx.doi.org/10.1016/j.bioelechem.2015.03.009] [PMID: 25862430]

[73]   L. Zhang, Y. Wang, C. Ma, P. Wang, and M. Yan, "Self-powered sensor for Hg2+ detection based on hollow-channel paper analytical devices", *RSC Advances,* vol. 5, pp. 24479-24485, 2015.
[http://dx.doi.org/10.1039/C4RA14154G]

[74]   D. Yu, L. Bai, J. Zhai, Y. Wang, and S. Dong, "Toxicity detection in water containing heavy metal ions with a self-powered microbial fuel cell-based biosensor", *Talanta,* vol. 168, pp. 210-216, 2017.
[http://dx.doi.org/10.1016/j.talanta.2017.03.048] [PMID: 28391844]

[75]   Y. Xiao, A.A. Rowe, and K.W. Plaxco, "Electrochemical detection of parts-per-billion lead via an electrode-bound DNAzyme assembly", *J. Am. Chem. Soc.,* vol. 129, no. 2, pp. 262-263, 2007.
[http://dx.doi.org/10.1021/ja067278x] [PMID: 17212391]

[76]   C. Ocaña, N. Malashikhina, M. del Valle, and V. Pavlov, "Label-free selective impedimetric detection of Cu2+ ions using catalytic DNA", *Analyst (Lond.),* vol. 138, no. 7, pp. 1995-1999, 2013.
[http://dx.doi.org/10.1039/c3an36778a] [PMID: 23423467]

[77]   Z. Lin, Y. Chen, X. Li, and W. Fang, "Pb2+ induced DNA conformational switch from hairpin to G-quadruplex: electrochemical detection of Pb2+", *Analyst (Lond.),* vol. 136, no. 11, pp. 2367-2372, 2011.
[http://dx.doi.org/10.1039/c1an15080d] [PMID: 21491024]

[78]   F. Li, Y. Feng, C. Zhao, and B. Tang, "Crystal violet as a G-quadruplex-selective probe for sensitive amperometric sensing of lead", *Chem. Commun. (Camb.),* vol. 47, no. 43, pp. 11909-11911, 2011.
[http://dx.doi.org/10.1039/c1cc15023e] [PMID: 21975421]

CHAPTER 8

Electrochemical, Sensing and Environmental Assessments of CuFe$_2$O$_4$/ZnO Nanocomposites Synthesized *via Azadirachta Indica* Plant Extract

B. S. Surendra[1,*], **N. Raghavendra**[1], **H. P. Nagaswarupa**[2,*], **T. R. Shashi Shekhar**[3] and **S. C. Prashantha**[1]

[1] *Department of Science, East West Institute of Technology, Bengaluru 560 091, Karnataka, India*

[2] *Department of Studies in Chemistry, Shivagangothri, Davangere University, Davangere - 577 007, Karnataka, India*

[3] *Department of Civil, East West Institute of Technology, Bengaluru 560 091, Karnataka, India*

Abstract: The CuFe$_2$O$_4$/ZnO nanocomposite (CZO NCs) has been successfully synthesized by a facile one-pot green assisted approach using *Azadirachtaindica* extract. These synthesized nanocomposites were characterized by different techniques; Powder X-ray diffractometer (PXRD) confirms the formation of nanocrystalline composite material with crystallite size in the range 15-30 nm. The surface morphology of prepared nanocomposite materials was changed due to the plant extraction action as reducing and capping agent studied by Scanning electron microscope (SEM), which shows porous and spongy like structure. The photocatalytic performance of CZO NCs was examined for their potential role in the photodegradation of Methylene Blue (MB) dye under UV light. The result shows that the 1:2 molar CZO NCs is suitable for excellent photocatalytic degradation performance of MB dye (98%). Electrochemical analysis of prepared material was conducted with graphite electrode paste in 0.1 M KCl electrolyte, which showed excellent performance in redox potential measured by Cyclic Voltammetry and extended their sensor activities towards chemical and bio-chemicals (Paracetamol and Glucose). These studies open a new platform to utilize a simple approach for the preparation of multifunctional CZO NCs for their potential applications in wastewater treatment and electrochemical studies.

Keywords: CuFe$_2$O$_4$/ZnO, Cyclic Voltammetry, Green combustion, Photo-catalysis, Sensor, Wastewater treatment.

---

* **Corresponding authors B.S. Surendra and H. P. Nagaswarupa:** Department of Science, East West Institute of Technology, Bengaluru 560 091, Karnataka, India AndDepartment of Studies in Chemistry, Shivagangothri, Davangere University, Davangere - 577 007, Karnataka, India; E-mail: surendramysore2010@gmail.com, nswarupa@davangereuniversity.ac.in,

J.G. Manjunatha (Ed.)

INTRODUCTION

Nowadays, wastewater treatment practices have gained more priority in the world by the effect gradually increasing contamination of water from various sources like industries, laboratories, paper and pulp production, *etc.* causes a serious adverse effect to the environmental issues [1 - 5]. The wide utilization of synthetic dye products such as garments, leather trimmings, furniture equipment, chemical laboratories, *etc*, excludes approximately 10-12% of these dyes and departs into the surroundings. Therefore the efficient degradation of such toxic chemicals is a great task for water treatment in different fields of science and technologies. The final degraded products should be eco-friendly and their experimental practices must be sustainable are the solution things that are anticipated [6]. Generally, dyes are the major organic pollutants present in wastewaters, creating severe problems and thus, lead to developing effective methods for wastewater treatment [7]. Nanomaterials are the modern field of science which plays a vital role in all aspects of human activity such as electronic devices, telecommunication equipment, magnetic fluids, drug delivery, sensors, *etc.* with low cost and environmentally friendly approaches. Presently, the semiconductor nanomaterials have been employed as Photocatalyst, which can be used to degrade the pollutants as an ecological approach [8 - 11]. The heterogeneous catalysis process is extensively employed for the bleaching of such dye pollutants [12].

The ferrite nanomaterials are a common type of spinel tetrahedral and octahedral crystalline structures, which has attracted great attention due to their extensive technological applications and key subject in environmental contamination remediation *via* photocatalysis examination [13 - 15]. The copper based ferrites and their composites are recognized as semiconductor photocatalysts for efficient degradation of toxic chemicals through photocatalysis [16, 17]. The exposure of semiconductor materials to light of energy not less than electrons and holes generated from the band gap. In the conduction band, oxygen molecules accept the electrons and transforming into superoxide radical ion. Hydroxyl radicals generated from the oxidation of adsorbed water molecules on the surface of semiconductor in the valence band, which converts the organic pollutants to carbon dioxide and water [18, 19].

Recently, few works confirmed the functionality of green preparation of nanoparticles/nanocomposites using various plant extracts for domestic/industrial waste-water treatment [20]. The non-edible biomass resources (Neem (*Azadirachtaindica*), Jatropha, Pongamia, *etc.*) are the most abundant and alternative source of fossil diesel, offers several benefits for the environmental and economical needs. Various synthesis methods are applied to prepare nano metal oxides [21, 22] and among those, solution combustion approach is a potential, economical and versatile methodology. This method involves a self-sustained reaction in homogeneous mixture of reactant oxidizers (*e.g.,* metal nitrates) and green fuels (Neem *(Azadirachtaindica)* extract). Thus, this process has more assistance due to the advantage of ecological friendliness, easy for synthesis with cost effective, energy efficient approach *etc* [23 - 28]. CZO NCs have been preferred for this research work due to its activity towards anodic–cathodic potential properties. Also, the important movement has been achieved for the prepared nanocomposite in the sensor measurements. Therefore, the sensor activities of prepared material monitoring chemical and biochemical sensing applications are of great importance for sustaining human health and also in different detection practices [29 - 35]. As a result, CV analysis was carried out for CZO NCs to illustrate the modified graphite-composite electrode paste and extended their sensor activities towards chemical and biochemical such as Paracetamol and Glucose, respectively. With such outcomes, the present research work reported that the excellent performance of CZO NCs towards UV light induced photocatalyst for degradation of MB dye. The composition of nanocomposites has been synthesized methodically by varying the ratios of both the CuFe$_2$O$_4$ and ZnO oxides.

EXPERIMENT

Synthesis of CuFe$_2$O$_4$ nanoparticles and ZnO oxide

CuFe$_2$O$_4$ nanomaterials were synthesized by green combustion method using Cu(NO$_3$)$_2$.3H$_2$O and Fe(NO$_3$)$_3$.9H$_2$O as oxidizers. The stoichiometric ratios of oxidizer and fuel were taken in silica crucible with a small amount of water and subjected to stirring to attain homogeneity. Further, this homogeneous mixture was subjected to combustion in a muffle furnace at 550±10 °C.

Similarly, the same procedure was adopted for the preparation of ZnO nanoparticles mentioned above; instead of Cu(NO$_3$)$_2$.3H$_2$O and Fe(NO$_3$)$_3$.9H$_2$O, Zn(NO$_3$)$_2$ was used.

Synthesis of CuFe$_2$O$_4$-ZnO Nanocomposite

The CZO NCs photocatalysts were synthesized from mixing up of Cu(NO$_3$)$_2$.3H$_2$O, Fe(NO$_3$)$_3$.9H$_2$O and Zn(NO$_3$)$_2$ (NICE chemical (P)Ltd.) with various ratios (1:1, 1:2 & 2:1) by green combustion method. [Cu(NO$_3$)$_2$.3H$_2$O, NICE chemical (P)Ltd.; Fe(NO$_3$)$_3$.9H$_2$O, NICE chemical (P)Ltd.]. Analytical grade was used without further purification. The specific amounts of copper and ferric nitrates in stoichiometric ratios were used as initial precursors and *Azadirachtaindica* extract was used as a green fuel. Finally, the product was obtained by placing a crucible containing homogeneous mixtures in a muffle furnace maintained at 470±10 °C.

Characterization

The phase analysis was carried out in Shimadzu Powder X-ray diffractometer (PXRD) using nickel filter in the 2θ range 20 - 70° with Cu Kα radiation. The surface morphology of prepared nanocomposites was found from SEM, Hitachi SU3500. A FT-IR study was performed in the range of 4000-400 cm^{-1} with the help of Spectrum-1000 (Perkin Elmer) spectrophotometer. UV-Visible absorption was recorded using Shimadzu UV 2600 UV – Visible Spectrophotometer.

Photocatalytic Activity Measurements of Organic Dyes by CZO NCs

The photocatalytic activity of synthesized CZO NCs was conducted under UV-light irradiation for degrading Methylene blue (MB) dye. The newly designed photocatalytic reactor is connected with a mercury vapor lamp for the supply of UV light with 125 W medium pressure. The glass container has 169.8 cm^2 surface area containing 20 mg of prepared CZO NCs and 250 ml of MB dye solution. Irradiate the UV-light directly on the solution at a distance of 23 cm in the normal atmosphere and at definite time intervals; the 5 ml solution was collected in aliquots from the suspensions. The concentration of MB dye was measured by observing the absorbance at 668 nm.

RESULTS AND DISCUSSION

PXRD Analysis

Phase composition and structure of the $CuFe_2O_4$, ZnO and CZO NCs obtained by *Azadirachtaindica* extract assisted combustion process were illustrated through PXRD study (Fig. **1**). The diffraction peaks in accordance with JCPDS file (card No. 034-0425) confirms the $CuFe_2O_4$ with the cubic spinel structure [36]. Additionally, diffraction peaks in the pattern were indexed to diffraction peaks at (100), (002), (101), (102), (110), (103), (200), and (112) are in accordance with JCPDS file (card no. 36-1451) having the hexagonal crystalline structure of ZnO [37]. The sample with 1:2 molar ratio of CZO NCs shows better crystallinity compared to 1:1 and 2:1. The crystal plane showed sharp and high intensity indicating that high crystallinity of the material and its average crystallite size was found to be ~ 15 nm measured through Scherer's formula (Eq.1). Additionally, its other physical parameters such as Micro strain, Dislocation density and Stress were calculated using the following equations (Eq.2, 3, 4).

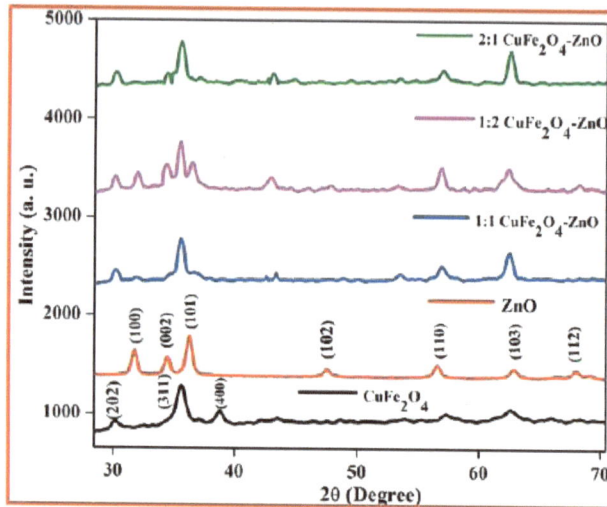

Fig. (1). PXRD studies of the synthesized CuFe₂O₄-ZnO and host Nanoparticles.

$$d = \frac{k\lambda}{\beta \cos \theta} \qquad (1)$$

$$\varepsilon = \beta \cos \frac{\theta}{4} \qquad (2)$$

$$\delta = \frac{1}{D2} \qquad (3)$$

$$\sigma = \varepsilon Y \qquad (4)$$

Where, D is the crystalline size, λ is the X-ray wavelength, β is the full-width half maximum of the peak. Smaller the particle size of CZO NCs shows, the higher the micro strain and lesser dislocation density representing well crystallinity of the material.

MORPHOLOGY EXAMINATION

SEM micrographs of CZO NCs with different molar ratios are shown in Fig. (2). The addition of ZnO into CuFe₂O₄ shows morphological changes of nanocomposites and the 1:2 composite sample exhibits more spongy like structures [38 - 40]. The SEM image of prepared pure nanomaterials [CuFe₂O₄ and ZnO] show voids and agglomeration caused by the chemical composition (fatty acids) of oil extract that reacts with cupric ion (Cu²⁺) and zinc ions (Zn²⁺) due to its reducing and capping agent properties [36]. The resulting that the coordination reaction occurs with fatty acids chains containing hydroxyl (OH⁻)

groups. This polymeric coordination binding was responsible for getting different structures due to the conjugation of all compounds present in the oil extract [8].

Fig. (2). SEM micrographs of CuFe$_2$O$_4$/ZnO and host Nanoparticles.

FT-IR Analysis

FT-IR studies evidence for the presence of functional groups, atomic/molecular vibrations and purity of the prepared composite in the region of 4000-400 cm^{-1} (Fig. **3**). The tetrahedral and octahedral complexes of metal-oxygen bond peaks are found at ~ 552 and ~ 415 cm^{-1}, respectively [41]. The stretching vibrational peaks of tetrahedral group Cu-O bond found at ~ 552 cm^{-1}, octahedral complex Fe-O vibration band showed at ~ 415 cm^{-1} and peak at 709 cm^{-1} shows the presence of Zn-O stretching vibration [42, 43]. *Azadirachtaindica* extract assisted prepared CZO NCs showed a peak at 2372 cm^{-1}. The broad peak appears at 3425 cm^{-1} representing the presence of water molecules and it's the bending vibrational peak found at ~1624 cm^{-1} [8]. The presence of OH$^-$ groups in CZO NCs showed the potential of the compound for photocatalytic activity.

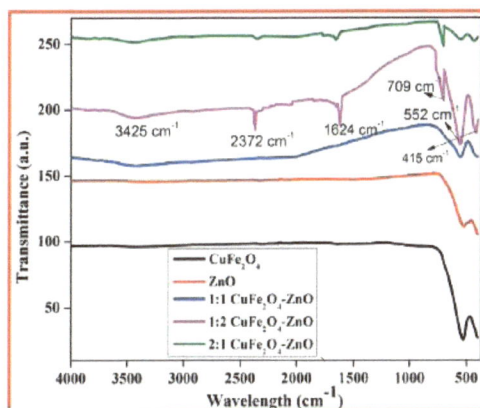

Fig. (3). FT-IR spectra of synthesized CuFe$_2$O$_4$-ZnO and host Nanoparticles.

Nitrogen Adsorption-desorption Isotherm

The pore diameter and surface area of prepared samples were measured by N_2 adsorption-desorption isotherms (Fig. **4**). The observed hysteresis loop of prepared CZO NCs have high relative pressure state (>0.8 P/P0) representing the macro porosity in nature. The result of this measurement was attributed to a class of the irreversible type-IV isotherm with H3 type hysteresis, indicating the existence of mesoporous size of nanoparticles.

Thus, the synthesized 1:1, 1:2 and 2:1 CZO NCs showed specific surface areas are 10.3, 28.2 and $13.2 m^2/g$, respectively. This consequence confirms that the obtained 1:2-CZO NCs are mesoporous in nature and assist the photocatalytic reactions.

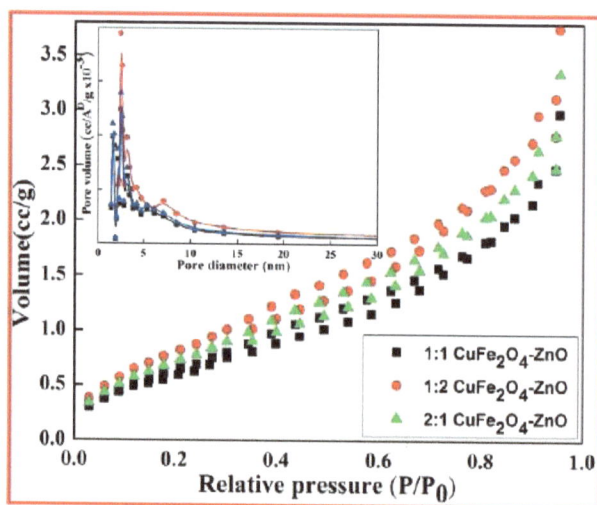

Fig. (4). N_2 adsorption and desorption isotherms and pore volume distribution curve (inset) of CZO NCs nanoparticles.

Cyclic Voltammetry (CV) Studies

The potentiodynamic measurements and super capacitive nature of the prepared nanomaterial (CuFe₂O₄/ZnO) modified with graphite paste were analyzed by CV studies. The CV testing employs an electrochemical cell associated with three electrode arrangements in 0.1M KClmedium in the potential range of 0.5 to −1.0 V (Fig. **5**). It is significant that the modified 1:2- CZO NCs electrode shows good redox peaks with enhanced peak current than the 1:1, 2:1- CZO NCs and host materials (CuFe₂O₄ and ZnO) attributed to its high photocatalytic performance [44 - 46].

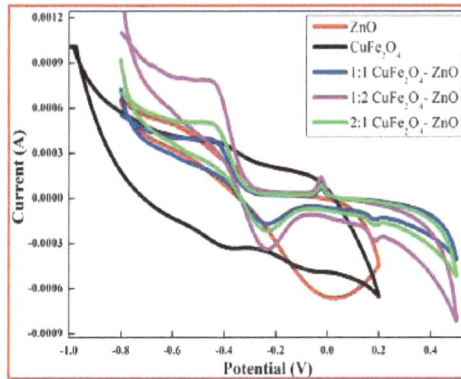

Fig. (5). Cyclic voltammogram of host and prepared nanomaterials (CuFe $_2$ O $_4$ /ZnO) modified with graphite paste in 0.1M KCl medium.

The EIS (Electrochemical Impedance spectra) Nyquist graph of the nano CZO NCs and nanoparticles (CuFe$_2$O$_4$ & ZnO) prepared by green combustion method as shown in Fig. (**6**). This plot shows that the electrochemical reactions of electrodes display the semicircle at the high frequency region and a line at the low frequency region comprised capacitive and resistive effects [47 - 51]. The arc radius for 1:1, 1:2, 2:1- CZO NCs was found to be 72, 46 and 120 Ω, respectively. This suggests that the 1:2- CZO NCs have lower charge transfer resistance due to the presence of ZnO in CuFe$_2$O$_4$ nanoparticles accelerates the interfacial charge-transfer process. Lesser charge-transfer resistance offers the high photocatalytic activity and hence, 1:2- CZO NCs having smaller arc radius performed with high photocatalytic action. The capacitance measurements of 1:2 CZO NCs electrode was observed for electrolyte with sensors of every 1mM Paracetamol and Glucose sensor chemical Figs. (**7a** and **7b**). Thus, the prepared material electrode showed good biosensor activity toward Glucose molecule.

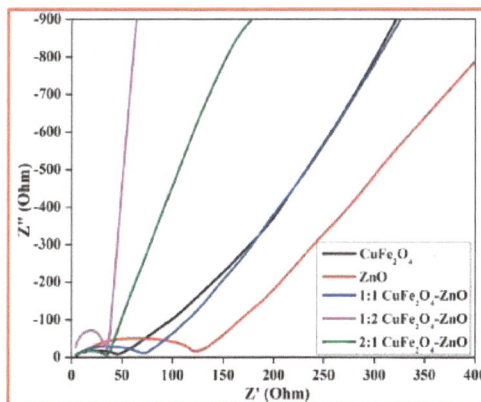

Fig. (6). Impedance plot of synthesized CuFe$_2$O$_4$-ZnO and host Nanoparticles.

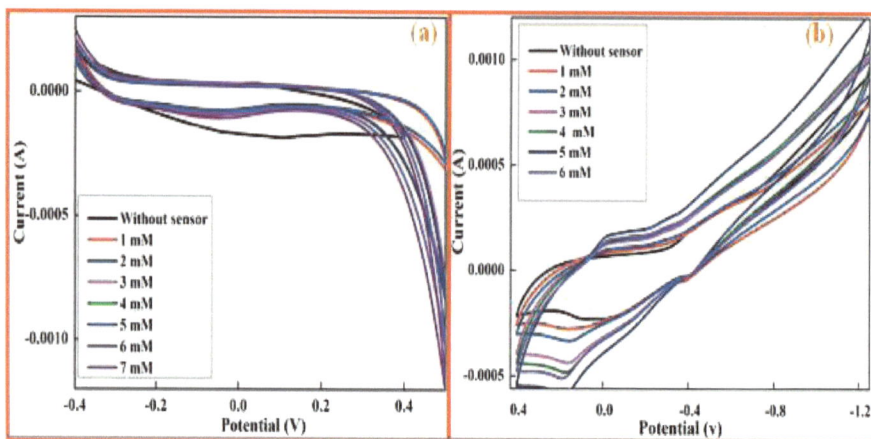

Fig. (7). Sensor activity of prepared CuFe $_2$ O $_4$ -ZnOnanocomposites for **(a)** Paracetamol and **(b)** Glucose.

Photocatalytic Degradation

In current investigation, photocatalytic degradation was carried out for CZO NCs prepared by the green combustion method. From experimental analysis, 1:2-CZO NCs sample showed good photocatalytic activity for degradation of MB dye under UV radiation due to small particle size, more active sites as indicated by the P-XRD data [52 - 56]. It was found that the photo-decolorization of MB for CZO NCs was show around 98% (Fig. **8**).

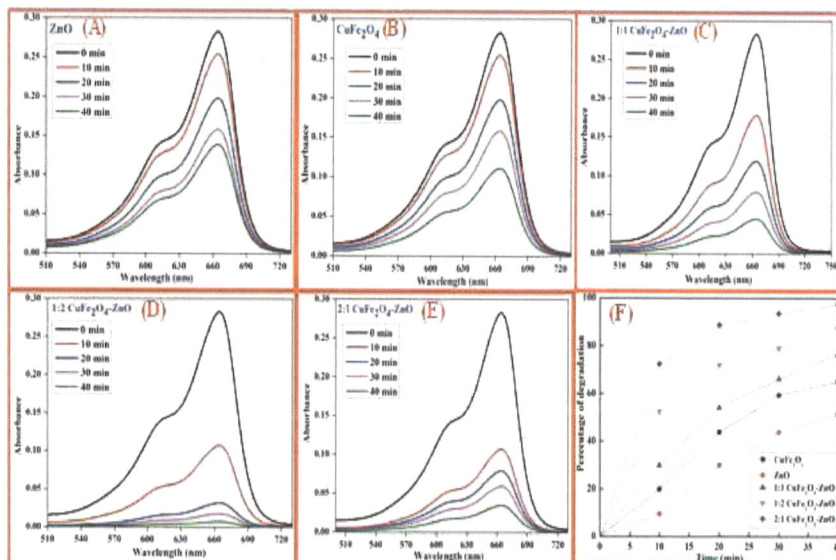

Fig. (8). Photocatalytic decomposition of MB under UV light for synthesized CuFe₂O₄-ZnO and host Nanoparticles.

Mechanism of Photocatalytic Degradation

The synthesized $CuFe_2O_4$ encapsulated ZnO nanoparticles effectively degraded the MB dye solution under UV light radiation. The mechanism involved in the photocatalytic activity of MB dye is shown in Fig. (**9**). The oxygen (O_2) and water (H_2O) molecules are easily absorbed on the outer surface and also the internal surface of photocatalyst under UV light irradiation. These Photocatalysts are capable of capturing the incident light with sufficient energy, released electrons (e^-) jump into the conduction band (CB) from the valence band (VB), leaving behind the same number of holes (h^+) in the VB. In MB dye degradation, chemical bonds are broken under light irradiation and jump to excited state or can be directly self-disintegrated.

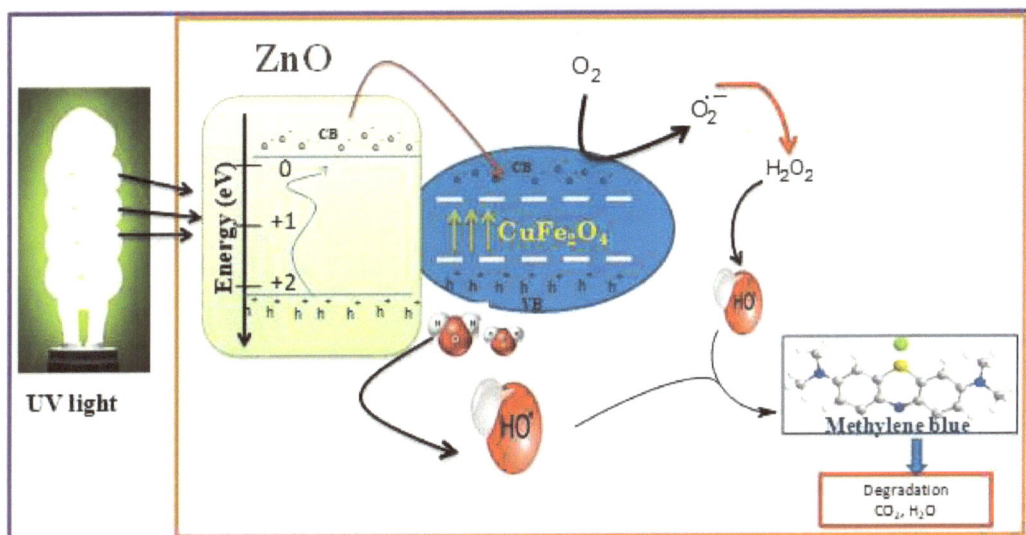

Fig. (9). Mechanism for the photocatalytic decomposition of MB under UV light irradiation.

ZnO photocatalyst in the dye solution shows the release of Zn^{2+} to the solution due to the photocatalytic process under UV light. The O_2 molecules absorbed photo-generated electrons and H_2O molecules interacted with holes, which leads produces a lot of superoxide radicals ($O^{2\cdot-}$) and hydroxyl radicals ($OH^{\cdot}$).These obtained active radicals react with the ionized MB molecules to decompose them into the nontoxic H_2O, CO_2, and mineral acids (NO^{2-}, NO^{3-} or SO_4^{2-}) [57 - 62].

Kinetic Studies

MB dye decomposition practice confirms the first-order kinetic model, which is generally matched since the original concentration of the pollutant is low and its kinetics can be measured (Eq. 5) as follows;

$$\ln\left(\frac{C}{C_0}\right) = - \, kt \qquad (5)$$

Where **k** is the apparent reaction rate constant, the original concentration of aqueous MB is C_0, and C is the concentration of aqueous MB at the reaction time **t**. Fig. (**10 A**) represents first-order plots for decomposition of MB for green combustion synthesized CZO NCs under UV light irradiation.

A general chemical 4-chlorophenol (4-CP) is used widely in several industrial relevancies. As a result, this compound is also preferred as a model pollutant for our research study. Fig. (**10 B**) illustrates the decomposition rate of 4-CP on effective green combustion synthesized CZO NCs under UV light irradiation. Thus, the decomposition of 4-CP was almost consistent with decomposition of MB dye for green combustion synthesized CZO NCs under UV light irradiation.

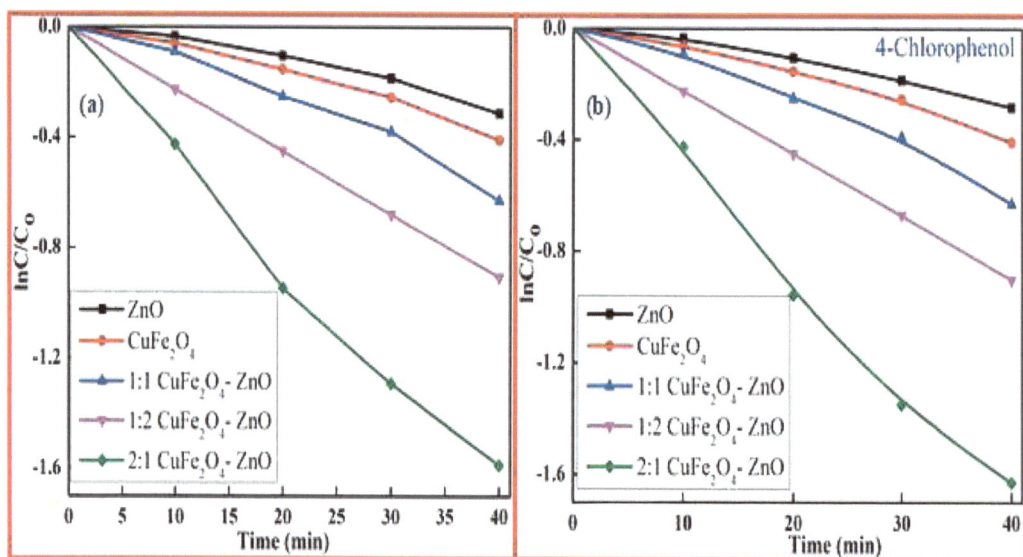

Fig. (10. (a) Plot of ln (C/C_0) *versus* time for the decomposition of MB under UV irradiation for synthesized $CuFe_2O_4$-ZnO and host Nanoparticles **(b)** Plot of ln (C/C_0) *versus* time for the decomposition of 4-chlorophenol under UV irradiation.

CONCLUSION

The $CuFe_2O_4$/ZnO nanocomposite was successfully synthesized *via* green assisted approach using green extract as a fuel *(Azadirachtaindica)* for the first time. P-XRD analysis confirmed the presence of cubic phase structure with particles size in the range of 15-30 nm. The electrochemical investigation results confirm that the prepared material revealed a promising material for the development of graphite electrode paste and showed excellent performance in reversible redox reaction confirmed by CV studies. Also, they extended their analysis for examining the sensor activities towards chemical and bio-chemical compounds 0.1 M KCl electrolyte. The synthesized 1:2- CZO NCs photocatalyst exhibit a potential substance for excellent photocatalytic activity towards the photo-degradation of MB (98%) dye. The reported results confirm that the prepared nanocomposite can be used as a promising substance for electrochemical, Sensor, Heavy metal detection of industrial wastewater treatments for environmental remediation by using several chemical and bio-sensors.

CONSENT FOR PUBLICATION

Not applicable.

CONFLICT OF INTEREST

The authors declare no conflict of interest, financial or otherwise.

ACKNOWLEDGEMENTS

The authors gratefully acknowledge the Department of Science and Technology (DST), New Delhi, for the financial support for this research, through DST-DDP Project [DST/TDT/DDP-36/2018, Dated 09/09/2020].

REFERENCES

[1]   G. Nabiyouni, D. Ghanbari, J. Ghasemi, and J. Yousofnejad, *Nano Struct.*, vol. 5, no. 3, pp. 289-295, 2015.

[2]   R.D. Rivera-Rangel, M.P. Gonzalez-Munoz, M. Avila-Rodriguez, T.A. Razo-Lazcano, and C. Solans, "Green synthesis of silver nanoparticles in oil-in-water microemulsion and nano-emulsion using geranium leaf aqueous extract as a reducing agent", *Colloids Surf. A Physicochem. Eng. Asp.*. [http://dx.doi.org/10.1016/j.colsurfa.2017.07.051]

[3]   *Anal. Bioanal. Electrochem.*, vol. 12, no. 7, pp. 893-903, 2020.

[4]   Naghikhani R., Nabiyouni, G., and Ghanbari D, Simple and green synthesis of $CuFe_2O_4$–CuO nanocomposite using some natural extracts: photo-degradation and magnetic study of nanoparticles. , *J Mater Sci: Mater Electron,* vol. 29, pp. 4689-4703, . [http://dx.doi.org/10.1007/s10854-017-8421-1]

[5]   C.P. Devathaa, and K. Jagadeesh, *Monitoring & Management,* vol. 9, pp. 85-94, 2018.

[6]   J.G. Girish Tigari, Manjunatha C. Raril, and N. Hareesha, "Appro", In: *Drug Des Dev* vol. 4. , 2019.

[7]   S.D. Edwin Souza, J.G. Manjunatha, and C. Raril, "Girish Tigari and P.A. Pushpanjali", *Anal. Bioanal. Chem. Res.,* vol. 7, no. 4, pp. 461-472, 2020.

[8]   *Surfaces,* vol. 3, no. 3, pp. 473-483, 2020. [http://dx.doi.org/10.3390/surfaces3030034]

[9]   "NajmehNajmoddin, Ali Beitollahi, Eamonn Devlin, Huseyin Kavas, Seyed Majid Mohseni, Johan Akerman, Dimitris Niarchos, HamidrezaRezaie, Mamoun Muhammed and S. MuhammetToprak", *Microporous Mesoporous Mater.,* vol. 190, pp. 346-355, 2014.

[10]   B.S. Surendra, "J. of Sci", *Advan. Mater. and Devices,* vol. 3, pp. 44-50, 2018.

[11]   *G. JamballiManjunatha.* vol. Vol. 14. The Open Chemical Engg. J., 2020, pp. 52-62.

[12]   N.R. Su, P. Lv, M. Li, and X. Zhang, "Minghui Li andJiashengNiu", *Mater. Lett.,* vol. 122, pp. 201-204, 2014.
[http://dx.doi.org/10.1016/j.matlet.2013.12.106]

[13]   B. S. Surendra, M. Veerabhdraswamy, K. S. Anantharaju, H. P. Nagaswarupa, and S. C. Prashantha, "J.ofNanostru", *in Chem,* vol. 8, pp. 45-59, 2018.

[14]   V.K. Pandit, S.S. Arbuj, Y.B. Pandit, S.D. Naik, S.B. Rane, U.P. Mulik, S.W. Gosavic, and B.B. Kale, "Nanocomp", *RSC Advances,* vol. 5, pp. 10326-10331, 2015.
[http://dx.doi.org/10.1039/C4RA11920G]

[15]   R.Q. Guo, L.A. Fang, W. Dong, F.G. Zheng, and M.R. Shen, *J. Phys. Chem. C,* vol. 114, pp. 21390-21396, 2010.
[http://dx.doi.org/10.1021/jp104660a]

[16]   "Pietro Altimari, EmanuelaMoscardini, Ida Pettiti and Luigi Toro Francesca Pagnanelli", *Chem. Eng. Trans.,* vol. 47, pp. 151-156, 2016.

[17]   N. Hareesha, J. G. Manjunatha, and C. Raril, *ChemistrySelect,* vol. 4, no. 15, pp. 4559-4567, 2019.
[http://dx.doi.org/10.1002/slct.201900794]

[18]   "JamballiGangadharappa Gowda Manjunatha and ChenthattilRaril", *Adv. Pharm. Bull.,* vol. 10, no. 2, pp. 247-253, 2020.
[PMID: 32373493]

[19]   "G. JamballiManjunatha, ChenthattilRaril and Girish Tigari", *Adv. Pharm. Bull.,* vol. 9, no. 1, pp. 132-137, 2019.
[PMID: 31011567]

[20]   N. Hareesha, and J.G. Manjunatha, *J. of Sci. Advan. Mater. and Devices,* vol. 5, no. 4, pp. 502-511, 2020.
[http://dx.doi.org/10.1016/j.jsamd.2020.08.005]

[21]   "Instru. and Experi", *Techni.,* vol. 63, pp. 750-757, 2020.

[22]   *J. Electrochem. Sci. Eng.,* vol. 10, no. 4, pp. 305-315, 2020.
[http://dx.doi.org/10.5599/jese.774]

[23]   "R. Jayaprakash, G. Sarala Devi andP. Siva Prasada Reddy", *J. Magn. Magn. Mater.,* vol. 355, pp. 87-92, 2014.

[24]   P. Jacob, "Binu, Kumar, Ashok, Pant, R.P. Singh andSukhvir, Mohammed", *Bull. Mater. Sci.,* vol. 34, pp. 1345-1350, 2011.
[http://dx.doi.org/10.1007/s12034-011-0326-7]

[25]   L. Sun, R. Shao, L. Tang, Z. Chen, and J. Al, *Com.,* vol. 564, pp. 55-62, 2013.

[26]   "K.M.Jadhav, D.V. Mane and SG Patil, The Phar", *Innov. J.,* vol. 7, no. 1, pp. 215-217, 2018.

[27]   *Desalination,* vol. 279, pp. 332-337, 2011.
[http://dx.doi.org/10.1016/j.desal.2011.06.027]

[28]   "Rakshitameta, C. Suresh. Ameta", *Int. J. Chem. Sci.,* vol. 14, no. 4, pp. 3256-3264, 2016.

[29]   Y. Choi, M.J. Choi, S.H. Cha, Y.S. Kim, S. Cho, and Y. Park, "Catechin-capped gold nanoparticles: green synthesis, characterization, and catalytic activity toward 4-nitrophenol reduction", *Nanoscale Res. Lett.,* vol. 9, no. 1, p. 103, 2014.

[http://dx.doi.org/10.1186/1556-276X-9-103] [PMID: 24589224]

[30] P. Sathishkumar, R. Sweena, and J. Wu, "Jerry, S. Anandan", *Chem. Eng. J.,* vol. 171, pp. 136-140, 2011.
[http://dx.doi.org/10.1016/j.cej.2011.03.074]

[31] B. Neppolian, Q. Wang, H. Yamashita, and H. Choi, *Appl. Catal. A,* vol. 333, pp. 264-271, 2007.
[http://dx.doi.org/10.1016/j.apcata.2007.09.026]

[32] S. Arvind Varma, "Alexander. Mukasyan, S. AlexanderRogachev and V.KhachaturManukyan", *Chem. Rev.,* vol. 116, no. 23, pp. 14493-14586, 2016.
[PMID: 27610827]

[33] B.S. Surendra, R. Gagan, and G.N. Soundarya, "Naresh, Appl", *Surface Sci. Advan.,* vol. 1, no. 1, p. 100017, 2020.
[http://dx.doi.org/10.1016/j.apsadv.2020.100017]

[34] "G. JamballiManjunatha, K. DoddarasinakereRavishankar, Santosh Fattepur, GurumallappaSiddaraju and LingappaNanjundaswamy", *Surg. Eng. Appl. Electrochem.,* vol. 56, no. 4, pp. 415-426, 2020.

[35] *Mater. Res. Innov.,* vol. 24, no. 6, pp. 349-362, 2020.
[http://dx.doi.org/10.1080/14328917.2019.1684657]

[36] G. Girish Tigari, "JamballiManjunatha", *ChenthatillRaril and NagarajappaHareesha,* vol. 4, no. 7, pp. 2168-2173, 2019.

[37] B.M. Amrutha, J.G. Manjunatha, A.S. Bhatt, C. Raril, and P.A. Pushpanjali, *Phys. Chem. Res.,* vol. 7, no. 3, pp. 523-533, 2019.

[38] B.S. Surendra, H.P. Nagaswarupa, and M.U. Hemashree, *Chem. Phys. Lett.,* vol. 739, p. 136980, 2020.
[http://dx.doi.org/10.1016/j.cplett.2019.136980]

[39] R. Dom, R. Subasri, K. Radha, and P.H. Borse, *Solid State Commun.,* vol. 151, pp. 470-473, 2011.
[http://dx.doi.org/10.1016/j.ssc.2010.12.034]

[40] E. Ranjith Kumar, and R. Jayaprakash, "G. Sarala Devi andP. Siva Prasada Reddy", *Sen. Act. B.,* vol. 191, pp. 186-191, 2014.
[http://dx.doi.org/10.1016/j.snb.2013.09.108]

[41] P.H. Borse, J. Hwichan, and S.H. Choi, "S. J. Hong andJ. S. Lee", *Appl. Phys. Lett.,* vol. 93, p. 173103, 2008.
[http://dx.doi.org/10.1063/1.3005557]

[42] "K.M.Jadhav, D.V. Mane and S.G.Patil, The Phar", *Innov. J.,* vol. 7, no. 1, pp. 215-217, 2018.

[43] S. Ahmed, and M.A. Saifullah, "Babu Lal Swami andSaiqaIkram", *J. of Radiation Resea. and Appl. Sci.,* vol. 9, pp. 1-7, 2016.

[44] S.H. Yu, and M. Yoshimura, *Adv. Funct. Mater.,* vol. 12, pp. 9-15, 2002.
[http://dx.doi.org/10.1002/1616-3028(20020101)12:1<9::AID-ADFM9>3.0.CO;2-A]

[45] S. Cho, S.H. Jung, and K.H. Lee, *J. Phys. Chem. C,* vol. 112, pp. 12769-12776, 2008.
[http://dx.doi.org/10.1021/jp803783s]

[46] Y. Tamaura, Y. Ueda, J. Matsunami, N. Hasegawa, and M. Nezuka, "T. Sano andM. Tsuji", *Sol. Energy,* vol. 65, pp. 55-57, 1999.
[http://dx.doi.org/10.1016/S0038-092X(98)00087-5]

[47] N. Hareesha, and J.G. Manjunatha, *J. of Sci. Advan. Mater. and Devices,* vol. 5, no. 4, pp. 502-511, 2020.
[http://dx.doi.org/10.1016/j.jsamd.2020.08.005]

[48] "G. JamballiManjunatha, LingappaNanjundaswamy, GurumallappaSiddaraju, K. DoddarasinakereRavishankar", *Santosh Fattepur and EshwarappaNiranjan,* vol. 10, no. 11, pp. 1479-1490, 2018.

[49]   A. Pemmatte, "Pushpanjali and Jamballi G", *Manjunatha,* vol. 32, no. 11, pp. 2474-2480, 2020.

[50]   " JamballiManjunatha and ChenthattilRaril, Anal", *Bioanal. Electrochem,* vol. 11, no. 6, pp. 742-756, 2019.

[51]   N. Harcesha, and J.G. Manjunatha, "J. of Electroanaly. Chem., Vol.878, pp.114533, 2020.R. Kalaiselvan, C. O. Augustin, L. John Berchmanns and R Saraswathi", *Mater. Res. Bull.,* vol. 38, pp. 41-54, 2003.

[52]   Y. Wang, H. Zhao, M. Li, J. Fan, and G. Zhao, "Appl", *Cat. B: Environ.,* vol. 147, pp. 534-545, 2014. [http://dx.doi.org/10.1016/j.apcatb.2013.09.017]

[53]   L.J. Zhuge, X.M. Wu, Z.F. Wu, and X.M. Yang, "X.M. Chen andQ. Chen", *Mater. Chem. Phys.,* vol. 120, pp. 480-483, 2010. [http://dx.doi.org/10.1016/j.matchemphys.2009.11.036]

[54]   "G. JamballiManjunatha and T. MellekatteShreenivas", *ChemistrySelect,* vol. 4, no. 46, pp. 13427-13433, 2019.

[55]   C. Raril, J. G. Manjunatha, and J. Microchemical,

[56]   *J. Electrochem. Sci. Eng.,* vol. 10, no. 4, pp. 305-315, 2020. [http://dx.doi.org/10.5599/jese.774]

[57]   A. Ren, C. Liu, Y. Hong, W. Shi, S. Lin, and P. Li, *Chem. Eng. J.,* vol. 258, pp. 301-308, 2014. [http://dx.doi.org/10.1016/j.cej.2014.07.071]

[58]   "XinyongLi, QidongZhao,YonghuaLi, CaizhiSun andYongqiang Cao", *Mater. Res. Bull.,* vol. 48, no. 8, pp. 2927-2932, 2013.

[59]   C.W. Nan, M.I. Bichurin, and S. Dong, "D. Viehland andG. Srinivasan", *J. Appl. Phys.,* vol. 103, p. 031101, 2008. [http://dx.doi.org/10.1063/1.2836410]

[60]   J. Yang, M. Gao, L. Yang, Y. Zhang, J. Lang, D. Wang, and Y. Wang, "H. Liu andH. Fan", *Appl. Surf. Sci.,* vol. 255, pp. 2646-2650, 2008. [http://dx.doi.org/10.1016/j.apsusc.2008.08.001]

[61]   Y. Wang, H. Zhao, and M. Li, "Jiaqi Fan andGuohua Zhao", *Appl. Catal. B,* vol. 147, pp. 534-545, 2014. [http://dx.doi.org/10.1016/j.apcatb.2013.09.017]

[62]   M.R. Anil Kumar, B. Mahendra, H.P. Nagaswarupa, B.S. Surendra, and C.R. Ravikumar, "Krushitha Shetty", *Mater. Today Proc.,* vol. 5, no. 10, pp. 22221-22228, 2018. [http://dx.doi.org/10.1016/j.matpr.2018.06.587]

CHAPTER 9

Advanced Sensor Materials for the Simultaneous Voltammetric Determination of Antihypertensive Drugs: An Overview

Carlos Alberto Rossi Salamanca-Neto[1], Bruna Coldibeli[1], Erison Pereira de Abreu[2], Gabriel Junquetti Mattos[1], Gabriel Rainer Pontes Manrique[1], Jessica Scremin[1], Natalia Sayuri Matunaga Campos[1], Débora Nobile Clausen[1,2] and Elen Romão Sartori[1,*]

[1] *Laboratório de Eletroanalítica e Sensores, Departamento de Química, Universidade Estadual de Londrina (UEL), Rodovia Celso Garcia Cid, PR 445 Km 380, Londrina – PR, C.P. 10.011, 86057-970, Brazil*

[2] *Departamento de Biomedicina, Centro Universitário Cesumar (UniCesumar), Avenida Santa Mônica, 450, Londrina – PR, 86027-610, Brazil*

Abstract: Antihypertensives are one of the most consumed drugs worldwide. To reduce pill burden, more than one class of antihypertensive drugs can be found in the same tablet in a fixed-dose combination. In this chapter, an overview of about 35 papers reporting on the advanced sensor materials for the voltammetric simultaneous determination of antihypertensive drugs will be presented. Special attention is given to the strategies the authors used for enabling simultaneous determination. Several types of carbon-based electrodes were considered as a voltammetric sensor platform for the development of simultaneous determination procedures. The samples analyzed were fixed-dose tablets, preferentially, but human fluids and water samples were also analyzed. Voltammetric methods used for the simultaneous determination of antihypertensive drugs revealed very sensitive approaches reaching limits of detection in the micro-and nanomolar levels with analysis lasting less than one minute and consuming very low or even no organic solvents.

Keywords: Antihypertensives, Carbon-based electrodes, Electroanalytical methods, Modified electrode, Simultaneous determination.

---

* **Corresponding authors Carlos A.R. Salamanca-Neto and Elen Romão Sartori:** Laboratório de Eletroanalítica e Sensores, Departamento de Química, Universidade Estadual de Londrina (UEL), Rodovia Celso Garcia Cid, PR 445 Km 380, Londrina – PR, C.P. 10.011, 86057-970, Brazil; Tel: +55 43 3371-4366; Fax: +55 43 3371-4286; E-mail:carlos.salamanca@uel.br and elensartori@uel.br

INTRODUCTION

Antihypertensive drugs are widely used by the population to reduce arterial hypertension on a daily and continuous basis [1]. Pharmaceutical formulations containing these drugs must undergo strict quality control in the pharmaceutical industry for human consumption.

Electroanalytical methods based on voltammetric techniques and advanced sensor materials have been considered for this purpose. The use of these techniques in the development of new analytical protocols for the simultaneous determination of antihypertensive drugs has increased in recent years [2 - 5]. The analytical advantages of the various voltammetric methods include high precision, accuracy, simple sample preparation, short analysis time, and minimal or no use of organic solvents, which makes them low cost and environmentally friendly procedures for large scale applications [2 - 7]. Additionally, several types of carbon-based materials, nanostructured or not, have been considered as a sensing platform for the simultaneous determination of different chemical species.

Considering these aspects, in this review, we have comprehensively summarized papers reporting the use of advanced sensors materials for the simultaneous determination of antihypertensives. The methods were categorized according to the type of material employed as the sensing element, and tables present the analytical characteristics towards the simultaneous determination of antihypertensives.

THE CURRENT APPLICATION OF ANTIHYPERTENSIVE AGENTS ON CARDIOVASCULAR EVENTS

Cardiovascular diseases are considered the major cause of death in the world, and hypertension is an important risk condition for the occurrence of cardiovascular events, as stroke, coronary disease, peripheral disease, cardiac insufficiency, and chronic kidney disease [8, 9]. Hypertension can be assigned to several causes, including behavior, genetic, metabolic, and unknown factors [10]. Besides the recommendation of a healthy lifestyle as the first strategy to reduce blood pressure, pharmacological intervention plays an essential role in the effectiveness of hypertension treatment. Antihypertensive agents consist of a diverse group of drugs that are classified according to their mechanisms and locals of action [1].

Blood pressure control with antihypertensives can be achieved by monotherapy of one drug or combined therapy with simultaneous administration of two or more drugs. Despite the divergence about the topic, the combined-therapy approach has become increasingly popular in the medical field, supported by the many favourable features. However, a suitable antihypertensive treatment must be prescribed, considering each patient individually [11, 12].

In combined therapy, low dosages result in reduced incidence of drug-related side effects, while the different drug properties act synergistically on different antihypertensive mechanisms for a more effective blood pressure lowering. Furthermore, this approach can be managed with separate products or a fixed-dosage combination product, which consists of a more appreciated way because it improves patient compliance with the treatment [11 - 13].

The antihypertensive classes available for medical practice nowadays are diuretics, β-blockers, angiotensin-converting enzyme (ACE) inhibitors, calcium antagonists, and angiotensin receptor blockers.

Diuretics have been an important class of antihypertensive agents since the 1950s. These drugs act on the renal system, increasing Na^+ and H_2O excretion, and are effective for reducing blood pressure in monotherapy or enhancing the efficiency of other antihypertensives when used as an association. They are subclassified in thiazide diuretics (*e.g.* hydrochlorothiazide), loop-diuretic (*e.g.* furosemide), and potassium-sparing diuretics (*e.g.* amiloride and triamterene). Thiazide-types are the primary choice in pharmacological treatment due to their efficacy, but differently from the other diuretics, potassium-sparing ones are prescribed in association with another diuretic to attenuate the excessive excretion of potassium (kaliuresis) [1, 8].

The β-blocker drugs were introduced in clinical medicine as antihypertensive agents in the 1960s. This group is divided into three generations, according to its β-adrenergic receptor selectivity and vasodilatory properties. The first-generation or non-selective β-blockers (*e.g.*, propranolol and pindolol) have similar affinities for β_1 and β_2-adrenergic receptors. The second-generation or selective β-blockers (*e.g.*, metoprolol, bisoprolol, and atenolol) present a higher affinity for β_1-receptors. Consequently, they reduce the incidence of adverse effects from the blockage of β_2-receptors. The third generation (*e.g.*, carvedilol and nebivolol) are vasodilators [14, 15].

The renin-angiotensin-aldosterone system (RAAS) expresses an important relationship with the homeostatic control of arterial pressure; therefore, hypertension can be related to dysregulation in this system. In this sense, RAAS blockers have been used as antihypertensive agents and include angiotensin-converting enzyme (ACE) inhibitors (*e.g.*, captopril, ramipril, enalapril, and lisinopril) and angiotensin receptor blockers (*e.g.*, valsartan, losartan, and telmisartan) [8, 16].

Calcium antagonists, approved in the 1980s, are now in widespread use for the long-term treatment of hypertension and are classified according to their molecular structures and modes of action. The dihydropyridine agents (*e.g.* nifedipine, amlodipine, nitrendipine, felodipine, lacidipine, lercanidipine, nimodipine, nisoldipine, and nicardipine) act as vasodilators. The non-dihydropyridine agents (*e.g.*, verapamil and diltiazem) act by reducing myocardial contractility, cardiac frequency, and impulse conductions [1, 8, 17].

Electrochemical Techniques and Instrumentation Applied to the Electroanalysis of Antihypertensive Drugs

Electrochemistry is the field of physical chemistry that associates the changing of electric parameters with a particular chemical reaction on the surface of an electrode. Its application in analytical chemistry generally involves the measurement of some electrical parameters (*e.g.*, potential, current, charge, impedance, *etc.*) under conditions that allow a direct or indirect association of the variable measured with the presence or concentration of a particular chemical species [18].

The electroanalysis is based on the application of electrochemical techniques on qualitative and quantitative analysis. This field of research has grown greatly for the development of several analytical methods over the last years due to the possibility of using a portable potentiostat device and flexible circuits (electrodes) as an electrochemical cell. The electroanalytical techniques can be categorized according to the parameters of the electrochemical cell that are being controlled and which are monitored by the electronic system. The main categories include electrogravimetric analysis, coulometry, amperometry, electrochemical impedance spectroscopy, and the most popular are the voltammetric techniques [19].

The analytical advantages of the voltammetric techniques include excellent sensitivity over both inorganic and organic species in aqueous and nonaqueous solutions, several useful solvents and electrolytes, rapid analysis time, in addition to the ability of simultaneous determination of many analytes in real samples. Although the description of voltammetric techniques comes from the same fundamental electrochemical theory, changes in the mode of the potential control and current measurement provide various types of voltammetry, including linear sweep, cyclic, square wave, stripping, and differential pulse voltammetry [6]. The configuration of the electrochemical cell where the experiment is carried out is the same for all of these operations, consisting of a working electrode, which provides the interface across which electrons are transferred during the reduction and/or oxidation of the analyte, in addition to an auxiliary electrode, and a reference electrode.

Cyclic voltammetry (CV) is rarely used for quantitative electroanalysis, but it is widely applied in the study of redox processes, and for understanding the electrochemical behavior of the analytes. In a CV measurement, the working electrode potential changes linearly as a function of time in cyclical phases. The scan rate of potential changing during each step of these cycles is a predetermined instrumental parameter and it is given in volts per second. The potential applied in the working electrode is based on the reference electrode, while the current from the reduction or oxidation of the analyte, which is called faradaic current, is monitored between the working electrode and the auxiliary electrode. These data are plotted as cathodic/anodic current (I) *versus* applied potential (E) and the resulting curves are called cyclic voltammograms. Similar to a CV operation system, in a linear sweep voltammetry experiment, the potential of the working electrode varies linearly in the time domain. However, after the limit potential is reached, there is no reverse step to get back to the initial potential. Redox processes of chemical species are registered as a peak (oxidation) or trough (reduction) in the current signal at the potential at which the electroactive species begins to be oxidized or reduced [20].

The quantitative electroanalysis is based on the measurement of faradaic currents that are generated when the analytes are oxidized or reduced at electrodes, as mentioned above. However, the interfaces of an electron conductor and a solution form a capacitor because charge can be accumulated on both sides, generating capacitive currents at this electrochemical double layer. When capacitive and faradaic currents flow at the same time, often the peak-shaped faradaic current (analytical signal) is disturbed by the capacitive current. Given this, the basis of all pulse voltammetric techniques is the difference in the rate of decay of the capacitive and the faradaic currents following a potential pulse. The capacitive current decays exponentially as a function of time, whereas the faradaic current decays considerably slower [21].

Among the different pulse techniques, differential pulse voltammetry (DPV) is one of the most important and the most widely applied in quantitative electroanalysis. In a DPV experiment, potential pulses are superimposed to a staircase ramp, the current is measured immediately before each potential change, and its value is subtracted from the current measured at the end of a potential step. With the procedure of sampling the current just before the potential is changed, the effect of the capacitive current can be decreased, since it decays faster than the faradaic current [22]. When a square-wave alternating potential is superimposed to a staircase ramp, the resulting voltammetric technique is called square-wave voltammetry (SWV). The current is sampled at two times within short periods of all pulses, at the end of the forward potential pulse, and at the end of the reverse pulse. The subtraction of the currents of the negative pulse from the following positive pulse (or vice versa), results in a minimal contribution of the current signal resulting from capacitive charging. Square-wave voltammetry is very popular in quantitative electroanalysis of several compounds, especially in stripping voltammetry, as described below [23].

Stripping electroanalysis has been currently employed for the monitoring of metal ions and organic structures. It is a useful analytical method to quantify analytes in low concentrations when the faradaic current is partially or completely hidden by the capacitive current. This technique involves the preconcentration of the analyte onto a solid electrode surface by applying a specific potential, and this procedure can be categorized according to the set potential for preconcentration. When analytes are electrochemically deposited by reduction and the analyte deposit is anodically dissolved, the technique is called anodic stripping voltammetry [24]. When the potential is initially held at an oxidizing value (preconcentration), and the oxidized species are stripped from the electrode surface by sweeping the potential negatively, the technique is called cathodic stripping voltammetry [25]. This method allows accumulating a considerable amount of the analyte on the electrode, providing analytical signals much higher than the undetectable oxidation or reduction currents. The voltammetric current is sampled during the stripping step, and it can be based on a linear, staircase, or pulse mode. Similar to the anodic and cathodic stripping voltammetry, the adsorptive stripping voltammetry is also based on the preconcentration of the analyte on the electrode, but this step is not controlled by electrolysis. The preconcentration is accomplished by direct physical-chemical adsorption or by chemical reactions on the surface of modified electrodes [26].

Advanced Sensor Platforms for Simultaneous Determination of Antihypertensives

The great success of electroanalytical techniques in analytical applications is strongly related to the electrode material. Currently, there is a wide variety of materials for electrode applications such as inert metals (*e.g.* platinum, gold, *etc.*), carbon-based materials (*e.g.* graphite, glassy carbon, diamond, fullerene, carbon fiber, nanotubes, *etc.*), and semiconductors (*e.g.* indium and tin oxides, *etc.*) [27 - 29].

The advantages of carbon-based electrodes include low cost, wide range of working potential, low background current, and chemical inertness [30]. Several unmodified and modified carbon electrodes were considered as a voltammetric platform for the simultaneous determination of antihypertensive drugs and they are presented below.

Glassy Carbon Electrode

Glassy carbon is widely employed as an electrode due to its excellent mechanical and electrical properties. The dense carbon structure of this material results from the random and tangled carbon plane networks, which confers some interesting features to glassy carbon such as high hardness, impermeability, and good electroconductivity [30, 31]. To enhance the analytical performance of the glassy carbon electrode (GCE), it is commonly summited to pretreatment by polishing its surface with smaller alumina particles (<0.05 μm) on a polishing cloth. This procedure guarantees the activation of the GCE surface improving the electron-transfer reactivity. Other treatments can be made for surface activation, such as electrochemical, chemical, heat, or laser treatments [28].

GCE has been employed in the development of voltammetric methods for the simultaneous determination of antihypertensive agents (Table **1**). Hydrochlorothiazide and bisoprolol were simultaneously determined using DPV and SWV. The pH study revealed a better separation of oxidation peaks for the active pharmaceutical ingredient at pH 3.0. The voltammetric method was successfully applied in the accurate analysis of pharmaceutical formulations, with a LOD of 2.3 and 1.1 μmol L^{-1} (DPV) and 2.4 and 0.47 μmol L^{-1} (SWV) for hydrochlorothiazide and bisoprolol, respectively [32]. The high working potential observed for the oxidation of bisoprolol (1.40 *vs*. Ag/AgCl) in this type of electrode can corrode it, resulting in microstructural and surface chemical changes that adversely affect the analytical response of these analytes and leads to the need for frequent time-consuming pretreatment and/or activation.

Amlodipine was the most determined active pharmaceutical ingredient using GCE. This calcium antagonist drug was determined with valsartan [33], telmisartan [34], and hydrochlorothiazide [35]. As noted, telmisartan is also oxidized at a high potential (1.55 V *vs*. Ag/AgCl), leading to damages in the electrode surface that affects the analytical response for these analytes. Erden *et al*. studied amlodipine and valsartan using CV and DPV. Despite the short 200 mV separation of the oxidation peaks of amlodipine and valsartan, there was no interference of one active in the oxidation current of its pair in pH 5.0. The DPV method achieved 0.31 and 0.36 μmol L^{-1} as LOD for amlodipine and valsartan, respectively, and was applied in the analysis of pharmaceutical formulations and human serum with recovery values over 90% and RSD lower than 15% [33]. In the intra-day repeatability study of peak current was observed RSD values ranging from 4.23% to 5.68%. Higher RSD values (8.97% to 12.47%) were obtained in the inter-day measurements, probably because AML adsorbs on this electrode material. However, the authors did not report any type of cleaning procedure performed on the electrode surface.

The chemically modified electrodes (CME) represent an interesting and increasingly field of study. Several inorganic and organic materials can be used as modifiers for electrodes for accelerating the electron-transfer reactions and enhance the sensitivity and reproducibility of the sensor [28]. Chemical modifications on the GCE surface were proposed for the simultaneous determination of antihypertensive drugs [36 - 39]. Multi-walled carbon nanotubes (MWCNTs) and metallic nanoparticles were the nanomaterials predominant in the modifications.

Kun *et al.* described the construction of a platinum nanoparticle-doped multiwalled carbon-nanotube-modified glassy carbon electrode (PtNPs/MWCNTs/GCE) for the determination of propranolol and atenolol. According to the results, the detection of the analytes was only possible due to the surface modification of the GCE, which proved to has excellent electrocatalytic activity towards the oxidation processes of propranolol and atenolol [36]. An iron metal-organic framework/mesoporous carbon nanocomposite-modified GCE (FeMOF/MC/GCE) was employed for the determination of a binary mixture of antihypertensives amlodipine and losartan. For both molecules, the oxidation processes were highly favorable on the modified surface of the sensor and higher peak currents were observed when compared to the bare CGE [37]. Hudari *et al.* reported the development of a sensor based on the modification of the GCE with MWCNTs (MWCNTs/GCE) for the determination of the diuretic hydrochlorothiazide and triamterene. On bare GCE, the oxidation processes take place at a very close potential, and overlapping occurs. On the other hand, the MWCNTs/CGE was efficient in separating the hydrochlorothiazide and triamterene peaks ($\Delta E = 160$ mV), enabling simultaneous quantification of the drugs, as well as providing a significant increase in the current intensity of the anodic peaks reporting LODs of 28 and 29 nmol L^{-1} for hydrochlorothiazide and triamterene, respectively [38]. In all three modification cases, the GCE surface modifications were performed by a single drop-casting step of the modifier over the alumina-polished surface of GCE. Doulache *et al.* proposed the voltammetric determination of amlodipine and ramipril employing a polyglycine modified GCE (Poly(Gly)/GCE). In this study, the modification was performed by electropolymerization of glycine on GCE and higher current and well-defined peaks were obtained for the antihypertensive when compared to the bare GCE. The method achieved LOD of 85 and 130 nmol L^{-1} for amlodipine and ramipril, respectively, and was applied to analyze human urine and serum [39].

The modifiers enhanced the electroanalytical performance of the bare GCE, and the modified electrodes were able to detect and quantify the antihypertensive drugs in more complex media as human urine, saliva and serum, and residual water.

Table 1. Voltammetric methods for the simultaneous determination of antihypertensives using GCE and modified GCE.

Sensor	Technique and Supporting Electrolyte	Analytes	Oxidation Potential (V)	Linear Range (μmol L⁻¹)	LOD (μmol L⁻¹)	Application	Refs.
GCE	DPV/SWV; 0.04 mol L⁻¹ BR buffer pH 3.0 with methanol (20%)	Hydrochlorothiazide	1.09[a] (DPV) 1.11[a] (SWV)	8.0 – 235	2.3 (DPV) 1.1 (SWV)	Pharmaceutical formulations	[32]
		Bisoprolol	1.40[a] (DPV) 1.44[a] (SWV)	11 – 181	2.4 (DPV) 0.47 (SWV)		
GCE	DPV; BR buffer solution pH 5.0	Amlodipine	0.95[a]	1.5 – 35.0	0.31	Pharmaceutical formulations and spiked human serum	[33]
		Valsartan	1.15[a]	1.5 – 32.0	0.36		
GCE	SWV; 0.5 mol L⁻¹ H₂SO₄	Amlodipine	0.87 and 1.40[a]	0.1 – 100	0.0006	Pharmaceutical formulations	[34]
		Telmisartan	1.55[a]	0.1 – 10	0.0226		
GCE	DPV; 0.04 mol L⁻¹ BR buffer pH 5.0	Amlodipine	0.80[a]	6.1 – 73	1.9		[35]
		Hydrochlorothiazide	1.03[a]	8.4 – 100	2.7		
PtNPs/MWCNTs/GCE	DPV; 0.1 mol L⁻¹ PBS pH 7.4	Atenolol	0.85[b]	5.8 – 116	1.17	Serum samples	[36]
		Propranolol	1.10[b]	0.2 – 50	0.15		
FeMOF/MC/GCE	DPV; 0.1 mol L⁻¹ PBS pH 6.0	Amlodipine	0.73[a]	0.009 – 500	0.00127	Pharmaceutical formulations, human blood serum, saliva, and urine	[37]
		Losartan	1.24[a]	0.009 – 500	0.00203		

(Table 1) cont.....

Sensor	Technique and Supporting Electrolyte	Analytes	Oxidation Potential (V)	Linear Range (µmol L⁻¹)	LOD (µmol L⁻¹)	Application	Refs.
MWCNTs/GCE	LSAdSV; BR buffer pH 4.0; E_{ac}= 0.8V; t_{ac} = 40s	Hydrochlorothiazide	1.01[a]	0.10 – 20	0.028	Artificial urine, tap water, water treatment plant water and hemodialysis sample	[38]
		Triamterene	1.17[a]	0.10 – 20	0.029		
Poly(Gly)/GCE	DPAdSV; 0.1 mol L⁻¹ H_2SO_4; E_{ac}= 0.1V; t_{ac}= 30s	Amlodipine	0.77[a]	0.5 – 25	0.085	Human serum and urine samples	[39]
		Ramipril	1.28[a]	0.5 – 25	0.130		

LSAdSV - Linear sweep adsorptive stripping voltammetry; DPAdSV – Differential Pulse Adsorptive Stripping Voltammetry; [a] E / V (*vs.* Ag/AgCl); [b] – E / V (*vs.* SCE)

Carbon Paste Electrodes

Carbon paste electrode (CPE) is also widely used in electroanalysis due to its special characteristics and benefits such as low cost, low residual current, ease of surface renewal, and reproducibility. The construction of this type of electrode consists of a mixture prepared from carbon powder (mostly graphite) and water-immiscible organic liquids of non-electrolytic character packed into a small inert holder with electrical contact. There are different pasting liquids available that can be used in the CPE preparation, such as mineral oil and paraffin, which have effects on electrode reactivity [30, 40].

Binary mixtures of antihypertensive agents were determined by voltammetry (Table **2**) employing MWCNTs paste electrode (MWCNTsPE) [41, 42], and graphite paste electrode (GPE) [43, 44]. The antihypertensives amlodipine and enalapril were simultaneously determined using an MWCNTsPE in the presence of cationic surfactant, with good reproducibility, sensitivity, and LODs of 49 and 810 nmol L⁻¹ for amlodipine and enalapril, respectively [41]. A GPE was used for the simultaneous determination of two binary mixtures of antihypertensives containing lisinopril: lisinopril-hydrochlorothiazide and lisinopril-amlodipine. These methods were not feasible if using GCE since enalapril was not oxidized under those specific conditions. The use of DPV and the GPE provided accurate results for their determination in pharmaceutical formulations and tap water samples [43].

Besides, the sensitivity of the CPE can be increased relatively ease with the incorporation of a variety of modifiers into the paste preparation or on the surface of the electrode [45 - 49]. The procedures for the simultaneous determination of antihypertensives were developed using modified CPEs, which were constructed by incorporating the modifier materials into the paste preparation (Table **2**). For example, Gholivand and Khodadadian incorporated graphene (Gr) and ferrocene (Fc) on the GPE (Gr/Fc/GPE) for the simultaneous determination of captopril and hydrochlorothiazide. This modification in the matrix of the GPE was responsible for an improvement in the current intensity of the anodic peaks of both analytes [50]. Common ions and excipients do not affect the analytical signals of captopril and hydrochlorothiazide on the Gr/Fc/GPE, enabling the direct determination of both analytes in pharmaceutical samples [50]. A GPE was also modified with the incorporation of 5-amino-20-ethyl-biphenyl-2-ol (5AEB) and CNTs (5AEB/CNTs/GPE) for the determination of methyldopa and hydrochlorothiazide in pharmaceutical sample and human urine [51]. The ternary mixture of pindolol, acebutolol, and metoprolol, was simultaneously determined by a sensor based on a GPE modified with amino-functionalized hexagonal mesostructured silica (NH_2-HMS-GPE). The proposed sensor was efficient in the simultaneous determination of the antihypertensive agents, since the oxidation processes of pindolol, acebutolol, and metoprolol exhibited three well-defined and separated peaks at +0.85 V, +1.11 V, and +1.45 V, respectively, which were not observed employing an unmodified GPE [52]. Beitollahi *et al.* modified a GPE with magnetic core-shell $Fe_3O_4@SiO_2$/MWCNT nanocomposite and an ionic liquid (n-hexyl-3-methylimidazolium hexafluoro phosphate) (ILFSCNPE) and studied the novel sensor for the simultaneous voltammetric oxidation of amlodipine and hydrochlorothiazide and its determination in pharmaceutical formulations and human urine [53]. The presence of ionic liquid in the GPE assured the enhancement of the peak current and a shift of the oxidation potential (decreasing the overpotential) when compared with the unmodified GCE [35].

Graphite, the ionic liquid octylpyridinium hexafluorophosphate, and copper hydroxide nanoparticles were mixed and formed an anti-fouling GPE ($Cu(OH)_2$NP/GILE) for the simultaneous determination of captopril and hydro-chlorothiazide in pharmaceutical formulations with a LOD of 12 and 60 nmol L^{-1}, respectively [54]. The agglutinating material mineral oil was replaced by polycaprolactone (PCL) in the construction of a GPE (PCL/GPE) for the determination of amlodipine and hydrochlorothiazide in synthetic urine samples. The sensor was used in an anionic surfactant-containing media and reached LOD at a nanomolar level [55].

Table 2. Voltammetric methods for the simultaneous determination of antihypertensives using CPE and modified CPE.

Sensor	Technique and supporting electrolyte	Analytes	Oxidation Potential (V)	Linear range ($\mu mol\ L^{-1}$)	LOD ($\mu mol\ L^{-1}$)	Application	Refs.
MWCTNsPE	SWV; BR buffer pH 6.0 with 10 $\mu mol\ L^{-1}$ CTAB	Amlodipine	0.79[a]	0.58 – 5.9	0.049	Pharmaceutical formulations	[41]
		Enalapril	1.21[a]	2.0 – 57	0.81		
	SWV; BR buffer pH 5.0	Hydrochlorothiazide	1.02[a]	0.49 – 45	0.014		[42]
		Enalapril	1.24[a]	5.0 – 83	0.041		
GPE	DPV; BR buffer pH 10.0	Lisinopril	0.65[a]	9.97 – 442	1.70	Pharmaceutical formulations and spiked tap water	[43]
		Hydrochlorothiazide	0.76[a]	1.99 – 70.8	0.78		
	DPV; BR buffer pH 10.0	Lisinopril	0.65[a]	19.9 – 450	2.08		
		Amlodipine	0.99[a]	1.99 – 54.1	0.784		
	DPV; acetate buffer pH 4.8 with 120 $\mu mol\ L^{-1}$ SDS	Ramipril	0.82[a]	50 – 800	4.5	Pharmaceutical formulations and human urine	[44]
		Felodipine	1.20[a]	50 – 800	2.7		
Gr/Fc/GPE	DPV; 0.2 mol L^{-1} PBS pH 7.0	Captopril	0.30[a]	10 – 430	0.87		[50]
		Hydrochlorothiazide	0.92[a]	0.5 – 390	0.38		
5AEB/CNTs/GPE	SWV; 0.1 mol L^{-1} PBS pH 8.0	Methyldopa	0.27[a]	5.0 – 190	–	Pharmaceutical formulations and urine	[51]
		Hydrochlorothiazide	0.73[a]	10.0 – 1200	–		
NH_2-HMS-GPE	DPV; 0.1 mol L^{-1} PBS pH 4.0	Pindolol	0.85[a]	0.5 – 40	0.11	Tap, mineral, river and sewage effluent waters	[52]
		Acebutolol	1.11[a]	0.5 – 70	0.046		
		Metoprolol	1.45[a]	1 – 60	0.23		

(Table 2) cont.....

Sensor	Technique and supporting electrolyte	Analytes	Oxidation Potential (V)	Linear range (µmol L⁻¹)	LOD (µmol L⁻¹)	Application	Refs.
ILFSCNPE	SWV; 0.1 mol L⁻¹ PBS pH 7.0	Amlodipine	0.73[a]	0.25 – 500	0.15	Pharmaceutical formulations and human urine	[53]
		Hydrochlorothiazide	0.84[a]	1.0 – 600	0.085		
Cu(OH)₂NP/GILE	SWV; 0.2 mol L⁻¹ PBS pH 8.0	Captopril	0.22[a]	0.7 – 70	0.012	Pharmaceutical formulations	[54]
		Hydrochlorothiazide	0.73[a]	3 – 600	0.060		
PCL/GPE	DPV; 0.04 mol L⁻¹ BR buffer pH 5.0 with 200 µmol L⁻¹ SDS	Amlodipine	0.77[a]	0.73 – 2.6	0.068	Commercial tablets and synthetic urine	[55]
		Hydrochlorothiazide	0.98[a]	2.4 – 22.0	0.27		

[a] E / V (*vs.* Ag/AgCl)

Screen Printed Electrodes

Screen Printed Electrode (SPE) has inherent advantages such as miniaturization, versatility, low cost, disposable, and the possibility of large-scale production. These devices were also considered as electrochemical platforms for the simultaneous determination of antihypertensives (Table **3**). Disposable SPEs were utilized in the DPV analysis of propranolol-hydrochlorothiazide and propranolol-amlodipine mixtures in anionic surfactant-containing media with nanomolar LOD values [56]. The addition of sodium dodecyl sulfate to the measurements enhanced the voltammetric signal of PROP mixtures, once SDS molecules adsorb on the GPE surface forming a negatively charged boundary layer related to sulfate groups, which promote the electrostatic interactions with the protonated molecules. However, in this type of electrode, hydrochlorothiazide presents a high oxidation potential (1.29 V *vs.* pseudo-Ag/AgCl). Additionally, the modification of SPE with metal oxide nanoparticles was also tested as voltammetric sensors for a mixture of antihypertensives. ZnO/Al_2O_3 nanocomposite modified graphite SPE (ZnO/Al_2O_3/SPE) was developed for the simultaneous determination of methyl-dopa and hydrochlorothiazide. The modified SPE showed excellent electrocatalytic activity towards the oxidation of methyldopa and was satisfactorily employed in the analysis of pharmaceutical formulations and urine samples with recovery values near 100% and RSD lower than 4%, however, no interferents were evaluated [57]. An SPE modified with MgO nanoplatelets (MgO/SPE) was fabricated and explored in the DPV determination of nifedipine and atenolol with a separation of 0.35 V in pH 9.0. The lab-made sensor proved to be accurate and suitable for routine analysis of pharmaceutical formulations and urine samples [58].

Table 3. Voltammetric methods for the simultaneous determination of antihypertensives using SPE and modified SPE.

Sensor	Technique and supporting electrolyte	Analytes	Oxidation Potential (V)	Linear range (µmol L⁻¹)	LOD (µmol L⁻¹)	Application	Ref.
SPE	DPV; 0.1 M H_2SO_4 with 277.4 µM SDS	Propranolol	1.11^c	0.33 – 14.98	0.0817	Pharmaceutical tablets and urine	[56]
		Hydrochlorothiazide	1.29^c	1.90 – 63.31	0.546		
	DPV; BR buffer pH 7.0 with 277.4 µM SDS	Propranolol	0.92^c	0.07 – 9.38	0.013		
		Amlodipine	0.59^c	0.33 – 4.98	0.075		
ZnO/Al_2O_3/SPE	DPV; 0.1 M PBS pH 7.0	Methyldopa	0.23^c	1.0 – 100	0.5	Pharmaceutical formulations and urine	[57]
		Hydrochlorothiazide	0.77^c	0.1 – 100	0.08		
MgO/SPE	DPV; BR buffer pH 9.0	Nifedipine	0.49^c	1.3 – 101	0.032	Pharmaceutical tablets and human urine	[58]
		Atenolol	0.70^c	6.67 – 500	1.76		

^{c} E / V (*vs.* pseudo-Ag/AgCl).

Nitrogen-Containing Tetrahedral Amorphous Carbon

The nitrogen-containing tetrahedral amorphous carbon (ta-C:N) belongs to the class of materials that contain a mixture of sp²- and sp³-hybridized carbon bonding. This material is very similar to the boron-doped diamond films and exhibits good resistance to passivation in a very wide working potential window with a low background current. This type of sensor was used for the simultaneous determination of hydrochlorothiazide and propranolol in artificial urine and serum samples, obtaining similar results when comparing with the use of boron-doped diamond electrode (BDDE). The simultaneous analytical curves were constructed employing SWV in 0.5 mol L⁻¹ H_2SO_4 and good linearity was obtained in the concentration ranges of 3.0 – 9.8 µmol L⁻¹ for hydrochlorothiazide and 0.9 – 9.8 µmol L⁻¹ for propranolol, with a LOD of 2.5 and 0.75 µmol L⁻¹, respectively [59]. On the ta-C:N thin-film electrode, hydrochlorothiazide presented oxidation potential at 1.40 V *vs.* Ag/AgCl. This electrode is quite useful for the electrooxidation of analytes that require high positive potentials for detection, due to a wide working potential window and microstructural stability at high potentials. No electrode fouling/response attenuation was detected in the complex matrices and the recovery percentages were very close to 100%.

Boron-Doped Diamond Electrode

BDD is a carbon material that has been used in very modern electroanalytical methods and is very applicable due to its extraordinary electrochemical properties [5, 60, 61]. Some of the best advantages of BDDE are the morphological stability at remarkable potentials and currents (due to the presence of sp^3-hybridized diamond carbon atoms), wide working potential window in aqueous solutions, low residual current, long term stability in acid and alkaline solutions, good resistance to passivation, high chemical stability, the easiness of the electrode surface cleaning when compared to GCE, low sensitivity to dissolved oxygen, and also its high performance in withstanding extreme potentials and a strong tendency to resist scale against conventional carbon-based electrodes [62, 63]. However, the analytical performance of BDDE for a given analyte greatly depends on their surface termination, *i.e.*, whether they are hydrogen or oxygen terminated and this notable dissimilarity between these two phases is due to electronic affinity and wettability of the diamond electrode surface [5, 61, 64]. Opposite to O-termination leading to a polar, hydrophilic surface with positive electron affinity [56], H-terminated BDD surfaces are negatively charged. That is due to the low polarity of the C-H bond showing a highly hydrophobic behavior whilst in the first case is the C-O dipole of the surface [57]. In the last decade, due to the superiority of the BDD material, many works reported in scientific publications describe its applications in the field of analytical chemistry and chemical materials. The properties of BDD were explored in the development of new analytical methods for organic compounds in fixed-dose formulations that reached LOD at a nanomolar level [65 - 70]. Over the past seven years, a considerable number of publications successfully demonstrate the versatility of the BDDE either anodically (Table **4**) or cathodically pretreated (Table **5**) as a voltammetric sensor for the simultaneous determination of antihypertensive drugs.

Anodically Pretreated BDDE

Santos *et. al.* described a method for simultaneous determination of hydrochlorothiazide and losartan in pharmaceutical formulations using DPV using an anodically pretreated BDDE (AP-BDDE). Two very well-resolved and reproducible oxidation peaks of hydrochlorothiazide and losartan, with separation of +0.23 V were obtained and the LOD values were 1.2 µmol L^{-1} and 0.95 µmol L^{-1}, respectively [71]. In this study, the BDDE was used without modification, showing its robustness and applicability for individual and simultaneous determination of hydrochlorothiazide and losartan in commercial formulations with a minimal pretreatment of samples and dispensing any use of organic reagents. The selectivity of the proposed methods was evaluated by adding the interfering species such as starch, povidone, lactose, polysorbate 80, microcrystalline cellulose, titanium dioxide, and magnesium stearate. The authors concluded that these compounds do not significantly interfere with the here proposed methods.

Nifedipine and atenolol were also simultaneously determined by using an AP-BDDE. The anodic pretreatment resulted in a smaller difference of potential for the oxidation peaks of the antihypertensives but considerably increased the current for atenolol. The AP-BDDE was compared to GCE and the diamond electrode provided a better reproducibility than GCE. Moreover, the authors showed that the GCE presents adsorption phenomena, mainly for atenolol, which makes their employment unfeasible for the simultaneous determination of these two drugs. However, BDDE provided good resistance and stability and simplicity of use. In this method, the authors sought to obtain not only a low detection limit but a method with chemical and instrumental characteristics for detecting both drugs with good linearity of the analytical curve and also capable of being applied to real samples. The method for the simultaneous determination of nifedipine and atenolol demonstrated good accuracy [72].

The electroanalytical method proposed by Moraes *et al.* employed BDDE and the SWV technique for simultaneous determination of a quaternary mixture of antihypertensive drugs (amlodipine, amiloride, hydrochlorothiazide, and atenolol). These drugs presented very well-resolved and reproducible oxidative peaks at 0.68, 0.89, 1.04, and 1.21 V (*vs* Ag/AgCl (3.0 mol L^{-1} KCl)), respectively, using the AP-BDDE. On the cathodically pretreated BDDE (CP-BDDE) otherwise, was observed an overlapping of peaks for amiloride, hydrochlorothiazide, and atenolol, attesting that the use of electrochemical pretreatment is crucial in the development of voltammetric methods using BDDE. Under the optimized experimental conditions, the corresponding peak heights increased linearly and were free of the interference of concomitant drugs. LODs were 0.30, 0.09, 0.08, and 0.06 μmol L^{-1} for amlodipine, amiloride, hydrochlorothiazide, and atenolol, respectively. The authors showed that BDDE is a robust electrochemical sensor in antihypertensive analysis capable of sensitive and selective simultaneous determination with a minimal sample pretreatment, without the use of considerable toxic organic reagent. This method was successfully applied in the simultaneous quantification of the drugs studied in pharmaceutical formulations, as well as in an enriched water sample, obtaining recovery values between 93.9 to 104% [73].

Table 4. Voltammetric methods for the simultaneous determination of antihypertensives using AP-BDDE.

Sensor	Technique and supporting electrolyte	Analytes	Oxidation Potential (V)	Linear range (μmol L⁻¹)	LOD (μmol L⁻¹)	Application	Ref.
AP-BDDE	DPV; BR buffer pH 9.5	Hydrochlorothiazide	1.00[a]	3.0 – 74	1.2		[71]
		Losartan	1.20[a]	3.0 – 74	0.95	Pharmaceutical formulations	
	DPV; TRIS buffer pH 8.0	Nifedipine	0.97[a]	3.98 – 107	0.612		[72]
		Atenolol	1.36[a]	1.99 – 47.2	0.999		
	SWV; ammonium buffer pH 9.0	Amlodipine	0.68[a]	0.90 – 31	0.30		[73]
		Amiloride	0.89[a]	8.7 – 125	0.09	Pharmaceutical formulations and tap water	
		Hydrochlorothiazide	1.04[a]	29 – 260	0.08		
		Atenolol	1.21[a]	11 – 91	0.06		

[a] E / V (*vs.* Ag/AgCl)

Cathodically Pretreated BDDE

The hydrogen-terminated BDDE was the platform most considered for the development of voltammetric methods for the simultaneous determination of antihypertensives (Table 5). Several methods have been described specially for hydrochlorothiazide and amlodipine in fixed-dose or combined with other antihypertensive drugs. An SWV method was the first one successfully applied in the simultaneous determination of antihypertensives hydrochlorothiazide and valsartan, reaching LOD values of 0.639 and 0.935 μmol L⁻¹, respectively [74]. The peak potentials difference of about 0.280 V between both oxidation peaks clearly allows the simultaneous determination of both antihypertensives on the cathodically pretreated BDDE (CP-BDDE). This electrode has a wide potential window in aqueous solutions, which allows for the detection of both antihypertensives without the interference of water decomposition and its resistance to molecular adsorption and electrode fouling. The proposed method was applied in the simultaneous determination of HCTZ and VAL in real samples (combined dosage forms) dissolved in methanol and showed no interference from the common additives and excipients. In another method proposed by Mansano *et al.*, it was possible to determine simultaneously the amlodipine and valsartan content in a synthetic urine sample and combined pharmaceutical formulations dissolved in ethanol using SWV. A comparison between pretreatments revealed a 2-fold higher response for valsartan on the CP-BDDE and the optimization of the experimental parameters yielded a LOD of 0.0764 μmol L^{-1} for amlodipine and 0.193 μmol L^{-1} for valsartan with no adsorption effects observed on the CP-BDDE. The linear range obtained in the analytical curves using this type of electrode is proper for the simultaneous quantification of these antihypertensives in pharmaceutical formulations with different dosages, which was not observed by using a glassy carbon material [33]. This facilitates sample preparation and reduces the number of reagents used and consequently, the waste generated, contributing to an important economic gain for the pharmaceutical industry. The effects of possible interferents, commonly present in the analyzed pharmaceutical formulations and urine, such as lactose, microcrystalline cellulose, silicon dioxide, titanium dioxide, magnesium stearate, ascorbic acid, and uric acid were studied. The authors showed that the interference may occur if the concentration of uric acid is 100 times higher than that of AML, because its oxidation peak becomes enlarged [75]. The simultaneous determination of hydrochlorothiazide and amlodipine in urine samples was reported twice [76, 77]. First, hydrochlorothiazide and amlodipine were simultaneously determined in BR buffer (pH 2.0) showing the difference in potentials of 0.42 V allowed the simultaneous determination of these two analytes, and LODs were 2.00 μmol L^{-1} and 60.0 nmol L^{-1}, respectively and the recoveries of these analytes ranged from 91.0 to 107% [76]. On the other hand, Mansano and Sartori [77] reported a simultaneous voltammetric method for hydrochlorothiazide and amlodipine in a BR buffer (pH 5.0). With the use of this pH, repetitive measurements were obtained and lower concentrations for both drugs were detected in the simultaneous analytical curve with LOD values of 0.160 μmol L^{-1} and 60.0 nmol L^{-1}, respectively. The authors showed no adsorption effects were observed on the BDD electrode, without the necessity of renovation of the surface of this electrode after each measurement. This method was simple, rapid, sensitive, precise and accurate, environmentally friendly, and a suitable alternative to chromatography methods in laboratories lacking the required facilities for these techniques [77].

Besides, Mansano *et al.* developed a method for the simultaneous determination of an antihypertensive ternary mixture of amlodipine, hydrochlorothiazide, and valsartan. In this study, three oxidative processes of amlodipine, hydrochlorothiazide, and valsartan were observed at 0.781, 1.14, and 1.41 V, respectively, which were well resolved and reproducible. For the simultaneous determination of these drugs, the linearity ranges were adapted to the dosages found commercially. The voltammetric method was of easy sample preparation and the LODs were 0.23, 0.75, and 6.20 μmol L^{-1}, respectively [78].

A convenient method for the simultaneous determination of antihypertensive drugs hydrochlorothiazide and metoprolol was described by Salamanca-Neto *et al.* [79]. The simultaneous determination was feasible after the pH study, where it was found that above pH 4 the oxidation peaks were sufficiently separated, +1.11 V for hydrochlorothiazide and +1.32 V for metoprolol. The method presented LODs of 0.376 and 0.077 µmol L^{-1}, respectively for hydrochlorothiazide and metoprolol, and was applied to detect the antihypertensives in human urine. In addition, adequate recovery results were obtained for the individual determination of MTP and the simultaneous determination of MTP and HCTZ in human urine samples, indicating that the method can be applied to this type of sample, since there is no significant interference from species present in the urine matrix [79].

The simultaneous voltammetric determination of antihypertensive drugs amlodipine and atenolol using a CP-BDDE was reported for the first time by Moraes *et al*. The anodic peaks of amlodipine and atenolol were found at +0.727 and +1.32 V on the BDDE using DPV and SWV [80]. DPV was more sensitive; however, SWV was chosen because of the repeatability of analytical signals. Both techniques provided similar LODs [80].

The voltammetric behavior of ramipril, an ACE inhibitor, was investigated for the first time on a BDDE by Mattos *et al*. Its simultaneous determination with hydrochlorothiazide was also reported. The cathodic pretreatment of BDDE increased the response for ramipril at +1.67 V in acid media. Due to the wide potential window in aqueous solutions of BDDE, it was possible to determine both drugs without the interference of water decomposition [81]. A simultaneous determination of antihypertensive drugs amlodipine and ramipril was also reported [82]. Under optimized instrumental parameters of DPV, the LOD values were 0.26 and 0.08 µmol L^{-1}, for amlodipine and ramipril, respectively, and the method was applied to pharmaceutical and rat serum samples without fouling [82].

As can be seen, the boron-doped diamond film may offer several particular advantages when used as an electrochemical sensor for the simultaneous determination of fixed-dose formulations because of the negligible adsorption of organic compounds enabling more reproducible measurements, wide potential window as well as the possibility of electroanalysis at very negative and very positive potentials. BDDE has proven as a robust electrochemical sensor in the antihypertensive analysis capable of sensitive and selective simultaneous determination with a minimal sample pretreatment, without the use of considerable toxic organic reagent, and with reliable results.

Table 5. Voltammetric methods for the simultaneous determination of antihypertensives using CP-BDDE.

Sensor	Technique and Supporting Electrolyte	Analytes	Oxidation Potential (V)	Linear range (µmol L⁻¹)	LOD (µmol L⁻¹)	Application	Ref.
CP-BDDE	SWV; BR buffer pH 5.0	Hydrochlorothiazide	1.14[a]	1.97 – 88.1	0.639	Pharmaceutical formulations	[74]
		Valsartan	1.42[a]	9.88 – 220	0.935		
	SWV; BR buffer pH 5.0	Amlodipine	0.79[a]	0.497 – 28	0.0764	Synthetic urine and pharmaceutical formulations	[75]
		Valsartan	1.37[a]	19.8 – 280	0.193		
	SWV; BR buffer pH 2.0	Amlodipine	0.88[a]	0.2 – 9.09	0.06	Synthetic urine	[76]
		Hydrochlorothiazide	1.30[a]	4.0 – 100	2.0		
	SWV; BR buffer pH 5.0	Amlodipine	0.78[a]	0.1 – 2.1	0.056		[77]
		Hydrochlorothiazide	1.14[a]	2.0 – 22	0.16		
	SWV; BR buffer pH 5.0	Amlodipine	0.78[a]	0.49 – 7.2	0.23	Pharmaceutical formulations	[78]
		Hydrochlorothiazide	1.14[a]	2.9 – 45	0.75		
		Valsartan	1.41[a]	9.7 – 130	6.2		
	DPV; lactate buffer pH 4.0	Hydrochlorothiazide	1.11[a]	0.51 – 18.7	0.376	Pharmaceutical formulations and human urine	[79]
		Metoprolol	1.32[a]	1.23 – 22.8	0.077		
	SWV; 0.1 mol L⁻¹ PBS pH 7.0	Amlodipine	0.73[a]	2.9 – 33	0.17		[80]
		Atenolol	1.32[a]	9.8 – 190	0.22		
	SWV; BR buffer pH 2.0	Ramipril	1.67[a]	1.96 – 36.7	0.027	Pharmaceutical formulations	[81]
		Hydrochlorothiazide	1.23[a]	2.46 – 36.7	0.018		
	DPV; BR buffer pH 6.0	Amlodipine	0.68[a]	0.99 – 14	0.26	Pharmaceutical formulations and rat blood serum	[82]
		Ramipril	1.70[a]	0.29 – 1.98	0.08		

[a] E / V (*vs.* Ag/AgCl)

ELECTROOXIDATION   MECHANISMS   OF   THE ANTIHYPERTENSIVES

Over this review, different carbon-based electrodes were reported in the electroanalytical studies of several antihypertensive agents simultaneously, and the oxidation mechanisms of some molecules were proposed. The participation of protons in the electrooxidation mechanisms is characterized by a displacement of the oxidation potential according to the pH value of the supporting electrolyte while in the no protons participating mechanisms, the oxidation potential remains constant by changing pH. Generally, the pKa value of the molecule determinates the pH where this behavior changes, considering proton-transfer reactions [27 - 29].

The antihypertensive classes are mostly represented by molecules with similar structures and electroactive groups, then it is possible to observe that the authors reported the same oxidation pathways for different antihypertensives belonging to the same class.

According to the literature, hydrochlorothiazide, a diuretic antihypertensive, can be oxidated to chlorothiazide with the participation of 2 protons and 2 electrons on the surface of different electrodes [42, 43, 55, 56, 71, 74, 78, 81].

The electrooxidation pathway described for β-blockers as atenolol [36, 72, 80], acebutalol [52], and propranolol [36, 56] occurs on the secondary alcoholic group in these molecules.

The renin-angiotensin-aldosterone system (RAAS) blockers include angiotensin-converting enzyme inhibitors (*e.g.* captopril, ramipril, enalapril, and lisinopril) and angiotensin receptor blockers (*e.g.* valsartan, losartan, and telmisartan). The possible mechanism for the electrochemical oxidation of the angiotensin-converting enzyme can be assigned to oxidation of the secondary amine group consuming one electron [39, 41 - 44, 81]. As for the angiotensin receptor blockers group, for example valsartan, the mechanism of the electrode reaction can be assigned to oxidation of tetrazolyl group –N– consuming one electron [33, 74, 75, 78]. The oxidation mechanism for telmisartan may occur from benzimidazole part [34]. On the other hand, the oxidation of losartan occurs by a two-electron mechanism [37, 71].

Also, the electrochemical mechanism of the calcium antagonist agents was consensually reported as been the oxidation of the 1,4-dyhidropyridine ring present in the amlodipine [34, 37, 39, 41, 43, 55, 56, 75, 78, 80], felodipine [44], and nifedipine [72], involving a two-protons and two-electrons mechanism.

CONCLUDING REMARKS

In this review, the authors' effort is focused on providing a summary of the use of advanced sensor materials in conjunction with voltammetry for the simultaneous determination of antihypertensive drugs. Using this approach and appropriate sensors or medium pH, the oxidation potential of a variety of antihypertensives was distinct, not requiring the use of time-consuming separation methods. The construction of analytical curves was a crucial point for the feasibility of simultaneous determination in real samples. In some papers, the authors reported that the first point of analytical curves was not the lowest detected by the sensors, but they had to be combined in agreement with the concentration of the co-formulated drug. In addition, these techniques allow the use of aqueous solvent, with buffers or mineral acids used as supporting electrolytes. Sample preparation before analysis is simplified and the measurements are precise and accurate. BDDE could be a good alternative to other sensors for the simultaneous determination of antihypertensives. The critical comparison showed this sensor as of greater applicability due to its excellent properties, such as minimization of the fouling effect, wide potential window, and good reproducibility. Finally, the authors' hope that this review will expand the view of the utilization of advanced sensor materials in conjunction with the voltammetry for the simultaneous determination of antihypertensives for application in quality control in the pharmaceutical industry and the broad scientific community.

CONSENT FOR PUBLICATION

Not applicable.

CONFLICT OF INTEREST

The authors declare no conflict of interest, financial or otherwise.

ACKNOWLEDGEMENTS

The authors gratefully acknowledge the financial support from the Brazilian funding agencies CNPq (grants no. 408591/2018-8 and 305320/2019-0 (E. R. Sartori); 153289/2019-8 (N. S. M. Campos)), Fundação Araucária do Paraná (ProPPG/DP/DIC 001/2020 (G.R.P. Manrique)) and CAPES (grant no. 88882.448537/2019-01 (C. A. R. Salamanca-Neto); 88882.448536/2019-01 (J. Scremin)).

REFERENCES

[1]   J.G. Hardman, L.E. Limbird, and A.G. Gilman, *Goodman & Gilman's – The Pharmacological Basis of Therapeutics.* 9th ed. McGraw-Hill: New York, 1996.

[2]   B. Uslu, and S.A. Ozkan, "Electroanalytical Methods for the Determination of Pharmaceuticals: A Review of Recent Trends and Developments", *Anal. Lett.,* vol. 44, no. 16, pp. 2644-2702, 2011.
[http://dx.doi.org/10.1080/00032719.2011.553010]

[3]   M. R. Siddiqui, Z. A. AlOthman, and N. Rahman, "Analytical Techniques in Pharmaceutical Analysis: A Review. ", *Arabian Journal of Chemistry,* Elsevier B.V. , pp. S1409-S1421, 2017.
[http://dx.doi.org/10.1016/j.arabjc.2013.04.016]

[4]   A.K. Gupta, R.S. Dubey, and J.K. Malik, "Application of modern electroanalytical techniques: recent trend in pharmaceutical and drug analysis", *Int. J. Pharm. Sci. Res.,* vol. 4, no. 7, pp. 2450-2458, 2013.
[http://dx.doi.org/10.13040/IJPSR.0975-8232.4(7).2450-58]

[5]   S. Baluchová, A. Daňhel, H. Dejmková, V. Ostatná, M. Fojta, and K. Schwarzová-Pecková, "Recent progress in the applications of boron doped diamond electrodes in electroanalysis of organic compounds and biomolecules - A review", *Anal. Chim. Acta,* vol. 1077, pp. 30-66, 2019.
[http://dx.doi.org/10.1016/j.aca.2019.05.041] [PMID: 31307723]

[6]   F. Scholz, "Voltammetric Techniques of Analysis: The Essentials", *ChemTexts.,* vol. 1, no. 4, p. 17, 2015.
[http://dx.doi.org/10.1007/s40828-015-0016-y]

[7]   C.A.R. Salamanca-Neto, A. Olean-Oliveira, J. Scremin, G.S. Ceravolo, R.F.H. Dekker, A.M. Barbosa-Dekker, M.F.S. Teixeira, and E.R. Sartori, "Carboxymethyl-botryosphaeran stabilized carbon nanotubes aqueous dispersion: A new platform design for electrochemical sensing of desloratadine", *Talanta,* vol. 210, p. 120642, 2020.
[http://dx.doi.org/10.1016/j.talanta.2019.120642] [PMID: 31987177]

[8]   D. E. Golan, *Princípios de Farmacologia: A Base Fisiopatológica Da Farmacoterapia* Guanabara Koogan: Rio de Janeiro, 2009.

[9]   "World Health Organization - A global brief on hypertension ", https://apps.who.int/iris/bitstream/handle/10665/79059/WHO_DCO_WHD_2013.2_eng.pdf;jsessionid=CE2A7A500DFFD02CF99B51EAB78F3828?sequence=1

[10]   *Q&As on hypertension.*http://www.who.int/features/qa/82/en/

[11]   G. Mancia, F. Rea, G. Corrao, and G. Grassi, "Two-Drug Combinations as First-Step Antihypertensive Treatment", *Circ. Res.,* vol. 124, no. 7, pp. 1113-1123, 2019.
[http://dx.doi.org/10.1161/CIRCRESAHA.118.313294] [PMID: 30920930]

[12]   P.A. van Zwieten, and C. Farsang, "Combination of antihypertensive drugs from a historical perspective", *Blood Press.,* vol. 14, no. 2, pp. 72-79, 2005.
[http://dx.doi.org/10.1080/08037050510008922] [PMID: 16036483]

[13]   C. Guerrero-García, and A.F. Rubio-Guerra, "Combination therapy in the treatment of hypertension", *Drugs Context,* vol. 7, p. 212531, 2018.
[http://dx.doi.org/10.7573/dic.212531] [PMID: 29899755]

[14]   "M. The Role of the New β-Blockers in Treating Cardiovascular Disease", *Am. J. Hypertens.,* vol. 18, no. 12, pp. 169-176, 2005.
[http://dx.doi.org/10.1016/j.amjhypcr.2005.09.009]

[15]   E. Oliver, F. Mayor Jr, and P. D'Ocon, "Beta-blockers: Historical Perspective and Mechanisms of Action", *Rev. Esp. Cardiol. (Engl. Ed.),* vol. 72, no. 10, pp. 853-862, 2019.
[http://dx.doi.org/10.1016/j.rec.2019.04.006] [PMID: 31178382]

[16]   S.A. Atlas, "The renin-angiotensin aldosterone system: pathophysiological role and pharmacologic inhibition", *J. Manag. Care Pharm.,* vol. 13, no. 8, suppl. Suppl. B, pp. 9-20, 2007.
[http://dx.doi.org/10.18553/jmcp.2007.13.s8-b.9] [PMID: 17970613]

[17]   E. Grossman, and F.H. Messerli, "Calcium antagonists", *Prog. Cardiovasc. Dis.,* vol. 47, no. 1, pp. 34-57, 2004.

[http://dx.doi.org/10.1016/j.pcad.2004.04.006] [PMID: 15517514]

[18] C.M. Hussain, and R. Keçili, "Electrochemical Techniques for Environmental Analysis", In: *Modern Environmental Analysis Techniques for Pollutants.* Elsevier, 2020, pp. 199-222.
[http://dx.doi.org/10.1016/B978-0-12-816934-6.00008-4]

[19] P. Kurzweil, HISTORY | Electrochemistry.*Encyclopedia of Electrochemical Power Sources.* Elsevier, 2009, pp. 533-554.
[http://dx.doi.org/10.1016/B978-044452745-5.00007-1]

[20] N. Elgrishi, K.J. Rountree, B.D. McCarthy, E.S. Rountree, T.T. Eisenhart, and J.L. Dempsey, "A Practical Beginner's Guide to Cyclic Voltammetry", *J. Chem. Educ.,* vol. 95, no. 2, pp. 197-206, 2018.
[http://dx.doi.org/10.1021/acs.jchemed.7b00361]

[21] F. Scholz, *Electroanalytical Methods.* 2nd ed. Springer Berlin Heidelberg: Berlin, 2010.
[http://dx.doi.org/10.1007/978-3-642-02915-8]

[22] Á. Molina, and J. González, *Pulse Voltammetry in Physical Electrochemistry and Electroanalysis,* 2016.
[http://dx.doi.org/10.1007/978-3-319-21251-7]

[23] V. Mirceski, S. Komorsky-Lovric, and M. Lovric, *Square-Wave Voltammetry* Monographs in Electrochemistry; Springer Berlin Heidelberg: : Berlin, Heidelberg, 2007.
[http://dx.doi.org/10.1007/978-3-540-73740-7]

[24] J. Wang, "Stripping Analysis at Bismuth Electrodes: A Review", *Electroanalysis,* vol. 17, no. 15–16, pp. 1341-1346, 2005.
[http://dx.doi.org/10.1002/elan.200403270]

[25] F. Scholz, and H. Kahlert, "The Calculation of the Solubility of Metal Hydroxides, Oxide-Hydroxides, and Oxides, and Their Visualisation in Logarithmic Diagrams", *ChemTexts.,* vol. 1, no. 1, p. 7, 2015.
[http://dx.doi.org/10.1007/s40828-015-0006-0]

[26] L.A.M. Baxter, A. Bobrowski, A.M. Bond, G.A. Heath, R.L. Paul, R. Mrzljak, and J. Zarebski, "Electrochemical and spectroscopic investigation of the reduction of dimethylglyoxime at mercury electrodes in the presence of cobalt and nickel", *Anal. Chem.,* vol. 70, no. 7, pp. 1312-1323, 1998.
[http://dx.doi.org/10.1021/ac9703616] [PMID: 21644728]

[27] C.M.A. Brett, and A.M.O. Brett, *Electrochemistry: Principles, Methods, and Applications.* 1st ed. Oxford University Press Inc: Oxford, 1993.
[http://dx.doi.org/10.1002/anie.199419892]

[28] J. Wang, *Analytical Electrochemistry.* 2nd ed. WILEY-VCH: New York, 2000.
[http://dx.doi.org/10.1002/0471228230]

[29] D.A. Skoog, D.M. West, F.J. Holler, and S.R. Crouch, *Fundamentos Da Química Analítica.* 8th ed. Thomson Learning: São Paulo, 2007.

[30] B. Uslu, and S. Ozkan, "Electroanalytical Application of Carbon Based Electrodes to the Pharmaceuticals", *Anal. Lett.,* vol. 40, no. 5, pp. 817-853, 2007.
[http://dx.doi.org/10.1080/00032710701242121]

[31] Y. Yi, G. Weinberg, M. Prenzel, M. Greiner, S. Heumann, S. Becker, and R. Schlögl, "Electrochemical Corrosion of a Glassy Carbon Electrode", *Catal. Today,* vol. 295, pp. 32-40, 2017.
[http://dx.doi.org/10.1016/j.cattod.2017.07.013]

[32] B. Bozal, M. Gumustas, B. Dogan-Topal, B. Uslu, and S.A. Ozkan, "Fully validated simultaneous determination of bisoprolol fumarate and hydrochlorothiazide in their dosage forms using different voltammetric, chromatographic, and spectrophotometric analytical methods", *J. AOAC Int.,* vol. 96, no. 1, pp. 42-51, 2013.
[http://dx.doi.org/10.5740/jaoacint.11-364] [PMID: 23513956]

[33] P.E. Erden, İ.H. Taşdemir, C. Kaçar, and E. Kılıç, "Simultaneous Determination of Valsartan and

Amlodipine Besylate in Human Serum and Pharmaceutical Dosage Forms by Voltammetry", *Int. J. Electrochem. Sci.,* vol. 9, pp. 2208-2220, 2014.

[34] N.K. Bakirhan, F. Kartal, and S.A. Ozkan, "Electroanalytical Studies and Simultaneous Validated Assay of Antihypertensive Compounds in Their Binary Mixtures Using Fast Electrochemical Technique", *Hacettepe J. Biol. Chem.,* vol. 47, no. 1, pp. 23-32, 2019.
[http://dx.doi.org/10.15671/HJBC.2019.272]

[35] B. Yilmaz, and U. Kocak, "Simultaneous Determination of Amlodipine and Hydrochlorothiazide in Pharmaceutical Preparations by Differential Pulse Voltammetry Method", *J. Adv. Pharm. Res,* vol. 0, no. 0, pp. 0-0, 2019.
[http://dx.doi.org/10.21608/aprh.2018.5851.1068]

[36] Z. Kun, C. Hongtao, Y. Yue, B. Zhihong, L. Fangzheng, and S. Li, "Platinum Nanoparticle-Doped Multiwalled Carbon-Nanotube-Modified Glassy Carbon Electrode as a Sensor for Simultaneous Determination of Atenolol and Propranolol in Neutral Solution", *Ionics (Kiel),* vol. 21, 2015.
[http://dx.doi.org/10.1007/s11581-014-1266-1]

[37] A.S. Rajpurohit, D.K. Bora, and A.K. Srivastava, "Simultaneous Determination of Amlodipine and Losartan Using an Iron Metal-Organic Framework/Mesoporous Carbon Nanocomposite-Modified Glassy Carbon Electrode by Differential Pulse Voltammetry", *Anal. Methods,* vol. 10, no. 45, pp. 5423-5438, 2018.
[http://dx.doi.org/10.1039/C8AY01553H]

[38] F.F. Hudari, J.C. Souza, and M.V.B. Zanoni, "Adsorptive stripping voltammetry for simultaneous determination of hydrochlorothiazide and triamterene in hemodialysis samples using a multi-walled carbon nanotube-modified glassy carbon electrode", *Talanta,* vol. 179, pp. 652-657, 2018.
[http://dx.doi.org/10.1016/j.talanta.2017.11.071] [PMID: 29310290]

[39] M. Doulache, N.K. Bakirhan, B. Saidat, and S.A. Ozkan, "Highly Sensitive and Selective Electrochemical Sensor Based on Polyglycine Modified Glassy Carbon Electrode for Simultaneous Determination of Amlodipine and Ramipril from Biological Samples", *J. Electrochem. Soc.,* vol. 167, no. 2, p. 027511, 2020.
[http://dx.doi.org/10.1149/1945-7111/ab68cd]

[40] A.A. Al-Rashdi, O.A. Farghaly, and A.H. Naggar, "Voltammetric Determination of Pharmaceutical Compounds at Bare and Modified Solid Electrodes: A Review", *J. Chem. Pharm. Res.,* vol. 10, no. 9, pp. 21-43, 2018.

[41] C.F. Valezi, E.H. Duarte, G.R. Mansano, L.H. Dall'Antonia, C.R.T. Tarley, and E.R. Sartori, "An Improved Method for Simultaneous Square-Wave Voltammetric Determination of Amlodipine and Enalapril at Multi-Walled Carbon Nanotubes Paste Electrode Based on Effect of Cationic Surfactant", *Sens. Actuators B Chem.,* vol. 205, pp. 234-243, 2014.
[http://dx.doi.org/10.1016/j.snb.2014.08.078]

[42] C.A.R. Salamanca-Neto, P.H. Hatumura, C.R.T. Tarley, and E.R. Sartori, "Electrochemical Evaluation and Simultaneous Determination of Binary Mixture of Antihypertensives Hydrochlorothiazide and Enalapril in Combined Dosage Forms Using Carbon Nanotubes Paste Electrode", *Ionics (Kiel),* vol. 21, no. 6, pp. 1615-1622, 2015.
[http://dx.doi.org/10.1007/s11581-014-1349-z]

[43] C.F. Valezi, A.P. Pires Eisele, and E. Romão Sartori, "Versatility of a Carbon Paste Electrode Coupled to Differential Pulse Voltammetry for Determination of Lisinopril with Its Associations (Hydrochlorothiazide and Amlodipine)", *Anal. Methods,* vol. 9, no. 31, pp. 4599-4608, 2017.
[http://dx.doi.org/10.1039/C7AY01455D]

[44] E.A. Taha, A.K. Attia, M.M. Fouad, and Z.M. Yousef, "Simultaneous Determination of Ramipril and Felodipine Using Carbon Paste Electrode in Micellar Medium", *Anal. Bioanal. Electrochem.,* vol. 11, no. 2, pp. 150-164, 2019.

[45] J.G. Manjunatha, "Fabrication of Efficient and Selective Modified Graphene Paste Sensor for the

Determination of Catechol and Hydroquinone", *Surfaces,* vol. 3, no. 3, pp. 473-483, 2020.
[http://dx.doi.org/10.3390/surfaces3030034]

[46] N. Hareesha, and J.G. Manjunatha, "Surfactant and Polymer Layered Carbon Composite Electrochemical Sensor for the Analysis of Estriol with Ciprofloxacin", *Mater. Res. Innov.,* vol. 24, no. 6, pp. 349-362, 2020.
[http://dx.doi.org/10.1080/14328917.2019.1684657]

[47] N. Hareesha, and J.G. Manjunatha, "Fast and Enhanced Electrochemical Sensing of Dopamine at Cost-Effective Poly(DL-Phenylalanine) Based Graphite Electrode", *J. Electroanal. Chem. (Lausanne),* vol. 878, p. 114533, 2020.
[http://dx.doi.org/10.1016/j.jelechem.2020.114533]

[48] M.M. Charithra, and J.G. Manjunatha, "Enhanced Voltammetric Detection of Paracetamol by Using Carbon Nanotube Modified Electrode as an Electrochemical Sensor", *J. Electrochem. Sci. Eng.,* vol. 10, no. 1, pp. 29-40, 2019.
[http://dx.doi.org/10.5599/jese.717]

[49] "Manjunatha, J. G.; Ravishankar, D. K.; Fattepur, S.; Siddaraju, G.; Nanjundaswamy, L. Validated Electrochemical Method for Simultaneous Resolution of Tyrosine, Uric Acid, and Ascorbic Acid at Polymer Modified Nano-Composite Paste Electrode", *Surg. Eng. Appl. Electrochem.,* vol. 56, no. 4, pp. 415-426, 2020.
[http://dx.doi.org/10.3103/S1068375520040134]

[50] M.B. Gholivand, and M. Khodadadian, "Simultaneous Voltammetric Determination of Captopril and Hydrochlorothiazide on a Graphene/Ferrocene Composite Carbon Paste Electrode", *Electroanalysis,* vol. 25, no. 5, pp. 1263-1270, 2013.
[http://dx.doi.org/10.1002/elan.201200665]

[51] S. Tajik, M.A. Taher, and H. Beitollahi, "First Report for Simultaneous Determination of Methyldopa and Hydrochlorothiazide Using a Nanostructured Based Electrochemical Sensor", *J. Electroanal. Chem. (Lausanne),* vol. 704, pp. 137-144, 2013.
[http://dx.doi.org/10.1016/j.jelechem.2013.07.008]

[52] M. Silva, S. Morante-Zarcero, D. Pérez-Quintanilla, and I. Sierra, "Simultaneous Determination of Pindolol, Acebutolol and Metoprolol in Waters by Differential-Pulse Voltammetry Using an Efficient Sensor Based on Carbon Paste Electrode Modified with Amino-Functionalized Mesostructured Silica", *Sens. Actuators B Chem.,* vol. 283, pp. 434-442, 2019.
[http://dx.doi.org/10.1016/j.snb.2018.12.058]

[53] H. Beitollahi, F. Ebadinejad, F. Shojaie, and M. Torkzadeh-Mahani, "A Magnetic Core–Shell Fe3O4@SiO2/MWCNT Nanocomposite Modified Carbon Paste Electrode for Amplified Electrochemical Sensing of Amlodipine and Hydrochlorothiazide", *Anal. Methods,* vol. 8, no. 32, pp. 6185-6193, 2016.
[http://dx.doi.org/10.1039/C6AY01438K]

[54] G. Absalan, M. Akhond, R. Karimi, and A.M. Ramezani, "Simultaneous determination of captopril and hydrochlorothiazide by using a carbon ionic liquid electrode modified with copper hydroxide nanoparticles", *Mikrochim. Acta,* vol. 185, no. 2, p. 97, 2018.
[http://dx.doi.org/10.1007/s00604-017-2630-4] [PMID: 29594606]

[55] E.M. da Silva, G.C. de Oliveira, A.B. de Siqueira, A.J. Terezo, and M. Castilho, "Development of a Composite Electrode Based on Graphite and Polycaprolactone for the Determination of Antihypertensive Drugs", *Microchem. J.,* vol. 158, p. 105228, 2020.
[http://dx.doi.org/10.1016/j.microc.2020.105228]

[56] M. Khairy, and A.A. Khorshed, "Simultaneous Voltammetric Determination of Two Binary Mixtures Containing Propranolol in Pharmaceutical Tablets and Urine Samples", *Microchem. J.,* vol. 159, p. 105484, 2020.
[http://dx.doi.org/10.1016/j.microc.2020.105484]

[57]   R. Zaimbashi, H. Beitollahi, and M. Torkzadeh-Mahani, "Simultaneous Electrochemical Sensing of Methyldopa and Hydrochlorothiazide Using a Novel ZnO/Al2O3 Nanocomposite Modified Screen Printed Electrode", *Anal. Bioanal. Electrochem,* vol. 9, no. 8, pp. 1008-1020, 2017.

[58]   M. Khairy, "Khorshed, A. A.; Rashwan, F. A.; Salah, G.; Abdel-Wadood, H. M.; Banks, C. E. Simultaneous Voltammetric Determination of Antihypertensive Drugs Nifedipine and Atenolol Utilizing MgO Nanoplatelet Modified Screen-Printed Electrodes in Pharmaceuticals and Human Fluids", *Sens. Actuators B Chem.,* vol. 252, pp. 1045-1054, 2017.
        [http://dx.doi.org/10.1016/j.snb.2017.06.105]

[59]   B.C. Lourencao, T.A. Silva, O. Fatibello-Filho, and G.M. Swain, "Voltammetric Studies of Propranolol and Hydrochlorothiazide Oxidation in Standard and Synthetic Biological Fluids Using a Nitrogen-Containing Tetrahedral Amorphous Carbon (Ta-C:N) Electrode", *Electrochim. Acta,* vol. 143, pp. 398-406, 2014.
        [http://dx.doi.org/10.1016/j.electacta.2014.08.008]

[60]   K. Pecková, J. Musilová, and J. Barek, "Boron-Doped Diamond Film Electrodes - New Tool for Voltammetric Determination of Organic Substances", *Crit. Rev. Anal. Chem.,* vol. 39, no. 3, pp. 148-172, 2009.
        [http://dx.doi.org/10.1080/10408340903011812]

[61]   B.C. Lourencao, R.F. Brocenschi, R.A. Medeiros, O. Fatibello☐Filho, and R.C. Rocha☐Filho, "Analytical Applications of Electrochemically Pretreated Boron☐Doped Diamond Electrodes", *ChemElectroChem,* vol. 7, no. 6, pp. 1291-1311, 2020.
        [http://dx.doi.org/10.1002/celc.202000050]

[62]   S.J. Cobb, Z.J. Ayres, and J.V. Macpherson, "Boron Doped Diamond: A Designer Electrode Material for the Twenty-First Century", *Annu. Rev. Anal. Chem. (Palo Alto, Calif.),* vol. 11, no. 1, pp. 463-484, 2018.
        [http://dx.doi.org/10.1146/annurev-anchem-061417-010107] [PMID: 29579405]

[63]   J. Xu, Y. Yokota, R.A. Wong, Y. Kim, and Y. Einaga, "Unusual Electrochemical Properties of Low-Doped Boron-Doped Diamond Electrodes Containing sp2 Carbon", *J. Am. Chem. Soc.,* vol. 142, no. 5, pp. 2310-2316, 2020.
        [http://dx.doi.org/10.1021/jacs.9b11183] [PMID: 31927922]

[64]   C.A.R. Salamanca-Neto, G.G. Marcheafave, G.J. Mattos, J.T. Moraes, K. Schwarzová-Pecková, and E.R. Sartori, "Boron-Doped Diamond Film and Multiple Linear Regression-Based Calibration Applied to the Simultaneous Electrochemical Determination of Paracetamol, Phenylephrine Hydrochloride, and Loratadine in Fixed-Dose Combinations", *Microchem. J.,* vol. 162, p. 105831, 2021.
        [http://dx.doi.org/10.1016/j.microc.2020.105831]

[65]   M. Brycht, K. Kaczmarska, B. Uslu, S.A. Ozkan, and S. Skrzypek, "Sensitive Determination of Anticancer Drug Imatinib in Spiked Human Urine Samples by Differential Pulse Voltammetry on Anodically Pretreated Boron-Doped Diamond Electrode", *Diamond Related Materials,* vol. 68, pp. 13-22, 2016.
        [http://dx.doi.org/10.1016/j.diamond.2016.05.007]

[66]   E.R. Sartori, D.N. Clausen, I.M.R. Pires, and C.A.R. Salamanca-Neto, "Sensitive Square-Wave Voltammetric Determination of Tadalafil (Cialis®) in Pharmaceutical Samples Using a Cathodically Pretreated Boron-Doped Diamond Electrode", *Diamond Related Materials,* vol. 77, pp. 153-158, 2017.
        [http://dx.doi.org/10.1016/j.diamond.2017.07.001]

[67]   K.Y.H. Nagao, C.A.R. Salamanca-Neto, B. Coldibeli, and E.R. Sartori, "A Differential Pulse Voltammetric Method for Submicromolar Determination of Antihistamine Drug Desloratadine Using an Unmodified Boron-Doped Diamond Electrode", *Anal. Methods,* vol. 12, no. 8, pp. 1115-1121, 2020.
        [http://dx.doi.org/10.1039/C9AY02785H]

[68]   C.A.R. Salamanca-Neto, M.L. Felsner, A. Galli, and E.R. Sartori, "In-House Validation of a Totally Aqueous Voltammetric Method for Determination of Diltiazem Hydrochloride", *J. Electroanal. Chem. (Lausanne),* vol. 837, 2019.
[http://dx.doi.org/10.1016/j.jelechem.2019.02.026]

[69]   B. Coldibeli, N.S.M. Campos, C.A.R. Salamanca-Neto, J. Scremin, G.J. Mattos, G.G. Marcheafave, and E.R. Sartori, "Feasibility of the Use of Boron-Doped Diamond Electrode Coupled to Electroanalytical Techniques for the Individual Determination of Pravastatin and Its Association with Acetylsalicylic Acid", *J. Electroanal. Chem. (Lausanne),* vol. 862, p. 113987, 2020.
[http://dx.doi.org/10.1016/j.jelechem.2020.113987]

[70]   G.R.P. Manrique, C.A.R. Salamanca-Neto, J. Tobias Moraes, and E.R. Sartori, "Fast Surface Water Quality Analysis Based on an Ultra-Sensitive Determination of the Antidepressant Drug Duloxetine Hydrochloride on a Diamond Electrode by Voltammetry", *Int. J. Environ. Anal. Chem.,* vol. •••, pp. 1-15, 2020.
[http://dx.doi.org/10.1080/03067319.2020.1802442]

[71]   M.C.G. Santos, C.R.T. Tarley, L.H. Dall'Antonia, and E.R. Sartori, "Evaluation of Boron-Doped Diamond Electrode for Simultaneous Voltammetric Determination of Hydrochlorothiazide and Losartan in Pharmaceutical Formulations", *Sens. Actuators B Chem.,* vol. 188, pp. 263-270, 2013.
[http://dx.doi.org/10.1016/j.snb.2013.07.025]

[72]   J. Scremin, and E.R. Sartori, "Simultaneous Determination of Nifedipine and Atenolol in Combined Dosage Forms Using a Boron-Doped Diamond Electrode with Differential Pulse Voltammetry", *Can. J. Chem.,* vol. 96, no. 1, pp. 1-7, 2018.
[http://dx.doi.org/10.1139/cjc-2017-0302]

[73]   J.T. Moraes, C.A.R. Salamanca-Neto, Ľ. Švorc, and E.R. Sartori, "Advanced Sensing Performance towards Simultaneous Determination of Quaternary Mixture of Antihypertensives Using Boron-Doped Diamond Electrode", *Microchem. J.,* vol. 134, pp. 173-180, 2017.
[http://dx.doi.org/10.1016/j.microc.2017.06.001]

[74]   A.P.P. Eisele, G.R. Mansano, F.M. De Oliveira, J. Casarin, C.R.T. Tarley, and E.R. Sartori, "Simultaneous Determination of Hydrochlorothiazide and Valsartan in Combined Dosage Forms: Electroanalytical Performance of Cathodically Pretreated Boron-Doped Diamond Electrode", *J. Electroanal. Chem. (Lausanne),* vol. 732, pp. 46-52, 2014.
[http://dx.doi.org/10.1016/j.jelechem.2014.08.033]

[75]   G.R. Mansano, A.P.P. Eisele, L.H. Dall'Antonia, S. Afonso, and E.R. Sartori, "Electroanalytical Application of a Boron-Doped Diamond Electrode: Improving the Simultaneous Voltammetric Determination of Amlodipine and Valsartan in Urine and Combined Dosage Forms", *J. Electroanal. Chem. (Lausanne),* vol. 738, pp. 188-194, 2015.
[http://dx.doi.org/10.1016/j.jelechem.2014.11.034]

[76]   L.P. Silva, B.C. Lourencao, and O. Fatibello-Filho, "Determinação Voltamétrica Simultánea de Besilato de Anlodipino e Hidroclorotiazida Em Amostras de Urina Sintética Utilizando Um Eletrodo de Diamante Dopado Com Boro", *Quim. Nova,* vol. 38, no. 6, pp. 801-806, 2015.
[http://dx.doi.org/10.5935/0100-4042.20150077]

[77]   G.R. Mansano, and E.R. Sartori, "Oxidação Eletroquímica de Anlodipino e Hidroclorotiazida Sobre o Eletrodo de Diamante Dopado Com Boro: Potencialidade de Determinação Simultânea Em Urina. Orbital - Electron", *J. Chem.,* vol. 7, no. 1, pp. 81-86, 2015.
[http://dx.doi.org/10.17807/orbital.v7i1.673]

[78]   G.R. Mansano, A.P.P. Eisele, and E.R. Sartori, "Electrochemical Evaluation of a Boron-Doped Diamond Electrode for Simultaneous Determination of an Antihypertensive Ternary Mixture of Amlodipine, Hydrochlorothiazide and Valsartan in Pharmaceuticals", *Anal. Methods,* vol. 7, no. 3, pp. 1053-1060, 2015.
[http://dx.doi.org/10.1039/C4AY02511C]

[79]   C.A.R. Salamanca-Neto, A.P.P. Eisele, V.G. Resta, J. Scremin, and E.R. Sartori, "Differential Pulse Voltammetric Method for the Individual and Simultaneous Determination of Antihypertensive Drug Metoprolol and Its Association with Hydrochlorothiazide in Pharmaceutical Dosage Forms", *Sens. Actuators B Chem.*, vol. 230, pp. 630-638, 2016.
       [http://dx.doi.org/10.1016/j.snb.2016.02.071]

[80]   J.T. Moraes, A.P.P. Eisele, C.A.R. Salamanca-Neto, J. Scremin, and E.R. Sartori, "Simultaneous Voltammetric Determination of Antihypertensive Drugs Amlodipine and Atenolol in Pharmaceuticals Using a Cathodically Pretreated Boron-Doped Diamond Electrode", *J. Braz. Chem. Soc.*, vol. 27, no. 7, pp. 1264-1272, 2016.
       [http://dx.doi.org/10.5935/0103-5053.20160023]

[81]   G.J. Mattos, J. Scremin, C.A.R. Salamanca-Neto, and E.R. Sartori, "The Performance of Boron-Doped Diamond Electrode for the Determination of Ramipril and Its Association with Hydrochlorothiazide", *Electroanalysis*, vol. 29, no. 4, 2017.
       [http://dx.doi.org/10.1002/elan.201600692]

[82]   J.T. Moraes, C.A.R. Salamanca-Neto, A.P.P. Eisele, B. Coldibeli, G.S. Ceravolo, and E.R. Sartori, "Fast and Sensitive Simultaneous Determination of Antihypertensive Drugs Amlodipine Besylate and Ramipril Using an Electrochemical Method: Application to Pharmaceuticals and Blood Serum Samples", *Anal. Methods,* vol. 11, no. 31, pp. 4006-4013, 2019.
       [http://dx.doi.org/10.1039/C9AY01232J]

CHAPTER 10

Voltammetric Applications in Anti-inflammatory Drug Detection

B. P. Sanjay[1], **D. N. Varun**[1], **S. Sandeep**[1], **C. S. Karthik**[1], **A. S. Santhosh**[2], **P. Mallu**[1] and **Nagaraja Sreeharsha**[3,4]

[1] *Department of Chemistry, JSS Science and Technology University, Mysuru- 570006, Karnataka, India*

[2] *Department of Chemistry (UG), NMKRV College for Women, Jayanagar, Bengaluru – 560011, India*

[3] *Department of Pharmaceutical Sciences, College of Clinical Pharmacy, King Faisal University, Al- Ahsa-31982, Saudi Arabia*

[4] *Department of Pharmaceutics, Vidya Siri College of Pharmacy, Off Sarjapura Road, Bengaluru - 560035, Karnataka, India*

Abstract: The past few years have seen substantial development in electrochemical sensors. The fundamental challenge in the construction of reliable analytical techniques for the quantification of anti-inflammatory drugs has been conveniently overcome by electrochemical methods. Voltammetric sensors have applications in biological, environmental, and chemical structures. Owing to their high sensitivity, selectivity, ease of preparation, and fast reaction, voltammetric sensors for detecting specific biomolecules have received significant attention in recent years. Polarography, cyclic voltammetry (CV), Differential Pulse Voltammetry (DPV), and Squarewave Voltammetry (SWV) are examples of electrochemical techniques that are classified. Non-steroidal anti-inflammatory drugs (NSAIDs) such as 4-amino antipyrine (4-AAP), Indomethacin (IND), and Diclofenac (DCF) are widely used in the treatment of inflammatory conditions such as musculoskeletal disorders, dental pain, menstrual pain, postoperative pain, and migraines. This chapter discusses the details and usages of the different voltammetric methods for the determination of NSAIDs.

Keywords: Anti-inflammatory drugs, Classifications of voltammetric techniques, Diclofenac, Indomethacin, Voltammetric applications, 4-Aminoantipyrine.

INTRODUCTION

The advancement of the human race can be witnessed with improved technologies, new inventions, and new paradigms being developed every day

---

* **Corresponding Author S. Sandeep:** Department of Chemistry, SJCE, JSS Science and Technology University, Mysuru, India; E-mail: sandeeps@jssstuniv.in

J.G. Manjunatha (Ed.)

using cutting-edge technology [1]. However, in contrast, there is an increase in the number of illicit drugs and trafficking uses, which causes problems for society, health, criminality, and the environment [2]. In addition, the number of unpredictable health issues in humans compels researchers worldwide to investigate accurate analytical techniques for drug detection in seized street samples, biological fluids, and wastewater [3]. Drug detection plays a vital role in the modern pharmaceutical industries involving tedious and cumbersome processes like formulation and stability studies, quality control (QC), and toxicology and pharmacological testing in animals and man. In clinical industries, drug detection is carried out using biological samples like blood, urine and saliva in support of clinical trials, *i.e.*, bioavailability and pharmacokinetic studies, and in the monitoring of therapeutic drugs and drug abuse [4]. These investigations require simple, novel, rapid, and reliable analytical methods to measure drugs in complex media such as formulation and biofluids. In this background, electrochemistry takes center stage by fulfilling all the criteria for an efficient analytical tool [5]. In particular, voltammetric methods for the determination of drugs have received enormous attention in recent years. This is due to the sensitivity and selectivity they provide on qualitative and quantitative aspects of the electroactive analyte under study [6]. This chapter, will explore the commonly used voltammetric techniques, electrodes utilized, modification procedures for the electrodes, solvents, and supporting electrolytes highlighting their electrochemical performances and limitations. Finally, we devote special attention to outlining the future trends of the electrochemical methods for drug analysis [7]. Voltammetry is part of electrochemistry, which was known after the discovery of polarography in 1922 by the Czech chemist Yaroslav Heyrovsky, for which he received the Nobel Prize in 1959 [8]. The techniques utilized in early voltammetry encountered several challenges, making them less applicable for routine analysis. However, around the 1960s and 1970s, considerable advances were made in all the fields of voltammetry, which expanded its application in different areas. Voltammetry gives data about oxidation, adsorption, and thermodynamic properties of solvated species [9].

CLASSIFICATION OF VOLTAMMETRY TECHNIQUES

Polarography

The concept of pulse polarography was introduced by the English chemist G.C. Barker (Fig. **1** A-C). Initially, the theory relied on the results of Barker's work on square wave polarography [6]. Barker and colleagues explored the principle of pulse polarography in various correspondences, adapted a portion of the hypothetical aspects of square wave polarography to pulse polarographic methods, and presented the overall instrumentation [10]. Polarography is also part of the electrochemistry procedure, which examines liquid solutions containing oxidizing or reducing substances. It is one of the most broadly utilized voltammetric methods. In this procedure, the estimations are directed based on diffusion mass transport [11]. It is accepted that mercury terminals are the most basic anodes with smooth surfaces and a high cathodic potential window, which is utilized to determine the reducible substances [12]. The ongoing investigation demonstrated noteworthy points of interest over the recently revealed strategies, including high affectability and brief time frame devouring. Moreover, they do not need a pretreatment of samples before analysis.

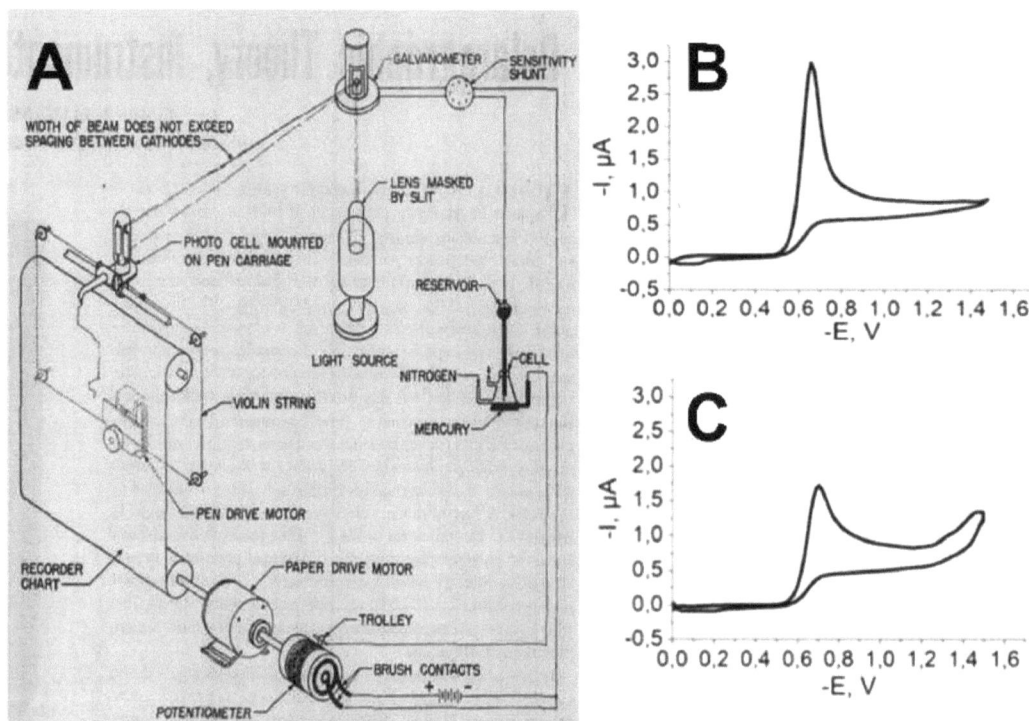

Fig. (1). (A) Schematic Diagram of Lykken-Pompeo-Weaver Polarograph. **(B)** and **(C)** graphs of Polarograms [13].

Cyclic Voltammetry

Cyclic voltammetry (CV) is one of the most efficient electroanalytical methods for the investigating of electroactive species. The use of CV is versatile, with applications in electrochemistry, organic chemistry, materials science, natural science, *etc.* CV is efficacious because of its ability to observe redox activity quickly across a large possible range. The resulting voltammogram is similar to a normal range because it propagates information as part of an energy channel [14].

The electrochemical setup is shown in Fig. (**2A**). It is a three-electrode system that consists of a working electrode, reference electrode, and counter electrode. Further, the cyclic voltammogram obtained for the simple reversible reaction is shown in Fig. (**2 B**).

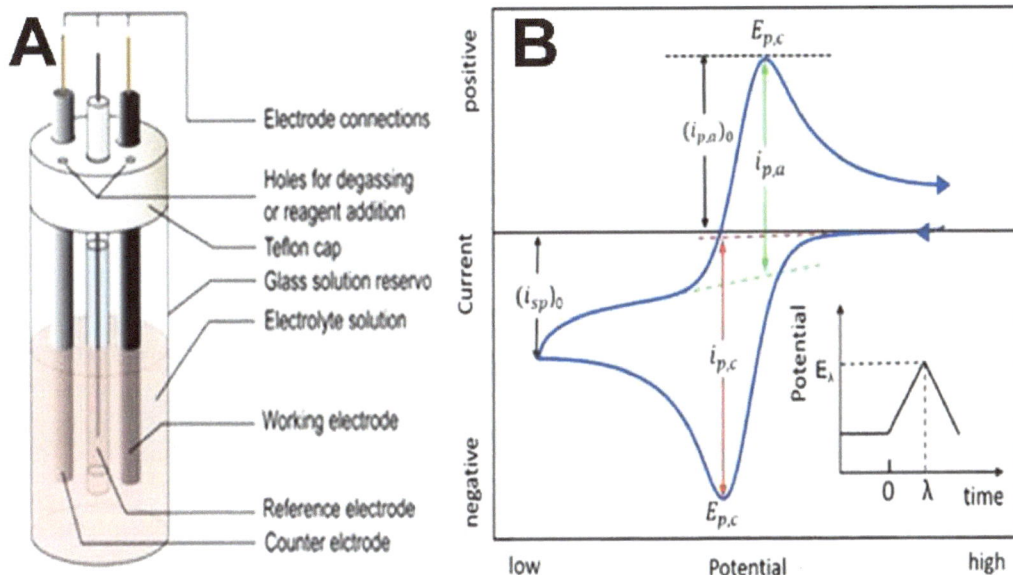

Fig. (**2 A**). Electrochemical setup and (**B**) Voltammogram for a simple reversible reaction [15].

Differential Pulse Voltammetry (DPV)

DPV is one of the most often used voltammetry methods [16 - 18] for distinguishing analytes in-stream frameworks, incorporating the *in vivo* examination of a patient's blood and DNA. Besides, DPV likewise suits the fast investigation of a solitary analyte. In DPV, the current is tested twice in each pulse period, and the distinction between these two current qualities is recorded to create a peaked voltammogram. The peak height is directly proportional to the concentration of the analyte. DPV has an advantage over the other analytical method as it can lessen the impact of capacitive current and improve the signal-to-noise proportion by attenuating the background current [19]. DPV joins the parts of chronoamperometry and linear sweep voltammetry and displays high selectivity and affectability [20]. The estimation of two-fold potential pulse differential strategies for electrochemical investigations of the active compound, speciation is examined inside the setting of liquid-liquid electrochemistry. It does not just give data about the speciation of thermodynamics and energy availability, but also the species lipophilicity. This gives a far-reaching perspective on their conduct in characteristic and natural media [21].

Square-Wave Voltammetry

Square-wave voltammetry (SWV) is a ground-breaking electrochemical procedure appropriate for the analytical application, mechanistic investigation of electrode measures, and electrokinetic estimations [22].

It is considered as one of the most developed voltammetric strategies, which binds together the benefits of pulse procedures, cyclic voltammetry, and impedance methods. Other than tending to the absolute highlights of SWV.

ELECTROCHEMICAL SENSORS

The detection of biomolecules is of great importance as each of the molecules plays a crucial part in the lives of human beings. Various conventional methods have been employed, such as spectrophotometric (enzymatic and nonenzymatic), titrimetry, chemiluminescence, *etc.* these techniques have complex procedures, low specificity, instability of reagents, and high cost. As a result, electrochemical sensors are receiving considerable attention due to their high sensitivity, excellent selectivity, long-term stability, quick studies and are amenable to miniaturization. Electrochemical sensors are devices that consist of a selective analyte interface or an immobilizing matrix on a transducer. The interface is selective and sensitive in recognizing its specific substrate. This interaction produces a chemical reaction that is transferred to the transducer. The transducer converts the chemical reaction to an electrochemical reaction. In an electrochemical sensor, the role of the transducer is to convert the chemical change into a measurable signal. This process is called signalization, and most of the transducers produce either optical or electrical signals that are usually proportional to the number of analyte interactions. The recognition element mainly determine the selectivity of the electrochemical sensor for the target analyte while, the transducer greatly influences the sensitivity of the sensor.

Nanotechnology plays a vital role in the development of electrochemical sensors. The nanoparticles (NPs) and nanomaterials exhibit unique chemical, physical, and electronic properties because of their small size (1-100 nm). NPs used for immobilization is gold (Au), silver (Ag), palladium (Pd), titanium dioxide (TiO_2), zirconium, and many more. Pure or metals combined with an alloy provide enhanced electron transfer. Apart from these NPs, nanomaterials such as graphene, multi-walled carbon nanotubes (MWCNTs) have been tremendously used in the development of biosensors due to their large number of unique and exceptional characteristics.

Application of the Electrochemical Method for the Detection of Non-steroidal Anti-inflammatory Drugs (NSAIDs)

Non-steroidal anti-inflammatory drugs (NSAIDs) have achieved a breakthrough role in the pharmaceutical industry since the isolation of salicylate from willow bark in about 1830s, followed by the discovery of aspirin (acetylsalicylate) by Felix Hoffman of the Bayer industry, Germany, in 1897 [23]. NSAIDs are among the most prescribed treatments and used to reduce symptoms such as pain and fever connected with respiratory infections [24]. Traditionally, NSAIDs are classified as major derivatives of salicylic acid, acetic acid, enolic acid, anthranilic acid, or propionic acid based on their chemical characteristics. Most of the common NSAIDs are categorized. However, the classification has also been relocated with the development of scientific knowledge based on their selectivity to inhibit cyclooxygenase/prostaglandins [25]. NSAIDs are mostly used to treat pain, and inflammatory disorders such as chronic pain, osteoarthritis, rheumatoid arthritis, postoperative surgery, menstrual cramps and are often commonly used as analgesics and antipyretics [26, 27]. In this chapter, we have discussed the electrochemical detection of some NSAIDs.

4-Aminoantipyrine (4-AAP)

4- AAP (Fig. **3**) is usually utilized as a pain-relieving, antipyretic, and non-steroid mitigating drug [1]. 4-AAP causes eye aggravation, skin, respiratory disturbance, *etc.*, can decrease the bloodstream [28]. Due to these reasons, it is crucial to quantify the amount of 4- AAP.

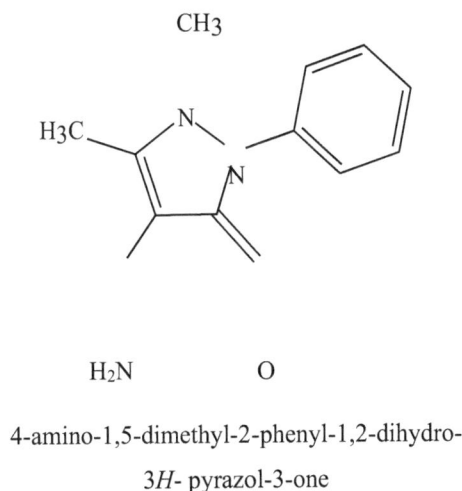

4-amino-1,5-dimethyl-2-phenyl-1,2-dihydro-

3*H*- pyrazol-3-one

Fig. (3). Structure of 4-Aminoantipyrine.

Many types of research have been reported on the construction of sensors for the detection of 4- SrCuO$_2$/GCN nanocomposite modified pencil graphite electrode. The SrCuO$_2$/GCN nanocomposite showed excellent electrocatalytic activity towards detecting 4-AAP with a low detection of 0.05 µM. This excellent property was attributed to the combined effect of SrCuO$_2$ and GCN nanocomposite [29]. The cyclic voltammogram and the DPV curves observed at the SrCuO$_2$/GCN nanocomposite modified pencil graphite electrode surface are shown in Fig. (**4A and B**). *Xiong et al.,* have fabricated the working electrode with Fe$_3$O$_4$@MnO$_2$ magnetic core-shell nanoflowers. The modified electrode showed a wide linear range from 1.0 µM to 120 µM with a detection limit of 0.15 µM [30]. Hu X *et al.,* prepared the CNT-modified electrode for the oxidation of 4-AAP. The modified electrode was employed to quantify 4-AAP in the range of 5 x 10^{-5} to 2 x 10^{-3} with the detection limit of 2.6 x 10^{-5} [31]. Palraj Ranganath *et al.,* have synthesized polystyrene:β-cyclodextrin embedded with yttrium oxide nanoparticles (PS:β-CD IC/Y$_2$O$_3$). The synthesized PS:β-CD IC/Y$_2$O$_3$ nanocomposite was modified on a glassy carbon electrode. The modified electrode was further used for the simultaneous detection of NSAIDs 4-AAP and anti-viral drug acyclovir. In the developed sensor, the anodic peak currents measured on the PS:β-CD IC/Y$_2$O$_3$/GCE are much higher than those on PS:β-CD IC/GCE, Y$_2$O$_3$/GCE, and bare GCE because of the functional group-rich PS:β-CD IC and the synergistic interaction between PS and Y$_2$O$_3$ *via* hydrophobic, covalent, or noncovalent interactions facilitate the fast electron- transfer activity. The very low LOD was found at PS:β-CD IC/Y$_2$O$_3$/GCE for 4-AAP and ACV. Last, the utility of this novel sensor has been successfully demonstrated by the detection of two crucial drugs 4-AAP and ACV levels in human urine and commercial ACV tablet samples with satisfactory results. *Xiong et al.* have fabricated the working electrode with Fe$_3$O$_4$@MnO$_2$ magnetic core-shell nanoflowers. The modified electrode showed a wide linear range from 1.0 µM to 120 µM with a detection limit of 0.15 µM [32]. The possible mechanism of 4-AAP at the electrode surface is given in scheme (**1**).

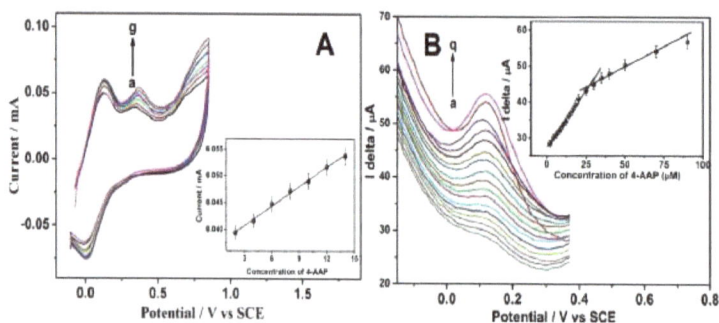

Fig. (4). (A) shows Cyclic voltammograms were investigated by measuring the CV response at different concentrations of 4-AAP and **(B)** shows DPVs recorded for the increasing concentration of 4-AAP [29].

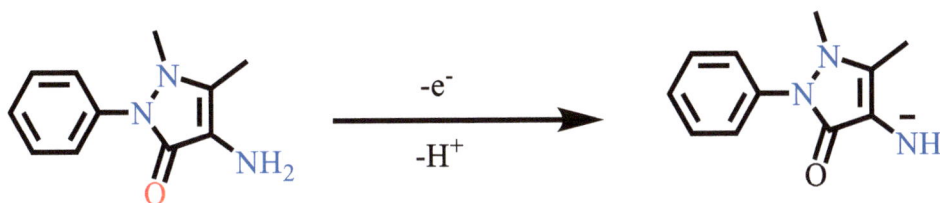

Scheme (1). The possible mechanism of oxidation of 4-AAP.

A compilation of improved performance of electrochemical sensors fabricated using different nanomaterials for the detection of 4-AAP is summerised in Table **1**.

Table 1. Comparison of different electrochemical sensors for the determination of 4- Aminoantipyrine.

Matrix	LinearRange (µM)	LOD (µM)	Oxidation Potentials	Stability mA	Reproducibility RSD %	Repeatability RSD %	Real Samples	References
f-CNF/LaCoO$_3$/GCE	0.001-1371	0.001	0.05V	---	1.15	---	urine and plasma	[33]
CNTs/GCE	5.0×10^{-5}–2.0×10^{-3}	2.6	200mv	---	---	---	Drugs in Pharmaceutical and Urine Samples	[34]
RF/GCE	0.02-0.4 µg mL^{-1}	0.016 µg mL^{-1}	0.5V	---	3.4	---	Different pharmaceutical samples	[35]
DLT	2-15 µmol l^{-1}	1.1 µmol l^{-1}	0.5V	---	1.15	---	urine and plasma	[36]
MWCNT/CTAB/GCE	5.0×10^{-9} - 4.0×10^{-8} M	1.63×10^{-10} M	0.512V	93.5%	2.21	2.38	urine	[37]

Indomethacin (IND)

Fig. (**5**) shows a schematic diagram of Ind. It was founded in 1963 [38] and endorsed by the US FDA in 1965 as a non-steroidal mitigating drug (NSAID). Drug and individual consideration items (PPCPs) are considered a developing toxin in oceanic conditions for as far back as barely any years, These accumulations have gotten a ton of interest in light of their expected damage to the human biological system [39 - 41]. As an application part of PPCPs for antipyretics and analgesics, IND generally exists in amphibian climates, for example, drinking water, groundwater, surface water, and sewage [42, 43]. Nephrotoxic prescription is much of the time given to Very low-birth-weight (VLBW) babies during the principal long periods of life [44, 45]. This incorporates non-steroidal mitigating drugs (NSAID) to conclude a patent ductus arteriosus (PDA). Since nephrogenesis is fragmented before the gestational age of

34–36 weeks, the juvenile kidney in VLBW newborn children is especially defenseless. Not with standing intense nephrotoxic impacts, long-haul antagonistic sequelae on renal structure and capacity following medication presentation are progressively perceived [46, 47]. The cyclic voltammogram and the DPV curves observed at the different electrodes in PBS (GCE, NiO/GCE, Gr/GCE, GrNiO/GCE) nanocomposite modified glassy carbon electrode surface is shown in Fig. (**6A** and **B**).

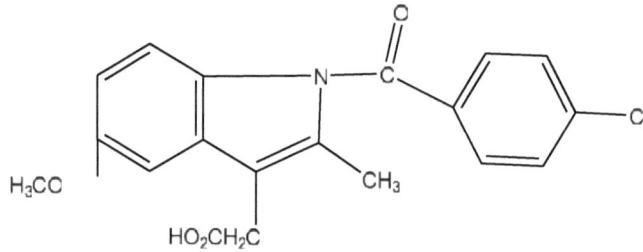

Fig. (5). Schematic diagram of Indomethacin [48].

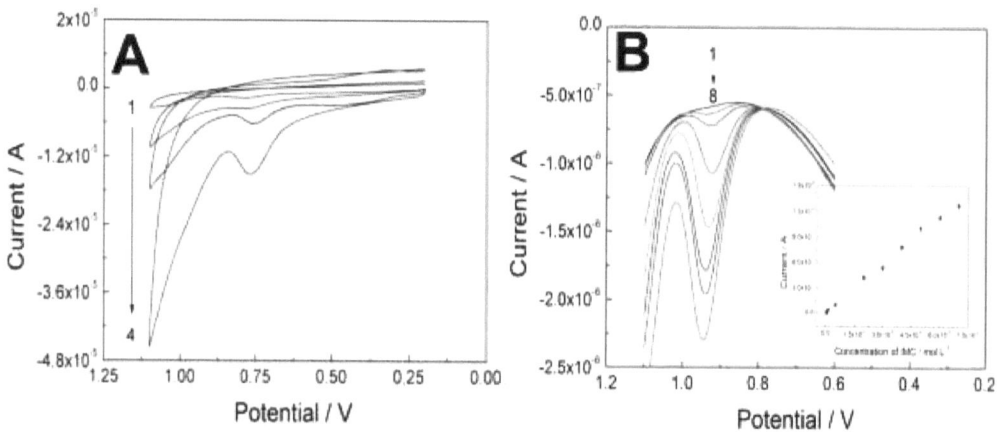

Fig. (6). Shows CVs of Indomethacin recorded at the different electrode in PBS and (B) Indomethacin recorded by DPV with different concentrations and Inset is the calibration curve [49].

The cyclic voltammogram and the DPV curves observed at the MnO_2Gr/GCE nanocomposite modified glassy carbon electrode surface is shown in Fig **A** and **B**. Baezzat *et al. 2020* has reported the electrochemical sensor for the detection of IND using a TiO_2 nanoparticle modified carbon ionic liquid electrode (CILE). The TiO_2 nanoparticles and the ionic liquid 1-hexylpyridinium hexafluorophosphate (HPFP) have been used fabricate the nanocomposite sensor. The TiO_2 and HPFP showed excellent electrocatalytic activity towards the detection of IND with the low detection of 2.1×10^{-8} M. The excellent property was attributed to the TiO_2 nanoparticle [50]. By CV methods, Electrochemical parameters of the electrode reaction of IND, including the electron transfer coefficient (α) and the electron-transfer number (n), were calculated. Under selected conditions, the anodic peak current was linear for the concentration of IND in the broad range is 1.0×10^{-7} to 1.0×10^{-4} M. *Heli et al. 2009* has reported the electrochemical sensor for the detection of IND using a nanoparticle of Ni–curcumin complex modified glassy carbon electrode (GCE). The development of the nanocomposite showed a low detection of 98 nM, and the Ni-curcumin complex displayed excellent electrocatalytic activity to detect IND. The excellent property was attributed to the Ni-curcumin complex [51]. *Petković etal.* has reported that the electrochemical sensor detecting IND using a nanoparticle of the boron-doped diamond electrode was employed as a sensitive sensor with a low detection of 57 nM, boron-doped diamond electrode exhibited excellent electrocatalytic activity towards IND [52]. *Sarhangzadeh et al. 2013* has reported the electrochemical sensor for detecting IND using a multiwalled carbon nanotube and ionic liquid-modified carbon ceramic electrode (MWCNT–IL/CCE). The sensor showed a low detection of 260 nM, and the MWCNT-IL reported the electrochemical sensor for the detection of IND using gold nanorods– graphene oxide nanocomposite incorporated carbon nanotube paste modified glassy carbon electrode (GNRs–GO–CNTP/GCE) was employed as a sensitive sensor. The GNRs–GO–CNTP nanocomposite has been used for the fabrication of nanocomposite sensors. The GNRs–GO–CNTP showed excellent electrocatalytic activity towards detection of IND with the low detection of 1.7×10^{-2} mM and two linear calibration ranges of 0.2–0.9 and 2.5–91.5 μM were obtained. The excellent property was attributed to the GNRs–GO–CNTP nanoparticle [54]. The possible mechanism of IND at the electrode surface is given in Scheme (**2**).

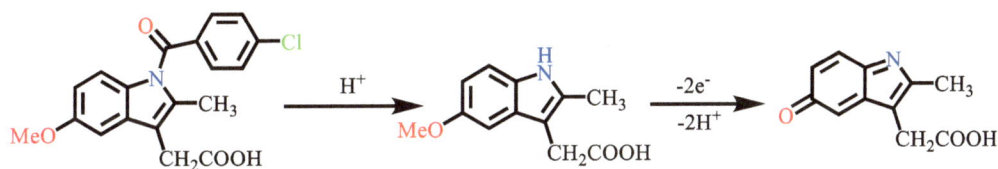

Scheme (2). Possible mechanism of oxidation of IND at electrode surface.

A compilation of improved performance of electrochemical sensors fabricated using different nanomaterials for the detection of IND is summerised in Table **2**.

Table 2. Comparison of different electrochemical sensors for the determination of Indomethacin.

Matrix	linear range (μM)	LOD (μM)	Oxidation Potentials	Stability mA	Reproducibility RSD%	Repeatability	Real Sample Used	Reference
GCOPE	8.5×10^{-8} -1.5×10^{-7}	5×10^{-8}	0.850 V	---	3	---	Urine	[48]

(Table 2) cont.....

Matrix	linear range (µM)	LOD (µM)	Oxidation Potentials	Stability mA	Reproducibility RSD%	Repeatability	Real Sample Used	Reference
TiO$_2$/CILE	1.0×10^{-7} -1.0×10^{-4}	2.1×10^{-8}	0.9 V	90.16%	1.91	---	Pharmaceutical drug	[49]
S,N/CQD	0.1–1.5 mg L^{-1}	2.6	0.61 V	---	---	---	Human plasma	[50]
MnO$_2$/Gr/GCE	1.0×10^{-7} -2.5×10^{-5}	3.2×10^{-8}	580 mV	---	---	---	Authentic drugs	[51]
GNRs/GO/CNTP/GCE	0.2–0.9 Mm	1.7×10^{-2}	0.9 V	S.D=3.5%	---	3.1-2.2%	Indomethacin tablets	[52]
CNTP/ILC/CCE	1 to 50 µM	0.26	800 mV	S.D=5%	3.2	---	Human blood	[53]
Gr/NiO/GCE	2.0×10^{-7} -7.0×10^{-5} mol L^{-1}	5.4×10^{-8}	0.52 V	---	2.46	2.56%	Human blood serum, urine, and pharmaceutical drug	[54]
S,N/CQD	0.1–1.5 mg L^{-1}	65 µg	---	---	---	---	Water sample and Human plasma	[55]
ZnFe$_2$O$_4$ /MWCNTs/CPE	5.0-200.0 µM	0.5	0.14 V	93.5%	3.06%	2.28%	Urine and Human blood	[56]

Diclofenac (DCF)

Fig. (**7**) shows a schematic diagram of DCF (Voltaren and different generics). An NSAID is primarily used to treat various rheumatoid issues, including osteoarthritis, rheumatoid joint irritation, ankylosing spondylitis, and severe muscle torture. Like various NSAIDs, in a patient's DCF causes remarkable events like a hepatic injury. Regardless of the way that the unmistakable event movement of genuine diclofenac induced hepatic threatening drug reactions (ADRs) is low (checks range from 1 to 2 cases for every million cures [57] to 6 to 18 cases/100,000- person-years [58], Worldwide, the number of hepatic cases appear to be incredible, in the patients who are treated with DCF. NSAIDs have relieving impacts due to the controlling of cyclooxygenase, the rate-limiting impetus in the difference in arachidonic acid to combustible mediators, for instance, prostaglandins, thromboxanes, and prostacyclins [59, 60]. Typically NSAIDs are convincing in the control of torture from extraordinary fragile tissue hurt. They similarly are valuable in controlling articular musculoskeletal desolation [61, 62]. DCF is one of the most extensively studied topical NSAIDs in treating of torture and bothering, including torture and irritation starting from myofascial jumble [63 - 65]. The cyclic voltammogram curves observed at the f-MWCNT/GCE nanocomposite modified glassy carbon electrode surface is shown in Fig. (**8**).

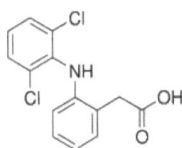

Fig. (7). Schematic diagram of Diclofenac [66].

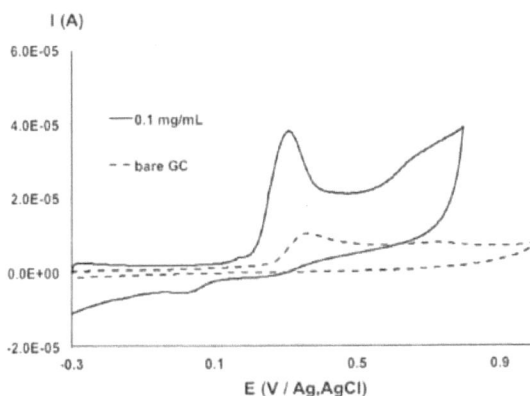

Fig. (8). Cyclic voltammograms of DCF on bare GCE (dotted line) and modified electrode (solid line) in 0.1 mol L−1 PBS [67].

Kassahun et al. 2020 has reported the electrochemiluminescent (ECL) immunosensor for the detection of DCF using a graphene oxide coupled graphite-like carbon nitride (GO-g-C_3N_4) as a transducer. The GO-g-C_3N_4 showed excellent electrocatalytic activity towards the detection of DCF with the low detection of 1.7 nM. The excellent property was attributed to the GO-g-C_3N_4 nanomaterial [68]. The calibration curves were linear over a wide range of concentrations of each species, including 0.005–1000 ng mL^{-1} for DCF. Elmbrook *et al.* has reported the electrochemical sensor for the detection of DCF using a functionalized multi-walled carbon nanotube (f-MWCNTs) modified glassy carbon electrode (f-MWCNTs/GCE). The nanoparticles of f-MWCNTs have been used for the development of nanocomposite sensors. With a low detection of 0.1 µM, the f-MWCNTs/GCE showed excellent electrocatalytic activity against detection of DCF. The excellent property has been attributed to the nanoparticle of GNRs-GO- CNTP [69]. *Deng et al. 2019* has reported the electrochemical sensor for detecting DCF using functionalized Ag, Ag@SiO_2, and AgMBA@SiO_2 nanoparticles. The Ag, Ag@SiO_2, and AgMBA@SiO_2 showed excellent electrocatalytic detection activity (DCF) with a low 0.07 pg mL^{-1} detection. The excellent property was attributed to the Ag, Ag@SiO_2, and AgMBA@SiO_2 nanoparticle [70]. Elmbrook *et al.* has reported the electrochemical sensor for the detection of DCF using a microporous silicon nitride film (SNF) modified GCE. The sensor exhibited a low detection of 1.5 µM, the f-SNF/GCE showed excellent electrocatalytic activity against using a microporous silicon nitride film (SNF) modified GCE. The sensor exhibited a low detection of 1.5 µM, the f- SNF/GCE showed excellent electrocatalytic activity against detection (DCF). The excellent property was attributed to the nanoparticle of the GNRs-GO-CNTP [71]. The calibration curves of each species are linear over a wide range of concentrations, including 8-56 µM. Zhang *et al.* 2020 has reported the electrochemical sensor for detecting DCF using an oxygen plasma-treated ZnO nanoflower modified SPE. The sensor showed excellent electrocatalytic activity towards the DCF with a detection limit of 0.11 µM [72]. Yang *et al.* has reported the electrochemical sensor for the detection of DCF using an AuNPs modified SPE. The AuNPs/SPE showed excellent electrocatalytic activity towards the detection of DCF with the low detection of 0.069 nM. The excellent property was attributed to the AuNPs nanoparticle [73]. The sensor also exhibited a wide linear range of 0.1–500-nM for DCF. Kimura *et al.* has reported the electrochemical sensor for the detection of DCF using platinum nanoflowers and reduced graphene oxide nanocomposite (PtNFs/rGO) modified SPE was employed as a sensitive sensor. The PtNFs/rGO/SPE showed excellent electrocatalytic activity towards the detection of DCF with the low detection of 40 nM with the wide linear range of 0.1−100 µM [74]. The possible mechanism for the oxidation of DFC at the electrode surface is given in scheme (**3**).

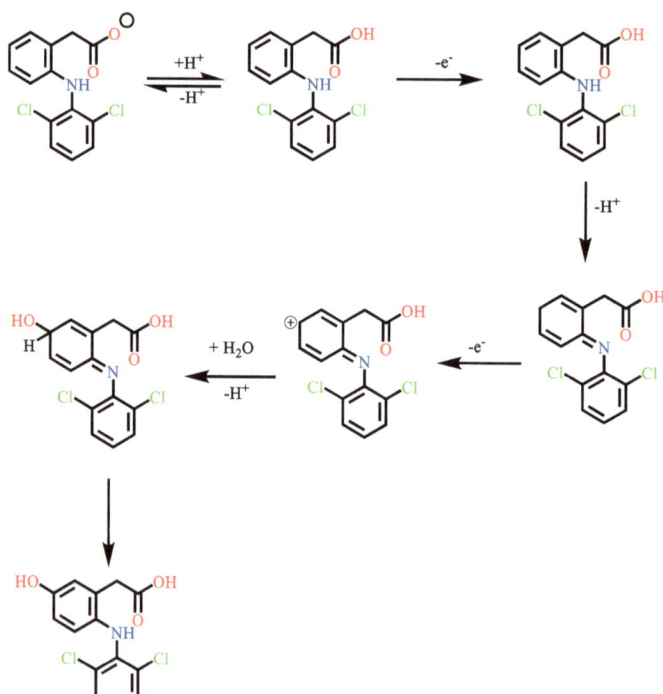

Scheme (3). Possible mechanism for the oxidation of DFC at the electrode surface.

A compilation of improved performance of electrochemical sensors fabricated using different nanomaterials for the detection of DFC is summerised in Table **3**.

Table **3**. Comparison of different electrochemical sensors for the determination of Diclofenac.

Matrix	linear range (μM)	LOD (μM)	Oxidation Potentials	Stability mA	Reproducibility RSD	Repeatability RSD	Real Sample Used	References
MWCNTs/AuNPs	0.005–1000 ng mL^{-1}	1.7 pg mL^{-1}	200 mV	S. D = 4.89%	1.79%	2.46%	Pharmaceutical formulation	[75]
Vinylferrocene/MWCNTs/CPE	5–600	2.0	520 mV	94%	1.9%	2.1%	Urine and Tablets	[76]
DCF/DBA/GCE	0-5.0	2.7×10^{-7}	0.25 V	---	4.7%	---	Human blood serum	[77]
Pt:Co/NL/1M3BIBr/CPE	0.1-900	50×10^{-9}	0.6 V	97.2%	3.13%	---	Pharmaceutical sample	[78]
MWCNTs/ EMIMPF$_6$/Cu(OH)$_2$/GCE	0.18 to 119 μM	0.04 μM	704 mV	S. D = 1.16%	2.13%	---	Human and fish blood serum and seawater	[79]
TCPE	10 μM and 140 μM	3.28 μM	0.130 mV	---	---	---	Pharmaceutical sample and Human Urine	[80]
MWCNTs/CTS-Cu/GCE	0.3 and 200 μmol L^{-1}	0.021 μmol L^{-1}	0.7 V	---	---	---	Pharmaceutical samples	[81]
Cu/ZEGE	2×10^{-5} - 3×10^{-7} M	5×10^{-8} M	0.2 V	---	---	---	Water samples	[82]
BFE	6.0–50.0 μmol L^{-1}	4.3 μmol L^{-1}	-1.4 V	S. D = 1.93%	3.71%	3.27%	Pharmaceutical formulations	[83]

Preparation of Real Sample

Generally, the blood sample is collected from the healthy volunteer, and it will be subjected to centrifugation. Then it will be treated with the reagents to precipitate the proteins and other components. Finally, the supernatant will be diluted for the desired concentration, which will be used for analysis. In the case of a urine sample, the urine will be collected from the volunteer, the sample is filtered using suitable filter paper, and then it will be frozen. Then the desired amount of urine sample will be assessed for the determination of the drug, or in another case the urine sample will be diluted, and the sample will be spiked with the known amount of drug. Whereas for the analysis of drugs in pharmaceutical tablets, the tablets containing the drug will be weighed accurately, and it is dissolved in a suitable solvent. The prepared tablet solution will be used for the quantification.

CONCLUSION

This chapter explains the current developments in the field of nanostructured materials to enhance the sensitivity of electrochemical sensors in the quantitative detection of anti-inflammatory drugs in several samples. To improve efficiency in the electrochemical sensing techniques, nanomaterials arc incorporated. A complete comparative investigation of various nanostructured materials has shown that sensor sensitivity can be significantly improved along with various strategies, such as hybrid composite fabrication, chemical functionalization, modification of surface *etc.*, The fundamental challenges associated with these advanced sensing instruments can be further solved by developing synergies between multifunctional nanomaterials, identification of components, and electrochemical approaches. However, effective connection and electrochemical sensor integration utilizing communication devices may be used to create the needed portable devices. Finally, this chapter will help to expand the view in the utilization of sensor materials in conjugation with the voltammetry for the determination of anti-inflammatory drugs.

CONSENT FOR PUBLICATION

Not applicable.

CONFLICT OF INTEREST

The authors declare no conflicts of interest, financial or otherwise.

ACKNOWLEDGMENTS

The authors acknowledge the great support from the Department of Chemistry, JSS Science and Technology University, Mysuru -570006, Karnataka, India.

REFERENCES

[1]   J.G. Manjunatha, "Surfactant modified carbon nanotube paste electrode for the sensitive determination of mitoxantrone anticancer drug", *J. Electrochem. Sci. & Eng.,* vol. 7, no. 1, pp. 39-49, 2017.
[http://dx.doi.org/10.5599/jese.368]

[2]   C. Raril, and J.G. Manjunatha, "A simple approach for the electrochemical determination of vanillin at ionic surfactant modified graphene paste electrode", *Microchem. J.,* vol. 154, p. 104575, 2020.
[http://dx.doi.org/10.1016/j.microc.2019.104575]

[3]   M.M. Charithra, and J.G. Manjunatha, "Enhanced voltammetric detection of paracetamol using carbon nanotube modified electrode as an electrochemical sensor", *J. Electrochem. Sci. & Eng.,* vol. 10, no. 1, pp. 29-40, 2020.
[http://dx.doi.org/10.5599/jese.717]

[4]   P.A. Pushpanjali, J.G. Manjunatha, B.M. Amrutha, and N. Hareesha, "Development of carbon nanotube-based polymer-modified electrochemical sensor for the voltammetric study of Curcumin", *Mater. Res. Innov.,* 2020.

[5]   T Girish, JG Manjunatha, PA Pushpanjali, NS Prinith, DK Ravishankar, and G Siddaraju, "Poly (DL-valine) electro-polymerized carbon nanotube paste sensor for determination of antihistamine drug cetirizine", *J. Electrochem. Sci. & Eng,* 2020.

[6]   J.G. Manjunatha, "A surfactant enhanced graphene paste electrode as an effective electrochemical sensor for the sensitive and simultaneous determination of catechol and resorcinol", *Chem. Data Collect.,* vol. 25, p. 100331, 2020.
[http://dx.doi.org/10.1016/j.cdc.2019.100331]

[7]   S. Sandeep, A.S. Santhosh, N.K. Swamy, G.S. Suresh, J.S. Melo, and N.A. Chamaraja, "A biosensor based on a graphene nanoribbon/silver nanoparticle/polyphenol oxidase composite matrix on a graphite electrode: application in the analysis of catechol in green tea samples", *New J. Chem.,* vol. 42, no. 20, pp. 16620-16629, 2018.
[http://dx.doi.org/10.1039/C8NJ02325E]

[8]   S. Sandeep, A.S. Santhosh, N.K. Swamy, G.S. Suresh, and J.S. Melo, "Detection of catechol using a biosensor based on biosynthesized silver nanoparticles and polyphenol oxidase enzymes", *Port. Electrochem. Acta,* vol. 37, no. 4, pp. 257-270, 2019.
[http://dx.doi.org/10.4152/pea.201904257]

[9]   S.R. Yashas, S. Sandeep, B.P. Shivakumar, and N.K. Swamy, "Potentiometric polyphenol oxidase biosensor for sensitive determination of phenolic micropollutant in environmental samples", *Environ. Sci. Pollut. Res. Int.,* vol. 27, no. 22, pp. 27234-27243, 2020.
[http://dx.doi.org/10.1007/s11356-019-05495-2] [PMID: 31134539]

[10]   N. Hareesha, and J.G. Manjunatha, "Fast and enhanced electrochemical sensing of dopamine at cost-effective poly (DL-phenylalanine) based graphite electrode", *J. Electroanal. Chem. (Lausanne),* vol. 878, p. 114533, 2020.
[http://dx.doi.org/10.1016/j.jelechem.2020.114533]

[11]   NS Prinith, and JG Manjunatha, "Polymethionine modified carbon nanotube sensor for sensitive and selective determination of L-tryptophan", *J. Electrochem. Sci. & Eng,* 2020.

[12]   E. Hammam, A. Tawfik, and M.M. Ghoneim, "Adsorptive stripping voltammetric quantification of the antipsychotic drug clozapine in bulk form, pharmaceutical formulation and human serum at a mercury electrode", *J. Pharm. Biomed. Anal.,* vol. 36, no. 1, pp. 149-156, 2004.
[http://dx.doi.org/10.1016/j.jpba.2004.04.012] [PMID: 15351059]

[13]   E.P. Parry, and R.A. Osteryoung, "Evaluation of analytical pulse polarography", *Anal. Chem.,* vol. 37, no. 13, pp. 1634-1637, 1965.
[http://dx.doi.org/10.1021/ac60232a001]

[14]   A.M. Stortini, L.M. Moretto, A. Mardegan, M. Ongaro, and P. Ugo, "Arrays of copper nanowire electrodes: Preparation, characterization and application as nitrate sensor", *Sens. Actuators B Chem.,* vol. 207, pp. 186-192, 2015.
[http://dx.doi.org/10.1016/j.snb.2014.09.109]

[15]   J. Gajdar, E. Horakova, J. Barek, J. Fischer, and V. Vyskocil, "Recent applications of mercury electrodes for monitoring of pesticides: A critical review", *Electroanalysis,* vol. 28, no. 11, pp. 2659-2671, 2016.
[http://dx.doi.org/10.1002/elan.201600239]

[16]   J.J. Lingane, "Polarographic Theory, Instrumentation, and Methodology", *Anal. Chem.,* vol. 21, no. 1, pp. 45-60, 1949.
[http://dx.doi.org/10.1021/ac60025a012]

[17]   P.T. Kissinger, and W.R. Heineman, "Cyclic voltammetry", *J. Chem. Educ.,* vol. 60, no. 9, p. 702, 1983.
[http://dx.doi.org/10.1021/ed060p702]

[18]   X. Huang, Z. Wang, R. Knibbe, B. Luo, S.A. Ahad, D. Sun, and L. Wang, "Cyclic voltammetry in lithium–sulfur batteries—challenges and opportunities", *Energy Technol. (Weinheim),* vol. 7, no. 8, p. 1801001, 2019.

[19]   N.A. Mansor, Z.M. Zain, H.H. Hamzah, M.S. Noorden, S.S. Jaapar, V. Beni, and Z.H. Ibupoto,

"Detection of breast cancer 1 (BRCA1) gene using an electrochemical DNA biosensor based on immobilized ZnO nanowires", *Open J. Appl. Biosens.*, vol. 3, no. 02, p. 9, 2014.
[http://dx.doi.org/10.4236/ojab.2014.32002]

[20]   M. Azimzadeh, M. Rahaie, N. Nasirizadeh, K. Ashtari, and H. Naderi-Manesh, "An electrochemical nanobiosensor for plasma miRNA-155, based on graphene oxide and gold nanorod, for early detection of breast cancer", *Biosens. Bioelectron.*, vol. 77, pp. 99-106, 2016.
[http://dx.doi.org/10.1016/j.bios.2015.09.020] [PMID: 26397420]

[21]   N.Z. Aqmar, W.F. Abdullah, Z.M. Zain, and S. Rani, "Embedded 32-bit Differential Pulse Voltammetry (DPV) Technique for 3-electrode Cell Sensing", In: *In IOP Conference Series: Materials Science and Engineering* vol. 340. , 2018, p. 012016. IOP Publishing

[22]   J. Wang, R. Polsky, and D. Xu, "Silver-enhanced colloidal gold electrochemical stripping detection of DNA hybridization", *Langmuir*, vol. 17, no. 19, pp. 5739-5741, 2001.
[http://dx.doi.org/10.1021/la011002f]

[23]   S. Crespi, and M. Fagnoni, "Generation of Alkyl Radicals: From the Tyranny of Tin to the Photon Democracy", *Chem. Rev.*, vol. 120, no. 17, pp. 9790-9833, 2020.
[http://dx.doi.org/10.1021/acs.chemrev.0c00278] [PMID: 32786419]

[24]   J.M. Olmos, E. Laborda, and A. Molina, "Differential double pulse voltammetry (DDPV) and additive differential pulse voltammetry (ADPV) applied to the study of the ACDT mechanism", *J. Solid State Electrochem.*, vol. 24, no. 11, pp. 2819-2831, 2020.
[http://dx.doi.org/10.1007/s10008-020-04619-w]

[25]   P.G. Pickup, and R.W. Murray, "Redox conduction: its use in electronic devices", *J. Electrochem. Soc.*, vol. 131, no. 4, p. 833, 1984.
[http://dx.doi.org/10.1149/1.2115709]

[26]   M.R. Montinari, S. Minelli, and R. De Caterina, "The first 3500 years of aspirin history from its roots - A concise summary", *Vascul. Pharmacol.*, vol. 113, pp. 1-8, 2019.
[http://dx.doi.org/10.1016/j.vph.2018.10.008] [PMID: 30391545]

[27]   A. Abdulla, N. Adams, M. Bone, A.M. Elliott, J. Gaffin, D. Jones, R. Knaggs, D. Martin, L. Sampson, and P. Schofield, "Guidance on the management of pain in older people", *Age Ageing*, vol. 42, suppl. Suppl. 1, pp. i1-i57, 2013.
[http://dx.doi.org/10.1093/ageing/afs199] [PMID: 23420266]

[28]   S. Bindu, S. Mazumder, and U. Bandyopadhyay, "Non-steroidal anti-inflammatory drugs (NSAIDs) and organ damage: A current perspective", *Biochem. Pharmacol.*, vol. 180, p. 114147, 2020.
[http://dx.doi.org/10.1016/j.bcp.2020.114147] [PMID: 32653589]

[29]   A. Gupta, and M. Bah, "NSAIDs in the treatment of postoperative pain", *Curr. Pain Headache Rep.*, vol. 20, no. 11, p. 62, 2016.
[http://dx.doi.org/10.1007/s11916-016-0591-7] [PMID: 27841015]

[30]   P.W. Budoff, "Use of mefenamic acid in the treatment of primary dysmenorrhea", *JAMA*, vol. 241, no. 25, pp. 2713-2716, 1979.
[http://dx.doi.org/10.1001/jama.1979.03290510021018] [PMID: 376875]

[31]   S.G. Sunderji, A. El Badry, E.R. Poore, J.P. Figueroa, and P.W. Nathanielsz, "The effect of myometrial contractures on uterine blood flow in the pregnant sheep at 114 to 140 days' gestation measured by the 4-aminoantipyrine equilibrium diffusion technique", *Am. J. Obstet. Gynecol.*, vol. 149, no. 4, pp. 408-412, 1984.
[http://dx.doi.org/10.1016/0002-9378(84)90155-8] [PMID: 6428232]

[32]   Y. Xiong, S. Chen, F. Ye, L. Su, C. Zhang, S. Shen, and S. Zhao, "Preparation of magnetic core–shell nanoflower Fe3O4@MnO2 as reusable oxidase mimetics for colorimetric detection of phenol", *Anal. Methods*, vol. 7, no. 4, pp. 1300-1306, 2015.
[http://dx.doi.org/10.1039/C4AY02687J]

[33] K.B. Narayanan, N. Sakthivel, and S.S. Han, "From Chemistry to Biology: Applications and Advantages of Green, Biosynthesized/Biofabricated Metal-and Carbon-based Nanoparticles", *Fibers Polym.,* no. 4, pp. 877-897, 2020.
[http://dx.doi.org/10.1007/s12221-021-0595-8]

[34] X. Hu, J. Yang, C. Yang, and J. Zhang, "UV/H2O2 degradation of 4-aminoantipyrine: a voltammetric study", *Chem. Eng. J.,* vol. 161, no. 1-2, pp. 68-72, 2010.
[http://dx.doi.org/10.1016/j.cej.2010.04.025]

[35] S. Sakthinathan, S. Palanisamy, S.M. Chen, P.S. Wu, L. Yao, and B.S. Lou, "Electrochemical detection of phenol in industrial pollutant absorbed molecular sieves by electrochemically activated screen printed carbon electrode", *Int. J. Electrochem. Sci.,* vol. 10, pp. 3319-3328, 2015.

[36] M. Sakthivel, S. Ramaraj, S.M. Chen, and B. Dinesh, "Synthesis of rose like structured LaCoO3 assisted functionalized carbon nanofiber nanocomposite for efficient electrochemical detection of anti-inflammatory drug 4-aminoantipyrine", *Electrochim. Acta,* vol. 260, pp. 571-581, 2018.
[http://dx.doi.org/10.1016/j.electacta.2017.11.122]

[37] G. Gopu, P. Manisankar, B. Muralidharan, and C. Vedhi, "Stripping voltammetric determination of analgesics in their pharmaceuticals using nano-riboflavin-modified glassy carbon electrode", *Int. J. Electrochem.,* vol. •••, p. 2011, 2011.

[38] T.R. Paixão, R. Camargo Matos, and M. Bertotti, "Diffusion layer titration of dipyrone in pharmaceuticals at a dual-band electrochemical cell", *Talanta,* vol. 61, no. 5, pp. 725-732, 2003.
[http://dx.doi.org/10.1016/S0039-9140(03)00334-5] [PMID: 18969237]

[39] J.I. Gowda, A.T. Buddanavar, and S.T. Nandibewoor, "Fabrication of multiwalled carbon nanotube-surfactant modified sensor for the direct determination of toxic drug 4-aminoantipyrine", *J. Pharm. Anal.,* vol. 5, no. 4, pp. 231-238, 2015.
[http://dx.doi.org/10.1016/j.jpha.2015.01.001] [PMID: 29403936]

[40] C. Chen, Y. Yao, W. Wang, L. Duan, W. Zhang, and J. Qian, "Selective bioimaging of cancer cells and detection of HSA with indomethacin-based fluorescent probes", *Spectrochim. Acta A Mol. Biomol. Spectrosc.,* vol. 241, p. 118685, 2020.
[http://dx.doi.org/10.1016/j.saa.2020.118685] [PMID: 32653821]

[41] J. Wang, and S. Wang, "Removal of pharmaceuticals and personal care products (PPCPs) from wastewater: A review", *J. Environ. Manage.,* vol. 182, pp. 620-640, 2016.
[http://dx.doi.org/10.1016/j.jenvman.2016.07.049] [PMID: 27552641]

[42] L. Yang, Y. Luo, L. Yang, S. Luo, X. Luo, W. Dai, T. Li, and Y. Luo, "Enhanced photocatalytic activity of hierarchical titanium dioxide microspheres with combining carbon nanotubes as "e-bridge"", *J. Hazard. Mater.,* vol. 367, pp. 550-558, 2019.
[http://dx.doi.org/10.1016/j.jhazmat.2019.01.016] [PMID: 30641425]

[43] L. Yang, Z. Chen, D. Cui, X. Luo, B. Liang, L. Yang, T. Liu, A. Wang, and S. Luo, "Ultrafine palladium nanoparticles supported on 3D self-supported Ni foam for cathodic dechlorination of florfenicol", *Chem. Eng. J.,* vol. 359, pp. 894-901, 2019.
[http://dx.doi.org/10.1016/j.cej.2018.11.099]

[44] A. Eslami, M.M. Amini, A.R. Yazdanbakhsh, N. Rastkari, A. Mohseni-Bandpei, E. Piroti, and A. Asadi, "Occurrence of non-steroidal anti-inflammatory drugs in Tehran source water, municipal and hospital wastewaters, and their ecotoxicological risk assessment", *Environ. Monit. Assess.,* vol. 187, no. 12, p. 734, 2015.
[http://dx.doi.org/10.1007/s10661-015-4952-1] [PMID: 26553436]

[45] N.H. Tran, and K.Y. Gin, "Occurrence and removal of pharmaceuticals, hormones, personal care products, and endocrine disrupters in a full-scale water reclamation plant", *Sci. Total Environ.,* vol. 599-600, pp. 1503-1516, 2017.
[http://dx.doi.org/10.1016/j.scitotenv.2017.05.097] [PMID: 28531959]

[46]   EA Lusth, "Therapeutic Atrial Natriuretic Peptide infusion in Acute Kidney Injury after surgery for Pediatric Congenital Heart Disease",

[47]   D.T. Selewski, J.R. Charlton, J.G. Jetton, R. Guillet, M.J. Mhanna, D.J. Askenazi, and A.L. Kent, "Neonatal acute kidney injury", *Pediatrics,* vol. 136, no. 2, pp. e463-e473, 2015.
[http://dx.doi.org/10.1542/peds.2014-3819] [PMID: 26169430]

[48]   A.L. Kent, L.E. Maxwell, M.E. Koina, M.C. Falk, D. Willenborg, and J.E. Dahlstrom, "Renal glomeruli and tubular injury following indomethacin, ibuprofen, and gentamicin exposure in a neonatal rat model", *Pediatr. Res.,* vol. 62, no. 3, pp. 307-312, 2007.
[http://dx.doi.org/10.1203/PDR.0b013e318123f6e3] [PMID: 17622959]

[49]   M.F. Schreuder, R.R. Bueters, M.C. Huigen, F.G. Russel, R. Masereeuw, and L.P. van den Heuvel, "Effect of drugs on renal development", *Clin. J. Am. Soc. Nephrol.,* vol. 6, no. 1, pp. 212-217, 2011.
[http://dx.doi.org/10.2215/CJN.04740510] [PMID: 21071516]

[50]   P. Purcell, D. Henry, and G. Melville, "Diclofenac hepatitis", *Gut,* vol. 32, no. 11, pp. 1381-1385, 1991.
[http://dx.doi.org/10.1136/gut.32.11.1381] [PMID: 1752473]

[51]   A. Radi, "Preconcentration and voltammetric determination of indomethacin at carbon paste electrodes", *Electroanalysis,* vol. 10, no. 2, pp. 103-106, 1998.
[http://dx.doi.org/10.1002/(SICI)1521-4109(199802)10:2<103::AID-ELAN103>3.0.CO;2-R]

[52]   Y. Liu, Z. Zhang, C. Zhang, W. Huang, C. Liang, and J. Peng, "Manganese dioxide□graphene nanocomposite film modified electrode as a sensitive voltammetric sensor of indomethacin detection", *Bull. Korean Chem. Soc.,* vol. 37, no. 8, pp. 1173-1179, 2016.
[http://dx.doi.org/10.1002/bkcs.10815]

[53]   M.R. Baezzat, F. Banavand, and S. Kamran Hakkani, Sensitive Voltammetric Detection of Indomethacin Using TiO$_2$ Nanoparticle Modified Carbon Ionic Liquid Electrode., *Anal. Bioanal. Chem. Res.,* vol. 7, no. 1, pp. 89-98, 2020.

[54]   H. Heli, A. Jabbari, S. Majdi, M. Mahjoub, A.A. Moosavi-Movahedi, and S. Sheibani, "Electrooxidation and determination of some non-steroidal anti-inflammatory drugs on nanoparticles of Ni–curcumin-complex-modified electrode", *J. Solid State Electrochem.,* vol. 13, no. 12, p. 1951, 2009.
[http://dx.doi.org/10.1007/s10008-008-0758-1]

[55]   BB Petković, M Ognjanović, M Krstić, V Stanković, L Babincev, M Pergal, and DM Stanković, "Boron-doped diamond electrode as efficient sensing platform for simultaneous quantification of mefenamic acid and indomethacin", *Diamond Relat. Mater,* 2020.

[56]   K. Sarhangzadeh, A.A. Khatami, M. Jabbari, and S. Bahari, "Simultaneous determination of diclofenac and indomethacin using a sensitive electrochemical sensor based on multiwalled carbon nanotube and ionic liquid nanocomposite", *J. Appl. Electrochem.,* vol. 43, no. 12, pp. 1217-1224, 2013.
[http://dx.doi.org/10.1007/s10800-013-0609-3]

[57]   M. Arvand, and T.M. Gholizadeh, "Gold nanorods–graphene oxide nanocomposite incorporated carbon nanotube paste modified glassy carbon electrode for voltammetric determination of indomethacin", *Sens. Actuators B Chem.,* vol. 186, pp. 622-632, 2013.
[http://dx.doi.org/10.1016/j.snb.2013.06.059]

[58]   T. Hallaj, M. Amjadi, J.L. Manzoori, and N. Azizi, "A novel chemiluminescence sensor for the determination of indomethacin based on sulfur and nitrogen co-doped carbon quantum dot-KMnO4 reaction", *Luminescence,* vol. 32, no. 7, pp. 1174-1179, 2017.
[http://dx.doi.org/10.1002/bio.3306] [PMID: 28524362]

[59]   M. Hassannezhad, M. Hosseini, M.R. Ganjali, and M. Arvand, "Electrochemical Sensor Based on Carbon Nanotubes Decorated with ZnFe2O4 Nanoparticles Incorporated Carbon Paste Electrode for

Determination of Metoclopramide and Indomethacin", *ChemistrySelect,* vol. 4, no. 25, pp. 7616-7626, 2019.
[http://dx.doi.org/10.1002/slct.201900959]

[60]   A.M. Walker, "Quantitative studies of the risk of serious hepatic injury in persons using nonsteroidal antiinflammatory drugs", *Arthritis Rheum.,* vol. 40, no. 2, pp. 201-208, 1997.
[http://dx.doi.org/10.1002/art.1780400204] [PMID: 9041931]

[61]   C.K. Ong, P. Lirk, C.H. Tan, and R.A. Seymour, "An evidence-based update on nonsteroidal anti-inflammatory drugs", *Clin. Med. Res.,* vol. 5, no. 1, pp. 19-34, 2007.
[http://dx.doi.org/10.3121/cmr.2007.698] [PMID: 17456832]

[62]   P.C. Gøtzsche, "Non-steroidal anti-inflammatory drugs", *BMJ,* vol. 320, no. 7241, pp. 1058-1061, 2000.
[http://dx.doi.org/10.1136/bmj.320.7241.1058] [PMID: 10764369]

[63]   M.J. Kao, T.I. Han, T.S. Kuan, Y.L. Hsieh, B.H. Su, and C.Z. Hong, "Myofascial trigger points in early life", *Arch. Phys. Med. Rehabil.,* vol. 88, no. 2, pp. 251-254, 2007.
[http://dx.doi.org/10.1016/j.apmr.2006.11.004] [PMID: 17270525]

[64]   L. Mason, R.A. Moore, J.E. Edwards, S. Derry, and H.J. McQuay, "Topical NSAIDs for acute pain: a meta-analysis", *BMC Fam. Pract.,* vol. 5, no. 1, p. 10, 2004.
[http://dx.doi.org/10.1186/1471-2296-5-10] [PMID: 15147585]

[65]   L. Mason, R.A. Moore, J.E. Edwards, S. Derry, and H.J. McQuay, "Topical NSAIDs for chronic musculoskeletal pain: systematic review and meta-analysis", *BMC Musculoskelet. Disord.,* vol. 5, no. 1, p. 28, 2004.
[http://dx.doi.org/10.1186/1471-2474-5-28] [PMID: 15317652]

[66]   R.A. Moore, M.R. Tramer, D. Carroll, P.J. Wiffen, and H.J. McQuay, "Quantitive systematic review", *BMJ,* vol. 316, no. 7128, pp. 333-338, 1998.
[http://dx.doi.org/10.1136/bmj.316.7128.333] [PMID: 9487165]

[67]   L. Di Rienzo Businco, A. Di Rienzo Businco, M. D'Emilia, M. Lauriello, and G. Coen Tirelli, "Topical versus systemic diclofenac in the treatment of temporo-mandibular joint dysfunction symptoms", *Acta Otorhinolaryngol. Ital.,* vol. 24, no. 5, pp. 279-283, 2004.
[PMID: 15871609]

[68]   G. Affaitati, F. Vecchiet, R. Lerza, S. De Laurentis, S. Iezzi, F. Festa, and M.A. Giamberardino, "Effects of topical diclofenac (DHEP plaster) on skin, subcutis and muscle pain thresholds in subjects without spontaneous pain", *Drugs Exp. Clin. Res.,* vol. 27, no. 2, pp. 69-76, 2001.
[PMID: 11392056]

[69]   R. Wolski, M. Półtorak, I. Rykowska, S. Kaczmarek, and P. Andrzejewski, Destruction of Selected Pharmaceuticals with Peroxydisulfate (PDS): An Influence of PDS Activation Methods.*Frontiers in Water-Energy-Nexus—Nature-Based Solutions, Advanced Technologies and Best Practices for Environmental Sustainability.* Springer: Cham, 2020, pp. 203-205.
[http://dx.doi.org/10.1007/978-3-030-13068-8_50]

[70]   C. Slim, N. Tlili, C. Richard, S. Griveau, and F. Bedioui, "Amperometric detection of diclofenac at a nano-structured multi-wall carbon nanotubes sensing films", *Inorg. Chem. Commun.,* vol. 107, p. 107454, 2019.
[http://dx.doi.org/10.1016/j.inoche.2019.107454]

[71]   G.S. Kassahun, S. Griveau, S. Juillard, J. Champavert, A. Ringuedé, B. Bresson, Y. Tran, F. Bedioui, and C. Slim, "Hydrogel Matrix-Grafted Impedimetric Aptasensors for the Detection of Diclofenac", *Langmuir,* vol. 36, no. 4, pp. 827-836, 2020.
[http://dx.doi.org/10.1021/acs.langmuir.9b02031] [PMID: 31910020]

[72]   E.M. Almbrok, N.A. Yusof, J. Abdullah, and R.M. Zawawi, "Electrochemical Behavior and Detection of Diclofenac at a Microporous Si3N4 Membrane Modified Water–1, 6- dichlorohexane Interface System", *Chemosensors (Basel),* vol. 8, no. 1, p. 11, 2020.

[http://dx.doi.org/10.3390/chemosensors8010011]

[73] D. Deng, H. Yang, C. Liu, K. Zhao, J. Li, and A. Deng, "Ultrasensitive detection of diclofenac in water samples by a novel surface-enhanced Raman scattering (SERS)-based immunochromatographic assay using AgMBA@ SiO2-Ab as immunoprobe", *Sens. Actuators B Chem.*, vol. 283, pp. 563-570, 2019.
[http://dx.doi.org/10.1016/j.snb.2018.12.076]

[74] P.K. Kalambate, J. Noiphung, N. Rodthongkum, N. Larpant, P. Thirabowonkitphithan, T. Rojanarata, M. Hasan, Y. Huang, and W. Laiwattanapaisal, "Nanomaterials-based Electrochemical Sensors and Biosensors for the Detection of Non-Steroidal anti-inflammatory Drugs", *Trends Analyt Chem TRAC-TREND ANAL CHEM.*, no. Oct, p. 116403, 2020.

[75] Y. Zhang, Z. Zhang, G. Qi, Y. Sun, Y. Wei, and H. Ma, "Detection of indomethacin by high-performance liquid chromatography with in situ electrogenerated Mn(III) chemiluminescence detection", *Anal. Chim. Acta*, vol. 582, no. 2, pp. 229-234, 2007.
[http://dx.doi.org/10.1016/j.aca.2006.09.028] [PMID: 17386497]

[76] L Yang, L Li, F Li, H Zheng, T Li, X Liu, J Zhu, Y Zhou, and S Alwarappan, "Ultrasensitive photoelectrochemical aptasensor for diclofenac sodium based on surface-modified TiO 2- FeVO 4 composite. Anal", *Bioanal. Chem. Res,* 2020.

[77] K. Kimuam, N. Rodthongkum, N. Ngamrojanavanich, O. Chailapakul, and N. Ruecha, "Single step preparation of platinum nanoflowers/reduced graphene oxide electrode as a novel platform for diclofenac sensor", *Microchem. J.*, vol. 155, p. 104744, 2020.
[http://dx.doi.org/10.1016/j.microc.2020.104744]

[78] R. Jain, and S. Shrivastava, "A graphene-polyaniline-Bi2O3 hybrid film sensor for voltammetric quantification of anti-inflammatory drug etodolac", *J. Electrochem. Soc.*, vol. 161, no. 4, p. 89, 2014.
[http://dx.doi.org/10.1149/2.043404jes]

[79] A. Mokhtari, H. Karimi-Maleh, A.A. Ensafi, and H. Beitollahi, "Application of modified multiwall carbon nanotubes paste electrode for simultaneous voltammetric determination of morphine and diclofenac in biological and pharmaceutical samples", *Sens. Actuators B Chem.*, vol. 169, pp. 96-105, 2012.
[http://dx.doi.org/10.1016/j.snb.2012.03.059]

[80] L. Kashefi-Kheyrabadi, and M.A. Mehrgardi, "Design and construction of a label free aptasensor for electrochemical detection of sodium diclofenac", *Biosens. Bioelectron.*, vol. 33, no. 1, pp. 184-189, 2012.
[http://dx.doi.org/10.1016/j.bios.2011.12.050] [PMID: 22265876]

[81] S. Cheraghi, M.A. Taher, H. Karimi-Maleh, and R. Moradi, "Simultaneous detection of nalbuphine and diclofenac as important analgesic drugs in biological and pharmaceutical samples using a Pt: Co nanostructure-based electrochemical sensor", *J. Electrochem. Soc.*, vol. 164, no. 2, p. B60, 2016.
[http://dx.doi.org/10.1149/2.0911702jes]

[82] M. Arvand, T.M. Gholizadeh, and M.A. Zanjanchi, "MWCNTs/Cu(OH)2 nanoparticles/IL nanocomposite modified glassy carbon electrode as a voltammetric sensor for determination of the non-steroidal anti-inflammatory drug diclofenac", *Mater. Sci. Eng. C,* vol. 32, no. 6, pp. 1682-1689, 2012.
[http://dx.doi.org/10.1016/j.msec.2012.04.066] [PMID: 24364977]

[83]   B.K. Chethana, S. Basavanna, and Y. Arthoba Naik, "Voltammetric determination of diclofenac sodium using tyrosine-modified carbon paste electrode", *Ind. Eng. Chem. Res.,* vol. 51, no. 31, pp. 10287-10295, 2012.
[http://dx.doi.org/10.1021/ie202921e]

CHAPTER 11

Voltammetric Applications in Drug Detection: Mini Review

G. Krishnaswamy[1,*], G. Shivaraja[1], S. Sreenivasa[2, 3,*] and Aruna Kumar D. B.[1]

[1] *Department of Studies and Research in Organic Chemistry, Tumkur University, Tumakuru-572 103, Karnataka, India*

[2] *Department of Studies and Research in Chemistry, University College of Science, Tumkur University, Tumakuru-572 103, Karnataka, India*

[3] *Deputy Adviser, National Assessment and Accreditation Council, Bengaluru-560 072, Karnataka, India*

Abstract: The application of electroanalytical techniques plays a very important role in the field of medicinal and pharmaceutical research. These techniques have been routinely used by analytical, inorganic, organic physical as well as biological chemists. In recent years the various voltammetric methods such as Differential pulse voltammetry, Cyclic voltammetry, Linear sweep voltammetry, Square wave voltammetry, Stripping voltammetry *etc* have gained enormous attention for the determination of different classes of drugs. Electroanalytical techniques gain supreme importance over other techniques in terms of the quantization of trace and ultra-trace components of biological fluids as well as pharmaceutical substances. Therefore, these techniques are alternative in detection and analysis due to their sensitivity, reproducibility and selectivity in terms of qualitative as well as quantitative aspects of the analyte under investigation. The present review focuses on the application of various voltammetric techniques for the analysis of analgesics, antibiotics, anthelmintic, anti-tubercular and anticancer drugs covering the period from 2015 to present.

Keywords: Analgesics, Anthelmintics, Antibiotic, Anticancer, Antitubercular, Cyclic voltammetry, Drug detection, Electroanalytical techniques.

INTRODUCTION

Voltammetry technique was developed in 1920s by Jaroslav Heyrovsky a Czechoslovakian chemist and in that year, an important electro analytical techni-

---

* **Corresponding authors G. Krishnaswamy and S. Sreenivasa:** Department of Studies and Research in Organic Chemistry, Tumkur University, Tumakuru-572 103, Karnataka, India and Deputy Adviser, National Assessment and Accreditation Council, Bengaluru-560 072, Karnataka, India; E-mail:drkrishna23org@gmail.com, drsreenivasa@yahoo.co.in

J.G. Manjunatha (Ed.)

que for the analysis of different pharmaceutical substances was developed [1]. Generally, Voltammetry technique is widely used because it provides both qualitative as well as quantitative information of electroactive sample based on the applied electric potential and current, respectively. Simple modified electrodes have been used as sensors for the electrochemical characterization of various bioactive moleculeslike riboflavin [2 - 5], Tyrosine [6], Dopamine [7], Uric acid, Ascorbic acid [8 - 10], Estriol [11, 12], Isatin [13], Phloroglucinol [14, 15], L-tryptophan [16], Vanillin [17] and also dyes such as Alizarin Red-S [18] and Indigotine [19].

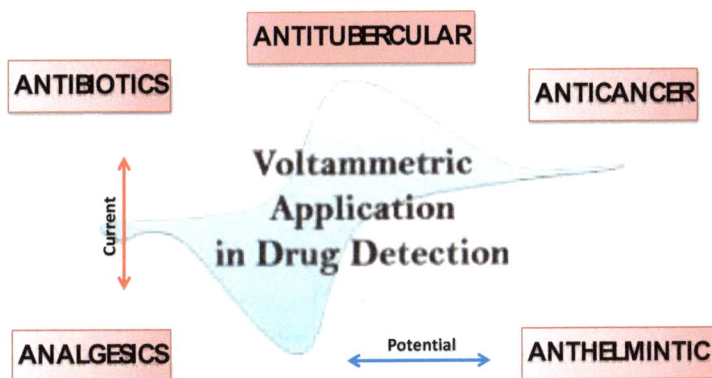

Voltammetry Techniques such as:

- Cyclic Voltammetry (CV) and Linear Sweep Voltammetry (LSV) were used to study the kinetics of electrochemical reactions, reversible and irreversible reactions as well as determination of formal redox potentials of the analyte since these techniques provide linear continuous changing potential with time.
- In Differential Pulse Voltammetry (DPV), the cell current is measured as a function of time and potential between the indicator and reference electrodes. DPV improves detection limits and is highly sensitive for the quantitative determination of electroactive drug compounds.
- Square Wave Voltammetry (SWV) is a large amplitude differential technique in which a waveform composed of symmetrical square wave compared with DPV. It requires a small amount of electroactive species and a very fast response is obtained due to an effective high scan rate.
- Stripping Voltammetry (SV) can be called as a sub division of voltammetric techniques and classified into four types namely, anodic (ASV), Cathodic (CSV), Adsorptive (AdSV), and Potentiometric Stripping Voltammetry (PSV). This technique uses a step of preconcentration to accumulate the analytes on the electrode surfaces for dilute samples, followed by the measurement of electroch-

emical response leading to trace level determination of pharmaceutical compounds.

Inspired by the important applications of the various voltammetric techniques. Here in the following section, we present the electrochemical characterization studies of antibiotics, analgesics, anthelmintic, anti-tuberculosis and anti-cancer drugs highlighting mainly electrode material, techniques employed and detection limit.

ELECTROCHEMICAL CHARACTERIZATION OF ANTIBIOTICS

Penicillin is a group of antibiotics, which are used to treat different bacterial infections in humans, livestock, and poultry.

Zhao *et al.* [20] in 2016 reported an electrochemical APTA sensor for highly sensitive detection of penicillin at GNS nanocomposite (GR-Fe$_3$O$_4$ NPs) and a poly (3, 4-ethylenedioxy-thiophene)-AuNP composite (PEDOT-AuNPs) using modified glassy carbon electrode (GCE) employing differential pulse voltammetry (DPV) method with a limit of detection (LOD) of 1.70×10^{-16} M. Li and co-workers [21] in 2015 reported an immuno sensor for detection of penicillin based on supported bilayer lipid membrane (s-BLM)-AuNP modified glassy carbon electrode (Au-sBLM/GCE) employing Cyclic Voltammetry (CV) with a limit of detection (LOD) of 8.07×10^{-16} M. Salihu *et al.* [22] in 2019 reported Nickel nanoparticle (Ni-NP) modified screen-printed carbon electrode (SPCE) for the electrochemical detection of penicillin in bovine milk samples with a limit of detection (LOD) of 0.00031 μM. Wizral *et al.* [23] in 2019 reported boron doped diamond electrode for determination of penicillin G in aquatic samples using differential pulse volatmmetric (DPV) method with detection limit of (LOD) 0.23 μM. The limit of detection for penicillin G detection using various voltammetric methods were compared in Table **1**.

Table 1. Comparison of different electrochemical sensors for voltammetric determination of pencillin G.

Chemical Structure		
Electrode	**Method of Analysis**	**LOD (M)**	**Refs.**
GR-Fe$_3$O$_4$ NPs/ PEDOT-AuNPs/ GCE	DPV	1.70×10^{-16}	[20]

(Table 1) cont.....

Electrode	Method of Analysis	LOD (M)	Refs.
Au-sBLM/GCE	CV	8.07×10^{-16}	[21]
Ni-NP/ SPCE	CV	$3.1x\ 10^{-10}$	[22]
Boron doped diamond	DPV	$2.3x10^{-7}$	[23]

Amoxicillin is a β-lactams antibiotic widely used to prevent and treat various bacterial infections in humans and also in the field of agriculture [24].

Hatamie and co-workers [25] in 2015 reported the sensor based on the fabrication of zinc oxide nanorods on a gold/glass electrode. The proposed sensor was found to be highly constructive and suitable for the voltammetric behavior of amoxicillin. Swave voltammetry (SWV) studies revealed a limit of detection (LOD) of 9.2×10^{-7} M. The electrochemistry of amoxicillin was investigated by the author Ağin [26] in 2016 using glassy carbon electrodes (GCEs) modified by electropolymerization of acridine orange and differential pulse voltammetry (DPV) study revealed a limit of detection (LOD) of 1.87×10^{-9} M. Nosuhi and Ejhieh [27] in 2017 reported a sensitive method for the voltammetric determination of amoxicillin on a modified carbon paste electrode using copper-doped nano-clinoptilolite as a modifier and adsorptive stripping square wave voltammetry (AdSSWV) study revealed a limit of detection (LOD) of 3.5×10^{-10} M. Valenga *et al.* [28] in 2020 developed and validated a voltammetric method for determination of amoxicillin in river water using a glassy carbon electrode modified with reduced grapheme oxide (rGO) and Nafion by square-wave voltammetry (SWV) with a limit of detection (LOD) 0.36 μmol L^{-1}. Yen *et al.* [29] in 2020 developed economical and simple graphite pencil electrode (GPE) for the determination of electrochemical properties of amoxicillin by square wave voltammetric method with limit of detection of (LOD) 0.224 μM. Wong *et al.* [30] in 2020 developed a electrochemical sensor composed of nanometrials printex 6L carbon and cadmium telluride quantum dots within a poly(3,4-ethylenedioxythiophene) polystyrene sulfonate film (QDs-P6LC-PEDOT: PSS) modified glassy carbon electrode (GCE) for the determination of amoxicillin in different matrices such as pharmaceutical products (commercial tablets), clinical samples (urine) and also to examine food quality (milk samples) by square wave voltammetric technique with a limit of detection (LOD) 0.05 μmol L^{-1}. From the above literature reports, graphite pencil electrode (GPE) using SWV technique was better for the detection of amoxicillin based on the limit of detection compared to other methods (Table **2**).

Table 2. Comparison of different electrochemical sensors for voltammetric determination of Amoxicilin.

Chemical Structure			
Electrode	**Method of Analysis**	**LOD (M)**	**Refs.**
ZnO NPs/ GGE	SWV	9.2×10^{-7}	[25]
Acridine/GCE	DPV	1.87×10^{-9}	[26]
Cu-NP-clinoptilolite/CPE	AdSSWV	3.5×10^{-10}	[27]
rGO/Nafion/GCE	SWV	3.6×10^{-7}	[28]
GPE	SWV	2.24×10^{-7}	[29]
QDs-P6LC-PEDOT: PSS/GCE	SWV	5×10^{-8}	[30]

General mechanism for the electrochemical oxidation of the amoxicilin is depicted in Fig. (**1**) and phenolic ring of the amoxicilin is responsible for the observed electrochemical oxidation reaction.

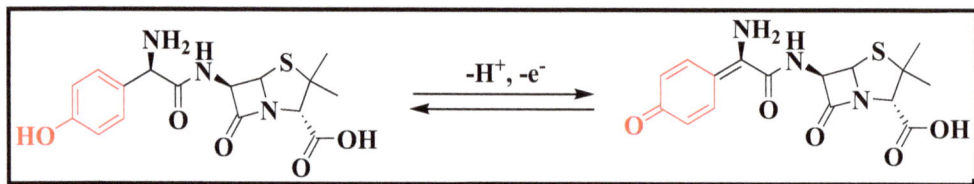

Fig. (1). Possible general mechanism for electrochemical oxidation of Amoxacilin.

Ciprofloxacin belongs to the second generation of fluoroquinolones and is used worldwide as an antimicrobial agent to treat infections in humans as well as livestock [31, 32].

Shan *et al*. [33] in 2015 reported a well-defined electrochemical method using graphene modified GCEs for determination of ciprofloxacin with good reproducibility and high selectivity with a detection limit of (LOD) 5.9×10^{-8} M. Faria *et al*. [34] in 2019 developed chemical reduction of graphene oxide modified glassy carbon electrode (rGO-GCE) for the determination of ciprofloxacin in pharmaceutical formulations and milk samples using SWV technique with a detection limit of (LOD) 0.21×10^{-6} M. Hosseini and co-workers [35] in 2019 developed a glassy carbon electrode modified with $CoFe_2O_4/$

MWCNT heterostructure for the determination of ciprofloxacin using differential pulse voltammetric technique with a limit of detection (LOD) of 0.036 μM. Comparison of the above methods with LOD is tabulated in Table **3**.

Table 3.   Comparison of different electrochemical sensors for voltammetric determination of Ciprofloxacin.

Chemical Structure			
Electrode	**Method of Analysis**	**LOD (M)**	**Refs.**
Graphene/ GCE	CV	5.9×10^{-8}	[33]
CRGO-GCE	SWV	0.21×10^{-6}	[34]
CoFe$_2$O$_4$/MWCNT /GCE	DPV	3.6×10^{-8}	[35]

The piperazine ring portion of Ciprofloxacin undergoes electrochemical oxidation and possible general mechanism is shown in Fig. (**2**).

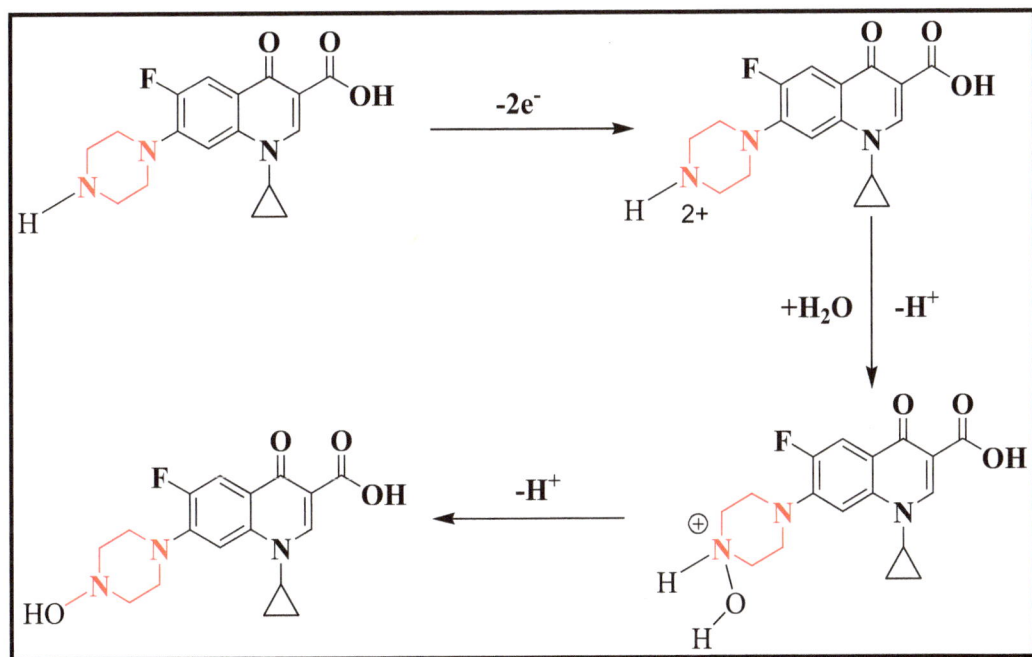

Fig. (2). Possible general mechanism for electrochemical oxidation of Ciprofloxacin.

Erythromycin is a macrolide antibiotic used for the treatment of bacterial infections in human and poultry chicks.

Voltammetric determination of erythromycin was reported by Vajdle *et al.* [36] in 2016 using a renewable silver amalgam film electrode (Ag/Hg FE) by square wave voltammetry (SWV) technique. This electrochemical study revealed that it has selectivity with a detection limit of (LOD) 1.85×10^{-6} M. Radicova *et al.* [37] in 2015 successfully carried out the determination of erythromycin in water sample on boron doped diamond electrode (BDDE) employing SWV technique with a detection limit of (LOD) 2.86×10^{-4} M. Veseli *et al.* [38] in 2019 developed on sensor-based, surface-modified, screen-printed carbon electrodes (SPCEs) which are modified by sodium dodecyl sulphate (SDS) for selective and sensitive determination of erythromycin in water samples in flow injection analysis (FIA) mode with a limit of detection of (LOD) 0.19×10^{-9} M. A Comparison of the methods with LOD is given in Table **4**.

Table 4. Comparison of different electrochemical sensors for voltammetric determination of Erythromycin.

Chemical Structure			
Electrode	**Method of Analysis**	**LOD (M)**	**Refs.**
Ag/Hg FE	SWV	1.85×10^{-6}	[36]
BDDE	SWV	2.86×10^{-4}	[37]
SDS/ SPCEs	FIA	0.19×10^{-9}	[38]

Levofloxacin is an antibacterial agent mainly used for the treatment of bacterial infections and bronchitis or pneumonia.

Rkik. M *et al.* [39] reported the determination of levofloxacin on boron-doped diamond electrode (BDDE) employing square wave voltammetry (SWV) technique with a detection limit of (LOD) 2.8×10^{-6} M. Voltammetric behaviour of levofloxacin was studied by Huang co-workers [40] on poly(o-aminophenol) and GNS quantum dots (PoAP/GQD) modified glassy carbon electrode by employing differential pulse voltammetry (DPV) technique. The detection limit for the experiment was found to be (LOD) 1.0×10^{-8} M. Ghanbari *et al.* [41] in 2019 reported reduced graphene oxide (rGO) and nanocomposite prepared from polymerized β-cyclodextrin (β-CD) modified glassy carbon electrode (GCE) for the determination of levofloxacin with a detection limit of (LOD) 30 pmol L^{-1}. Farias *et al.* [42] in 2020 reported the determination of levofloxacin in pharmaceutical formulations and urine at reduced graphene oxide and carbon nanotube modified electrode employing differential pulse voltammetry (DPV), square wave voltammetry (SWV) and flow injection analysis with amperometric detection (FIA-AMP) methods with a limit of detection (LOD) of 1.45, 6.70 and 1.90 μmol L^{-1}, respectively. A comparative table of the methods with LOD is given in Table **5**.

Table 5.   **Comparison of different electrochemical sensors for voltammetric determination of Levofloxacin.**

Chemical Structure			
Electrode	**Method of Analysis**	**LOD (M)**	**Refs.**
BDDE	SWV	2.8×10^{-6}	[39]
PoAP/GQD/GCE	DPV	1.0×10^{-8}	[40]
rGO- β-CD/ GCE	CV	3×10^{-11}	[41]
rGO/ GCE	DPV SWV FIA-AMP	1.45 6.70 1.90	[42]

Levofloxacin undergoes electrochemical oxidation at N-methyl piperazine ring portion as depicted in Fig. (**3**).

Ofloxacin is an antibiotic used for the treatment of bacterial infections of the urinary tract, lungs and skin in human beings.

Its determination has been reported by Jiang *et al* [43] in 2017 using rGO/Pt/Au modified glassy carbon electrode (GCE) employing differential pulse voltammetric (DPV) technique with a limit of detection of (LOD) 5.0×10^{-8} M. Wong *et al.* [44] in 2016 reported the determination of ofloxacin using GNS oxide and ionic liquid (1-butyl-3-methyl imidazolium tetrafluroborate) modified carbon paste electrode (CPE) employing adsorptive stripping square wave voltammetric

(AdSSWV) technique with a detection limit of (LOD) 2.8×10^{-10} M. Feng *et al.* [45] in 2020 reported voltammetric determination of ofloxacin in water samples using laser modified carbon glassy electrode (LGCE), employing differential pulse voltammetry (DPV) with a limit of detection of (LOD) 75 nM. Comparison of different electrochemical sensors for ofloxacin is tabulated in Table **6**.

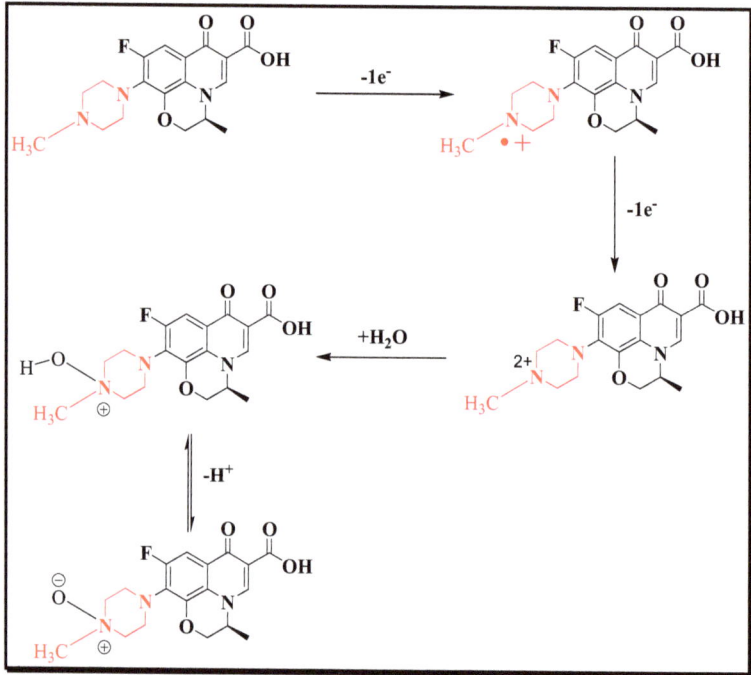

Fig. (3). Possible mechanism for electrochemical oxidation of Levofloxacin.

Table 6. Comparison of different electrochemical sensors for voltammetric determination of Ofloxacin.

Chemical Structure			
Electrode	**Method of Analysis**	**LOD (M)**	**Refs.**
rGO/Pt/Au/ GCE	DPV	5.0×10^{-8}	[43]
GNS oxide/ CPE	AdSSWV	2.8×10^{-10}	[44]
LGCE	DPV	7.5×10^{-8}	[45]

Fig. (**4**). depicts the possible general electrochemical oxidation of Ofloxacin along with portion of responsible for the observed electrochemical oxidation process.

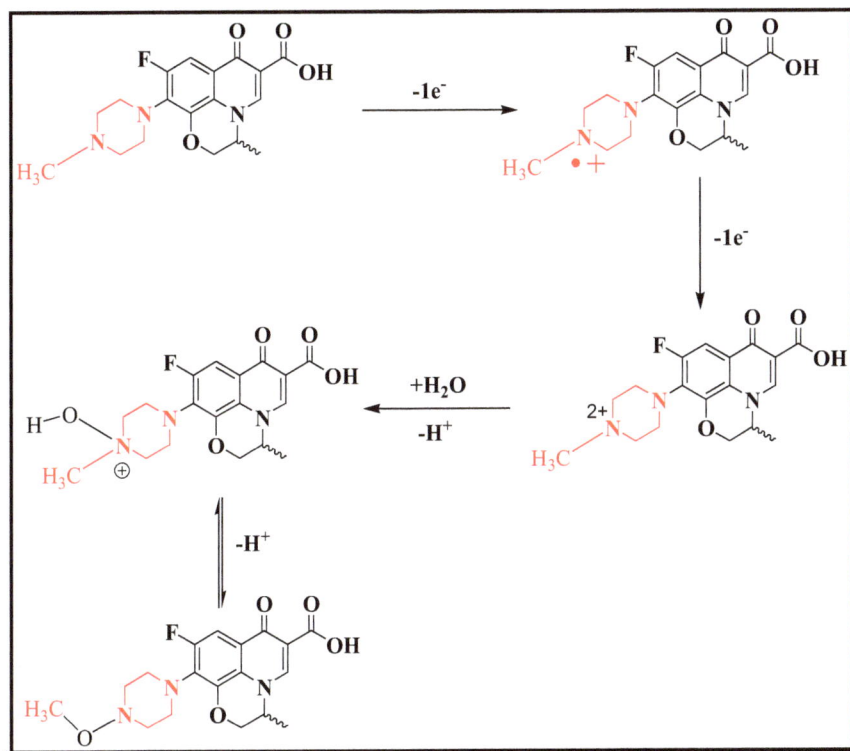

Fig. (4). Possible mechanism for electrochemical oxidation of Ofloxacin.

ELECTROCHEMICAL CHARACTERIZATION OF ANALGESICS

Paracetamol (Acetaminophen) is a widely used analgesic for relieving pain associated with headache, backache, arthritis and for reducing viral or bacterial fever.

Dai *et al*. [46] in 2016 reported the determination of paracetamol using Prussian blue (PB) and molecularly imprinted polymer (MIP) modified glassy carbon electrode (GCE) employing differential pulse voltammetric (DPV) technique with a detection limit of (LOD) 5.3×10^{-10}M. Kalambate *et al*. [47] in 2015 reported simultaneous determination of paracetamol and domperidone with a detection limits of 1.1×10^{-10} and 4.4×10^{-10} M, respectively using nafion-platinum nanoparticles and GNS modified glassy carbon electrode (GCE) employing adsorptive stripping square wave voltammetric (AdSSWV) technique. Amare *et al*. [48] in 2019 reported a simpler, relatively cheaper and more sensitive method of determination of paracetamol using anthraquinone modified carbon paste electrode (AQM/CPE) by square wave voltammetric (SWV) method with a limit of detection (LOD) of 0.13 μM. Table **7** compares the method of analysis and LOD of different methods for the detection of Paracetamol.

Table 7. **Comparison of different electrochemical sensors for voltammetric determination of Paracetamol.**

Chemical Structure			
Electrode	**Method of Analysis**	**LOD (M)**	**Refs.**
PB/MIP/GCE	DPV	5.3×10^{-10}	[46]
NAF/PtNP/GCE	AdSSWV	1.1×10^{-10}	[47]
AQM/CPE	SWV	1.3×10^{-7}	[48]

General electrochemical oxidation mechanism of Paracetamol is depicted in Fig. (**5**).

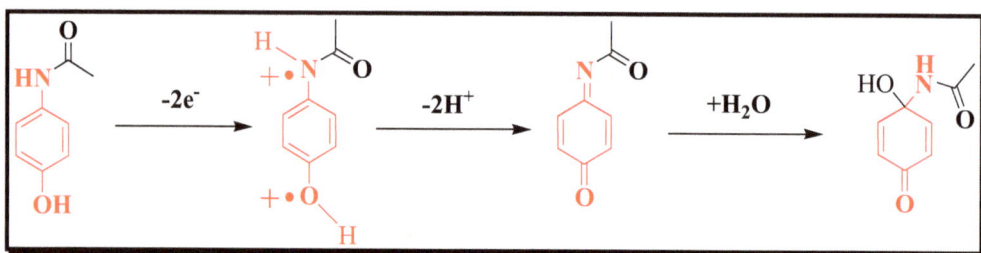

Fig (5). Possible mechanism for electrochemical oxidation of Paracetamol.

Aspirin (Acetylsalicylic acid) is a multi-quality pharmaceutical possessing anti-inflammatory, analgesic, antioxidant and antipyretic properties. Hence, widely used to cure headaches, fever and is also effective in Alzheimer's and Cardiovascular diseases.

Iacob *et al.* [49] in 2015 reported the determination of aspirin using multi-walled carbon nanotubes (MWCNTs) within epoxy composite employing differential pulse voltammetric (DPV) technique with 4.0×10^{-6} M limit of detection (LOD). Zhu *et al.* [50] in 2016 reported the voltammetric determination of aspirin using poly(diallyldimethylammonium chloride) functionalised on reduced GNS oxide modified glassy carbon electrode (GCE) employing differential pulse voltammetric (DPV) technique with a detection limit of (LOD) 1.2×10^{-6} M. Comparative method and LOD for determination of aspirin is tabulated in Table **8**.

Table 8. Comparison of different electrochemical sensors for voltammetric determination of Aspirin.

Chemical Structure			
Electrode	**Method of Analysis**	**LOD (M)**	**Refs.**
EP-MWCNTs	DPV	4.0×10^{-6}	[49]
PDDA-RGO/GCE	DPV	1.2×10^{-6}	[50]

Tramadol is a synthetic monoamine uptake inhibitor used to treat moderate to severe pain.

Fathirad and co-workers [51] in 2016 reported the voltammetric determination of tramadol using Pd nanoparticle loaded on vulcan carbon/conductive polymeric ionic liquid composite nanofibres modified glassy carbon electrode (GCE) employing square wave voltammetric technique with a detection limit of (LOD) 2.2×10^{-6} M. Deiminiat *et al.* [52] in 2017 reported the determination of Tramadol using MIP and functionalised multi-walled carbon nanotubes (f-MWCNTs) modified glassy carbon electrode (GCE) by square wave technique (SWV) techniques with detection limit of (LOD) 3.0×10^{-11} M. Mohamed *et al.* [53] in 2017 used graphene oxide/multi-walled carbon nanotubes (GO/MWCNT) modified carbon paste electrode (CPE) employing differential pulse voltammetric (DPV) technique with a detection limit of (LOD) 1.5×10^{-10} M. Esra *et al.* [54] in 2019 reported a novel composite for the determination of tramadol using antimony oxide nanoparticles (Sb_2O_3NPs) and multi walled carbon nanotubes (MWCNTs) glassy carbon electrode with a limit of detection (LOD) of 9.5×10^{-9} M employing cyclic voltammetric (CV) technique (Table **9**).

Table 9. Comparison of different electrochemical sensors for voltammetric determination of Tramadol.

Chemical Structure		
Electrode	**Method of Analysis**	**LOD (M)**	**Refs.**
PdVCPILNF/GCE	SWV	2.2×10^{-6}	[51]
MIP/f-MWCNTs/GCE	SWV	3.0×10^{-11}	[52]
GO/MWCNT/ CPE	DPV	1.5×10^{-10}	[53]
Sb$_2$O$_3$NPs/ MWCNTs/GCE	CV	9.5×10^{-9}	[54]

General electrochemical oxidation mechanism of Tramadol and the portion responsible for the observed oxidation process is depicted in Fig. (**6**).

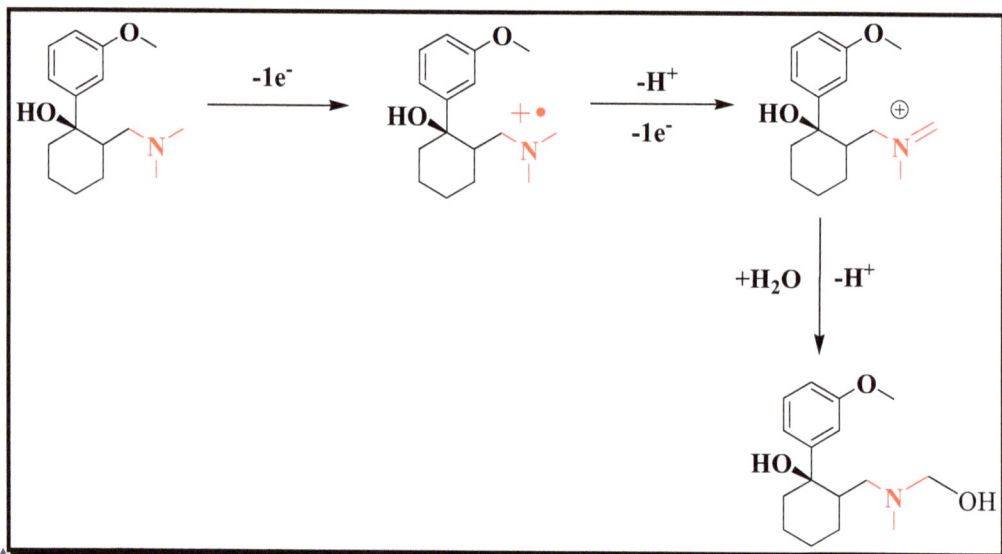

Fig. (6). Possible general mechanism for electrochemical oxidation of Tramadol.

Ibuprofen is a widely used anti-inflammatory and antipyretic agent used for the treatment of Rheumatoid arthritis, degenerative joint disease and other inflammatory rheumatic diseases.

Suresh and co-workers [55] in 2016 reported the determination of Ibuprofen with a detection limit of 9.7×10^{-7} M using multi-walled carbon nanotubes (MWCNTs) modified glassy carbon electrode (GCE) employing square wave voltammetric (SWV) technique. Hernandez *et al.* [56] in 2016 used multi-walled carbon nanotubes (MWCNTs) modified carbon paste electrode (CPE) for the determination of ibuprofen with a detection limit of (LOD) 9.1×10^{-6} M employing differential pulse voltammetric (DPV) method (Table **10**).

Table 10. Comparison of different electrochemical sensors for voltammetric determination of Ibupufen.

Chemical Structure			
Electrode	**Method of Analysis**	**LOD (M)**	**Refs.**
MWCNTs/ GCE	SWV	9.7×10^{-7}	[55]
MWCNTs/ CPE	DPV	9.1×10^{-6}	[56]

ELECTROCHEMICAL CHARACTERIZATION OF ANTHELMINTIC DRUGS

Albendazole is one of the most important and extensively used anthelmintic drug active against most of the helminths.

Srivastava and co-worker [57] in 2016 developed an eco-friendly electrochemical sensor for the determination of albendazole using Au electrode deposited with nanoparticles of biopolymer chitosan in the presence of albendazole template with a detection limit of (LOD) 1.0×10^{-8} M employing cyclic voltammetry technique (Table **11**).

Table 11. Electrochemical sensors for voltammetric determination of Albendazole.

Chemical Structure

Electrode	Method of Analysis	LOD (M)	Refs.
Au/Chitosan	CV	1.0×10^{-8}	[57]

Mebendazole is also an anthelmintic drug used to treat against ascaris, thread worms, hook worms and whip worms.

Munusamy *et al.* [58] in 2015 reported the differential pulse voltammetric (DPV) technique for the determination of mebendazole using gallium nitride poly (3,4-ethylenedioxythiophene)-co-polypyrrole (GaN/PEDOT-PPY) modified glassy carbon electrode (GCE) with a detection limit of (LOD) 2.4×10^{-6} M (Table **12**).

Table 12. Electrochemical sensors for voltammetric determination of Mebendazole.

Chemical Structure

Electrode	Method of Analysis	LOD (M)	Refs.
GaN/PEDOT-PPY/GCE	DPV	2.4×10^{-6}	[58]

The possible general mechanism of electrochemical oxidation of Mebendazole is shown in Fig. (**7**) and benzimidazole portion is involved in the electrochemical oxidation process.

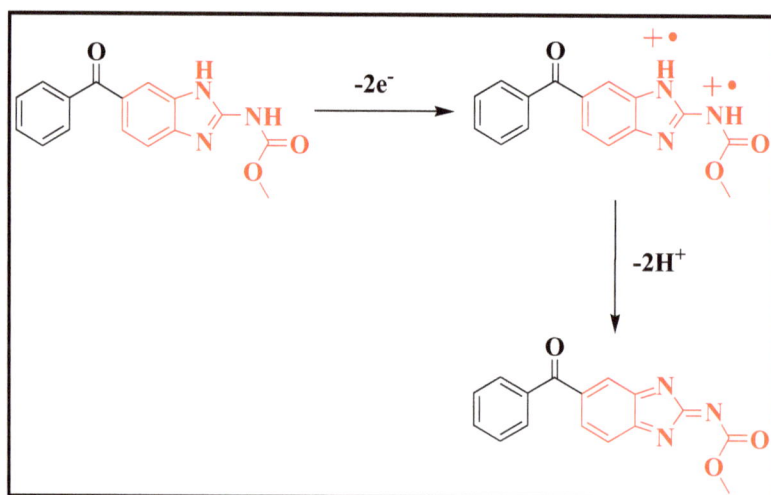

Fig. (7). Possible general mechanism for electrochemical oxidation of Mebendazole.

ELECTROCHEMICAL CHARACTERIZATION OF ANTI-TUBERCULOSIS DRUGS

Isoniazid is an effective anti-TB drug used in the prevention and treatment of pulmonary Tuberculosis.

Couto *et al.* [59] in 2015 reported the determination of Isoniazid using NAF coated screen-printed carbon electrode (SPCE) with a detection limit of 1.4×10^{-5} M employing square wave voltammetry (SWV) method. Absalan and co-workers [60] in 2016 developed palladium nanoparticle modified carbon ionic liquid electrode having N-octylpyridinium hexafluorophosphate binder as a working electrode for the detection of isoniazid with a detection limit of 4.7×10^{-7} M by cyclic voltametric (CV) method. Zhu *et al.* [61] in 2015 reported the determination of isoniazid using electrochemically reduced GNS oxide modified glassy carbon (GCE) employing linear sweep voltammetric (LSV) method with 1.5×10^{-8} M detection limit. Rastogi *et al.* [62] in 2016 designed sliver nanoparticles incorporated copolymer of methyl methacrylate and 2-acrylamid--2-methylpropane sulfonic acid modified glassy carbon electrode (GCE) as a working electrode for the quantitative determination of isoniazid with a detection limit of 1.0×10^{-8} M by cyclic voltammetric (CV) method (Table **13**).

Table 13. Comparison of different electrochemical sensors for voltammetric determination of Isoniazid.

Chemical Structure			
Electrode	**Method of Analysis**	**LOD (M)**	**Refs.**
NAF/ SPCE	SWV	1.4×10^{-5}	[59]
Pd-NP/CILE	CV	4.7×10^{-7}	[60]
ERGO/GCE	LSV	1.5×10^{-8}	[61]
Ag-P(MMA-co-AMPS)/GCE	CV	1.0×10^{-8}	[62]

The possible general mechanism of electrochemical oxidation of Isoniazid is shown in Fig. (**8**).

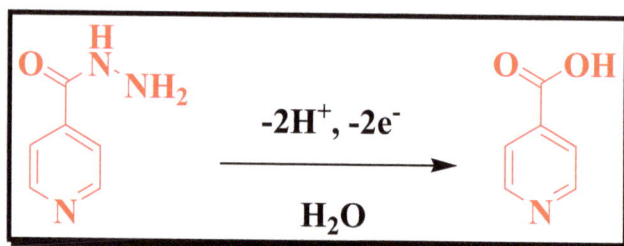

Fig. (8). Possible general mechanism for electrochemical oxidation of Isoniazid.

Pyrazinamide also known as pyrazine carboxamide is a first-line oral drug used to treat tuberculosis caused by *Mycobacterium tuberculosis*.

Mani and Co-worker [63] in 2015 have used multi-walled carbon nanotube (MWCNT) and GNS oxide (GO) modified glassy carbon electrode (GCE) for determination of pyrazinamide employing differential pulse voltammetric (DPV) method with a detection limit of 5.5×10^{-6} M. Simioni *et al.* [64] in 2017 reported determination of pyrazinamide using nanodiamond modified glassy carbon electrode (GCE) as a working electrode with a detection limit of 2.2×10^{-7} M employing square wave voltammetric (SWV) technique. Kalambate *et al.* [65] in 2016 developed GNS-ZnO modified carbon paste electrode (CPE) working electrode for determination of pyrazinamide employing cyclic voltammetric (CV) method with a detection limit of 4.3×10^{-17} M (Table **14**).

Table 14. Comparison of different electrochemical sensors for voltammetric determination of Pyrazinamide.

Chemical Structure			
Electrode	**Method of Analysis**	**LOD (M)**	**Refs.**
MWCNT/GO/GCE	DPV	5.5×10^{-6}	[63]
ND/GCE	SWV	2.2×10^{-7}	[64]
GNS/ZnO/CPE	CV	4.3×10^{-17}	[65]

ELECTROCHEMICAL CHARACTERIZATION OF ANTI-CANCER DRUGS

Nilutamide is an oral anticancer drug used to treat men with advanced prostate cancer.

Karthik *et al.* [66] in 2017 reported the determination of anticancer drug nilutamide using a disposable screen-printed carbon electrode (SPCE) modified with nanocomposite of β-cyclodextrin, gold nanoparticle and graphene oxide employing differential pulse voltammetric (DPV) technique with a detection limit of 0.4×10^{-9} M (Table **15**).

Table 15. Electrochemical sensors for voltammetric determination of Pyrazinamide.

Chemical Structure

Electrode	Method of Analysis	LOD (M)	Refs.
β-CD/Ag-NP/GNS/SPCE	DPV	0.4×10^{-9}	[66]

The possible general mechanism of electrochemical oxidation of Nilutamide is depicted in Fig. (**9**).

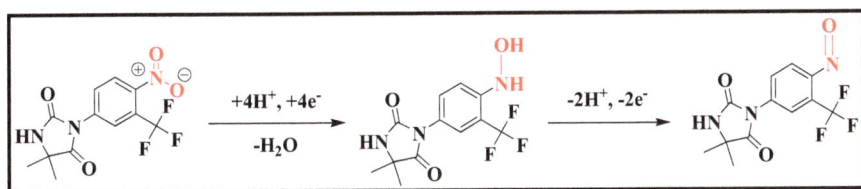

Fig. (9). Possible general mechanism for electrochemical oxidation of Nilutamide.

Cis platin (Cis-dichlorodiammineplatinum(II)) is electrochemically characterized as alkylating agent that binds to DNA and leads to cancer cell death which is also used to treat advanced bladder, metastatic ovarian, metastatic testicular cancer.

Materon *et al.* [67] in 2015 designed screen-printed electrode functionalised with multi-walled carbon nanotubes (MWCNT-COOH/SPCEs) for the detection of cisplatin by differential pulse voltammetric (DPV) technique with a detection limit of 4.6×10^{-6} M (Table **16**).

Table 16. Electrochemical sensors for voltammetric determination of Cis platin.

Chemical Structure

Electrode	Method of Analysis	LOD (M)	Refs.
MWCNT-COOH/SPCEs	DPV	4.6×10^{-6}	[67]

5-Fluorouracil is an excellent antineoplastic agent used to treat various kinds of cancer, namely colorectal, breast, stomach, pancreatic and cervical.

Bukkitgar *et al.* [68] in 2016 reported the electrochemical sensor for the determination of 5-Fluorouracil using glucose modified carbon paste electrode by differential pulse voltammetric method with a detection limit of 5.2×10^{-9} M. Pattar *et al.* [69] in 2015 developed a chemically reduced graphene oxide and chitosan composite film modified glassy carbon electrode (GCE) for the determination of 5-fluorouracil using cyclic, staircase and square wave voltammetric (SWV) techniques with a detection limit of 1.2×10^{-9} M (Table **17**).

Table 17. Comparison of different electrochemical sensors for voltammetric determination of 5-Fluorouracil.

Chemical Structure

Electrode	Method of Analysis	LOD (M)	Refs.
Glu/CPE	DPV	5.2×10^{-9}	[68]
rGO/GCE	SWV	1.2×10^{-9}	[69]

The possible general mechanism of electrochemical oxidation of Nilutamide is depicted in Fig. (**10**).

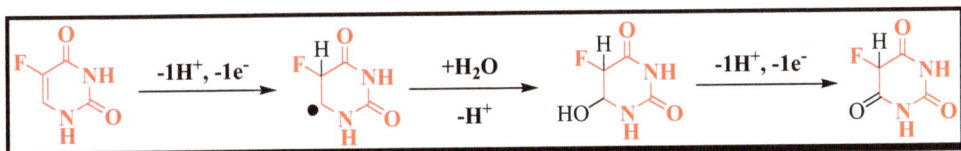

Fig. (10). Possible general mechanism for electrochemical oxidation of 5-Fluorouracil

CONCLUSION

In the present chapter, an attempt is made to investigate the recent developments and applications of the voltammetry technique in the detection of few important drugs that have been taken up based on the works published from 2015 till date.

CONSENT FOR PUBLICATION

Not applicable.

CONFLICT OF INTEREST

The authors declare no conflict of interest, financial or otherwise.

ACKNOWLEDGEMENTS

The authors are thankful to Department of Organic Chemistry, Tumkur University, Tumakuru.

REFERENCES

[1]   "Heyrovsky", *J. Chem. Listy.*, vol. 16, pp. 256-264, 1922.

[2]   N. Hareesha, and J.G. Manjunatha, "A simple and low-cost poly (DL-phenylalanine) modified carbon sensor for the improved electrochemical analysis of Riboflavin", *Journal of Science: Advanced Materials and Devices.*, vol. 5, pp. 502-511, 2020.

[3]   G. Tigari, J.G. Manjunatha, C. Raril, and N. Hareesha, "Determination of Riboflavin at Carbon Nanotube PasteElectrodes Modified with an Anionic Surfactant", *ChemistrySelect,* vol. 4, pp. 2168-2173, 2019.
        [http://dx.doi.org/10.1002/slct.201803191]

[4]   N. Hareesha, and J. G. Manjunatha, "Elevated and rapid voltammetric sensing of riboflavinat poly(helianthin dye) blended carbon paste electrodewith heterogeneous rate constant elucidation", *Journal of the Iranian Chemical Society,* 2020.
        [http://dx.doi.org/10.1007/s13738-020-01876-4]

[5]   E.S. D'Souzaa, J.G. Manjunathaa, C. Rarila, G. Tigaria, and P.A. Pushpanjalia, "Polymer Modified Carbon Paste Electrode as a Sensitive Sensor for the Electrochemical Determination of Riboflavin and Its Application in Pharmaceutical and Biological Samples", *Anal. Bioanal. Chem. Res.,* vol. 7, pp. 461-472, 2020.

[6]   N. Hareesha, J.G.G. Manjunatha, C. Raril, and G. Tigari, "Design of novel Surfactant Modified Carbon Nanotube Paste Electrochemical Sensor for the Sensitive Investigation of Tyrosine as a Pharmaceutical Drug", *Adv. Pharm. Bull.,* vol. 9, no. 1, pp. 132-137, 2019.
        [http://dx.doi.org/10.15171/apb.2019.016] [PMID: 31011567]

[7]   N. Hareesha, and J.G. Manjunatha, "Fast and enhanced electrochemical sensing of dopamine at cost-effectivepoly(DL-phenylalanine) based graphite electrode", *J. Electroanal. Chem. (Lausanne),* vol. 878, p. 114533, 2020.
[http://dx.doi.org/10.1016/j.jelechem.2020.114533]

[8]   N. Hareesha, J.G. Manjunatha, C. Raril, and G. Tigari, "Sensitive and Selective Electrochemical Resolution of Tyrosine with Ascorbic Acid through the Development of Electropolymerized Alizarin Sodium Sulfonate Modified Carbon Nanotube Paste Electrodes", *ChemistrySelect,* vol. 4, pp. 4559-4567, 2019.
[http://dx.doi.org/10.1002/slct.201900794]

[9]   N.S. Prinith, J.G. Manjunatha, and C. Raril, "Electrocatalytic Analysis of Dopamine, Uric Acid and Ascorbic Acid at Poly(Adenine) Modified Carbon Nanotube Paste Electrode: A Cyclic Voltammetric Study", *Anal. Bioanal. Electrochem.,* vol. 11, pp. 742-756, 2019.

[10]   C. Raril, J.G. Manjunatha, D.K. Ravishankar, F. Santosh, G. Siddaraju, and L. Nanjundaswamy, "Validated Electrochemical Method for Simultaneous Resolution of Tyrosine, Uric Acid, and Ascorbic Acid at Polymer Modified Nano-Compositse Paste Electrode", *Surg. Eng. Appl. Electrochem.,* vol. 56, pp. 415-426, 2020.
[http://dx.doi.org/10.3103/S1068375520040134]

[11]   N. Hareesha, and J.G. Manjunatha, "Surfactant and polymer layered carbon composite electrochemical sensor for the analysis of estriol with ciprofloxacin", *Mater. Res. Innov..*
[http://dx.doi.org/10.1080/14328917.2019.1684657]

[12]   P.A. Pushpanjali, J.G. Manjunatha, and M.T. Shreenivas, "The Electrochemical Resolution of Ciprofloxacin, Riboflavin and Estriol Using Anionic Surfactant and Polymer-Modified Carbon Paste Electrode", *ChemistrySelect,* vol. 4, pp. 13427-13433, 2019.
[http://dx.doi.org/10.1002/slct.201903897]

[13]   A.B. Monnappa, J.G. Manjunatha, A.S. Bhatt, and K. Malini, "Sodium Dodecyl Sulfate Modified Carbon Nano Tube Paste Electrode for Sensitive Cyclic Voltammetry Determination of Isatin", *Adv. Pharm. Bull.,* vol. 11, no. 1, pp. 111-119, 2021.
[http://dx.doi.org/10.34172/apb.2021.012] [PMID: 33747858]

[14]   G. Tigari, and J.G. Manjunatha, "Optimized Voltammetric Experiment for the Determination of Phloroglucinol at Surfactant Modified Carbon Nanotube Paste Electrode", *Instrum. Exp. Tech.,* vol. 63, pp. 750-757, 2020.
[http://dx.doi.org/10.1134/S0020441220050139]

[15]   J.G. Manjunatha, "A Promising Enhanced Polymer Modified Voltammetric Sensor for the Quantification of Catechol and Phloroglucinol", *Anal. Bioanal. Electrochem.,* vol. 12, pp. 893-903, 2020.

[16]   N.S. Prinith, and J.G. Manjunatha, "Polymethionine modified carbon nanotube sensor for sensitive and selective determination of L-tryptophan", *J. Electrochem. Sci. Eng.,* vol. 10, no. 4, pp. 305-315, 2020.
[http://dx.doi.org/10.5599/jese.774]

[17]   C. Raril, and J.G. Manjunatha, "A simple approach for the electrochemical determination of vanillin at ionicsurfactant modified graphene paste electrode", *Microchem. J.,* vol. 154, p. 104575, 2020.
[http://dx.doi.org/10.1016/j.microc.2019.104575]

[18]   B.M. Amruthaa, J.G. Manjunatha, A.S. Bhatt, C. Raril, and P.A. Pushpanjali, "Electrochemical Sensor for the Determination of Alizarin Red-S at Non-ionicSurfactant Modified Carbon Nanotube Paste Electrode", *Phys. Chem. Res.,* vol. 7, pp. 523-533, 2019.

[19]   C. Raril, J.G. Manjunatha, L. Nanjundaswamy, and G. Siddaraju, "Ravishankar, D. K.; Santosh, F.; Niranjan, E. Surfactant Immobilized Electrochemical Sensor for the Detection of Indigotine", *Anal. Bioanal. Electrochem.,* vol. 10, pp. 1479-1490, 2018.

[20]   J. Zhao, W. Guo, M. Pei, and F. Ding, "GR-Fe3O4NPs and PEDOTAuNPs composite based

electrochemical aptasensor for the sensitive detection of penicillin", *Anal. Methods,* vol. 8, pp. 4391-4396, 2016.
[http://dx.doi.org/10.1039/C6AY00555A]

[21] H. Li, B. Xu, D. Wang, Y. Zhou, H. Zhang, W. Xia, S. Xu, and Y. Li, "Immunosensor for trace penicillin G detection in milk based on supported bilayer lipid membrane modified with gold nanoparticles", *J. Biotechnol.,* vol. 203, pp. 97-103, 2015.
[http://dx.doi.org/10.1016/j.jbiotec.2015.03.013] [PMID: 25840366]

[22] S. Suleiman, Y. Nor Azah, M. Faruq, A. Jaafar, and A. Hamad, "Al-Lohedan. Nickel Nanoparticle-Modified Electrode for the Electrochemical Sensory Detection of Penicillin G in Bovine Milk Samples", *J. Nanomater.,* vol. •••, p. 1784154, 2019.

[23] D.H.W. Mohd, "Abdull Rahim Mohd Yusoff.; Munawar Saeed Qureshi.; Riffat Javeria.; Roil Bilad M.; Ghulam Zuhra Memon.; Ali Osman Solak. Differential Pulse Voltammetric Determination of Penicillin G at Boron Doped Diamond Electrode", *Eurasian Journal of Analytical Chemistry.,* vol. 14, no. 1, pp. 12-20, 2019.

[24] T. Wang, H. Yin, Y. Zhang, L. Wang, Y. Du, Y. Zhuge, and S. Ai, "Electrochemical aptasensor for ampicillin detection based on the protective effect of aptamer-antibiotic conjugate towards DpnII and Exo III digestion", *Talanta,* vol. 197, pp. 42-48, 2019.
[http://dx.doi.org/10.1016/j.talanta.2019.01.010] [PMID: 30771956]

[25] A. Hatamie, A. Echresh, B. Zargar, O. Nur, and M. Willander, "Fabrication and characterization of highly-ordered zinc oxide nanorods on gold/glass electrode, and its application as a voltammetric sensor", *Electrochim. Acta,* vol. 174, pp. 1261-1267, 2015.
[http://dx.doi.org/10.1016/j.electacta.2015.06.083]

[26] F. Agın, "Electrochemical determination of amoxicillin on a poly (acridine orange) modified glassy carbon electrode", *Anal. Lett.,* vol. 49, pp. 1366-1378, 2016.
[http://dx.doi.org/10.1080/00032719.2015.1101602]

[27] M. Nosuhi, and A. Nezamzadeh-Ejhieh, "Comprehensive study on the electrocatalytic effect of copper - doped nano-clinoptilolite towards amoxicillin at the modified carbon paste electrode - solution interface", *J. Colloid Interface Sci.,* vol. 497, pp. 66-72, 2017.
[http://dx.doi.org/10.1016/j.jcis.2017.02.055] [PMID: 28268183]

[28] "Eryza Guimar~aes de Castro.; Maria Lurdes Felsner, Carolina Ferreira de Matos, Andressa Galli, Development and validation of voltammetric method for determination of amoxicillin in river waterAnalytica Chimica Acta",

[29] T.H.Y. Pham, N.H. Anh, T.T.H. Vu, L.Q. Hung, P.H. Phong, and T.T.H. Chu, "Electrochemical properties of amoxicillin on an economical, simple graphite pencil electrode and the ability of the electrode in amoxicillin detection", *Vietnam J. Chem.,* vol. 58, no. 2, pp. 201-205, 2020.
[http://dx.doi.org/10.1002/vjch.201900158]

[30] A. Wong, A.M. Santos, F.H. Cincotto, F.C. Moraes, O. Fatibello-Filho, M.D.P.T. Sotomayor, and T. Sotomayor, "A new electrochemical platform based on low cost nanomaterials for sensitive detection of the amoxicillin antibiotic in different matrices", *Talanta,* vol. 206, p. 120252, 2020.
[http://dx.doi.org/10.1016/j.talanta.2019.120252] [PMID: 31514822]

[31] R. Islam, "Huy Tran Le Luu.; Sabine Kus. Review-Electrochemical Approaches and Advances towards the Detection of Drug Resistance", *J. Electrochem. Soc.,* vol. 167, p. 045501, 2020.
[http://dx.doi.org/10.1149/1945-7111/ab6ff3]

[32] A.A. Ensafi, M. Taei, T. Khayamian, and F. Hasanpour, "Simultaneous voltammetric determination of enrofloxacin and ciprofloxacin in urine and plasma using multiwall carbon nanotubes modified glassy carbon electrode by least-squares support vector machines", *Anal. Sci.,* vol. 26, no. 7, pp. 803-808, 2010.
[http://dx.doi.org/10.2116/analsci.26.803] [PMID: 20631443]

[33] J. Shan, Y. Liu, R. Li, C. Wu, L. Zhu, and J. Zhang, *J. Electroanal. Chem. (Lausanne),* vol. 738, p.

123, 2015.
[http://dx.doi.org/10.1016/j.jelechem.2014.11.031]

[34]   V. Lucas, "Faria, Jian F. S. Pereira, Gustavo C. Azevedo, Maria A. C. Matos, Lucas Rodrigo A. A. Munoz, and Renato C. Matos, Square-Wave Voltammetry Determination of Ciprofloxacin in Pharmaceutical Formulations and Milk Using a Reduced Graphene Oxide Sensor", *J. Braz. Chem. Soc.,* vol. 9, pp. 1947-1954, 2019.

[35]   H. Asieh, S. Esmail, and G. Maryam, Ali, Sobhani.; Seyed Ali M. Electrochemical Determination of Ciprofloxacin Using Glassy Carbon Electrode Modified with $CoFe_2O_4$-MWCNT., *Anal. Bioanal. Electrochem.,* vol. 8, pp. 996-1008, 2019.

[36]   O. Vajdle, V. Guzsvany, D. Skoric, J. Anojcic, P. Jovanov, M.A. Ivic, J. Janos Csanadi, Z. Konyad, S. Petrovic, and A. Bobrowski, "Voltammetric behavior of erythromycin ethyl succinate at a renewable silver-amalgam film electrode and its determination in urine and in a pharmaceutical preparation", *Electrochim. Acta,* vol. 191, pp. 44-54, 2016.
[http://dx.doi.org/10.1016/j.electacta.2015.12.207]

[37]   M. Radicova, M. Behul, M. Vojs, R. Bodor, and A.V. Stanova, "Voltammetric determination of erythromycin in water samples using a boron-doped diamond electrode", *Phys. Status Solidi, B Basic Res.,* vol. 252, pp. 2608-2613, 2015.
[http://dx.doi.org/10.1002/pssb.201552245]

[38]   A. Veseli, F. Mullallari, and F. Balidemaj, "Electrochemicaldetermination of erythromycin in drinking water resources by surface modified screenprintedcarbon electrodes", *Microchemical,* vol. 148, pp. 412-418, 2019.
[http://dx.doi.org/10.1016/j.microc.2019.04.086]

[39]   M. Rkik, M.B. Brahim, and Y. Samet, "Electrochemical determination of levofloxacin antibiotic in biological samples using boron doped diamond electrode", *J. Electroanal. Chem. (Lausanne),* vol. 794, pp. 175-181, 2017.
[http://dx.doi.org/10.1016/j.jelechem.2017.04.015]

[40]   J.Y. Huang, T. Bao, T.X. Hu, W. Wen, X.W. Zhang, and S.F. Wang, "Voltammetric determination of levofloxacin using a glassy carbon electrode modified with poly (o-aminophenol) and grapheme quantum dots", *Mikrochim. Acta,* vol. 184, pp. 127-135, 2017.
[http://dx.doi.org/10.1007/s00604-016-1982-5]

[41]   M.H. Ghanbari, F. Shahdost-Fard, A. Khoshroo, M. Rahimi-Nasrabadi, M.R. Ganjali, M. Wysokowski, T. Rębiś, S. Żółtowska-Aksamitowska, T. Jesionowski, P. Rahimi, Y. Joseph, and H. Ehrlich, "A nanocomposite consisting of reduced graphene oxide and electropolymerized β-cyclodextrin for voltammetric sensing of levofloxacin", *Mikrochim. Acta,* vol. 186, no. 7, p. 438, 2019.
[http://dx.doi.org/10.1007/s00604-019-3530-6] [PMID: 31197468]

[42]   D. Marques de Farias, "Lucas Vinícius de Faria.; Thalles Pedrosa Lisboa.; Maria Auxiliadora Costa Matos.; Rodrigo Alejandro Abarza Muñoz.; Renato Camargo Matos. Determination of levofloxacin in pharmaceutical formulations and urine at reduced graphene oxide and carbon nanotube-modified electrodes", *J. Solid State Electrochem.,* vol. 24, pp. 1165-1173, 2020.
[http://dx.doi.org/10.1007/s10008-020-04589-z]

[43]   Z. Jiang, Q. Liu, Y. Tang, and M. Zhang, "Electrochemical sensor based on a novel Pt@Au bimetallic nanoclusters decorated on reduced grapheme oxide for sensitive detection of ofloxacin", *Electroanalysis,* vol. 29, pp. 602-608, 2017.
[http://dx.doi.org/10.1002/elan.201600408]

[44]   A. Wong, T.A. Silva, F.C. Vicentini, and O. Fatibello-Filho, "Electrochemical sensor based on graphene oxide and ionic liquid for ofloxacin determination at nanomolar levels", *Talanta,* vol. 161, pp. 333-341, 2016.
[http://dx.doi.org/10.1016/j.talanta.2016.08.035] [PMID: 27769415]

[45]   L. Feng, Q. Xue, F. Liu, Q. Cao, J. Feng, L. Yang, and F. Zhang, "Voltammetric determination of

ofloxacin by using a laser-modified carbon glassy electrode", *Mikrochim. Acta,* vol. 187, no. 1, p. 86, 2020.
[http://dx.doi.org/10.1007/s00604-019-4065-6] [PMID: 31897751]

[46]   Y. Dai, X. Li, X. Lu, and X. Kan, "Voltammetric determination of paracetamol using a glassy carbon electrode modified with Prussian blue and a molecularly imprinted polymer, and ratiometric read-out of two signals", *Mikrochim. Acta,* vol. 183, pp. 2771-2778, 2016.
[http://dx.doi.org/10.1007/s00604-016-1926-0]

[47]   P. Kalambate, B. Sanghavi, S. Karna, and A. Srivastava, "Simultaneous voltammetric determination of paracetamol and domperidone based on a graphene/platinum nanoparticles/nafion composite modified glassy carbon electrode", *Sens. Actuators B Chem.,* vol. 213, pp. 285-294, 2015.
[http://dx.doi.org/10.1016/j.snb.2015.02.090]

[48]   Meareg Amare, and Welday Teklay, "Voltammetric determination of paracetamol in pharmaceutical tablet samples using anthraquinone modified carbon paste electrode", *Cogent Chemistry,* vol. 5, 2019.

[49]   A. Iacob, F. Manea, N. Vaszilcsin, S. Picken, and J. Schoonman, "Anodic determination of acetylsalicylic acid at multiwall carbon nanotubes-epoxy composite electrode", *Int. J. Electrochem. Sci.,* vol. 10, pp. 5661-5672, 2015.

[50]   H. Zhu, "Electrochemical determination of acetylsalicylic acid in human urine samples based on poly (diallyldimethylammonium chloride) functionalized reduced graphene oxide sheets", *Int. J. Electrochem. Sci.,* vol. 11, pp. 4007-4017, 2016.
[http://dx.doi.org/10.20964/110395]

[51]   F. Fathirad, A. Mostafavi, and D. Afzali, "Electrospun Pd nanoparticles loaded on Vulcan carbon/ conductive polymeric ionic liquid nanofibers for selective and sensitive determination of tramadol", *Anal. Chim. Acta,* vol. 940, pp. 65-72, 2016.
[http://dx.doi.org/10.1016/j.aca.2016.08.051] [PMID: 27662760]

[52]   B. Deiminiat, G. Rounaghi, and M. Zavar, "Development of a new electrochemical imprinted sensor based on poly-pyrrole, sol-gel and multiwall carbon nanotubes for determination of tramadol", *Sens. Actuators B Chem.,* vol. 238, pp. 651-659, 2017.
[http://dx.doi.org/10.1016/j.snb.2016.07.110]

[53]   M. Mohamed, S. Atty, N. Salama, and C. Banks, "Highly selective sensing platform utilizing graphene oxide and multiwalled carbon nanotubes for the sensitive determination of tramadol in the presence of co-formulated drugs", *Electroanalysis,* vol. 29, pp. 1-12, 2017.
[http://dx.doi.org/10.1002/elan.201600668]

[54]   E. Çidem, T. Teker, and M. Aslanoglu, "sensitive determination of tramadol using a voltammetric platform based on antimony oxide nanoparticles", *Microchem. J.,* vol. 147, pp. 879-885, 2019.
[http://dx.doi.org/10.1016/j.microc.2019.04.018]

[55]   E. Suresh, K. Sundaram, B. Kavitha, and N. Kumar, "Square wave voltammetry sensing of ibuprofen on glassy carbon electrode", *Int. J. Pharm. Tech. Res.,* vol. 9, pp. 182-188, 2016.

[56]   S. Hernandez, G. Romero, S. Avendaño, M. Hernández, C. Vidal, and M. Romo, "Voltammetric determination of ibuprofen using a carbon paste-multiwalled carbon nanotube composite electrode", *Instrum. Sci. Technol.,* vol. 44, pp. 483-494, 2016.
[http://dx.doi.org/10.1080/10739149.2016.1173061]

[57]   J. Srivastava, and M.A. Singh, "Biopolymeric nano-receptor for sensitive and selective recognition of albendazole", *Anal. Methods,* vol. 8, pp. 1026-1033, 2016.
[http://dx.doi.org/10.1039/C5AY03048J]

[58]   S. Munusamy, R. Suresh, K. Giribabu, R. Manigandan, S. Kumar, S. Muthamizh, C. Bagavath, A. Stephen, J. Kumar, and V. Narayanan, "Synthesis and characterization of GaN/PEDOT-PPY nanocomposites and its photocatalytic activity and electrochemical detection of mebendazole", *Arab. J. Chem.,* 2015.
[http://dx.doi.org/10.1016/j.arabjc.2015.10.012]

[59]   R. Couto, J. Lima, and M. Quinaz, "Screen-printed electrode based electrochemical sensor for the detection of isoniazid in pharmaceutical formulations and biological fluids", *Int. J. Electrochem. Sci.,* vol. 10, pp. 8738-8749, 2015.

[60]   G. Absalan, M. Akhond, M. Soleimani, and H. Ershadifar, "Efficient electrocatalytic oxidation and determination of isoniazid on carbon ionic liquid electrode modified with electrodeposited palladium nanoparticles", *J. Electroanal. Chem. (Lausanne),* vol. 761, pp. 1-7, 2016.
[http://dx.doi.org/10.1016/j.jelechem.2015.11.041]

[61]   X. Zhu, J. Xu, X. Duan, L. Lu, K. Zhang, Y. Yub, H. Xing, Y. Gao, L. Donga, H. Sun, and T. Yanga, "Controlled synthesis of partially reduced graphene oxide: Enhance electrochemical determination of isoniazid with high sensitivity and stability", *J. Electroanal. Chem. (Lausanne),* vol. 757, pp. 183-191, 2015.
[http://dx.doi.org/10.1016/j.jelechem.2015.09.038]

[62]   P. Rastogi, V. Ganesan, and U. Azad, "Electrochemical determination of nanomolar levels of isoniazid in pharmaceutical formulation using silver nanoparticles decorated copolymer", *Electrochim. Acta,* vol. 188, pp. 818-824, 2016.
[http://dx.doi.org/10.1016/j.electacta.2015.12.058]

[63]   S. Mani, S. Cheemalapati, S. Chen, and B. Devadas, "Antituberculosis drug pyrazinamide determination at multiwalled carbon nanotubes/graphene oxide hybrid composite fabricated electrode", *Int. J. Electrochem. Sci.,* vol. 10, pp. 7049-7062, 2015.

[64]   N. Simioni, T. Silva, G. Oliveira, and O. Filho, "A nanodiamondbased electrochemical sensor for the determination of pyrazinamide antibiotic", *Sens. Actuators B Chem.,* vol. 250, pp. 315-323, 2017.
[http://dx.doi.org/10.1016/j.snb.2017.04.175]

[65]   P. Kalambate, C. Rawool, and A. Srivastava, "Voltammetric determination of pyrazinamide at graphene-zinc oxide nanocomposite modified carbon paste electrode employing differential pulse voltammetry", *Sens. Actuators B Chem.,* vol. 237, pp. 196-205, 2016.
[http://dx.doi.org/10.1016/j.snb.2016.06.019]

[66]   R. Karthik, N. Karikalan, and S-M. Chen, "Gnanaprakasam, P.; Karuppiah, C. Voltammetric determination of the anti-cancer drug nilutamide using a screen-printed carbon electrode modified with a composite prepared from β-cyclodextrin, gold nanoparticles and graphene oxide", *Mikrochim. Acta,* vol. 184, pp. 507-514, 2017.
[http://dx.doi.org/10.1007/s00604-016-2037-7]

[67]   E.M. Materon, A. Wong, S.I. Klein, J. Liu, and M.D.P.T. Sotomayor, "Multi-walled carbon nanotubes modified screen-printed electrodes for cisplatin detection", *Electrochim. Acta,* vol. 158, p. 271, 2015.
[http://dx.doi.org/10.1016/j.electacta.2015.01.184]

[68]   S.D. Bukkitgar, and N.P. Shetti, "Electrochemical Sensor for the Determination of Anticancer Drug 5-Fluorouracil at Glucose Modified Electrode", *ChemistrySelect,* vol. 1, no. 4, pp. 771-777, 2016.
[http://dx.doi.org/10.1002/slct.201600197]

[69]   V.P. Pattar, and S.T. Nandibewoor, "Electroanalytical method for the determination of 5-fluorouracil using a reduced graphene oxide/chitosan modified sensor", *RSC Advances,* vol. 5, p. 34292, 2015.
[http://dx.doi.org/10.1039/C5RA04396D]

CHAPTER 12

Voltammetric Application in Detection of Anticancer Drug

Shankar Ashok Itagi[1], M. H. Chethana[1], P. Manikanta[1], S. Sandeep[1], C. S. Karthik[1], A. S. Santhosh[2], K. S. Nithin[3], Kumara Swamy N. K.[1] and P. Mallu[1]

[1] *Department of Chemistry, Sri Jayachamarajendra College of Engineering, JSS Science and Technology University, Mysuru, Karnataka 570006, India*

[2] *Department of Chemistry, NMKRV College for Womens, Jayanagar, Bengaluru*

[3] *Department of Chemistry, The National Institute of Engineering, Mysuru, Karnataka 570008, India*

Abstract: From the past few years the development of electrochemical sensor has gained more attention due to the fact that these methods have high sensetivity, less expensive and easy to handle when compared to other conventional methods. Cancer is one among the serious disease faced by the human race in the current era. Presently, It is treated with various treatments like chemotherapy, surgery and radiotherapy. Since the impressive growth of these prime treatments, ample of anticancer drugs have been developed. Alongside in recent years, the synthesis of nanomaterials and their application in medicinal (biomedical) field has gained quite a focus in the research field. Due to smaller size, improved solubility. surface tailor ability, electrocatalytic activity, sensible detection and multifunctionality the nanomaterials are found to possess greater applications in the growing field of pharmaceutical research. The present chapter gives the reprots on the electrochemical detection of various anticancer drugs like Methotrexate (MTX), 5-Flourouracil, Doxorubicin and Etoposide.

Keywords : Anticancer drugs , Doxorubicin , Etoposide , Methotrexate , Nano materials , Voltammetric methods , 5-Flourouracil .

INTRODUCTION

In the present era, cancer is termed as a major global issue causing many annual deaths. The impact of this issue has led to the development of variety of expensive drugs and treatments [1]. One such drug is methotrexate which is a derivative of folic acid antagonists, that finds application as an anticancer and anti-inflammatory agent [2].

* **Corresponding author S. Sandeep:** Department of Chemistry, SJCE, JSS Science and Technology University, Mysuru, India; E-mail: sandeeps@jssstuniv.in

J.G. Manjunatha (Ed.)
All rights reserved-© 2022 Bentham Science Publishers

In the last few decades, chemotherapy is employed as one among the best main major medical advancements in cancer care [3]. However, the medications employed for this practice possesses a limited therapeutic index and the response obtained are often soothing and unpredictable [4]. Though biomacromolecules are the components which are pointed by these approaches, the drawback found by these are failure in discriminating among swiftly splitting non-malignant and tumour cells [5]. In recent years, targeted therapy has been aimed toward tumour-specific targets and signalling pathways, and therefore has relatively more limited nonspecific mechanisms [6]. A number of studies have shown a key role for epigenetic pathways in the development of cancer. Genetic alteration alone is incapable of explaining carcinogenesis; thus, it includes epigenetic processes (DNA methylation, histone modifications and non-coding RNA deregulation) [7]. The chromatin decondensation which is asserted by condensins and interactions between histonesis caused by H3 and H4 histones lysine deacetylation which are part of the Histone decondensation [8].

The advancement of chemotherapy has led to the maximal cure of many cancer cells by the utility of numerous anti-cancer drugs [9]. However, this progress has attributed to the further establishment of innovative treatment techniques *i.e.*, surgery and radiation therapy. The keen draw back of the chemotherapeutic treatment is their cause of harmfulness on normal tissues of the body [10]. The bone marrow, gastrointestinal tract, hair follicle *etc* are the vital points that are majorly affected by the acute toxicities of chemotherapeutic treatment [11].

The common toxicities encountered are haematological, gastrointestinal, skin and hair follicle toxicity, nervous system toxicity, local toxicity, metabolic disturbances, hepatic toxicity, urinary tract toxicity, cardiac toxicity, pulmonary toxicity, gonadal toxicity, *etc* [12].

The cancer victims are frequently examined with various drug therapies which not only include chemotherapeutic agents but also includes non-chemotherapeutic drugs. The care should be taken in such a way that, the treatment should not encourage the growth of the tumour or cause toxicity to other tissues [13].

The anticancer drugs that can be detected using various methods are as follows:

1. Methotrexate (MTX)

2. 5-Flourouracil

3. Doxorubicin

4. Etoposide

METHOTREXATE

MTX differs structurally from folic acid by two sites:

- hydroxyl group replaces amino group in one site.
- methyl group replaces an amino at another [14].

MTX (amethopterin, 2,4-diamine-N10-methylpteroyl-glutamic acid) being a folic acid counterpart varies from that by an amino group being replaced by the hydroxyl group at the C4 region on the pyrimidine ring (Fig. **1**). It finds its application in treatment of various illness. However, the usage of this drugs results in effects like marrow suppression, gastro-intestinal lesions, renal insufficiency, hepatic failure, hypoalbuminemia, and pancytopenia [15]. It is therefore, very important to establish a responsive and versatile method for the assessment of MTX [16]. The structure of MTX is shown in below Fig. (**1**).

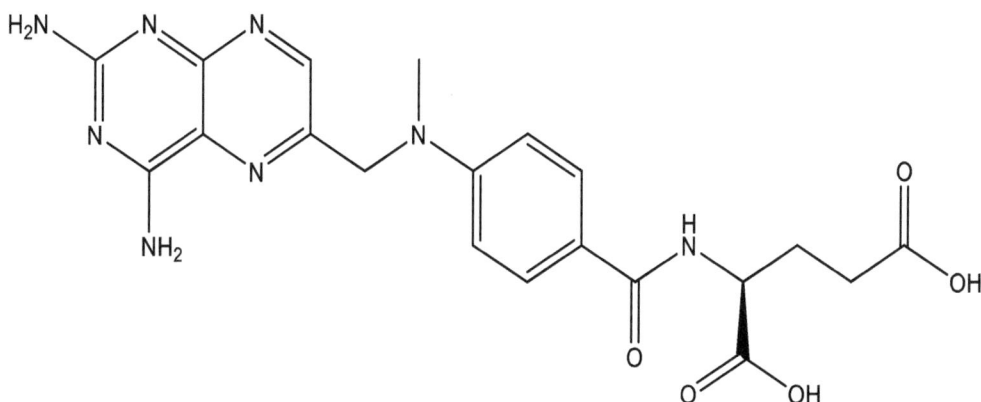

Fig. (1). The structure of Methotrexate.

Deng, Z et al. synthesised stearyltrimethylammoniumbromide (STAB)/novel acetylene black (AB). The modified sensor showed lower detection limit of 3.07 nM with linear range of 0.005 μM to 7.0 μM which showed application in determination of MTX [17]. *Deng, Z et al.* fabricated acetylene black (AB) on glassy carbon electrode which resulted in the development of novel sensor. The so designed sensor showed linear relationship within range of .005 μM to 3.0 μM with detection limit of 3.81 nM for MTX [18]. *Chen, J et al.* reported glassy carbon electrode (GCE) modified with graphene oxide (GO) for sensitive detection of MTX. This showed higher range of 5.5×10^{-8} to 2.2×10^{-6} mol L^{-1} with a lower limit of detection 7.6×10^{-9} mol L^{-1} [19]. *Ensafi, et al.* constructed an electrochemical sensor by modifying glassy carbon electrode with $CoFe_2O_4/$

reduced graphene oxide (GrO)/ ionic liquid that showed lower detection limit of 0.01 nmol mL^{-1} and linear range of 0.05–7.5 nmol mL^{-1} [20]. *Guo, Y et al.* reported potent sensor that is developed for application in detection of methotrexate compound by using cyclodextrin-graphene hybrid nanosheets modified glassy carbon electrode (CD-GNs/GCE). The sensor possessed linear range of 0.1 mM–1.0 mM, along with limits of detection of 20 nM [21]. *Lima, H.R.S et al.* developed an (indium tin oxide) ITO electrode modified by nanostructured films for detecting of MTX. The sensitivity of the sensor was enhanced by merging PVS (polyvinyl sodium sulfonate) with BM from 2.49 (ITO) to 5.55 µAµmol L^{-1} (ITO/PVS/BM). The detection limit possessed was on lower range of 5.95 × 10^{-7} mol L^{-1} for MTX [22]. *Ghadimi, H et al.* designed a sensor by fabricating bimetallic palladium–silver alloy (PdAg/NG–GCE)/N-graphene composite on glassy carbon electrode for the inspection of MTX. A significant linear range of 0.02–200 µM with a limit of detection of 1.32 nM was attained [23]. By utilizing screen printed electrode (SPE) *Wang, Y et al.* fabricated with nano-Au/MWNTs-ZnO developed an effective sensor that showed application in determination of MTX in blood samples. The linear ranges of MTX were estimated to be 0.02–1.00 µM, with the lower detection limits of 10 nM [24]. *Asbahr, D et al.* established an bismuth film modified electrode (BiFE) based sensor for the measurement of methotrexate (MTX). The linear range for MTX was observed to be in the range of 12 nM to 1650 nM along with detection limit of 0.9 nM [25]. *Zhao, Y et al.* integrated an fluorescent N, S co-doped carbon quantum dots (N, S co-doped CQDs) probe for the detection of methotrexate. The lower detection limit of 12 ng/mL (S/N = 3) with linear range of 0.4 µg/mL to 41.3 µg/mL was displayed by fluorescent probe [26]. *Wang, Y et al.* established a effective sensor by fabricating DNA-functionalized single-walled carbon nanotube (DNA/SWCNT) on a glassy carbon electrode and also assembles nafion composite film for the detectability of methotrexate. The detection limit of 8.0 × 10^{-9} mol L^{-1} with linear voltametric range of 2.0 × 10^{-8}–1.5 × 10^{-6} mol L^{-1} was illustrated [27]. With the use of Langmuir–Blodgett (LB) technique *Wang, F et al.* established a sensor by immobilizing DNA molecules onto a pre-treated glassy carbon electrode (GCE (ox)) surface. The sensor presented a good linear range of 2.0 × 10^{-8} to 4.0 × 10^{-6} mol L^{-1}, along with a lower detection of limit 5.0 × 10^{-9} mol L^{-1} [28].

Some of the electrochemical methotrexate sensors fabricated utilizing different nanomaterials is shown in Table **1**.

Table 1. Methotrexate sensor depending on various nano materials reported in literature.

Matrix	Linear Range	LOD	Potential	Stability	Reproducibility	Application to Real Sample	% Recovery	Ref.
MWCNT – SPE	0.5 µM–100 µM	100.0	0.82 V	3 months	-	Biological samples	96.04% and 116.57%	[29]
BDDE	50 nM–20 µM	10.0	1050mV	-	2%	Drugs and Spiked Human Urine	99.2 and 101.4	[30]
PLL/GCE	5 nM–200 nM	1.7	0.890V	-	-	Tablets	98% to 105%	[31]
MWCNTs-DHP/GC	50 nM–5 µM	33.0	0.96 V	45days	2.42%	Biologial Formulation	-	[32]
Ce-doped ZnO nano-flowers	0.01–500.0 µM	6.3 nM	0.76 V	-	-	Human Serum and Urine	97.5% to 102.5%	[33]
GCE/GO	0.055–2.2 µM	7.6 nM	1.12 V	7 days	-	Human Serum and Urine	97.4–102.5%	[34]
Gold nanoparticles and l-cystei	0.04–2.0 µM	0.01 µM	0.938 V	4 weeks	-	Tablets and Spiked human blood serum samples	95% to 101%	[35]

5-FLOUROURACIL

Based on the knowledge of tumour biochemistry, 5-Flourouracil has been rationally designed as a clinically useful antineoplastic drug [36]. Various cancers like colon, esophageal, stomach, breast, pancreas and cervix are presently been treated with 5-Fluorouracil (5-FU) as a medicine in an IV form. It is also used in the form of cream in treating actinic keratosis and basal cell carcinoma [37]. The drug in order to be more effective in treatment, it requires to be converted to the nucleotide level. It can also result in the disturbance in the maturation of nuclear RNA with its incorporation into the RNA [38]. The 5-Fluorouracil exhibits the following structure as shown in Fig. (2).

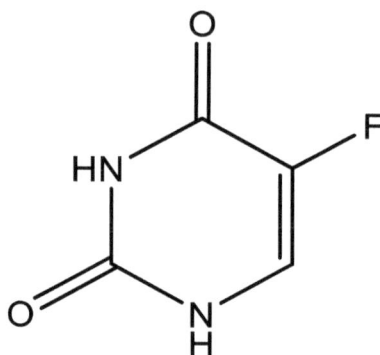

Fig. (2). The structural representation of 5-Flourouracil.

MojtabaHadi et al. modified Screen-Printed Carbon Electrode (SPC) by introducing Graphene oxides (GOs)/Carbon nanotubes (CNTs). The developed sensor exhibited lower detection limit of 0.05–5, 5–1200 µM and limit of detection of 16nM due to the new combination matrix of GOs, CNTs [39]. *Norouzi, P et al.* developed biosensor by altering gold electrode decorated with polyaniline and gold nanoparticles. The formed biosensor showed linear range of 1×10^{-9} to 1×10^{-6} M (r^2=0.996) along with lower limits of detection of 2.5×10^{-10} M [40]. *Zahed, F.M et al.* synthesised and fabricated silver nanoparticles-polyaniline nanotube (AgNPs@PANINTs) sensor for the detection of 5-FU. They reported lower limiting range 1.0 to 300.0 µM with a low detection limit (0.06 µM) which can be attributed to the incorporation of AgNPs [41]. *Emamian, R et al.* fabricated carbon paste electrode (GQD/BPBr/CPE) biosensor for 5-FU determination using graphene quantum dots (GQD) and 1-butylpyridinium bromide (BPBr) as a matrix. The developed biosensor showed lower detection limit of 0.5 nM with a linear range of 0.001–400 µM the developed biosensor showed excellent selectivity in presence of potent interferents [42]. *Jyothi C. Abbar et al.* developed multi-walled carbon nanotubes-paraffin oil paste electrode as a potent sensor. The sensor showed the linear range of 0.1×10^{-6} to 5.0×10^{-6} M with a detection limit of 3.94×10^{-8} M [43]. *Hatamluyi, B et al.* synthesised Graphene quantum dots-polyaniline/zinc oxide nanocomposites (GQDs-PANI/ZnO-NCs) whose detection limit was detected lower 0.023 µM with liner response range of 0.1–50.0 µM [44]. *Hua, X e al.* utilized carbon electrode of glass which was modified with blue bromothymol and nanotube of Multi walled carbon to develop a sensor. The sensor resulted in lower detection limit of 30.69nM with linear range of 1.9×10^5nM due to the excellent electrocatalytic activity of transducer element [45]. *Lima, D et al.* fabricated porphyran-capped gold nanoparticles on to the carbon paste electrodes and developed a potent sensor. The sensor so developed exhibited liner relationship amongst DPV peak currents and 5-FU concentration. The limit of detection and quantification was found in lower limit of 2.22 and 0.66 mol L^{-1} with limiting range up to 29.9 to 234 mol L^{-1} [46]. The synthesised gold nanoparticle decorated multiwall carbon nanotube (GNP-MWCNT) sensor developed by *Satyanarayana, M et al.* showed high sensitivity in detection of 5-FU. The sensor resulted in lower detection limit of 20 nM with wide concentration range of 0.03–10 μM due to the presence of gold nanoparticles [47]. *Roushani, M et al.* constructed a sensor using chitosan composite decorated with multi-walled carbon nanotube/copper-nanoparticles. The sensor showed a wide range of concentration of 1-110 μM with a low detection limit of 0.15 μM [48]. A few of the electrochemical 5-Flourouracil sensors constructed using distinct nanomaterials are represented in Table 2.

Table 2. 5-flourouracil sensor based on the relative nano materials reported in the literature.

Matrix	Linear Range	LOD	Potential	Stability	Reproducibility	Application to Real Sample	% Recovery	Ref.
CP/Glucose	0.1–40μM	5.17 nM	1.12 V	-	-	Urine	97.12% to 104.1%	[49]
MTB/CPE	0.2– 10 μM	2.04 nM	1.16 V	-	2.4%	Tablets	95.6% to 99.4%	[50]
GC/Reduced GOs/chitosan	0.01–0.11μM	4.93 nM	1.321 V	-	-	Tablets and Spiked human blood serum samples	97.3% to 100.7%	[51]
CPE/ Pr:Er	0.01–50 μM	0.98 nM	1.15 V	1 month	-	Injections, Urine and Blood samples	96% to 102%	[52]
GQD/BPBr/CPE	0.001–400 μM	0.5 nM	1006 m V	14 days	-	Pharmaceutical Samples		[53]

DOXORUBICIN

Doxorubicin is considered as one of the potent drug in the present world. However, its application is found to be limited in treating few tumours like colorectal cancers due to their cardiotoxicity and toxicity to healthy haematopoietic clonogenic cells [54]. Doxorubicin shows an effective result on wide range of malignancies like, leukaemias, sarcomas, breast cancer, small cell lung cancer and ovarian cancer [55 - 57].

Doxorubicin is available as dry powder and it can be easily dissolved in water. It is found more stable at a mild acidic pH 4 and unstable at a far more acidic or basic pH. It is also found to be stable in presence of illumination at room temperature for about 24 hours though stability may be found less in plasma and culture media [58, 59]. Doxorubicin was prescribed as a recurrent (weekly) small dosage, a solitary high dosage, and as a continuous infusion.The ideal protocol for tumour cytotoxic activity and dose-limiting health consequences, for example, myelosuppression or cardiotoxicity, has never been looked over in an imminent, randomized way.The Structure of Doxorubicin is as shown in the following Fig. (3).

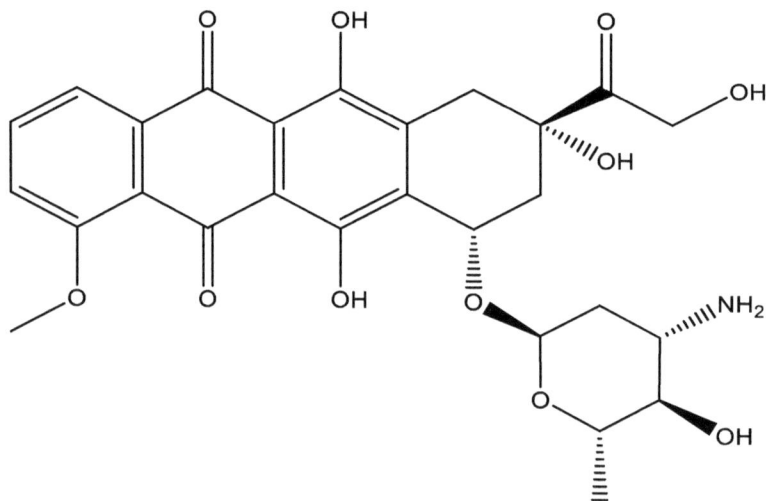

Fig. (3). The structural representation of Doxorubicin.

Jian Liu et al. utilized new combination matrix like mesoporous carbon spheres (MCS) and graphene oxide (GO) which resulted in the reduction of graphene oxide (rGO) in the development of sensor. This altered GCE presented broad linear parameter range between 10 nM - 10 µM along with lower limit of detection of 1.5 nM [60]. *Jahandari, S et al.* created an effective sensor by modifying carbon paste electrode (CPE) with bis(1,10-phenanthroline) (1,10-phenanthroline-5,6-dione) nickel (II) hexafluorophosphate (BPPDNi) as a new electro-catalyst and Pt:Co nanoparticle (Pt:CO-NPs) as highly conductive mediator. The linear range of 0.5–300 was observed along with limit of detection of 0.1 µM [61]. *Er, E et al.* decorated metallic phase MoS_2 ($1T-MoS_2$) with shape-dependent gold nanostructures in the development of electrochemical sensor for the sensitive detection of doxorubicin (DOX). The sensor exhibited lower detection limit of 2.5 nM along with linear range of 0.01–9.5 µM [62]. Sensor consisting of ternary nanocomposite composed of silver nanoparticles (AgNPs), carbon dots (CDS) and reduced graphene oxide (rGO) was constructed by *Guo, H et al.* The synthesised sensor exhibited good range of linearity of 1.0×10^{-8} to 2.5×10^{-6} M ($R^2 = 0.9956$) along with detection limit of 2 nM [63]. For the detection of doxorubicin, new electrochemical sensor composed of glassy carbon electrode (GCE) fabricated with nitrogen-doped carbon nano onions (N-CNOs) was designed by *Ghanbari, M.H et al.* The sensitivity and limit of detection was found to be 1.13 µA µmol L^{-1} cm^{-2} and 60.0 pmol L^{-1} (an at S/N ratio of 3) with range of detection between from 0.2 nmol L^{-1} to 10.0 µmol L^{-1} [64]. By modified chemical coprecipitation method *Taei, M et al.* developed an electrochemical sensor composed of Spinel-structured cobalt ferrite ($CoFe_2O_4$) magnetic nanoparticles (MNPs) for the sensitivity of doxorubicin. The designed sensor exhibited linearity of 0.05–1150 nM range with detection limit of 10 pM at a signal-to-noise ratio of 3 [65]. *Alavi-Tabari et al.* utilized ZnO nanoparticle/1-butyl-3-methylimidazolium tetrafluoroborate modified carbon paste electrode (ZnO-NPs/BMTFB/CPE) for crafting an effective sensor. The linear relation of doxorubicin with oxidation currents was found in the linearity range of 0.07–500 µM along with lower detection limit of 9.0 nM for doxorubicin [66]. Some of the electrochemical Doxorubicin sensors fabricated utilizing different nanomaterials is depicted in Table **3**.

Table 3. Sensor of Doxorubicin based on the different nano materials that are reported in the literature.

Matrix	Linear Range	LOD	Potential	Stability	Reproducibility	Application to Real Sample	% Recovery	Refs.
OMWCNT/GCE	0.04–90 µmol	9.4 × 10^{-3} nM	-444V	20days	2.6%	Blood Serum and Urine	-	[67]
HMDE	0.5–10 µmol	0.1 nM	-410 V	-	-	Pharamaceutical Preparations	-	[68]
nano-TiO_2/nafion/GCE	0.005–2 µmol	0.001 nM	-0.60V	2 weeks	-	Plasma samples	94.9% - 104.4%	[69]
MCS/rGO/GCE	0.01–10 µmol	1.5 nM	-0.45V	30 days	-	Serum	96% - 103%	[70]
GQD/GCE	0.018–3.6 µmol	16nM	-0.56 V	7 days	-	Plasma samples	90% - 118%	[71]
MWNTs/Polt-L-lysine SPE	0.0025–0.25 µmol	1nM	-0.69 V	-	-	Blood serum	92.60% - 106.87%	[72]
AgNPs-CDs-RGO/GCE	0.01–2.5 µmol	2 nM	-	15 days	-	Serum	98% - 102%	[73]
MWCNTs/Pt	0.09–7.36 µmol	3 nM	-0.522V	-	-	Blood Sample	97.5% - 105%	[74]

ETOPOSIDE

In the present world, the most prescribed anticancer drug is Etoposide. It is widely utilized in treating various cancers like small cell lung cancer, germ-line malignancies, sarcomas, leukaemia's, and lymphomas. The topoisomerase II is the main cellular target of etoposide. This significant enzyme establishes transient double-stranded splits in the DNA spine by eliminating ties and strands from the genome [75]. Cell death pathways are caused by the production of irreversible cracks in the genetic material established by the aggregation of cleavage complexes with in treated cells. As a result, etoposide transforms this important enzyme into a lethal cell toxin that results in the fragmentation of the genome and finally results in cell death [76].

Despite its application as a keen chemotherapeutic agent for human cancer, it is also found to exhibit a negative impact on human health by triggering chromosomal translocation leading to the specific types of leukaemia [77, 78]. The etoposide is found to possess the following structure as shown in Fig. (**4**).

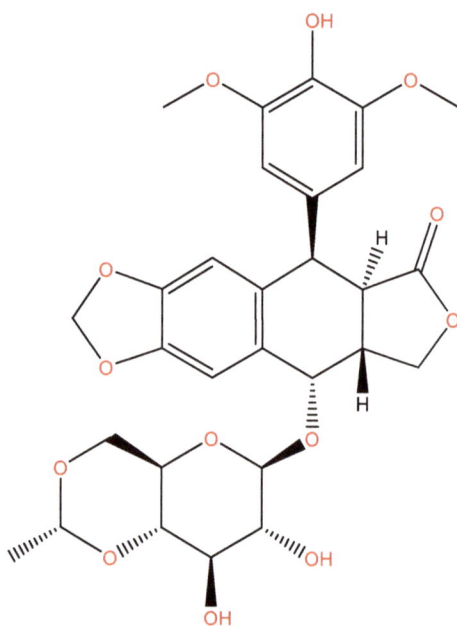

Fig. (4). The structural representation of Etoposide.

Hrichi, H et al. developed a sensor glassy carbon electrode fabricated by a pyrrole in the existence of ETP molecules. The sensor presented lower detection limit of 2.8×10^{-9} M with linear range of 5.0×10^{-7} M – 1.0×10^{-5} M [79]. *Sadat Vajedi et al.* fabricated high-sensitive electrochemical DNA biosensor with the ZnAl/layered double hydroxide modified cobalt ferrite-graphene oxide nanocomposite $(GO/CoFe_2O_4/ZnAl\text{-}LDH)$ which showed the potential electrochemical activity with detection limit of 0.0010 µM in the lower linear range of 0.2–10 µM [80]. *Bozal-Palabiyik et al.* constructed the multi-walled carbon nanotube-modified glassy carbon electrode (MWCNT-modified GCE) sensor. The calibration curve obtained was found linear in the range of $2.0 \times 10^{-8}-2.0 \times 10^{-6}$ M with the detection limit of 5.4×10^{-9} M [81]. Ozkan, *A et al.* constructed an SPR nanosensor based on core-shell nanoparticles (Ag@AuNPs) incorporated with hexagonal boron nitride (HBN) nanosheets and molecularly imprinted polymer (MIP) for the detection of etoposide (ETO). The results of the developed sensor showed linearity range of $0.001-1.00$ ng mL^{-1} ($1.70 \times 10^{-12}-1.70 \times 10^{-9}$ M) with lower detection limit of 0.00025 ng mL^{-1} (4.25×10^{-13} M) [82]. *Hatamluyi, B et al.* developed the Au/Pd@rGO@p(L-Cys) modified electrode sensor that found application in the sensing of ETO behaviour. The manufactured sensor displayed enhanced performance in simultaneous determination of ETO over a wide linear range of 0.01 to 40.0 µM with lower detection limits of 0.718 nM [83]. A few of the electrochemical Etoposide sensors constructed using distinct nanomaterials are represented in Table **4**.

Table 4. Etoposide sensor depending on various nano materials reported in literature.

Matrix	Linear Range	LOD	Potential	Stability	Reproducibility	Application to Real Sample	% Recovery	Ref.
CPE	0.25-25.0 µM	0.10 µM	0.660 V	-	-	Human Serum	90%	[84]
SPCE/DNA	0.1-10 µM	0.005 µM	0.56 V	-	-	Serum	91%	[85]
GCE/CQDs	0.02-10.0 µM	0.005 µM	0.42 V	-	-	Drugs	-	[86]

Preparation of Real Sample

Generally, the blood sample is collected from the healthy volunteer and it will be subjected to centrifugation. Then, it will be treated with the reagents to precipitate the proteins and other components. Finally, the supernatant will be diluted for the desired concentration and this will be used for analysis. In case of urine sample, the urine will be collected from the volunteer, the sample is filtered using suitable filter paper and then it will be frozen. Then, the desired amount of urine sample will be assessed for the determination of drug or in other case, the urine sample will be diluted and the sample will be spiked with the known amount of drug. Whereas for the analysis of drug in pharmaceutical tablets, the tablets containing the drug will be weighed accurately and it is dissolved in suitable solvent. The prepared tablet solution will be used for the quantification.

CONCLUSION

In attempt to elucidate current advancements in the field of nanostructured materials and in order to enhance the performance of electrochemical sensors in the quantitative detection of anti-cancer drugs in various matrices, the present review was coordinated. With rapid, accurate and responsive anti-cancer detection, electrochemical sensing techniques developed with the aid of nanomaterial technologies are seen to have improved efficiency. A detailed comparative study of different nanostructured materials has shown that sensor sensitivity can be dramatically enhanced along different routes such as, chemical functionalization, hybrid composite fabrication, and modification of surface architecture of these nanomaterials. However, further advances in this field of research need to be made in order to be able to overcome the major obstacles in this field of research which include, controlled synthesis, design of novel functional nanometric interfaces, understanding the relationships between structure, composition, and reactivity of nanomaterials, biocompatibility of nanomaterials with the biorecognition molecules, and cross interference within the structurally similar anti-cancer drugs (*e.g.*, methotrexate, 5-flourouracil, doxorubicin and etoposide). As such, it is anticipated that the use of NM-based electrochemical sensing systems can provide highly accurate on-the-spot evaluation of anti-cancer drugs in complex biological environments and under real-world conditions. By achieving synergies between multifunctional nanomaterials, recognition components, and electrochemical methods, the conceptual challenges associated with these advanced sensing tools can be further overcome. In addition, the desired portable devices can be built through successful connectivity and electrochemical sensor integration using communication devices. A successful convergence of two distinct research fields is presumed to have tremendous potential to effectively revolutionize all available anticancer diagnostic options.

CONSENT FOR PUBLICATION

Not applicable.

CONFLICT OF INTEREST

The authors declare that there are no conflicts of interest.

ACKNOWLEDGMENTS

The authors acknowledge the great support from the Department of Chemistry, JSS Science and Technology University, Mysuru -570006, Karnataka, India

REFERENCES

[1]   J.G. Manjunatha, "Surfactant modified carbon nanotube paste electrode for the sensitive determination of mitoxantrone anticancer drug", *J. Electrochem. Sci. Eng.,* vol. 7, no. 1, pp. 39-49, 2017.
        [http://dx.doi.org/10.5599/jese.368]

[2]   A.I. Barbosa, S.R. Fernandes, S. Machado, P. Sousa, O.Y. Sze, E.M. Silva, L. Barreiros, S.A. Lima, S. Reis, and M.A. Segundo, "Fast monolith-based chromatographic method for determination of methotrexate in drug delivery studies", *Microchem. J.,* vol. 148, pp. 185-189, 2019.
        [http://dx.doi.org/10.1016/j.microc.2019.04.075]

[3]   J.G. Manjunatha, M. Deraman, N.H. Basri, N.S.M. Nor, I.A. Talib, and N. Ataollahi, "Sodium dodecyl sulfate modified carbon nanotubes paste electrode as a novel sensor for the simultaneous determination of dopamine, ascorbic acid, and uric acid", *C. R. Chim.,* vol. 17, no. 5, pp. 465-476, 2014.
        [http://dx.doi.org/10.1016/j.crci.2013.09.016]

[4]   J.G. Manjunatha, "A novel voltammetric method for the enhanced detection of the food additive tartrazine using an electrochemical sensor", *Heliyon,* vol. 4, no. 11, p. e00986, 2018.
        [http://dx.doi.org/10.1016/j.heliyon.2018.e00986] [PMID: 30761373]

[5]   N. Hareesha, J.G. Manjunatha, C. Raril, and G. Tigari, "Sensitive and selective electrochemical resolution of tyrosine with ascorbic acid through the development of electropolymerized alizarin sodium sulfonate modified carbon nanotube paste electrodes", *ChemistrySelect,* vol. 4, no. 15, pp. 4559-4567, 2019.
        [http://dx.doi.org/10.1002/slct.201900794]

[6]   J.G. Manjunatha, "Electroanalysis of estriol hormone using electrochemical sensor", *Sens. Biosensing Res.,* vol. 16, pp. 79-84, 2017.
        [http://dx.doi.org/10.1016/j.sbsr.2017.11.006]

[7]   J.G. Manjunatha, B.K. Swamy, G.P. Mamatha, O. Gilbert, B.N. Chandrashekar, and B.S. Sherigara, "Electrochemical studies of dopamine and epinephrine at a poly (tannic acid) modified carbon paste electrode: a cyclic voltammetric study", *Int. J. Electrochem. Sci.,* vol. 5, pp. 1236-1245, 2010.

[8]   G. Li, R. Margueron, G. Hu, D. Stokes, Y.H. Wang, and D. Reinberg, "Highly compacted chromatin formed in vitro reflects the dynamics of transcription activation in vivo", *Mol. Cell,* vol. 38, no. 1, pp. 41-53, 2010.
        [http://dx.doi.org/10.1016/j.molcel.2010.01.042] [PMID: 20385088]

[9]   G. Tigari, and J.G. Manjunatha, "A surfactant enhanced novel pencil graphite and carbon nanotube composite paste material as an effective electrochemical sensor for determination of riboflavin", *J. Sci.: Adv. Mater. Devices,* vol. 5, no. 1, pp. 56-64, 2020.
        [http://dx.doi.org/10.1016/j.jsamd.2019.11.001]

[10]   J.G. Manjunatha, M. Deraman, N.H. Basri, and I.A. Talib, "Fabrication of poly (Solid Red A) modified carbon nano tube paste electrode and its application for simultaneous determination of epinephrine, uric acid and ascorbic acid", *Arab. J. Chem.,* vol. 11, no. 2, pp. 149-158, 2018.
        [http://dx.doi.org/10.1016/j.arabjc.2014.10.009]

[11]   G. Tigari, and J.G. Manjunatha, "Electrochemical preparation of poly (arginine)-modified carbon nanotube paste electrode and its application for the determination of pyridoxine in the presence of riboflavin: an electroanalytical approach", *J. Anal. Test.,* vol. 3, no. 4, pp. 331-340, 2019.
        [http://dx.doi.org/10.1007/s41664-019-00116-w]

[12]   G.M. Brenner, and C.W. Stevens, "Antineoplastic drugs", In: *Text book of Pharmacology* Saunders Elsevier, 2010, p. 493511.

[13]   H.J. Kondryn, C.L. Edmondson, J. Hill, and T.O. Eden, "Treatment non-adherence in teenage and young adult patients with cancer", *Lancet Oncol.,* vol. 12, no. 1, pp. 100-108, 2011.
        [http://dx.doi.org/10.1016/S1470-2045(10)70069-3] [PMID: 20580606]

[14]   T.G. Benedek, "Methotrexate: from its introduction to non-oncologic therapeutics to anti-TNF-α", *Clin. Exp. Rheumatol.*, vol. 28, no. 5, suppl. Suppl. 61, pp. S3-S8, 2010.
[PMID: 21044425]

[15]   L. Gao, Y. Wu, J. Liu, and B. Ye, "Anodic voltammetricbehaviors of methotrexate at a glassy carbon electrode and its determination in spiked human urine", *J. Electroanal. Chem. (Lausanne),* vol. 610, no. 2, pp. 131-136, 2007.
[http://dx.doi.org/10.1016/j.jelechem.2007.07.030]

[16]   S. Eksborg, F. Albertioni, C. Rask, O. Beck, C. Palm, H. Schroeder, and C. Peterson, "Methotrexate plasma pharmacokinetics: importance of assay method", *Cancer Lett.*, vol. 108, no. 2, pp. 163-169, 1996.
[http://dx.doi.org/10.1016/S0304-3835(96)04394-7] [PMID: 8973590]

[17]   Z. Deng, H. Li, Q. Tian, Y. Zhou, X. Yang, Y. Yu, B. Jiang, Y. Xu, and T. Zhou, "Electrochemical detection of methotrexate in serum sample based on the modified acetylene black sensor", *Microchem. J.,* vol. •••, p. 105058, 2020.
[http://dx.doi.org/10.1016/j.microc.2020.105058]

[18]   Z. Deng, and Q. Tian, "YinghuaXu, L., SukaiSun, X.Y., Liu, R., Zhou, Y., Li, H. and Zhou, T., Rapid and sensitive detection of methotrexate using acetylene black electrode", *Int. J. Electrochem. Sci.,* vol. 15, pp. 5058-5066, 2020.
[http://dx.doi.org/10.20964/2020.06.84]

[19]   E. Asadian, S. Shahrokhian, A.I. Zad, and F. Ghorbani-Bidkorbeh, "Glassy carbon electrode modified with 3D graphene–carbon nanotube network for sensitive electrochemical determination of methotrexate", *Sens. Actuators B Chem.,* vol. 239, pp. 617-627, 2017.
[http://dx.doi.org/10.1016/j.snb.2016.08.064]

[20]   A.A. Ensafi, F. Rezaloo, and B. Rezaei, "CoFe2O4/reduced graphene oxide/ionic liquid modified glassy carbon electrode, a selective and sensitive electrochemical sensor for determination of methotrexate", *J. Taiwan Inst. Chem. Eng.,* vol. 78, pp. 45-50, 2017.
[http://dx.doi.org/10.1016/j.jtice.2017.05.031]

[21]   Y. Guo, Y. Chen, Q. Zhao, S. Shuang, and C. Dong, "Electrochemical Sensor for Ultrasensitive Determination of Doxorubicin and Methotrexate Based on Cyclodextrin☐Graphene Hybrid Nanosheets", *Electroanalysis,* vol. 23, no. 10, pp. 2400-2407, 2011.
[http://dx.doi.org/10.1002/elan.201100259]

[22]   H.R.S. Lima, E.A. de Oliveira Farias, P.R.S. Teixeira, C. Eiras, and L.C.C. Nunes, "Blend films based on biopolymers extracted from babassu mesocarp (Orbignya phalerata) for the electrochemical detection of methotrexate antineoplastic drug", *J. Solid State Electrochem.,* vol. 23, no. 11, pp. 3153-3164, 2019.
[http://dx.doi.org/10.1007/s10008-019-04406-2]

[23]   H. Ghadimi, B. Nasiri-Tabrizi, P.M. Nia, W.J. Basirun, R.M. Tehrani, and F. Lorestani, "Nanocomposites of nitrogen-doped graphene decorated with a palladium silver bimetallic alloy for use as a biosensor for methotrexate detection", *RSC Advances,* vol. 5, no. 120, pp. 99555-99565, 2015.
[http://dx.doi.org/10.1039/C5RA18109G]

[24]   Y. Wang, J. Xie, L. Tao, H. Tian, S. Wang, and H. Ding, "Simultaneous electrochemical determination of epirubicin and methotrexate in human blood using a disposable electrode modified with nano-Au/MWNTs-ZnO composites", *Sens. Actuators B Chem.,* vol. 204, pp. 360-367, 2014.
[http://dx.doi.org/10.1016/j.snb.2014.07.099]

[25]   D. Asbahr, L.C.S. Figueiredo-Filho, F.C. Vicentini, G.G. Oliveira, O. Fatibello-Filho, and C.E. Banks, "Differential pulse adsorptive stripping voltammetric determination of nanomolar levels of methotrexate utilizing bismuth film modified electrodes", *Sens. Actuators B Chem.,* vol. 188, pp. 334-339, 2013.
[http://dx.doi.org/10.1016/j.snb.2013.07.027]

[26]   Y. Zhao, S. Zou, D. Huo, C. Hou, M. Yang, J. Li, and M. Bian, "Simple and sensitive fluorescence sensor for methotrexate detection based on the inner filter effect of N, S co-doped carbon quantum dots", *Anal. Chim. Acta,* vol. 1047, pp. 179-187, 2019.
[http://dx.doi.org/10.1016/j.aca.2018.10.005] [PMID: 30567648]

[27]   Y. Wang, H. Liu, F. Wang, and Y. Gao, "Electrochemical oxidation behavior of methotrexate at DNA/SWCNT/Nafion composite film-modified glassy carbon electrode", *J. Solid State Electrochem.,* vol. 16, no. 10, pp. 3227-3235, 2012.
[http://dx.doi.org/10.1007/s10008-012-1763-y]

[28]   F. Wang, Y. Wu, J. Liu, and B. Ye, "DNA Langmuir–Blodgett modified glassy carbon electrode as voltammetric sensor for determinate of methotrexate", *Electrochim. Acta,* vol. 54, no. 5, pp. 1408-1413, 2009.
[http://dx.doi.org/10.1016/j.electacta.2008.09.027]

[29]   S. Wang, Z. Qi, H. Huang, and H. Ding, "Electrochemical determination of methotrexate at a disposable screen-printed electrode and its application studies", *Anal. Lett.,* vol. 45, no. 12, pp. 1658-1669, 2012.
[http://dx.doi.org/10.1080/00032719.2012.677790]

[30]   R. Šelešovská, L. Janíková☐Bandžuchová, and J. Chýlková, "Sensitive Voltammetric Sensor Based on Boron☐Doped Diamond Electrode for Determination of the Chemotherapeutic Drug Methotrexate in Pharmaceutical and Biological Samples", *Electroanalysis,* vol. 27, no. 1, pp. 42-51, 2015.
[http://dx.doi.org/10.1002/elan.201400326]

[31]   Y. Wei, L. Luo, Y. Ding, X. Si, and Y. Ning, "Highly sensitive determination of methotrexate at poly (l-lysine) modified electrode in the presence of sodium dodecyl benzene sulfonate", *Bioelectrochemistry,* vol. 98, pp. 70-75, 2014.
[http://dx.doi.org/10.1016/j.bioelechem.2014.03.005] [PMID: 24727063]

[32]   G. Oliveira, B. Janegitz, V. Zucolotto, and O. Fatibello-Filho, "Differential pulse adsorptive stripping voltammetric determination of methotrexate using a functionalized carbon nanotubes-modified glassy carbon electrode", *Open Chem.,* vol. 11, no. 11, pp. 1837-1843, 2013.
[http://dx.doi.org/10.2478/s11532-013-0305-5]

[33]   N. Jandaghi, S. Jahani, M. Kazemipour, M.M. Foroughi, K. Mohammadi, and F. Borhani, "One-pot synthesis of cerium doped 3D ZnO nano-flowers modified on glassy carbon electrode as portable electro-chemical sensing platform for sensitive detection of methotrexate as an anticancer drug", *Synth. Met.,* vol. 256, p. 116119, 2019.
[http://dx.doi.org/10.1016/j.synthmet.2019.116119]

[34]   J. Chen, B. Fu, T. Liu, Z. Yan, and K. Li, "A graphene oxide-DNA electrochemical sensor based on glassy carbon electrode for sensitive determination of methotrexate", *Electroanalysis,* vol. 30, no. 2, pp. 288-295, 2018.
[http://dx.doi.org/10.1002/elan.201700615]

[35]   F. Wang, Y. Wang, K. Lu, X. Wei, and B. Ye, "Sensitive determination of methotrexate at nano-Au self-assembled monolayer modified electrode", *J. Electroanal. Chem. (Lausanne),* vol. 674, pp. 83-89, 2012.
[http://dx.doi.org/10.1016/j.jelechem.2012.04.010]

[36]   R.B. Diasio, and B.E. Harris, "Clinical pharmacology of 5-fluorouracil", *Clin. Pharmacokinet.,* vol. 16, no. 4, pp. 215-237, 1989.
[http://dx.doi.org/10.2165/00003088-198916040-00002] [PMID: 2656050]

[37]   S.B. Brown, E.A. Brown, and I. Walker, "The present and future role of photodynamic therapy in cancer treatment", *Lancet Oncol.,* vol. 5, no. 8, pp. 497-508, 2004.
[http://dx.doi.org/10.1016/S1470-2045(04)01529-3] [PMID: 15288239]

[38]   H.M. Pinedo, and G.F. Peters, "Fluorouracil: biochemistry and pharmacology", *J. Clin. Oncol.,* vol. 6, no. 10, pp. 1653-1664, 1988.

[http://dx.doi.org/10.1200/JCO.1988.6.10.1653] [PMID: 3049954]

[39]   M. Hadi, T. Mollaei, and A. Ehsani, "Graphene oxides/multi-walled carbon nanotubes hybrid-modified carbon electrodes for fast and sensitive voltammetric determination of the anticancer drug 5-fluorouracil in spiked human plasma samples", *Chem. Pap.*, vol. 72, no. 2, pp. 431-439, 2018.
[http://dx.doi.org/10.1007/s11696-017-0295-4]

[40]   P. Norouzi, M.A. Eshraghi, and M. Ebrahimi, "DNA Biosensor for Determination of 5-Fluorouracil based on Gold Electrode Modified with Au and Polyaniline Nanoparticles and FFT Square Wave Voltammetry", *Journal of Applied Chemical Research,* vol. 13, no. 1, pp. 24-35, 2019.

[41]   F.M. Zahed, B. Hatamluyi, F. Lorestani, and Z. Es'haghi, "Silver nanoparticles decorated polyaniline nanocomposite based electrochemical sensor for the determination of anticancer drug 5-fluorouracil", *J. Pharm. Biomed. Anal.*, vol. 161, pp. 12-19, 2018.
[http://dx.doi.org/10.1016/j.jpba.2018.08.004] [PMID: 30142492]

[42]   P.A. Pushpanjali, J.G. Manjunatha, and N. Hareesha, "An overview of recent developments of carbon-based sensors for the analysis of drug molecules", *J. Electrochem. Sci. Technol.*, vol. 11, no. 3, pp. 161-177, 2021.

[43]   J.C. Abbar, N.P. Shetti, and S.T. Nandibewoor, "Development of Voltammetric Method for the Determination of an Anticancer Drug, 5-Flurouracil, at a Multiwalled Carbon Nanotubes Paste Electrode", *Synth. React. Inorg. Met.-Org. Nano-Met. Chem.*, vol. 46, no. 6, pp. 814-820, 2016.
[http://dx.doi.org/10.1080/15533174.2014.989586]

[44]   B. Hatamluyi, "Es' haghi, Z., Zahed, F.M. and Darroudi, M. A novel electrochemical sensor based on GQDs-PANI/ZnO-NCs modified glassy carbon electrode for simultaneous determination of Irinotecan and 5-Fluorouracil in biological samples", *Sens. Actuators B Chem.*, vol. 286, pp. 540-549, 2019.
[http://dx.doi.org/10.1016/j.snb.2019.02.017]

[45]   X. Hua, X. Hou, X. Gong, and G. Shen, "Electrochemical behavior of 5-fluorouracil on a glassy carbon electrode modified with bromothymol blue and multi-walled carbon nanotubes", *Anal. Methods,* vol. 5, no. 10, pp. 2470-2476, 2013.
[http://dx.doi.org/10.1039/c3ay40149a]

[46]   D. Lima, G.N. Calaça, A.G. Viana, and C.A. Pessôa, "Porphyran-capped gold nanoparticles modified carbon paste electrode: a simple and efficient electrochemical sensor for the sensitive determination of 5-fluorouracil", *Appl. Surf. Sci.*, vol. 427, pp. 742-753, 2018.
[http://dx.doi.org/10.1016/j.apsusc.2017.08.228]

[47]   M. Satyanarayana, K.Y. Goud, K.K. Reddy, and K.V. Gobi, "Biopolymer stabilized nanogold particles on carbon nanotube support as sensing platform for electrochemical detection of 5-fluorouracil in-vitro", *Electrochim. Acta*, vol. 178, pp. 608-616, 2015.
[http://dx.doi.org/10.1016/j.electacta.2015.08.036]

[48]   M. Roushani, Z. Saeidi, S. Hemati, and M. Hosseini, "Highly sensitive electrochemical determination of 5-fluorouracil using CuNPs/MWCNT/IL/Chit composite modified glassy carbon electrode", *Advances in Nanochemistry,* vol. 1, no. 2, pp. 73-77, 2019.

[49]   S.D. Bukkitgar, and N.P. Shetti, "Electrochemical Sensor for the Determination of Anticancer Drug 5-Fluorouracil at Glucose Modified Electrode", *ChemistrySelect,* vol. 1, no. 4, pp. 771-777, 2016.
[http://dx.doi.org/10.1002/slct.201600197]

[50]   S.D. Bukkitgar, and N.P. Shetti, "Electrochemical behavior of an anticancer drug 5-fluorouracil at methylene blue modified carbon paste electrode", *Mater. Sci. Eng. C,* vol. 65, pp. 262-268, 2016.
[http://dx.doi.org/10.1016/j.msec.2016.04.045] [PMID: 27157751]

[51]   V.P. Pattar, and S.T. Nandibewoor, "Electroanalytical method for the determination of 5-fluorouracil using a reduced graphene oxide/chitosan modified sensor", *RSC Advances,* vol. 5, no. 43, pp. 34292-34301, 2015.
[http://dx.doi.org/10.1039/C5RA04396D]

[52] M. Rahimi-Nasrabadi, F. Ahmadi, H. Beigizadeh, M.S. Karimi, A. Sobhani-Nasab, Y. Joseph, H. Ehrlich, and M.R. Ganjali, "A modified sensitive carbon paste electrode for 5-fluorouracil based using a composite of praseodymium erbium tungstate", *Microchem. J.,* vol. 154, p. 104654, 2020. [http://dx.doi.org/10.1016/j.microc.2020.104654]

[53] R. Emamian, M. Ebrahimi, and H. Karimi-Maleh, "A sensitive sensor for nano-molar detection of 5-fluorouracil by modifying a paste sensor with graphene quantum dots and an ionic liquid", *Journal of Nanostructures,* vol. 10, no. 2, pp. 230-238, 2020.

[54] P.A.J. Speth, Q.G.C.M. van Hoesel, and C. Haanen, "Clinical pharmacokinetics of doxorubicin", *Clin. Pharmacokinet.,* vol. 15, no. 1, pp. 15-31, 1988. [http://dx.doi.org/10.2165/00003088-198815010-00002] [PMID: 3042244]

[55] J.L. Speyer, M.D. Green, J. Bottino, J. Wernz, R.H. Blum, and F.M. Muggia, "A Phase I/II Study of 6-Hour and 24-Hour Intravenous Infusions of Doxorubicin", In: *Anthracycline Antibiotics in Cancer Therapy.* Springer: Dordrecht, 1982, pp. 398-410. [http://dx.doi.org/10.1007/978-94-009-7630-6_37]

[56] T. Suda, J. Suda, and M. Ogawa, "Disparate differentiation in mouse hemopoietic colonies derived from paired progenitors", *Proc. Natl. Acad. Sci. USA,* vol. 81, no. 8, pp. 2520-2524, 1984. [http://dx.doi.org/10.1073/pnas.81.8.2520] [PMID: 6585813]

[57] R.C. Young, R.F. Ozols, and C.E. Myers, "The anthracycline antineoplastic drugs", *N. Engl. J. Med.,* vol. 305, no. 3, pp. 139-153, 1981. [http://dx.doi.org/10.1056/NEJM198107163050305] [PMID: 7017406]

[58] J.H. Beijnen, H. Lingeman, H.A. Van Munster, and W.J.M. Underberg, "Mitomycin antitumour agents: a review of their physico-chemical and analytical properties and stability", *J. Pharm. Biomed. Anal.,* vol. 4, no. 3, pp. 275-295, 1986. [http://dx.doi.org/10.1016/0731-7085(86)80050-4] [PMID: 16867594]

[59] J.A. Benvenuto, S.C. Adams, H.M. Vyas, and R.W. Anderson, "Pharmaceutical issues in infusion chemotherapy stability and compatibility", In: *Cancer chemotherapy by infusion.* Springer: Dordrecht, 1987, pp. 100-113. [http://dx.doi.org/10.1007/978-94-009-3193-0_7]

[60] C. Zou, J. Hu, Y. Su, Z. Zhou, B. Cai, Z. Tao, T. Huo, N. Hu, and Y. Zhang, "Highly repeatable and sensitive three-dimensional γ-Fe2O3@ reduced graphene oxide gas sensors by magnetic-field assisted assembly process", *Sens. Actuators B Chem.,* vol. 306, p. 127546, 2020. [http://dx.doi.org/10.1016/j.snb.2019.127546]

[61] S. Jahandari, M.A. Taher, H. Karimi-Maleh, and G. Mansouri, "Simultaneous voltammetric determination of glutathione, doxorubicin and tyrosine based on the electrocatalytic effect of a nickel(II) complex and of Pt:Co nanoparticles as a conductive mediator", *Mikrochim. Acta,* vol. 186, no. 8, p. 493, 2019. [http://dx.doi.org/10.1007/s00604-019-3598-z] [PMID: 31267341]

[62] E. Er, and N. Erk, "Construction of a sensitive electrochemical sensor based on 1T-MoS2 nanosheets decorated with shape-controlled gold nanostructures for the voltammetric determination of doxorubicin", *Mikrochim. Acta,* vol. 187, no. 4, p. 223, 2020. [http://dx.doi.org/10.1007/s00604-020-4206-y] [PMID: 32166596]

[63] F. Yan, J. Chen, Q. Jin, H. Zhou, A. Sailjoi, J. Liu, and W. Tang, "Fast one-step fabrication of a vertically-ordered mesoporous silica-nanochannel film on graphene for direct and sensitive detection of doxorubicin in human whole blood", *J. Mater. Chem. C Mater. Opt. Electron. Devices,* vol. 8, no. 21, pp. 7113-7119, 2020. [http://dx.doi.org/10.1039/D0TC00744G]

[64] M.H. Ghanbari, and Z. Norouzi, "A new nanostructure consisting of nitrogen-doped carbon nanoonions for an electrochemical sensor to the determination of doxorubicin", *Microchem. J.,* p. 105098, 2020.

[http://dx.doi.org/10.1016/j.microc.2020.105098]

[65]   M. Taei, F. Hasanpour, H. Salavati, and S. Mohammadian, Fast and sensitive determination of doxorubicin using multi-walled carbon nanotubes as a sensor and $CoFe_2O_4$ magnetic nanoparticles as a mediator., *Microchimica acta,* vol. 183, no. 1, pp. 49-56, 2016.

[66]   S.A. Alavi-Tabari, M.A. Khalilzadeh, and H. Karimi-Maleh, "Simultaneous determination of doxorubicin and dasatinib as two breast anticancer drugs uses an amplified sensor with ionic liquid and ZnO nanoparticle", *J. Electroanal. Chem. (Lausanne),* vol. 811, pp. 84-88, 2018.
[http://dx.doi.org/10.1016/j.jelechem.2018.01.034]

[67]   E. Haghshenas, T. Madrakian, and A. Afkhami, "Electrochemically oxidized multiwalled carbon nanotube/glassy carbon electrode as a probe for simultaneous determination of dopamine and doxorubicin in biological samples", *Anal. Bioanal. Chem.,* vol. 408, no. 10, pp. 2577-2586, 2016.
[http://dx.doi.org/10.1007/s00216-016-9361-y] [PMID: 26873212]

[68]   Y. Hahn, and H.Y. Lee, "Electrochemical behavior and square wave voltammetric determination of doxorubicin hydrochloride", *Arch. Pharm. Res.,* vol. 27, no. 1, pp. 31-34, 2004.
[http://dx.doi.org/10.1007/BF02980041] [PMID: 14969334]

[69]   J. Fei, X. Wen, Y. Zhang, L. Yi, X. Chen, and H. Cao, "Voltammetric determination of trace doxorubicin at a nano-titania/nafion composite film modified electrode in the presence of cetyltrimethylammonium bromide", *Microchimica acta,* vol. 164, no. 1-2, pp. 85-91, 2009.

[70]   J. Liu, X. Bo, M. Zhou, and L. Guo, "A nanocomposite prepared from metal-free mesoporous carbon nanospheres and graphene oxide for voltammetric determination of doxorubicin", *Mikrochim. Acta,* vol. 186, no. 9, p. 639, 2019.
[http://dx.doi.org/10.1007/s00604-019-3754-5] [PMID: 31440837]

[71]   N. Hashemzadeh, M. Hasanzadeh, N. Shadjou, J. Eivazi-Ziaei, M. Khoubnasabjafari, and A. Jouyban, "Graphene quantum dot modified glassy carbon electrode for the determination of doxorubicin hydrochloride in human plasma", *J. Pharm. Anal.,* vol. 6, no. 4, pp. 235-241, 2016.
[http://dx.doi.org/10.1016/j.jpha.2016.03.003] [PMID: 29403988]

[72]   A. Peng, H. Xu, C. Luo, and H. Ding, "Application of a disposable doxorubicin sensor for direct determination of clinical drug concentration in patient blood", *Int. J. Electrochem. Sci.,* vol. 11, pp. 6266-6278, 2016.
[http://dx.doi.org/10.20964/2016.07.38]

[73]   H. Guo, H. Jin, R. Gui, Z. Wang, J. Xia, and F. Zhang, "Electrodeposition one-step preparation of silver nanoparticles/carbon dots/reduced graphene oxide ternary dendritic nanocomposites for sensitive detection of doxorubicin", *Sens. Actuators B Chem.,* vol. 253, pp. 50-57, 2017.
[http://dx.doi.org/10.1016/j.snb.2017.06.095]

[74]   R. Hajian, Z. Tayebi, and N. Shams, "Fabrication of an electrochemical sensor for determination of doxorubicin in human plasma and its interaction with DNA", *J. Pharm. Anal.,* vol. 7, no. 1, pp. 27-33, 2017.
[http://dx.doi.org/10.1016/j.jpha.2016.07.005] [PMID: 29404015]

[75]   E.L. Baldwin, and N. Osheroff, "Etoposide, topoisomerase II and cancer", *Curr. Med. Chem. Anticancer Agents,* vol. 5, no. 4, pp. 363-372, 2005.
[http://dx.doi.org/10.2174/1568011054222364] [PMID: 16101488]

[76]   E.N. Chaney Jr, and R.P. Baldwin, "Electrochemical determination of adriamycin compounds in urine by preconcentration at carbon paste electrodes", *Anal. Chem.,* vol. 54, no. 14, pp. 2556-2560, 1982.
[http://dx.doi.org/10.1021/ac00251a034] [PMID: 7158780]

[77]   C.A. Felix, ""Secondary leukemias induced by topoisomerase-targeted drugs." Biochimica et Biophysica Acta (BBA)-", *Gene Structure and Expression,* vol. 1400, no. 1-3, pp. 233-255, 1998.
[http://dx.doi.org/10.1016/S0167-4781(98)00139-0]

[78]   C.A. Felix, "Leukemias related to treatment with DNA topoisomerase II inhibitors", *Medical and*

Pediatric Oncology: The Official Journal of SIOP—International Society of Pediatric Oncology (Societé Internationale d'OncologiePédiatrique, vol. 36, no. 5, pp. 525-535, 2001.
[http://dx.doi.org/10.1002/mpo.1125]

[79]   H. Hrichi, L. Monser, and N. Adhoum, "Selective Electrochemical Determination of Etoposide Using a Molecularly Imprinted Overoxidized Polypyrrole Coated Glassy Carbon Electrode", *Int. J. Electrochem.,* 2019.
[http://dx.doi.org/10.1155/2019/5394235]

[80]   "sadatVajedi, F. and Dehghani, H. A high-sensitive electrochemical DNA biosensor based on a novelZnAl/layered double hydroxide modified cobalt ferrite-graphene oxide nanocomposite electrophoretically deposited onto FTO substrate for electroanalytical studies of etoposide", *Talanta,* vol. 208, p. 120444, 2020.
[http://dx.doi.org/10.1016/j.talanta.2019.120444] [PMID: 31816745]

[81]   B. Bozal-Palabiyik, B. Dogan-Topal, B. Uslu, A. Can, and S.A. Ozkan, "Sensitive voltammetric assay of etoposide using modified glassy carbon electrode with a dispersion of multi-walled carbon nanotube", *J. Solid State Electrochem.,* vol. 17, no. 11, pp. 2815-2822, 2013.
[http://dx.doi.org/10.1007/s10008-013-2184-2]

[82]   A. Özkan, N. Atar, and M.L. Yola, "Enhanced surface plasmon resonance (SPR) signals based on immobilization of core-shell nanoparticles incorporated boron nitride nanosheets: Development of molecularly imprinted SPR nanosensor for anticancer drug, etoposide", *Biosens. Bioelectron.,* vol. 130, pp. 293-298, 2019.
[http://dx.doi.org/10.1016/j.bios.2019.01.053] [PMID: 30776616]

[83]   B. Hatamluyi, F. Lorestani, and Z. Es'haghi, "Au/Pd@rGO nanocomposite decorated with poly (L-Cysteine) as a probe for simultaneous sensitive electrochemical determination of anticancer drugs, Ifosfamide and Etoposide", *Biosens. Bioelectron.,* vol. 120, pp. 22-29, 2018.
[http://dx.doi.org/10.1016/j.bios.2018.08.008] [PMID: 30144642]

[84]   A.E. Radi, N. Abd-Elghany, and T. Wahdan, "Electrochemical study of the antineoplastic agent etoposide at carbon paste electrode and its determination in spiked human serum by differential pulse voltammetry", *Chem. Pharm. Bull. (Tokyo),* vol. 55, no. 9, pp. 1379-1382, 2007.
[http://dx.doi.org/10.1248/cpb.55.1379] [PMID: 17827766]

[85]   A.E. Radi, H.M. Nassef, and A. Eissa, "Voltammetric and ultraviolet–visible spectroscopic studies on the interaction of etoposide with deoxyribonucleic acid", *Electrochim. Acta,* vol. 113, pp. 164-169, 2013.
[http://dx.doi.org/10.1016/j.electacta.2013.09.046]

[86]   H.V. Nguyen, L. Richtera, A. Moulick, K. Xhaxhiu, J. Kudr, N. Cernei, H. Polanska, Z. Heger, M. Masarik, P. Kopel, M. Stiborova, T. Eckschlager, V. Adam, and R. Kizek, "Electrochemical sensing of etoposide using carbon quantum dot modified glassy carbon electrode", *Analyst (Lond.),* vol. 141, no. 9, pp. 2665-2675, 2016.
[http://dx.doi.org/10.1039/C5AN02476E] [PMID: 26882954]

CHAPTER 13

Voltammetric Studies of Dyes and their Role as Carbon Electrode Modifiers

Shefali Sharma[1], Anup Pandith[2], Shankramma Kalikeri[3], Nagaraja Sreeharsha[4,5] and Gururaj Kudur Jayaprakash[1,*]

[1] *School of Advanced School of Chemical Science, Shoolini University, Bajhol, Himachal Pradesh 173229, India*

[2] *Department of Chemistry, School of Natural Sciences, Kyung Hee University, Seoul, South Korea*

[3] *Division of Nanoscience and Technology, Department of Water and Health (Faculty of Life Sciences) JSS Academy of Higher Education & Research (Deemed to be University), Mysuru, Karnataka 570004, India*

[4] *Department of Pharmaceutical Sciences, College of Clinical Pharmacy, King Faisal University, Al-Ahsa-31982, Saudi Arabia*

[5] *Department of Pharmaceutics, Vidya Siri College of Pharmacy, Off Sarjapura Road, Bengaluru – 560035, Karnataka, India*

Abstract: Synthetic dyes are very commonly seen in commercial products nowadays. Synthetic dyes are commonly used to give a variety of colors and shades in several commercial products. The occurrence of synthetic dyes in water after certain levels can significantly affect living beings by causing toxicity and mutagenicity. Electroanalytical methods can detect these dyes with high sensitivity and selectivity. It is also interesting to note that, commercial dyes like calmagite, rhodamine, Congo red, and many more have shown significant applications as electrode modifiers. In the current chapter, we are going to discuss the use of electrochemical methods to sense commercial dyes and applications of dye-modified carbon paste electrodes in the voltammetric analysis.

Keywords: Carbon paste, Dyes, Real samples, Sensors, Voltammetry.

INTRODUCTION

In the past century, synthetic dyes have become more common in industrialization and they have been extensively used as colors and shades in many commercial products like textile, foods, paper, and cosmetics [1-5]. The organic dyes are typically complex aromatic compounds with substantial structural diversity.

** **Corresponding author Gururaj Kudur Jayaprakash:** School of Advanced School of Chemical Science, Shoolini University, Bajhol, Himachal Pradesh 173229, India; rajguru97@gmail.com

J.G. Manjunatha (Ed.)

Most organic compounds, dyes possess color because they absorb light in the visible spectrum (400-700 nm); have at least one chromophore (color-bearing group); have a conjugated system *i.e.*, a structure with alternating double and single bond and exhibit resonance of electrons, which is a stabilizing force in organic compounds. When any one of these features is lacking from the molecular structure the color is lost. In addition to chromophores, most dyes also contain groups known as auxochromes (color helpers), examples of which are carboxylic acid, sulfonic acid, amino, and hydroxyl groups. Increasing the number of electron-attracting groups conjugated with the electron-donor has a bathochromic effect.

The utilization and release of dyes into the ecosystem have received great attention since approximately 9% (40,000 tons) of the dyes produced worldwide (450,000 tons) are discharged as textile waste water [6-7]. The occurrence of synthetic dyes in water after certain levels can significantly affect living beings by causing toxicity and mutagenicity. Therefore, analytical methods able to determine these synthetic dyes in industrial effluents, food, and water are of great interest to the research community.

Methods such as spectroscopy and chromatography [8-10] are tested to detect the dyes in the commercial samples. Unfortunately, these methods are laborious and time-consuming. In this context, electrochemical techniques, especially the voltammetric techniques, have been used as an alternative method due to the high selectivity, sensitivity, low cost, use of low-quantity sample, little or no sample treatment, and low waste generation, which contribute to reduced environmental impact [11-24].

Carbon paste electrodes (CPEs) belong to promising electrochemical or bio-electrochemical sensors with wide applicability for detecting various organic and inorganic molecules. In 1958, it was exactly 63 years before Ralph Norman Adams from the University of Kans published a short one-page report in which he introduced CPE, initially, experiments were designed to develop a new electrode as an alternative to the dropping mercury electrode [11]. The first carbon paste was prepared by mixing 1 gram of carbon with 7 ml of bromoform1. In the last decade variety of research had been done to improve the electrochemical performance of CPE and expand its application to detect a wide variety of analytes. In the below-mentioned table (Table **1**), we are mentioning different working electrodes with their advantages and disadvantages.

The current form of CPE is quite different from Adams's work. The most widely used carbon material to prepare CPE is graphite powder apart from its variety of materials like graphene, carbon nanotube, and fullerene was also tested.

Currently, researchers are modifying CPE to further improve their properties like kinetics and stability. CPE can be modified mainly using three different methods like grinding, surface immobilization, and electro-deposition [19-24]. In the current chapter, we will mainly discuss the utilities of organic dyes modified (electro-deposited) CPE for sensing utilities.

Table 1. Advantages and disadvantages of different electrodes.

Electrode Materials	Advantage	Limitation
Pt	Available in dimension (as wire flat plate & tube). Large range of sizes. The rigidity can be improved (Pt-Rh alloy).	Expensive. Low hydrogen overvoltage socathodic potential range limited.
Au	Configurations same as Pt. The larger cathodic potential range.	The larger cathodic potential range. The anodic window is limited by surface oxidation.Expensive.
Carbon	Many types and configurations. The good cathodic potential range.	Quality varies greatly.Hard to give shape.
Carbon-paste	Wide potential range. Low background currentInexpensive.	Unstable in flow cellsCannot use in organic solvents.
Hg	Excellent cathodic window. Easy to "refresh"Forms amalgams.	Limited anodic window due tomercuryoxidationToxic.

PREPARATION OF ELECTRODES

To prepare an organic modified carbon paste modified electrode, CPE was placed in some electrolyte (buffer or bases like NaOH and KOH) solution. Electro-deposition can be (polymerization) was achieved by applying a potential of between to fixed potential (negative and positive) at a suitable scan rate for a few cycles using a method like cyclic voltammetry. After the electro-deposition, modified CPE was thoroughly washed with distilled water before electroanalysis. The electro-deposition process was pictorially represented in Fig. (**1**).

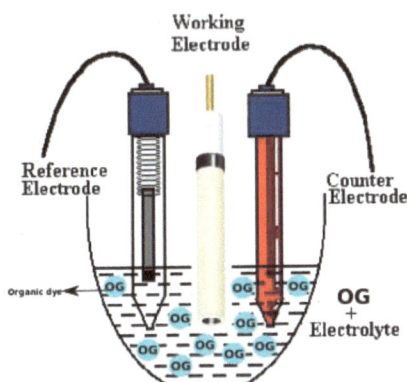

Fig. (1). Pictorial representation of voltammetric cell for organic dye deposition.

ELECTROANALYSIS OF TEXTILE DYES

It is possible to analyze the number of textile dyes from electroanalytical methods. In the current chapter, we will discuss how voltammetry can be used to detect textile dye Rhodamine B (RhB). Fig. (**2**) shows the atomic structure of RhB. RhB is also widely used in leather, paper, and porcelain productions. RhB is a carcinogenic chemical: ingestion, inhalation, and skin contact of RhB may cause acute and chronic poisoning and injury. Thus, due to the perilous nature and harmful effects of RhB, researchers making efforts to develop a simple method for the determination of RhB. Various methods have been proposed to detect RhB, such as chromatography [25] and spectrometry [26]. Unfortunately, the above-mentioned techniques are not economical, time-consuming, less available, difficult to handle thus making them less convenient in practice. In contrast, electrochemical methods have been widely used in chemo/biosensing due to their low cost, rapid analysis, easy operation, and high selectivity [11 - 24].

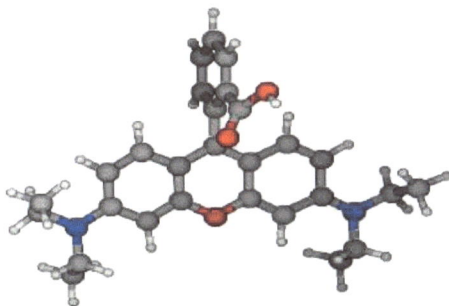

Fig. (2). Ball and stick atomic model of RhB-14 (C=Grey; O=Red; H=White; N=Blue).

Recently, some electrochemical methods have been developed to detect RhB. For instance, Y. Yi *et al.*, modified the glassy carbon electrode from the 6-thio-β-cyclodextrin functionalized nanogold/hollow carbon nanospheres, they found that after modification the oxidation peak potential of RhB shifted in a negative direction with an increase in peak current [27]. This indicates 6-thio-β-cyclodextrin functionalized nanogold/hollow carbon nanospheres acts as an electro-catalyst for the detection of RhB. They have also analyzed the determination of RhB in water and tomato juice. They got very good results with recovery range from 92 to 104%. D. Sung and X. Yang modified the glassy carbon electrode (GCE) from the graphene nanosheets for the determination of RhB [28]. These results prove that surface area, the number of active sites, and accumulation efficiency of RhB increases at the graphene nanosheet modified GCE as a result sensitivity of the electrode will be improved and the electrode can detect the RhB in real samples (Soy sauce) effectively. The above-mentioned research confirms that electrochemical methods are promising for the detection of RhB in real samples. H. D. Ngo and his colleagues modified electrodes based on the composite of zeolite imidazolate framework-67/reduced graphene oxide (ZIF-67/rGO) [29]. Their results showed that modifications are useful for improving the electrode efficiency for determining the RhB (Fig. 3).

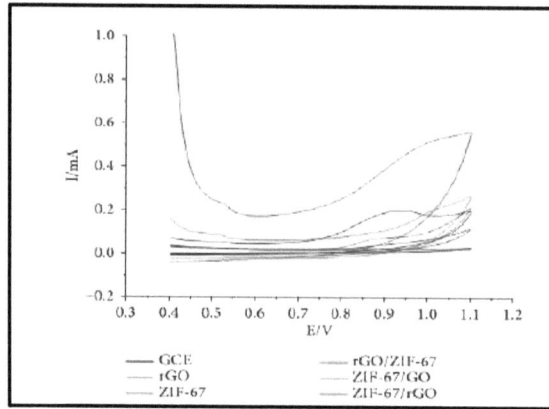

Fig. (3). CVs is obtained in an aqueous solution containing 24 mg/L RhB (0.1 M BR-BS pH 6). Reprinted with the permission of Reference 29. Copyright 2020, Hindawi.

rGO showed the synergetic effects with good electroconductivity properties and ZIF-67 with excellent textural properties, the use of ZIF-67/rGO modified electrode in differential pulse voltammetry (DPV) technique provides the broad linear response range and low LOD of RhB detection. Furthermore, they prove that the ZIF-67/rGO-GCE has good stability and reproducibility making the proposed method a prospective application for the determination of RhB in foodstuffs.

ELECTROANALYSIS OF MARKER DYES

The marker dye is commonly used for water and fuel applications. These dyes are used in fuel coloring due to their higher solubility in organic solvents, nonpolar and slightly polar. They are commonly used as marker solvents in petroleum derivatives, such as waxes, lubricants, plastics, and other nonpolar materials. The marker dye usually has the chemical structure protected and only the manufacturer can detect it by specific methods. Analytical methods, able to determine these types of dyes and help in the protection of the quality of fuel, are highly demanded. For these purposes electrochemical methods are useful. One of the most used anthraquinone dyes is 1,4-Bis(pentylamino)anthraquinone also called Solvent Blue 14(SB-14). The structure of SB-14. is showed below (Fig. **4**).

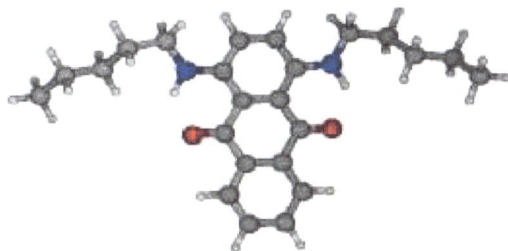

Fig. (4). Ball and stick atomic model of SB-14 (C=Grey; O=Red; H=White; N=Blue).

M. A. G. Tindade *et al.* determined the SB-14 by cyclic voltammetry and square wave voltammetry [30]. For their studies, they have used the screen-printed carbon electrode as a working electrode. In their method, they have proved that voltammetric methods can be used to detect fuel marker dyes in fuel samples like alcohol and kerosene without any pretreatments. It is interesting to note that in their work relative standard deviations are higher in kerosene (8.30%) when compared to alcohol (2.40%). It will be interesting to findout the reasons why detection of marker dyes in kerosene is difficult when compared to alcohols.

BENEFITS OF ORGANIC DYE MODIFICATIONS

As we have discussed in the Introduction section, CPE is a more suitable candidate as a working electrode material in electrochemical analysis. Carbon materials have been utilized in both analytical and industrial electrochemistry, and in fields, they showed superior performance than the noble metals. However, CPE properties further can be improved by electro-deposition of organic dyes. Organic dye at the CPE interface improves active surface area for electron transfer reactions, selectivity, stability, conductivity, detection limits, accuracy, and reproducibility of the working electrodes. Therefore, organic dye deposited CPE is very common in electrochemical analysis.

Organic Dyes Modification for Voltammetric Analysis of Dihydroxybenzene

Dihydroxybenzene has two positional isomers catechol (CC) and hydroquinone (HQ). These two positional isomers are widely present in the modern industrial effluents like plastic, cosmetic, paint, and leather industries. These two molecules are considered pollutants these two molecules can significantly affect the health of animals. Therefore, there is a huge demand to develop a sensor for these molecules.

HQ and CC generally coexist in the raw sample. Both possess almost similar oxidation potential because of their isomorphic nature. HQ and CC both have slower electron transfer possess at the CPE interface. Therefore, it is important to modify the CPE for the simultaneous detection of HQ and CC. Organic dyes can be used to detect HQ and CC.

K. Chetankumar *et al.* modified the pencil graphite electrode (PGE) from direct yellow by the electro-polymerization method [31]. After the modification of PGE, it exhibits an excellent electrochemical response for the analytes using cyclic voltammetry (CV) and differential pulse voltammetric (DPV) technique. They have analyzed the kinetic property of a modified electrode. By the scan rate analysis, they found that electron transfer properties after modification were both adsorption and diffusion controlled. The detection limit of the modified electrode for catechol (CC) (0.11 µM) and hydroquinone (HQ) (0.16 µM) respectively. They also did the simultaneous electroanalysis, bare PGE was failed to depict the oxidation peaks, whereas modified electrode gives well separated three anodic peaks for CC, HQ, and resorcinol (RS) with superior enhancement in peak current. Their proposed modified electrode was applied for the quantification of CC, HQ, and RS in tap water and obtained a consistent recovery in-between 96 and 100%.

Jamballi G. Manjunatha modified the graphene paste electrode from rosaniline. He has modified the electrode by electropolymerization method using the cyclic voltammetric method [32]. After the modification electrode capacity was improved for detection of CC as showed in Fig. (**5**).

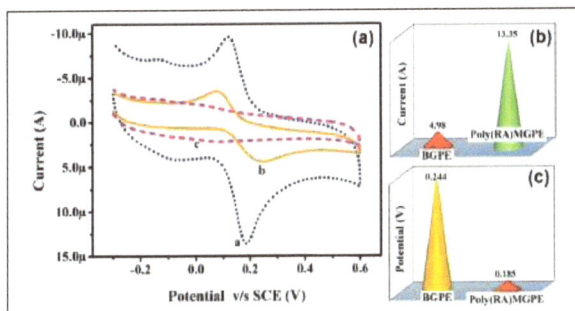

Fig. (5). (a) Cyclic voltammograms for the the presence absence (curve c) of 0.1 mM CC at Poly(RA)MGPE (curve a) and BGPE (curve b) in 0.1 M PBS (pH 7.0) at a 0.1 V/s scan rate, **(b)** relation between currentcurrent response BGPE and Poly(RA)MGPE, **(c)** relationship between current obtained at BGPE and Poly(RA)MGPE. Reprinted with the permission Ref 32. Copyright 2020, MDPI.

The proposed electrode by the author has some imperious applications such as fine reproducibility, repeatability, high stability, environmentally friendly, easily approachable, low price, and these applications provide a good platform in the sensor field. Additionally, the proposed sensor exhibits enhanced stability for the electro-reduction of CC in the occurrence of HQ.

Poly(RA)MGPE displays a superb linear response, reliability, selectivity, sensitivity, and rapid response, and a lower LOD towards CC detection. The suggested voltammetrictechnique was efficiently applied.for CC analysis in water samples. The inset in Fig. **(6)** picturizes the surface morphology of BGPE (Fig. **6a**) and Poly(RA)MGPE (Fig. **6b**). In Fig. **(6a)**, the unsystematically distributed flakes of graphene represent the surface of the BGPE material. Conversely, Fig. **(6b)** displays the regularly arranged electrochemical polymer layers on the surface of graphene flakes, characterizing the Poly(RA)MGP material.

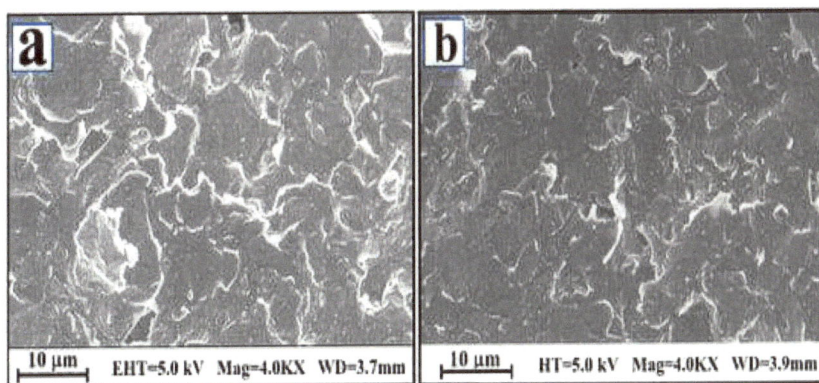

Fig. (6). a) FE-SEM image of BGPE. **(b)** FE-SEM image of Poly(RA)MGPE. Reprinted with the permission Ref 32. Copyright 2020, MDPI.

Organic Dyes Modification for Voltammetric Analysis of Dopamine

Dopamine (DA) is the neurochemical present in the mammalian nervous system; it plays a crucial role in maintaining the functional activities of cardiovascular and hormonal systems. Abnormal levels of DA are linked to various diseases like schizophrenia, Parkinson's, gout, scurvy, and cardiovascular diseases. Therefore, there is a great scope for developing DA sensors and researchers who are working on voltammetry are interested to develop sensors for DA. Fig. (**7**) shows the atomic structure of DA.

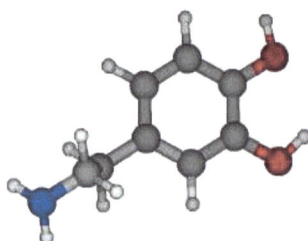

Fig. (7). Ball and stick atomic model of DA (C=Grey; O=Red; H=White; N=Blue).

Umesh Chandra *et al.* have modified CPE by calmagite. They have deposited the calmagite film on the CPE surface using CV [33]. The calmagite modified electrode showed a better electrochemical resolution when compared to bare CPE. Calmagite reduced the potentials difference (ΔEp) of DA with increased peak current. This indicates that calmagite modified CPE is very advantageous for determining DA. The authors also showed analytical applications of calmagite modified CPE for detecting DA in commercial injection samples. As a part of the work authors also proved that calmagite can sense DA selectively in presence of ascorbic acid and uric acid. C. M. Kuskur *et al.* modified the CPE from the naphthol green B, after modification sensitivity, selectivity, kinetics of the CPE was greatly improved [34]. T. Thomas *et al.* modified the CPE from the Rhodamine B [35]. They have applied modified electrode to detect DA, they found that peak currents of DA are higher almost double at the Rhodamine B modified CPE surface. The increase in peak currents is attributed to the electrostatic attraction of the DA cation towards the negatively charged Rhodamine B film.

Organic Dyes Modification for Voltammetric Analysis of Paracetamol

Paracetamol (PA) is used for reducing flue which originates from viruses and bacteria. Fig. (**8**) shows the atomic structure of PA. PA is used as an antipyretic and analgesic drug. Comparatively, PA is an effective and safe analgesic agent effectively used for the relief of mild to moderate pain associated with headache, backache, arthritis, and postoperative pain. Overdose of paracetamol leads to hepatic toxicity and in some cases associated with liver and kidney damage and even death. There is a huge demand for developing sensors for detecting PA.

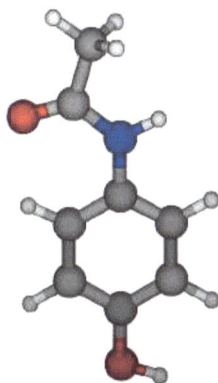

Fig. (8). Ball and stick atomic model of Paracetamol (C=Grey; O=Red; H=White; N=Blue).

C. M. Kuskur *et al.* modified the CPE from congo red after electrodeposition of congo red sensitivity, and selectivity of the electrode has improved [36]. The improved sensitivity is improved mainly due to the increase in active surface area of the electrode by congo red film. C. Madhuri *et al.* Modified the CPE from the orange dye [37]. They have deposited poly-orange on CPE by electro-polymerization and applied the modified electrode for determining PA. They found that at the modified electrode anodic peak current increased three folds with a significant shift in oxidation potential which indicates that dyes act as a catalyst for the electron transfer reactions at the CPE interface. S. Chitravathi and N.Munichandraiah [38] modified the GCE from the Nile blue. They found that modification of Nile blue will increase the active surface area of the GCE by decreasing the overpotential of PA. As a part of the work, they also find that Nile blue modified electrode can be applied to determine the PA in commercial tablet samples. Thus, organic dyes modified electrode is also useful to determine the PA.

CONCLUSION

Organic dyes have wide applications in the modern industrial era. Unfortunately, these dyes can also harm health. Therefore, it is very important to determine them. The electrochemical (voltammetric) methods have shown to be an interesting and promising tool for dye determination in real samples like food and fuels. In the current chapter, we have discussed the importance of developing sensors for RhB and SB-14. Voltammetric methods-based sensors are very effective to determine them in food samples like sauce and juice. These sensors can also be used to determine dyes in fuels. In the current chapter, we have also shown how dyes can be useful as modifiers to improve electron transfer properties of working electrodes. Dyes modified electrode applications in PA and DA determination were also discussed in the chapter. Dyes increase the active surface area of the electrode, as a result the sensing properties of the electrode improve. Therefore, we believe that dyes are very useful in electrode modifications.

CONSENT FOR PUBLICATION

Not applicable.

CONFLICT OF INTEREST

The authors declare no conflict of interest, financial or otherwise.

ACKNOWLEDGEMENT

Declared none.

REFERENCES

[1]   E. Forgacs, T. Cserháti, and G. Oros, "Removal of synthetic dyes from wastewaters: a review", *Environ. Int.,* vol. 30, no. 7, pp. 953-971, 2004.
[http://dx.doi.org/10.1016/j.envint.2004.02.001] [PMID: 15196844]

[2]   H. Ali, "Biodegradation of Synthetic Dyes-", *RE:view,* vol. 213, pp. 251-273, 2010.

[3]   T. Robinson, G. McMullan, R. Marchant, and P. Nigam, "Remediation of dyes in textile effluent: a critical review on current treatment technologies with a proposed alternative", *Bioresour. Technol.,* vol. 77, no. 3, pp. 247-255, 2001.
[http://dx.doi.org/10.1016/S0960-8524(00)00080-8] [PMID: 11272011]

[4]   R.G. Saratale, G.D. Saratale, J.S. Chang, and S.P. Govindwara, "Bacterial decolorization and degradation of azo dyes: A review", *J. Taiwan Inst. Chem. Eng.,* vol. 42, pp. 138-157, 2011.
[http://dx.doi.org/10.1016/j.jtice.2010.06.006]

[5]   C.S. DeRoo, and R.A. Armitage, "Direct identification of dyes in textiles by direct analysis in real time-time of flight mass spectrometry", *Anal. Chem.,* vol. 83, no. 18, pp. 6924-6928, 2011.
[http://dx.doi.org/10.1021/ac201747s] [PMID: 21846126]

[6]   C. O' Neill, F.R. Hawkes, D.L. Hawkes, N.D. Lourenço, H.M. Pinheiro, and W. Delée, "Colour in textile effluents—sources, measurement, discharge consents, and simulation: a review", *J. Chem. Technol. Biotechnol.,* vol. 74, pp. 1009-1018, 1999.
[http://dx.doi.org/10.1002/(SICI)1097-4660(199911)74:11<1009::AID-JCTB153>3.0.CO;2-N]

[7]   B. Lellis, and C. Zani, "Fávaro-Polonio; Pamphile J.A.; Polonio J.C.; Effects of textile dyes on health and the environment and bioremediation potential of living organisms", *Biotechnology Research and Innovation.,* vol. 3, pp. 275-290, 2019.
[http://dx.doi.org/10.1016/j.biori.2019.09.001]

[8]   S.P. Alves, D.M. Brum, É.C. Branco de Andrade, and A.D.P. Netto, "Remediation of dyes in textile

effluent: a critical review on current treatment technologies with a proposed alternative", *Bioresource Technology,* vol. 77, pp. 247-255, 2008.
[http://dx.doi.org/10.1016/j.foodchem.2007.07.054]

[9]   R. Noguerol-Cal, J.M. López-Vilariño, G. Fernández-Martínez, L. Barral-Losada, and M.V. González-Rodríguez, "High-performance liquid chromatography analysis of ten dyes for control of safety of commercial articles", *J. Chromatogr. A,* vol. 1179, no. 2, pp. 152-160, 2008.
[http://dx.doi.org/10.1016/j.chroma.2007.11.099] [PMID: 18093605]

[10]   M.V.B. Zanoni, W.R. Sousa, J.P.D. Lima, P.A. Carneiro, and A.G. Fogg, "Application of voltammetric technique to the analysis of indanthrene dye in alkaline solution", *Dyes Pigments,* vol. 68, pp. 19-25, 2006.
[http://dx.doi.org/10.1016/j.dyepig.2004.12.008]

[11]   R.N. Adams, "Carbon Paste Electrodes", *Anal. Chem.,* vol. 30, p. 1576, 1958.
[http://dx.doi.org/10.1021/ac60141a600]

[12]   C. Raril, and J.G. Manjunatha, "Sensitive and Selective Analysis of Nigrosine Dye at Polymer Modified Electrochemical Sensor", *Anal. Bioanal. Electrochem.,* vol. 10, pp. 372-382, 2018.

[13]   C. Akkapinyo, K. Subannajui, Y. Poo-Arporn, and R.P. Poo-Arporn, "Disposable Electrochemical Sensor for Food Colorants Detection by Reduced Graphene Oxide and Methionine Film Modified Screen Printed Carbon Electrode", *Molecules,* vol. 26, no. 8, p. 2312, 2021.
[http://dx.doi.org/10.3390/molecules26082312] [PMID: 33923482]

[14]   C. Raril, and J.G. Manjunatha, "A simple approach for the electrochemical determination of vanillin at ionic surfactant modified graphene paste electrode", *Microchem. J.,* vol. 154, p. 104575, 2020.
[http://dx.doi.org/10.1016/j.microc.2019.104575]

[15]   P.A. Pushpanjali, J.G. Manjunatha, and M.T. Shreenivas, "The electrochemical resolution of ciprofloxacin, riboflavin, and estriol using anionic surfactant and polymer-modified carbon paste electrode", *ChemistrySelect,* vol. 4, pp. 13427-13433, 2019.
[http://dx.doi.org/10.1002/slct.201903897]

[16]   J.G. Manjunatha, B.E.K. Swamy, R. Deepa, V. Krishna, G.P. Mamatha, U. Chandra, S.S. Shankar, and B.S. Sherigara, "Electrochemical studies of dopamine at eperisone and cetyl trimethyl ammonium bromide surfactant modified carbon paste electrode: a cyclic voltammetric study", *Int. J. Electrochem. Sci.,* vol. 4, pp. 662-667, 2009.

[17]   G.K. Jayaprakash, B.K. Swamy, S. Rajendrachari, S.C. Sharma, and R. Flores-Moreno, "Dual descriptor analysis of cetylpyridinium modified carbon paste electrodes for ascorbic acid sensing applications", *J. Mol. Liq.,* vol. 334, p. 116348, 2021.
[http://dx.doi.org/10.1016/j.molliq.2021.116348]

[18]   J.G. Manjunatha, M. Deraman, N.H. Basri, and I.A. Talib, "Fabrication of poly (Solid Red A) modified carbon nano tube paste electrode and its application for simultaneous determination of epinephrine, uric acid and ascorbic acid", *Arab. J. Chem.,* vol. 11, pp. 149-158, 2018.
[http://dx.doi.org/10.1016/j.arabjc.2014.10.009]

[19]   J.G.G. Manjunatha, "A novel poly (glycine) biosensor towards the detection of indigo carmine: A voltammetric study", *J. Food Drug Anal.,* vol. 26, no. 1, pp. 292-299, 2018.
[http://dx.doi.org/10.1016/j.jfda.2017.05.002] [PMID: 29389566]

[20]   G.K. Jayaprakash, B.E.K. Swamy, B.N. Chandrashekar, and R. Flores-Moreno, "Simultaneous determination of epinephrine, ascorbic acid and folic acid using TX-100 modified carbon paste electrode: A cyclic voltammetric study", *J. Mol. Liq.,* vol. 240, pp. 395-401, 2017.
[http://dx.doi.org/10.1016/j.molliq.2017.05.093]

[21]   G.K. Jayaprakash, B.E.K. Swamy, N. Casillas, and R. Flores-Moreno, "Analytical Fukui and cyclic voltammetric studies on ferrocene modified carbon electrodes and effect of Triton X-100 by immobilization method", *Electrochim. Acta,* vol. 258, pp. 1025-1034, 2017.
[http://dx.doi.org/10.1016/j.electacta.2017.11.154]

[22]   G.K. Jayaprakash, B.E.K. Swamy, H.N.G. Ramirez, M.T. Ekanthappa, and R. Flores-Moreno, "Quantum chemical and electrochemical studies of lysine modified carbon paste electrode surfaces for sensing dopamine", *New J. Chem.,* vol. 42, pp. 4501-4506, 2018.
[http://dx.doi.org/10.1039/C7NJ04998F]

[23]   G.K. Jayaprakash, B.E.K. Swamy, S.C. Sharma, and J.J. Santoyo-Flores, "Analyzing electron transfer properties of ferrocene in gasoline by cyclic voltammetry and theoretical methods", *Microchem. J.,* vol. 158, p. 105116, 2020.
[http://dx.doi.org/10.1016/j.microc.2020.105116]

[24]   G.K. Jayaprakash, B.E.K. Swamy, J.P.M. Sanchez, X. Li, S.C. Sharma, and S.L. Lee, "Electrochemical and quantum chemical studies of cetylpyridinium bromide modified carbon electrode interface for sensor applications", *J. Mol. Liq.,* vol. 315, p. 113719, 2020.
[http://dx.doi.org/10.1016/j.molliq.2020.113719]

[25]   T-L. Chiang, Y-C. Wang, and W-H. Ding, "Trace Determination of Rhodamine B and Rhodamine 6G Dyes in Aqueous Samples by Solid□phase Extraction and High□performance Liquid Chromatography Coupled with Fluorescence Detection", *J. Chin. Chem. Soc. (Taipei),* vol. 59, pp. 515-519, 2012.
[http://dx.doi.org/10.1002/jccs.201100318]

[26]   C.C. Wang, A.N. Masi, and L. Fernández, "On-line micellar-enhanced spectrofluorimetric determination of rhodamine dye in cosmetics", *Talanta,* vol. 75, no. 1, pp. 135-140, 2008.
[PMID: 18371858]

[27]   Y. Yi, H. Sun, G. Zhu, Z. Zhanga, and X. Wu, "Sensitive electrochemical determination of rhodamine B based on cyclodextrin-functionalized nanogold/hollow carbon nanospheres", *Anal. Methods,* vol. 7, pp. 4965-4970, 2015.
[http://dx.doi.org/10.1039/C5AY00654F]

[28]   D. Sun, and X. Yang, "Rapid Determination of Toxic Rhodamine B in Food SamplesUsing Exfoliated Graphene-Modified Electrode", *Food Anal. Methods,* vol. 10, pp. 2046-2052, 2017.
[http://dx.doi.org/10.1007/s12161-016-0773-2]

[29]   H.T. Ngo, V.T. Nguyen, T.D. Manh, T.T.T. Toan, N.T.M. Triet, N.T. Binh, N.T.V. Hoan, T.V. Thien, and D.Q. Khieu, "Voltammetric Determination of Rhodamine B Using a ZIF-67/Reduced Graphene Oxide Modified Electrode", *Journal of Nanomaterials,* pp. 1-7, 2020.

[30]   M.A.G. Trindade, and M.V.B. Zanoni, "Voltammetric sensing of the fuel dye marker Solvent Blue 14 by screen-printed electrodes", *Sens. Actuators B Chem.,* vol. 138, pp. 257-263, 2009.
[http://dx.doi.org/10.1016/j.snb.2009.01.043]

[31]   K. Chetankumar, B.E.K. Swamy, and S.C. Sharma, "Electrochemical preparation of poly (direct yellow 11) modified pencil graphite electrode sensor for catechol and hydroquinone in presence of resorcinol: A voltammetric study", *Microchem. J.,* vol. 156, p. 104979, 2020.
[http://dx.doi.org/10.1016/j.microc.2020.104979]

[32]   J.G. Manjunatha, "Fabrication of Efficient and Selective Modified Graphene Paste Sensor for the Determination of Catechol and Hydroquinone", *surfaces,* vol. 3, pp. 1-10, 2020.

[33]   U. Chandra, B.E.K. Swamy, O. Gilbert, and B.S. Sherigara, "Voltammetric resolution of dopamine in the presence of ascorbic acid and uric acid at poly (calmagite) film coated carbon paste electrode", *Electrochim. Acta,* vol. 55, pp. 7166-7174, 2010.
[http://dx.doi.org/10.1016/j.electacta.2010.06.091]

[34]   C.M. Kuskur, B.E.K. Swamy, and H. Jayadevappa, "Poly (naphthol green B) modified carbon paste electrode for the analysis of paracetamol and norepinephrine", *Ionics,* vol. 25, pp. 1845-1855, 2019.
[http://dx.doi.org/10.1007/s11581-018-2606-3]

[35]   T.T. Ronald, J. Mascarenhas, and B.E.K. Swamy, "Poly(Rhodamine B) modified carbon paste electrode for the selective detection of dopamine", *J. Mol. Liq.,* vol. 174, pp. 70-75, 2012.

[http://dx.doi.org/10.1016/j.molliq.2012.07.022]

[36]   C.M. Kuskur, B.E.K. Swamy, H. Jayadevappa, and K. Shivakumar, "Electropolymerized Congo Red Film based Sensor for Dopamine: A Voltammetric Study", *Insights in Analytical Electrochemistry,* vol. 3, pp. 1-9, 2017.

[37]   C. Madhuri, S. Kiranmai, D. Saritha, V.P. Rao, G. Madhavi, and H.S. Kusuma, "Electrochemical determination of paracetamol at poly(orange dye) modified carbon paste electrode by using cyclic voltammetry", *Journal of Chemical Technology and Metallurgy,* vol. 54, pp. 758-764, 2019.

[38]   S. Chitravathi, and N. Munichandraiah, "Voltammetric determination of paracetamol, tramadol and caffeine using poly(Nile blue) modified glassy carbon electrode", *J. Electroanal. Chem. (Lausanne),* vol. 764, pp. 93-103, 2016.
[http://dx.doi.org/10.1016/j.jelechem.2016.01.021]

CHAPTER 14

Advanced Voltammetric Devices for Disease Detection

P. Sophiya[1], Deepadarshan Urs[1], Rajkumar S. Meti[1], H. A. Nagma Banu[2], J Shankar[3] and K. K. Dharmappa[1,*]

[1] *Department of Studies and Research in Biochemistry, Jnana Kaveri Post Graduate Centre, Mangalore University, Chikka Aluvara, Kodagu, Karnataka, India*

[2] *Course in Chemistry, Jnana Kaveri Post Graduate Centre, Mangalore University, Chikka Aluvara, Kodagu, Karnataka, India*

[3] *Department of Studies and Research in Food Technology, Davangere University, Shivagangothri, Davangere, Karnataka, India*

Abstract: Electroanalytical methods are beneficial in investigating the electrochemical characteristics of chemicals and macromolecules. The innovation of electrochemical sensors has been of tremendous interest owing to their high sensitivity, fast analysis and capability to analyse intricate samples; these properties make electrochemistry significant for its use in medical applications, especially in disease detection. The appropriate target analytes must be identified to recognize any diseases by electrochemical techniques and their interplay with biological cells needs to be investigated. In this chapter, we have explained an electrochemical approach and techniques for several disease detection.

Keywords: Biosensors, Disease detection, Electrochemical sensors, Voltammetry.

INTRODUCTION

Voltammetry is an electroanalytical technique derived from voltamperometry, which gives information about the analyte that helps measure the current by varying the voltage, *i.e.*, the electrode's potential. A voltammogram produces the plot by the change in the current produced by voltage variance. An electrochemical cell requires two electrodes, one electrode is made considerably smaller than the other so that the moving current is restricted only by this electrode, which is called the active electrode, and the second (larger) electrode is

---
[*] **Corresponding author K. K. Dharmappa:** Department of Studies and Research in Biochemistry, Jnana Kaveri Post Graduate Centre, Mangalore University, ChikkaAluvara, Kodagu, Karnataka, India-571232; Email: dharmappa@gmail.com

[#] First two authors have contributed equally.

called the auxiliary electrode [1]. Voltammograms vary largely with measurement time and thus, it is important to pay attention to time variables. The voltage is a controlled variable in conventional voltammetry, and the current is measured as a function of the applied voltage at a specific point in time [2].

Types of Voltammetry: Linear sweep, cyclic, square wave, stripping, alternating current (AC), pulse, steady-state microelectrode, and hydrodynamic voltammetry are the various types of voltammetry, characterised based on the mode of potential control. The cyclic voltammetry technique is widely used and operates on a time scale of seconds (CV). In contrast, the voltammetry currently in use is AC voltammetry at only a time scale of milliseconds. Here, we describe the hypothesis and tips for the practical application of mainly the two types of voltammetry (Fig. **1**) [1].

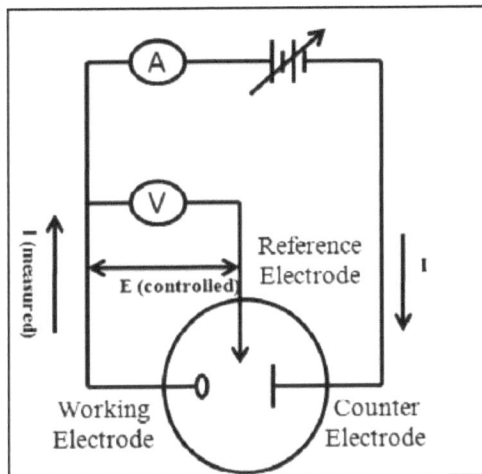

Fig (1). Schematic representation of cyclic voltammetry (Joshi PS *et al*, 2018) [3].

Biosensors: Analytical devices that transform biological reactions into electrical signals are termed as biosensors. Biosensors should typically possess several properties such as reusability, precision and should not be affected by pH and temperature. The word "biosensor" was coined by Cammann [4] and IUPAC has introduced its definition [5 - 7].

The application of biosensors in the detection of blood biomarkers and proteins has received a lot of attention in diagnosing multiple diseases over the last decade [8, 9]. Biosensors with inexpensive advanced technology and low consumption of clinical samples are considered along with laboratory-based devices as a simpler clinical diagnostic tool [10, 11]. The invention of a genosensor for the detection of extremely sensitive DNA is currently of enormous interest in the discovery of pathogenic diseases, clinical diagnostics and forensics with genomic DNA sequences/probes.

Voltammetric biosensors for cancer biomarkers detection: The electrochemical biosensors and the sort of electrochemical technique are employed to detect cancer biomarkers.

For example, 3D gold nanowire electrodes embedded in polypyrrole are used in Differential Pulse Voltammetry to detect prostate specific antigen concentrations of less than one femtogram per ml. Neuron-Specific Enolase (NSE) is a biomarker of lung cancer that can be detected by Glassy Carbon Electrodes (GCE) modified with gold nanocrystals at a concentration of 0.3 picograms per ml using cyclic voltammetry [12].

To detect HER2, a breast cancer biomarker by using Electrochemical Impedance Spectroscopy (EIS) at nanogram per ml, with a Carbon Ionic Liquid Electrode (CILE) [12]. This electrode is updated with multi-walled carbon nanotubes which enhances the biological recognition element's stability. Redox mediators such as the ferro/ferricyanide ion are used to maximise the efficiency of the biosensors. For a better prognosis, multiple biomarkers have to be analysed concurrently. Therefore, the use of multiple working electrodes or a single electrode with multiple labeling of antibodies is convenient [13].

The Human Epidermal growth factor Receptor 2- Extra Cellular Domain (HER2-ECD) is a breast cancer biomarker, which can be quantified by electrochemical molecularly imprinted polymer (MIP) sensor. The MIP is developed by electropolymerisation with a solution containing phenol and HER2-ECD on a screen-printed gold electrode (AuSPE) by using cyclic voltammetry. In addition, analysis of human serum samples by cyclic voltammetry and electrochemical impedance spectroscopy for the precise result is mandated. Thus MIP sensor could be useful in the analysis of HER2-ECD in breast cancer patients [14].

Lung cancer is gene associated disease; the genosensor is required to detect single point EGFR (Epidermal Growth Factor Receptor) gene variation. In the detection of DNA hybridisation, voltammetry detection was recognised as one of the strongest detective methods. When the genomic biomarkers bear, the charge was translated into a readable signal. In comparison to other sensing techniques, voltammetry genosensor is chosen because of its fast response, minimal effort in the invention with affordable prices and is less time consuming [15, 16].

The limit of detection and enhanced sensitivity by immobilising the least quantity of DNA samples on the sensing surface are the two primary problems in the development of the genosensor. For the excellent hybridisation and immobilisation of efficiency, nanomaterial has been studied for the surface modification of genosensor. In previous decades, gold, silver and titanium oxide nanoparticles have been applied to the sensor surface to increase the surface area for DNA immobilisation and hybridisation due to their physicochemical properties, resulting in a notable improvement in genosensor sensitivity [17 - 22].

Alzheimer's disease (AD): This is a very well known neurodegenerative disease caused by the deposition of amyloid-beta (Aβ) peptide, tau phosphorylation, and the growth of neurofibrillary tangles in the brain [23, 24]. Aβ peptide plaques and its aggregation play a significant role in the development of Alzheimer's disease and is one of the disease's biomarkers. The quantification of Aβ peptide concentration in cerebrospinal fluid can provide significant information in the diagnosis of AD [25 - 28]. Some of the studies showed that the variation in Aβ concentration is not just enough to confirm the AD. Therefore, Apolipoprotein E (APOE) is considered, which regulates the metabolism, aggregation, and deposition of Aβ [29]. The three APOE isoforms (E2, E3 and E4) are reported; among these three isoforms, the people who carry APOE 4 genes have a higher risk of AD. Therefore, APOE 4 gene can offer useful early diagnosis and precise medication for the treatment of AD [30, 31]. Electrochemical impedance spectroscopy (EIS) and cyclic voltammetry (CV) can be used to detect the APOE 4 gene, which consists of three electrodes: working electrode as a gold electrode with a diameter of 2mm, counter electrode as a platinum wire and reference electrode as Ag/AgCl electrode.

The APOE4 gene can be detected by the sequence specific restriction enzyme by immobilising the allele specific probe on the electrode to localise the particular target. To accomplish sufficient sensitivity for the detection of non-amplified DNA, signal amplification is a mandate which could be achieved by synthesizing ferrocene-capped (Fc) streptavidin/gold nanoparticles. With diluted genomic DNA samples, successful APOE4 gene detection was demonstrated. This experiment is simple with high sensitivity and appropriate to detect amplification-free APOE 4 genes [32].

Black disease: *Clostridium novyi* lives as a spore in the air and is responsible for the Black disease, which causes death in cattle and sheep [33]. *C. novyi* and related species of *Clostridium* causes wound infection that leads to gas gangrene in a later stage [34]. The onset of Black disease and death is so sudden that treatment is less effective or worthwhile [35]. An electrochemical and loop-mediated isothermal amplification (LAMP)-based cyclic voltammetry methods are recommended for the detection of *C. novyi* [36]. The DNA of *C. novyi* is obtained from the sample and amplified by LAMP. Further, based on the oxidation and reduction current peaks measured by a solely developed electrode which enable the quantitative detection of DNA. To test the accuracy and sensitivity of the cyclic voltammetry method is confirmed by the polymerase chain reaction method [37].

HBV: Hepatitis B viruses are highly virulent localized mainly in the liver [38]. HBV boosts IgG antibodies due to the presence of hepatitis B surface antigen. The specific electrochemical biosensors are used to detect antibodies of HBV surface antigens. An electrochemical biosensor dependent on a sandwich immunoassay is employed to detect anti- HBV antibodies [39]. An electrochemical DNA biosensor has been evolved to detect HBV DNA. The biosensor is based on the covalent immobilisation of the single-stranded DNA (ssDNA), which is linked to HBV genes, on the modified glassy carbon electrode (GCE). The differential pulse voltammetry is used to examine the hybridization between the probe and its complementary ssDNA. HBV DNA can be quantified in the range from 1.01×10^{-8} to 1.62×10^{-6} mol/L with good linearity ($\gamma = 0.9962$) [40].

Diabetes: Diabetes is a lifestyle disease where glucose metabolism is reduced due to a lack of insulin production [41]. Analytical methods for insulin detection include bioassays and immunoassays. These methods require the derivatisation of insulin with isotopes or fluorogenic labels [42, 43]. There are biosensors for direct insulin determination using disposable electrochemical sensors [43, 44]. The modified multi wall carbon nanotubes (MWCNTs) are used to detect insulin by the direct voltammetric method.

Voltammetric immune sensor consists of electrodes modified with multi-walled carbon nanotubes pyrenyl units are used for the direct diagnosis of type 1 and type 2 diabetic conditions based on the simple redox signals. The concentration range of insulin in the blood serum under fasting conditions is approximately 50 picomole (pM) [45], whereas the concentration range of the insulin is less than 50 pM in type 1 diabetes and greater than 70 pM during the onset of type 2 diabetes [46 - 48]. The immune biosensors can measure serum insulin at picomolar level by simple voltammetric methods by monitoring current signals. The nanomolar range of insulin can also be detected by aptamer-based immunoassay using graphene oxide (GO)-modified electrodes [49]. Modified voltammetric methods are beneficial for multiple independent detection features to assess and diagnose disease markers. Thus, voltammetric immune sensors are designed to diagnose and differentiate type 1 and type 2 diabetes in patients.

COVID-19: Coronavirus disease 2019 (COVID-19) is a pandemic worldwide, most of the research efforts are focused on this issue. The research on COVID-19 can be classified into the following, 1. Diagnosis of infection, 2. Discovery and development of successful therapies [50]. As a result, the research community attempting to develop rapid, highly sensitive devices and sensing systems (*e.g.*, nano biosensors) for COVID-19 diagnosis.

The most used diagnostic tests are Reverse Transcriptase-Polymerase chain reaction (RT-PCR) for the detection of Corona infection, chest computed tomography (CT), and Enzyme-Linked Immunosorbent Assay (ELISA) [51]. The detection of COVID-19 by RT-PCR is a most used technique, but time-consuming, costly and not always reproducible. As a result, alternative diagnostic techniques such as cyclic voltammetry could be used to diagnose the infection. The cyclic voltammetry-based nano biosensors need only simple instrumentation, rapid detection of virus and most of the biosensors are at low cost [52].

The nano biosensor developed using a fluorine-doped tin oxide (FTO) electrode and with gold nanoparticles (AuNPs), is immobilised with a monoclonal COVID-19 (nCOVID-19Ab) antibody to test the change in electrical conductivity. Cyclic voltammetry (CV) and differential pulse voltammetry (DPV) are commonly used for excellent sensitivity. The sensitivity limit for the detection of nCovid-19 antigen is 10 femtomolar (fM) [53] (Table **1**).

Table 1. Comparative tables with different methods/sensors.

Sl. No	Diseases	Method/Sensor	References
1	**CANCER** For detecting a cancer biomarker protein in serum at femtomolar quantities. To detect epidermal growth factor receptor (EGFR) mutation causes non-small cell lung cancer (NSCLC). To detect the breast cancer (BRCA1) biomarker.	Glassy carbon electrode Genosensor Electrochemical DNA (E–DNA) sensor based on polymer/reduced graphene–oxide biosensor.	[54] [55] [56]
2	**ALZHEIMER'S** Early detection of Alzheimer's biomarkers like micro RNA -137 A label-free immunosensor with great sensitivity for detecting the key biomarker of Alzheimer's disease (AD)) amyloid beta 1–42 (A (1–42))	Cyclic voltammetry and electro-chemical impedance spectroscopy Scanning electron microscopy, square-wave voltammetry, and electrochemical impedance spectro-scopy were utilised to analyse the biosensor's fabrication.	[57] [58]
3	**DIABETES** Diabetic HbA1c detection in undiluted human serum. Voltammetric analysis of two putative diabetic indicators, Methylglyoxal MGO as well asglyoxalase GLO	Electrochemical immunosensor by differential pulse voltammetry. CuO/Au/GCE nanostructured sensor	[59] [60]

(Table 1) cont.....

Sl. No	Diseases	Method/Sensor	References
4.	**HBV** Detection of viral genome of Hepatitis B virus. Detection and isolation of a single-stranded DNA (ssDNA) oligonucleotide with a sequence obtained from the hepatitis B virus genom (HBV)	Cathodic stripping voltammetry method Differential pulse voltammetry.	[61] [62]
5	**COVID -19** Early detection of COVID -19 disease of RNA-dependent RNA polymerase (RdRP) sequence as a specific target of novel coronavirus based on the CPE HT18C6(Ag)/chitosan / SiQDs @ PAMAM/ probe Sequence Detection of SARS-CoV-2 antigen.	Nanoscale genosensors Cyclic voltammetry Differential pulse voltammetry. Electrochemical biosensor designed with screen-printed carbon electrodes coated with gold nanoparticles.	[63] [64]

PERSPECTIVE ON THE USAGE OF SENSORS IN DISEASE DETECTION

The development of sensors is active research in all disciplines of science. The electrochemical biosensors have proven useful in clinical diagnosis, agriculture, food and pharma industries and forensics. The voltammetric biosensors are portable and their electrodes are modified with gold nanoparticles, multi-walled carbon nanotubes, *etc.*, enhancing their functional efficacy. The discovery of biosensors in diagnosis and therapy has a great demand because of their sensitivity, accuracy and cost-effectiveness.

Recent advancements in the field of voltammetry include the chemical processing of highly volatile organic compounds (VOCs) and the use of flexible sensors, ultra-small sensors and related artificial intelligence biosensors to study protein synthesis, gene expression, metabolic pathways and drug interaction in medical research. This may open up an opportunity in disease detection and drug discovery.

CONCLUSION

Voltammetry has a wide application in research, innovations, and diagnosis and covers the processes such as synthesis, characterisation, mechanism and analysis, *etc.* The main features of this technique are speedy, accuracy, sensitivity and cost-effectiveness. This chapter is focused on aspects of the detection of a few important diseases infection by using different biosensors. The diagnostic approach of diseases such as different cancers (breast, prostate and lung), Alzheimer's disease, hepatitis, black disease, and COVID-19 by cyclic voltammetry is discussed. Voltammetry technique is used in disease diagnosis based on reaction mechanism that involved the thermodynamic and kinetic analysis of electroactive species. Electrochemical techniques allow rapid screening of many samples and preparation and therefore have several advantages over other techniques such as chromatography, spectrophotometry. In the medical sciences, voltammetry is significant not only in diagnosis but also in other fields such as studies of the mechanism of infection, drug development, *etc*. This could be used as a starting point for further research in this area.

CONSENT FOR PUBLICATION

Not applicable.

CONFLICT OF INTEREST

The authors declare no conflict of interest, financial or otherwise.

ACKNOWLEDMENTS

The authors are thankful to Dr. J. G. Manjunatha, Assistant Professor, Department of Chemistry, FMKMC College Madikeri, for his support.

REFERENCES

[1]   F. Scholz, "Voltammetric techniques of analysis: the essentials", *ChemTexts.,* vol. 1, no. 4, pp. 1-24, 2015.
[http://dx.doi.org/10.1007/s40828-014-0001-x]

[2]   KJ Aoki, and J Chen, "Tips of Voltammetry", *InVoltammetry 2018 Nov 5*. IntechOpen

[3]   P.S. Joshi, and D.S. Sutrave, "A brief study of cyclic voltammetry and electrochemical analysis", *Int. J. Chemtech Res.,* vol. 11, no. 9, p. 77, 2018.
[http://dx.doi.org/10.20902/IJCTR.2018.110911]

[4]   K. Cammann, "Biosensors based on ion-selective electrodes", *Fresenius Z. Anal. Chem.,* vol. 287, pp. 1-9, 1977.
[http://dx.doi.org/10.1007/BF00539519]

[5]   D.R. Thevenot, K. Toth, R.A. Durst, and G.S. Wilson, "Electrochemical biosensors: recommended definitions and classification", *Pure Appl. Chem.,* vol. 71, pp. 2333-2348, 1999.
[http://dx.doi.org/10.1351/pac199971122333]

[6]   D.R. Thévenot, K. Toth, R.A. Durst, and G.S. Wilson, "Electrochemical biosensors: recommended definitions and classification", *Biosens. Bioelectron.,* vol. 16, no. 1-2, pp. 121-131, 2001.
[PMID: 11261847]

[7]   D.R. Thévenot, K. Toth, R.A. Durst, and G.S. Wilson, "Electrochemical biosensors: recommended definitions and classification", *Biosens. Bioelectron.,* vol. 16, no. 1-2, pp. 121-131, 2001.
[PMID: 11261847]

[8]   I. Letchumanan, M.K. Md Arshad, S.R. Balakrishnan, and S.C.B. Gopinath, "Gold-nanorod enhances dielectric voltammetry detection of c-reactive protein: A predictive strategy for cardiac failure", *Biosens. Bioelectron.,* vol. 130, pp. 40-47, 2019.
[http://dx.doi.org/10.1016/j.bios.2019.01.042] [PMID: 30716591]

[9]   I. Letchumanan, S.C.B. Gopinath, M.K. Md Arshad, P. Anbu, and T. Lakshmipriya, "Gold nano-

urchin integrated label-free amperometric aptasensing human blood clotting factor IX: A prognosticative approach for "Royal disease"", *Biosens. Bioelectron.,* vol. 131, pp. 128-135, 2019.
[http://dx.doi.org/10.1016/j.bios.2019.02.006] [PMID: 30826647]

[10]   Q. Liu, S.Y. Lim, R.A. Soo, M.K. Park, and Y. Shin, "A rapid MZI-IDA sensor system for EGFR mutation testing in non-small cell lung cancer (NSCLC)", *Biosens. Bioelectron.,* vol. 74, pp. 865-871, 2015.
[http://dx.doi.org/10.1016/j.bios.2015.07.055] [PMID: 26233643]

[11]   K.Z. Chen, F. Lou, F. Yang, J.B. Zhang, H. Ye, W. Chen, T. Guan, M.Y. Zhao, X.X. Su, R. Shi, L. Jones, X.F. Huang, S.Y. Chen, and J. Wang, "Circulating tumor DNA detection in early-stage non-small cell lung cancer patients by targeted sequencing", *Sci. Rep.,* vol. 6, no. 1, p. 31985, 2016.
[http://dx.doi.org/10.1038/srep31985] [PMID: 27555497]

[12]   M. Freitas, H.P. Nouws, and C. Delerue☐Matos, "Electrochemical Biosensing in Cancer Diagnostics and Follow☐up", *Electroanalysis,* vol. 30, no. 8, pp. 1584-1603, 2018.
[http://dx.doi.org/10.1002/elan.201800193]

[13]   C. Mercer, A. Jones, J.F. Rusling, and D. Leech, "Multiplexed electrochemical cancer diagnostics with automated microfluidics", *Electroanalysis,* vol. 31, no. 2, pp. 208-211, 2019.
[http://dx.doi.org/10.1002/elan.201800632] [PMID: 32390709]

[14]   M.J. Shiddiky, P.H. Kithva, S. Rauf, and M. Trau, "Femtomolar detection of a cancer biomarker protein in serum with ultralow background current by anodic stripping voltammetry", *Chem. Commun. (Camb.),* vol. 48, no. 51, pp. 6411-6413, 2012.
[http://dx.doi.org/10.1039/c2cc32810k] [PMID: 22618633]

[15]   Y. Ye, J. Ji, F. Pi, H. Yang, J. Liu, Y. Zhang, S. Xia, J. Wang, D. Xu, and X. Sun, "A novel electrochemical biosensor for antioxidant evaluation of phloretin based on cell-alginate/L-cysteine/gold nanoparticle-modified glassy carbon electrode", *Biosens. Bioelectron.,* vol. 119, pp. 119-125, 2018.
[http://dx.doi.org/10.1016/j.bios.2018.07.051] [PMID: 30121423]

[16]   J. Peng, T. He, Y. Sun, Y. Liu, Q. Cao, Q. Wang, and H. Tang, "An organic electrochemical transistor for determination of microRNA21 using gold nanoparticles and a capture DNA probe", *Mikrochim. Acta,* vol. 185, no. 9, p. 408, 2018.
[http://dx.doi.org/10.1007/s00604-018-2944-x] [PMID: 30097715]

[17]   L. Coudron, M.B. McDonnell, I. Munro, D.K. McCluskey, I.D. Johnston, C.K.L. Tan, and M.C. Tracey, "Fully integrated digital microfluidics platform for automated immunoassay; A versatile tool for rapid, specific detection of a wide range of pathogens", *Biosens. Bioelectron.,* vol. 128, pp. 52-60, 2019.
[http://dx.doi.org/10.1016/j.bios.2018.12.014] [PMID: 30634074]

[18]   M.H. Ghalehno, M. Mirzaei, and M. Torkzadeh-Mahani, "Double strand DNA-based determination of menadione using a Fe3O4 nanoparticle decorated reduced graphene oxide modified carbon paste electrode", *Bioelectrochemistry,* vol. 124, pp. 165-171, 2018.
[http://dx.doi.org/10.1016/j.bioelechem.2018.07.014] [PMID: 30064048]

[19]   S. Ramanathan, and S.C. Gopinath, "Potentials in synthesising nanostructured silver particles", *Microsyst. Technol.,* vol. 23, no. 10, pp. 4345-4357, 2017.
[http://dx.doi.org/10.1007/s00542-017-3382-0]

[20]   S. Ramanathan, S.C. Gopinath, P. Anbu, T. Lakshmipriya, F.H. Kasim, and C.G. Lee, "Eco-friendly synthesis of Solanumtrilobatum extract-capped silver nanoparticles is compatible with good antimicrobial activities", *J. Mol. Struct.,* vol. 1160, pp. 80-91, 2018.
[http://dx.doi.org/10.1016/j.molstruc.2018.01.056]

[21]   B. Kaur, K. Malecka, D.A. Cristaldi, C.S. Chay, I. Mames, H. Radecka, J. Radecki, and E. Stulz, "Approaching single DNA molecule detection with an ultrasensitive electrochemical genosensor based on gold nanoparticles and cobalt-porphyrin DNA conjugates", *Chem. Commun. (Camb.),* vol. 54, no. 79, pp. 11108-11111, 2018.

[http://dx.doi.org/10.1039/C8CC05362F] [PMID: 30101270]

[22]   D. Uhlirova, M. Stankova, M. Docekalova, B. Hosnedlova, M. Kepinska, B. Ruttkay-Nedecky, J. Ruzicka, C. Fernandez, H. Milnerowicz, and R. Kizek, "A Rapid Method for the Detection of Sarcosine Using SPIONs/Au/CS/SOX/NPs for Prostate Cancer Sensing", *Int. J. Mol. Sci.*, vol. 19, no. 12, p. 3722, 2018.
[http://dx.doi.org/10.3390/ijms19123722] [PMID: 30467297]

[23]   J Hardy, and DJ Selkoe, "The amyloid hypothesis of Alzheimer's disease: progress and problems on the road to therapeutics", *science*, vol. 297, no. 5580, pp. 353-6, 2002.

[24]   D.J. Selkoe, "Alzheimer's disease: genes, proteins, and therapy", *Physiol. Rev.*, vol. 81, no. 2, pp. 741-766, 2001.
[http://dx.doi.org/10.1152/physrev.2001.81.2.741] [PMID: 11274343]

[25]   L. Liu, Q. He, F. Zhao, N. Xia, H. Liu, S. Li, R. Liu, and H. Zhang, "Competitive electrochemical immunoassay for detection of β-amyloid (1-42) and total β-amyloid peptides using p-aminophenol redox cycling", *Biosens. Bioelectron.*, vol. 51, pp. 208-212, 2014.
[http://dx.doi.org/10.1016/j.bios.2013.07.047] [PMID: 23962708]

[26]   N. Xia, X. Wang, J. Yu, Y. Wu, S. Cheng, Y. Xing, and L. Liu, "Design of electrochemical biosensors with peptide probes as the receptors of targets and the inducers of gold nanoparticles assembly on electrode surface", *Sens. Actuators B Chem.*, vol. 239, pp. 834-840, 2017.
[http://dx.doi.org/10.1016/j.snb.2016.08.079]

[27]   N. Xia, X. Wang, B. Zhou, Y. Wu, W. Mao, and L. Liu, "Electrochemical detection of amyloid-β oligomers based on the signal amplification of a network of silver nanoparticles", *ACS Appl. Mater. Interfaces*, vol. 8, no. 30, pp. 19303-19311, 2016.
[http://dx.doi.org/10.1021/acsami.6b05423] [PMID: 27414520]

[28]   N. Xia, B. Zhou, N. Huang, M. Jiang, J. Zhang, and L. Liu, "Visual and fluorescent assays for selective detection of beta-amyloid oligomers based on the inner filter effect of gold nanoparticles on the fluorescence of CdTe quantum dots", *Biosens. Bioelectron.*, vol. 85, pp. 625-632, 2016.
[http://dx.doi.org/10.1016/j.bios.2016.05.066] [PMID: 27240009]

[29]   H.H. H Adams, S. A Swanson, A. Hofman, and M.A. Ikram, "Amyloid-β transmission or unexamined bias?", *Nature*, vol. 537, no. 7620, pp. E7-E9, 2016.
[http://dx.doi.org/10.1038/nature19086] [PMID: 27629649]

[30]   Y. Liu, L.P. Xu, S. Wang, W. Yang, Y. Wen, and X. Zhang, "An ultrasensitive electrochemical immunosensor for apolipoprotein E4 based on fractal nanostructures and enzyme amplification", *Biosens. Bioelectron.*, vol. 71, pp. 396-400, 2015.
[http://dx.doi.org/10.1016/j.bios.2015.04.068] [PMID: 25950934]

[31]   A.M. Saunders, W.J. Strittmatter, D. Schmechel, P.H. George-Hyslop, M.A. Pericak-Vance, S.H. Joo, B.L. Rosi, J.F. Gusella, D.R. Crapper-MacLachlan, M.J. Alberts, and C. Hulette, "Association of apolipoprotein E allele ∊ 4 with late-onset familial and sporadic Alzheimer's disease", *Neurology*, vol. 43, no. 8, pp. 1467-1472, 1993.
[http://dx.doi.org/10.1212/WNL.43.8.1467] [PMID: 8350998]

[32]   H. Lu, L. Wu, J. Wang, Z. Wang, X. Yi, J. Wang, and N. Wang, "Voltammetric determination of the Alzheimer's disease-related ApoE 4 gene from unamplified genomic DNA extracts by ferrocene-capped gold nanoparticles", *Mikrochim. Acta*, vol. 185, no. 12, p. 549, 2018.
[http://dx.doi.org/10.1007/s00604-018-3087-9] [PMID: 30426239]

[33]   A.I. Aronson, and P. Fitz-James, "Structure and morphogenesis of the bacterial spore coat", *Bacteriol. Rev.*, vol. 40, no. 2, pp. 360-402, 1976.
[http://dx.doi.org/10.1128/br.40.2.360-402.1976] [PMID: 786255]

[34]   C.L. Hatheway, "Toxigenic clostridia", *Clin. Microbiol. Rev.*, vol. 3, no. 1, pp. 66-98, 1990.
[http://dx.doi.org/10.1128/CMR.3.1.66] [PMID: 2404569]

[35]   I. Parsonson, "Epidemic catarrh of sheep: the possible role of black disease", *Aust. Vet. J.,* vol. 85, no. 1-2, pp. 78-83, 2007.
[http://dx.doi.org/10.1111/j.1751-0813.2006.00073.x] [PMID: 17300469]

[36]   L.L. Liu, D.N. Jiang, G.M. Xiang, C. Liu, J.C. Yu, and X.Y. Pu, "Development of a cyclic voltammetry method for the detection of Clostridium novyi in black disease", *Genet. Mol. Res.,* vol. 13, no. 1, pp. 1724-1734, 2014.
[http://dx.doi.org/10.4238/2014.January.14.9] [PMID: 24446342]

[37]   Y.J. Xu, X.H. Tian, X.Z. Zhang, X.H. Gong, H.H. Liu, H.J. Zhang, H. Huang, and L.M. Zhang, "Simultaneous determination of malachite green, crystal violet, methylene blue and the metabolite residues in aquatic products by ultra-performance liquid chromatography with electrospray ionization tandem mass spectrometry", *J. Chromatogr. Sci.,* vol. 50, no. 7, pp. 591-597, 2012.
[http://dx.doi.org/10.1093/chromsci/bms054] [PMID: 22542891]

[38]   V. Barnaba, A. Franco, A. Alberti, C. Balsano, R. Benvenuto, and F. Balsano, "Recognition of hepatitis B virus envelope proteins by liver-infiltrating T lymphocytes in chronic HBV infection", *J. Immunol.,* vol. 143, no. 8, pp. 2650-2655, 1989.
[PMID: 2477450]

[39]   A. de la Escosura-Muñiz, M. Maltez-da Costa, C. Sánchez-Espinel, B. Díaz-Freitas, J. Fernández-Suarez, Á. González-Fernández, and A. Merkoçi, "Gold nanoparticle-based electrochemical magnetoimmunosensor for rapid detection of anti-hepatitis B virus antibodies in human serum", *Biosens. Bioelectron.,* vol. 26, no. 4, pp. 1710-1714, 2010.
[http://dx.doi.org/10.1016/j.bios.2010.07.069] [PMID: 20724135]

[40]   S. Zhang, Q. Tan, F. Li, and X. Zhang, "Hybridization biosensor using diaquabis [N-(--pyridinylmethyl) benzamide-κ2N, O]-cadmium (II) dinitrate as a new electroactive indicator for detection of human hepatitis B virus DNA", *Sens. Actuators B Chem.,* vol. 124, no. 2, pp. 290-296, 2007.
[http://dx.doi.org/10.1016/j.snb.2006.12.040]

[41]   V. Singh, and S. Krishnan, "Voltammetric immunosensor assembled on carbon-pyrenyl nanostructures for clinical diagnosis of type of diabetes", *Anal. Chem.,* vol. 87, no. 5, pp. 2648-2654, 2015.
[http://dx.doi.org/10.1021/acs.analchem.5b00016] [PMID: 25675332]

[42]   M. Zhang, C. Mullens, and W. Gorski, "Insulin oxidation and determination at carbon electrodes", *Anal. Chem.,* vol. 77, no. 19, pp. 6396-6401, 2005.
[http://dx.doi.org/10.1021/ac0508752] [PMID: 16194105]

[43]   L. Fujcik, R. Prokop, J. Prasek, J. Hubalek, and R. Vrba, "New CMOS potentiostat as ASIC for several electrochemical microsensors construction", *Microelectron. Int.,* vol. 27, pp. 3-10, 2010.
[http://dx.doi.org/10.1108/13565361011009450]

[44]   J. Prasek, D. Huska, O. Jasek, L. Zajickova, L. Trnkova, V. Adam, R. Kizek, and J. Hubalek, "Carbon composite micro- and nano-tubes-based electrodes for detection of nucleic acids", *Nanoscale Res. Lett.,* vol. 6, no. 1, p. 385, 2011.
[http://dx.doi.org/10.1186/1556-276X-6-385] [PMID: 21711910]

[45]   G. Freckmann, S. Hagenlocher, A. Baumstark, N. Jendrike, R.C. Gillen, K. Rössner, and C. Haug, "Continuous glucose profiles in healthy subjects under everyday life conditions and after different meals", *J. Diabetes Sci. Technol.,* vol. 1, no. 5, pp. 695-703, 2007.
[http://dx.doi.org/10.1177/193229680700100513] [PMID: 19885137]

[46]   D.M. Muoio, and C.B. Newgard, "Mechanisms of disease:Molecular and metabolic mechanisms of insulin resistance and β-cell failure in type 2 diabetes", *Nat. Rev. Mol. Cell Biol.,* vol. 9, no. 3, pp. 193-205, 2008.
[http://dx.doi.org/10.1038/nrm2327] [PMID: 18200017]

[47]   C. Weyer, R.L. Hanson, P.A. Tataranni, C. Bogardus, and R.E. Pratley, "A high fasting plasma insulin concentration predicts type 2 diabetes independent of insulin resistance: evidence for a pathogenic role

of relative hyperinsulinemia", *Diabetes,* vol. 49, no. 12, pp. 2094-2101, 2000.
[http://dx.doi.org/10.2337/diabetes.49.12.2094] [PMID: 11118012]

[48]   F.C. Goetz, L.R. French, W. Thomas, R.L. Gingerich, and J.P. Clements, "Are specific serum insulin levels low in impaired glucose tolerance and type II diabetes?: measurement with a radioimmunoassay blind to proinsulin, in the population of Wadena, Minnesota", *Metabolism,* vol. 44, no. 10, pp. 1371-1376, 1995.
[http://dx.doi.org/10.1016/0026-0495(95)90045-4] [PMID: 7476300]

[49]   Y. Pu, Z. Zhu, D. Han, H. Liu, J. Liu, J. Liao, K. Zhang, and W. Tan, "Insulin-binding aptamer-conjugated graphene oxide for insulin detection", *Analyst (Lond.),* vol. 136, no. 20, pp. 4138-4140, 2011.
[http://dx.doi.org/10.1039/c1an15407a] [PMID: 21874167]

[50]   B. Alhalaili, I.N. Popescu, O. Kamoun, F. Alzubi, S. Alawadhia, and R. Vidu, "Nanobiosensors for the Detection of Novel Coronavirus 2019-nCoV and Other Pandemic/Epidemic Respiratory Viruses: A Review", *Sensors (Basel),* vol. 20, no. 22, p. 6591, 2020.
[http://dx.doi.org/10.3390/s20226591] [PMID: 33218097]

[51]   F. Cui, and H.S. Zhou, "Diagnostic methods and potential portable biosensors for coronavirus disease 2019", *Biosens. Bioelectron.,* vol. 165, p. 112349, 2020.
[http://dx.doi.org/10.1016/j.bios.2020.112349] [PMID: 32510340]

[52]   SI Kaya, L Karadurmus, G Ozcelikay, NK Bakirhan, and SA Ozkan, "Electrochemical virus detections with nanobiosensors", *InNanosensors for Smart Cities,* pp. 303-326, 2020.
[http://dx.doi.org/10.1016/B978-0-12-819870-4.00017-7]

[53]   S. Mahari, A. Roberts, D. Shahdeo, and S Gandhi, "eCovSens-ultrasensitive novel in-house built printed circuit board based electrochemical device for rapid detection of nCovid-19 antigen, a spike protein domain 1 of SARS-CoV-2", *BioRxiv,* .
[http://dx.doi.org/10.1101/2020.04.24.059204]

[54]   J.G. Pacheco, P. Rebelo, M. Freitas, H.P. Nouws, and C. Delerue-Matos, "Breast cancer biomarker (HER2-ECD) detection using a molecularly imprinted electrochemical sensor", *Sens. Actuators B Chem.,* vol. 273, pp. 1008-1014, 2018.
[http://dx.doi.org/10.1016/j.snb.2018.06.113]

[55]   S. Ramanathan, S.C.B. Gopinath, M.K.M. Arshad, P. Poopalan, P. Anbu, T. Lakshmipriya, and F.H. Kasim, "Aluminosilicate nanocomposite on genosensor: a prospective voltammetry platform for epidermal growth factor receptor mutant analysis in non-small cell lung cancer", *Sci. Rep.,* vol. 9, no. 1, p. 17013, 2019.
[http://dx.doi.org/10.1038/s41598-019-53573-9] [PMID: 31745155]

[56]   S. Shahrokhian, and R. Salimian, "Ultrasensitive detection of cancer biomarkers using conducting polymer/electrochemically reduced graphene oxide-based biosensor: Application toward BRCA1 sensing", *Sens. Actuators B Chem.,* vol. 266, pp. 160-169, 2018.
[http://dx.doi.org/10.1016/j.snb.2018.03.120]

[57]   M. Azimzadeh, N. Nasirizadeh, M. Rahaie, and H. Naderi-Manesh, "Early detection of Alzheimer's disease using a biosensor based on electrochemically-reduced graphene oxide and gold nanowires for the quantification of serum microRNA-137", *RSC Advances,* vol. 7, no. 88, pp. 55709-55719, 2017.
[http://dx.doi.org/10.1039/C7RA09767K]

[58]   P. Carneiro, J. Loureiro, C. Delerue-Matos, S. Morais, and M. do Carmo Pereira, "Alzheimer's disease: Development of a sensitive label-free electrochemical immunosensor for detection of amyloid beta peptide", *Sens. Actuators B Chem.,* vol. 239, pp. 157-165, 2017.
[http://dx.doi.org/10.1016/j.snb.2016.07.181]

[59]   A. Molazemhosseini, L. Magagnin, P. Vena, and C.C. Liu, "Single-use disposable electrochemical label-free immunosensor for detection of glycated hemoglobin (HbA1c) using differential pulse voltammetry (DPV)", *Sensors (Basel),* vol. 16, no. 7, p. 1024, 2016.

[http://dx.doi.org/10.3390/s16071024] [PMID: 27376299]

[60]   A.S. Rajpurohit, N.S. Punde, and A.K. Srivastava, "An electrochemical sensor with a copper oxide/gold nanoparticle-modified electrode for the simultaneous detection of the potential diabetic biomarkers methylglyoxal and its detoxification enzyme glyoxalase", *New J. Chem.*, vol. 43, no. 42, pp. 16572-16582, 2019.
[http://dx.doi.org/10.1039/C9NJ03553B]

[61]   K. Fatemi, H. Ghourchian, A.A. Ziaee, S. Samiei, and H. Hanaee, "Paramagnetic nanoparticle-based detection of hepatitis B virus using cathodic stripping voltammetry", *Biotechnol. Appl. Biochem.*, vol. 52, no. Pt 3, pp. 221-225, 2009.
[http://dx.doi.org/10.1042/BA20070199] [PMID: 18573093]

[62]   S. Huang, M. Feng, J. Li, Y. Liu, and Q. Xiao, "Voltammetric determination of attomolar levels of a sequence derived from the genom of hepatitis B virus by using molecular beacon mediated circular strand displacement and rolling circle amplification", *Mikrochim. Acta*, vol. 185, no. 3, p. 206, 2018.
[http://dx.doi.org/10.1007/s00604-018-2744-3] [PMID: 29594734]

[63]   L. Farzin, S. Sadjadi, A. Sheini, and E. Mohagheghpour, "A nanoscale genosensor for early detection of COVID-19 by voltammetric determination of RNA-dependent RNA polymerase (RdRP) sequence of SARS-CoV-2 virus", *Mikrochim. Acta*, vol. 188, no. 4, p. 121, 2021.
[http://dx.doi.org/10.1007/s00604-021-04773-6] [PMID: 33694010]

[64]   S. Eissa, H.A. Alhadrami, M. Al-Mozaini, A.M. Hassan, and M. Zourob, "Voltammetric-based immunosensor for the detection of SARS-CoV-2 nucleocapsid antigen", *Mikrochim. Acta*, vol. 188, no. 6, p. 199, 2021.
[http://dx.doi.org/10.1007/s00604-021-04867-1] [PMID: 34041585]

CHAPTER 15

Advanced Materials for Immune Sensors

A.H. Sneharani[1,*]

[1] *DoS in Biochemistry, Jnana Kaveri P.G. Center, Mangalore University, Karnataka 571232, India*

Abstract: The chapter outlines the trends in the rapid development of biofunctionalized hybrid nanomaterials for application as immune sensors to detect biomarkers, thehallmarks of diseases. Biofunctioanalized advanced materials incorporate biological macromolecules/biomaterials with high specificity and selectivity, such as proteins with unique binding properties which are combined with the electronic, optical or photonic properties of nanostructures, enabling them to function as sensors. Functionalization materialises due to similarity in dimension scale allowing functional coupling leading to an effective signal transduction in these sensors. A variety of nano objects metals, carbon nanotubes are coupled with proteins/immunoglobulins or antigens to yield biofunctionalized advanced materials leading to the development of promising novel immune biosensors with many advanced features, like ergonomic, time saving, sensitivity, point-of-care cost economy, feasibility and reproducibility.

Keywords: Antibody, Antigen, Diagnostics, Functionalization, Immunosenors, Nanomaterials.

INTRODUCTION

Biosensors are advanced diagnostic tools that measure and give quantitative output signals about the concentration of the target analyte in the sample. Analytes are components to be detected in the biological sample. They have a biological origin, range from genetic material (DNA or RNA), have proteins of pathogenic bacteria or viruses, proteins or other biological molecules secreted during pathological conditions and immunoglobulins (proteins) synthesized from the immune response of the infected person. Simple molecules like glucose and ions also can be detected. In some cases, the analyte is also known by the term biomarker. The recognition and detection of an analyte in a biosensor occur due to the specific recognition of the target analyte with the immobilised biomolecule often termed as bioreceptor. The bioreceptor is usually the complementary comp-

---

* **Corresponding author A.H. Sneharani:** DoS in Biochemistry, Jnana Kaveri P.G. Center, Mangalore University, Karnataka 571232, India; Tel: +91 9945566321; Email: sneharaniah@gmail.com

J.G. Manjunatha (Ed.)

onent of the analyte which is often an enzyme, monoclonal antibody, DNA, RNA, lectin, glycan or whole cell. Bioreceptors are very crucial elements in designing biosensors, as they provide high sensitivity and specificity in recognising the biomarker present in the sample. Bioreceptor specifically interacts with the target analyte either through covalent or non-covalent interactions generating a signal in the form of heat, charge or change in pH or mass change. The signal generated by these biorecognition events between the biomarker/ analyte and the bioreceptor is converted into a quantitatively measurable signal by transducers. The recorded signal allows the detection and identification of pathogen or disease conditions. A schematic diagram of the principle and design of biosensor is given in Fig. (**1**). Biosensors are very sensitive devices allowing the analyte to detect sub-micro molar concentration, hence requireing a sample in minute amounts. A sensor that is specifically used to detect the immune components comprises the immunosensor. These are ligand based sensing devices which recognise biological molecules by specific affinity interactions. Immunosensors work on the general principle based on the specific immunochemical recognition of antigens or antibodies immobilized on the transducer to antibodies or antigens respectively present in the sample.

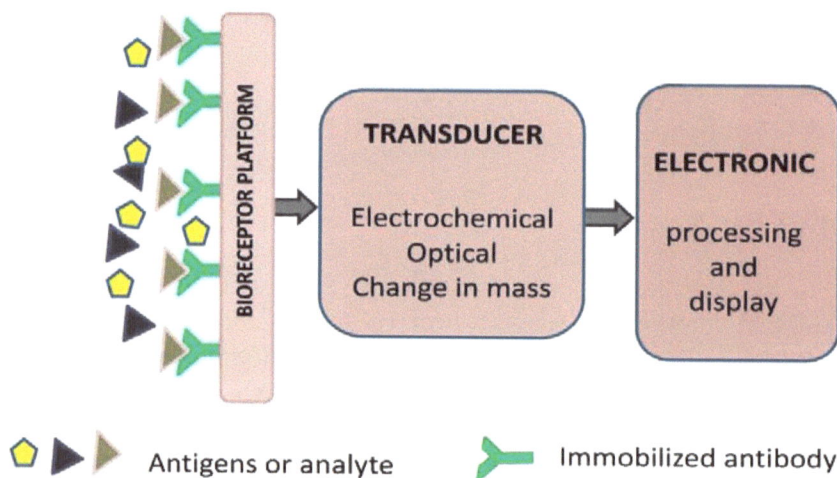

Fig. (1). Schematic diagram of the principle and design of biosensor.

IMMUNE COMPLEX

The cells and molecules responsible for protecting from foreign substances or infectious diseases constitute the immune system. The collective and co-ordinated physiological response of the immune system against foreign substances is called an immune response. Immune response mechanism normally protects individuals from infectious agents and eliminates foreign substances; however, under certain conditions, the same mechanism may cause tissue injury and disease. Defense against foreign substances by early response comprise 'innate immunity' and later responses called 'adaptive immunity'. Innate immunity provides early defense against microorganisms. The mechanisms are in place before even the onset of infection and the response is faster. The immune response in adaptive immunity is stimulated by exposure to infectious agents in response to infection which increases in magnitude with each successive exposure. Cellular and chemical barriers in case of innate immunity include skin, mucosal epithelia and antimicrobial chemicals, whereas, in adaptive immunity, lymph cells and their secreted products such as antibodies bring about the immune response [1, 2]. Adaptive immunity is characterised by exquisite specificity towards a large number of different or closely related molecules and microbes. Foreign substances that elicit immune response are called antigens. The adaptive immune response is mediated *via* humoral and cell mediated immunity. Humoral immunity is mediated by molecules in the blood called antibodies. Antibodies very specifically recognise microbial antigens and help in eliminating the microbes and toxins from the body. Humoral immune response is the principal defence mechanism in eliminating the extracellular microbes by secreting the antibodies. Intracellular microbes such as viruses and some bacteria survive and proliferate in host cells such as phagocytes where they are inaccessible to circulating antibodies. Defense against such infectious agents is mediated by cell mediated immunity which acts in eliminating the cells which are reservoirs of infectious agents.

Antibodies are glycoproteins that belong to the immunoglobulin (Ig) superfamily. The distinctive structure of an antibody is its Y-shape, comprising two pairs of polypeptide chains known as heavy and light chains which are identical. The light and heavy chains are linked through the disulphide linkage. The antibody structure is distinguished typically into three regions; two Fabs (fragment antigen binding) and one Fc (fragment crystallisable). The recognition and binding of antibodies on a specific site on the antigen is called an epitope. The specific binding of the anti body towards epitope is through Fab fragment. The antibody with a wide range of specificities for different epitopes of the same antigen is called a polyclonal antibody. On the other hand, antibodies' displaying high specificity and affinity for a particular epitope is called a monoclonal antibody. A key milestone in the development of immunoassays was developed by the advent of hybridoma method of antibody production. By this method, monoclonal antibodies are developed to detect antigens with specific and desired epitopes. The continuous production of monoclonal antibodies through hybridoma technology opened avenues in the field of immunodiagnostics.

IMMUNOSENSORS

The interaction between antigen-antibody components produces analytical signals which change in proportion with the concentration of analytes of the test. The immunosensor uses the antibody (bioreceptor) or epitopes of antibodies (peptide) for the molecular biorecognition of antigens forming a stable antigen-antibody complex. The biorecognition or interaction involves different types of weak forces; hydrogen, hydrophobic, van der Waals and electrostatic interactions (Fig. 2). Hence, the interaction between antigen and antibody is reversible. The interaction between them is very specific, though, due to relatively weak interactive forces the antigen-antibody complex dissociates with changes in the reaction conditions viz., ionic strength, pH or presence of detergents or chaotropic agents. The intensity of the binding of antigen with an antibody is characterized by its affinity constant (K_a), which is in the range of 5×10^4 to 1×10^{15} L mol^{-1}. High value of K_a represents high affinity between antigen and antibody, *i.e.*, more antigen-antibody complex than unbound antigen or antibody [3]. When K_a is low, the affinity between them is less with fewer binding sites occupied. These properties of specific and reversible interaction make immunosensors unique, as analytical tools.

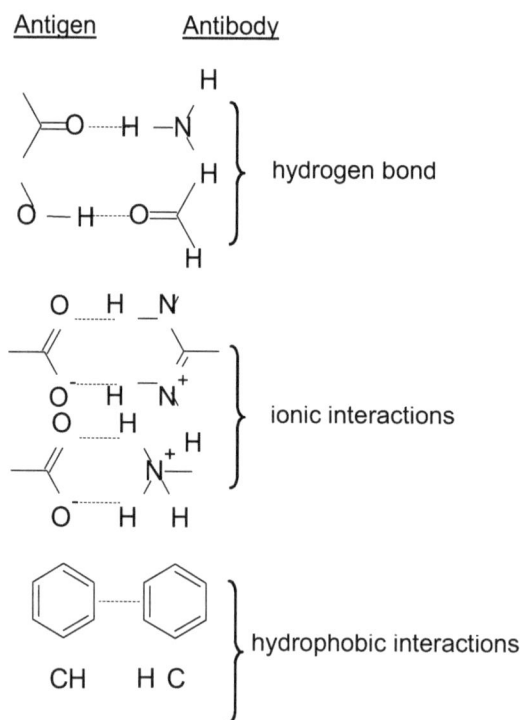

Fig. (2). Major contribution of forces in antigen-antibody interaction.

Immunosensors are designed with three parts; a biological recognition platform, a physicochemical transducer and an electronic part. The biological element (enzymes, antigen, nucleic acids, antibody or hapten) are integrated very closely with the physicochemical transducer and this in turn with the electronic part.. The high selectivity and sensitivity of the enzyme-substrate or antigen-antibody reaction defines the recognition by the transducer, and the electronic component amplifies and digitalises the signal. The formation of an immune complex is measured by coupling the response to a transducer. The transducer detects the chemical reaction and converts it to a processed signal. In immunosensors, transduction of signals can be carried out by different ways depending on the type of signal generation generated during the events of antigen-antibody complex formation [4]. The main transducers employed in immunosensors measure either electrons, photons or change in mass. Different types of transducers are classified based on signal output (*e.g.* electrical) or alteration in properties (change in mass) to detect the formation of an immune complex. To meet the need to detect different types of analytes under different operating conditions, various types of immunosensors are developed. Depending on the type of transducers employed in immunosensors, they are categorised as electrochemical, optical and piezoelectric.

An electrochemical immunosensor measures electrical signal as the output with the generation of antigen-antibody complex. This is a surface based method which is sensitive, selective and easy to operate [5]. Depending on different modes of measurement, the main transducers are amperometric, potentiometric or conductometric, which measures current, potential or difference in voltage, and impedance or conductance, respectively [6]. One major application of electrochemical immunosensor is in the field of clinical diagnostics [7]. In optical transducers, an optical signal is generated or a change in optical property is observed upon the formation of an antigen-antibody complex. Optical immunosensors detect the analytes based on the emission of chemical properties like reflectance, refractive index, fluorescence, phosphorescence, absorbance, chemiluminescence and surface plasmon resonance. Piezoelectric transducers work on the principle of measuring the increase in mass of antigen-antibody complex, as compared with mass of antigen or antibody alone, using quartz crystal balance or cantilever [8, 9].

An ideal immunosensor is specified by the following factors: ability to detect and quantify specific antigen or antibody, to form the immune complex without the additional requirement of reagents, identify the antigen in a short duration of time with high reproducibility and capacity to identify the analyte in unprocessed or actual sample [10].

The immunosensors are designed to possess all essential features of immunoassay technique, like high selectivity, specificity, sensitivity, reversibility and efficacious use of reagents. Immunosensors are automated and generally easy to operate which makes them user friendly over the traditional immunoanalytical technique. Hence, immunosensors are expanding as an immunoanalytical technique owing to its potential to detect a wide range of compounds. Environmental analysis and biological process monitoring utilises immunosensors for the detection of pollutants and toxins in ecological bodies. Immunosensors or immune biosensors are used in clinical diagnostics as smart point-of care devices, forensic analysis, agricultural, pharmaceutical and food sectors. Biomarkers in various disease conditions such as cancer, infectious diseases, pathogenic microorganisms and food contaminants and food pathogens are commonly detected using immunosensors. Considerable interest is devoted to the development of sensitive, reliable and rapid immunosensors by measuring the biomarkers in clinical diagnosis. Initial-stage screening of cancer is usually asymptomatic until it progresses to advanced stages, where the survival rate is poor or in case of Alzheimer's disease the quality of life becomes devastating on a daily basis. Detection of biomolecular markers or any other specific biological agents (such as tumor-associated antigens, circulating cells) in biological fluid is very crucial to detect the onset and progression of any pathological condition. The biotechnology approach to produce and purify antibodies (monoclonal antibodies, mab) against a particular antigen or mass production of enzymes has opened up avenues in biomedical science and clinical diagnosis. For the purpose of clinical diagnosis, immune sensors are developed using advanced biofunctionalized materials. The advantage of these immune sensors employing the nanomaterials over conventional immunoassays (radial immunodiffusion, immunonephelometry, immunoturbidimetry, western blotting, ELISA. immunochemiluminescence) by their exquisite specificity of the biomolecular recognition of antigens by monoclonal antibody, forming a stable immune complex which gives signal with high sensitivity for the presence of the analyte biomarker in question. Apart from this, disadvantages associated with conventional immunoassays are long reaction time involving multiple steps, skilled laboratory personnel and pricey chemicals and reagents. With the advancement of nanotechnology and the advent of biofunctionalized materials, focus on smart point-of-care immunodiagnostics has been shifted in developing more powerful and non-invasive immune sensing methods for detection of biomolecular markers. For example, hepatitis B surface antigen is detected by developing a potentiometric immunosensor. Human epidermal growth factor receptor 2 (HER2) is a biomarker for breast cancer. Gold nanoparticles (AuNPs) coated electrochemical sensors are developed to detect HER2. Likewise, C-reactive protein (CRP) is a clinical biomarker for detecting inflammation. Peptide sequence specific to HER2 epitope was immobilised on

AuNPs, resulting in a large electrochemical current upon sensing. Phospholipase A2 receptor is a biomarker in primary membranous nephropathy. A sensitive and low cost immunoassay based on customised open-source quartz crystal microbalance coupled with grapheme biointerface sensors (G-QCM) is developed to quantify the antibodies (anti-PLA2R) in patients with primary membranous nephropathy. Low concentration of denatured bovine serum albumin (dBSA) protein is adsorbed on the surface of a reduced graphene oxide sensor to create a graphene-protein biointerface. The dBSA functions as an armour, preventing the denaturation of PLA2 and acts as a cross-linker for immobilization of PLA2R on the surface for the binding of anti-PLA2R. The detection limit of this G-QCM immunoassay is reported to be 100 ng/mL with good specificity and sensitivity. The multifunctional dBSA-rGO platform provides a promising biofunctional-ization method for universal immunoassay and developing biosensors [11]. The COVID-19 pandemic has adversely affected people around the world, creating economic and social disruption across the globe. The available methods to detect the disease are by real-time reverse transcriptase-polymerase chain reaction (RT-PCR) and rapid antigen test. These methods have their advantages and disadvantages. For example, RT-PCR is more sensitive and accurate which detects the ribonucleic acid (RNA) of the SARS-CoV-2 virus. However, the RT-PCR method requires the RNA preparation method, which eventually decreases the accuracy. In addition, this method requires skilled technicians and diagnosis is expensive. Likewise, rapid antigen tests have a higher false negative rate. To overcome such disadvantages biosensors based on epitaxial graphene for rapid detection of COVID-19 are developed. Epitaxial graphene-based biosensor detects the SARS-CoV-2 spike protein. Epitaxial graphene-based biosensor allows rapid detection of one attogram (10^{-18} g) to one microgram (10^{-6} g) of SARS-CoV-2 spike protein in a few seconds. After each use, these sensors can be rinsed in sodium chloride solution which allows them to be reused. Epitaxial graphene is a single to few layers of carbon atoms with very surface area, high conductivity, which makes them a very material for developing the biological sensors [11]. A major challenge in developing the biosensor is the capture of biological transducing events. Analytes present in biological samples subjected to tests are present in highly complex environments containing a variety of biological molecules in a broad range of concentration. As a result signal to noise ratio poses a threat of sensitivity, wherein the analyte may be present in low concentration. Hence, immunosensors are designed with improved detection limits by taking advantage of the intrinsic properties of nanoparticles. Nanomaterials offer superior characteristics in their optical, photoluminescence, fluorescence, photostability, surface plasmon resonance band, magnetic and catalytic properties in comparison with their bulk counterparts [12 - 15]. These superior properties are dependent on size, shape and method of fabrication.

Advanced synthetic procedures, which are operated under highly controlled and tuneable methods have allowed the synthesis of fabricated and surface functionalized advanced materials. These desirable qualities of nanomaterials have made them illustrious for their applications in immunoassay, which allows the detection of highly specific and sensitive clinically relevant biomarkers or any other analyte of biological origin. The biological recognition element (either antibody or antigen) are immobilised on the surface of nanoparticle and their binding with its counterpart (antigen or antibody respectively, or any other biomolecular marker) produces a signal that is recorded. Henceforth, tailoring and immobilising the nanomaterials becomes a key factor in immunosensing applications as the nanomaterials are the interface in presenting the biological moiety. Hence the structure which reflects the activity of the biomolecule should be upheld, allowing maximum accessibility of binding sites during the process of immobilization.

Immunosensors measure the signals arising from specific immunochemical reactions between analytes and their immobilised counter component which may be an antigen or antibody. It is now made clear that the immobilization component plays a major role in the development of immunosensors. Numerous immobilization techniques are employed to capture the transducing events, utilising the nanomaterials including carbon (graphene, CNT), metal (inorganic or organic), hybrid conjugate system. This is because of their similar dimension (surface to volume) properties and excellent conductive properties [16]. The immobilization technique involves non-covalent or covalent based interactions which include entrapment, cross-linking, hydrophobic adsorption, electrostatic interaction, or covalent bonding.

The fusion of nanotechnology and biomolecular science has led to a very fast development of biofunctionalized advanced materials leading to advancement in nanotechnology. The advanced nanomaterials are chemical substances which are tailor made to possess specific physical, chemical, biological and functional properties. Fabrication and fine tuning of nanomaterials allow them to achieve photonic, optical, magnetic and luminescence properties giving the tag to nanomaterials as designer biofuctionalized nanomaterials. The biofunctionalized advanced materials find their applications in sensing or detection of pathogens, diagnosis of diseases and bioimaging. Biosensors are a popular analytical instrument used for the detection of pathogens and pathologies like cancer and autoimmune diseases.

Nanoparticles immobilised or functionalized with ligands exhibit preferential binding towards proteins or cell surface receptors that have been used for their detection. For example, nanoparticles through electrostatic interaction complexed with green fluorescence protein (GFP) were used to differentiate between cancerous and metastatic cells from the healthy cells in human and rodent cell lines. GFP is quenched when it is bound to the surface of a gold nanoparticle. Upon addition of cells, GFP is displaced from the surface of a gold nanoparticle, restoring the fluorescence. This process of sensing is driven by the specific affinity between the cell and the ligand head groups on functionalized gold nanoparticles [17]. In a similar approach, antibodies coated on the gold nanoparticle provide a useful tool as immunosensors [18]. Herceptin, is an antibody specific towards the Her-2 receptor overexpressed on the ovarian and breast tumor cells. Multivalent gold and silver nanoparticles were surface functionalized to immobilise Herceptin. The specificity between the Herceptin antibody towards the cancer cells expressing the HER-2 receptor allows the immunosensing of the cancer cells with the HER-2 receptor.

Immobilization of antibodies on the surface of nanoparticles is an important factor to be considered during the design of nanoparticle based immunosensors [19]. The performance of the immunosensors depend on the positioning or the orientation of the antibody on the nanoparticle with its antigen binding fragment exposed for easy accessibility and binding of the analyte component. Alongside this factor, the number or density of the antibody molecules immobilised per unit surface area of the nanoparticles also is a key feature for the performance of immunosensors. The successful implementation of these two major parameters is dependent on the choice of immobilisation strategy. When nanomaterials are subjected to the process of chemical modification to possess one or more functional groups to link or conjugate a biomolecule constitute the surface functionalization. This is achieved by adopting different immobilization or coupling strategies which allows synthesising advanced functional materials with tailored properties. The synthesis of designer biofunctionalized materials is synthesized by adopting either the covalent coupling strategy or through non-covalent interactions. However, irrespective of the coupling strategies adopted, the immobilization procedure should not alter the conformation, physico-chemical or kinetic properties which is a characteristic of the biomolecule [20]. Most commonly used surface functionalization technique involving chemistry used in the modification of substrates (antibody) for developing immunosensors is discussed in the following section (Table **1**). Physical adsorption, electrostatic binding through ionic interactions, affinity binding through specific interaction and covalent coupling are the few techniques commonly employed to functionalise the nanoparticles with biomolecules. Affinity interaction is an effective method for the bioconjugation of targeting ligands to nanoparticles because of specific and strong association between the antibody and antigen interaction.

Table 1. Common types of functionalization of advanced materials.

Coupling Strategies for Functionalization	Uses
Carbodiimide coupling	Covalently links -COOH to -NH$_2$ group, on the surface of nanoparticle to biological molecule respectively
Maleimide coupling	Conjugation of proteins to nanoparticles
Ionic coupling	Coupling of nanoparticles and biological species *via* oppositely charged electrostatic interactions
Click chemistry	cuprous catalysed terminal alkyne-azide cycloaddition reaction
Affinity interaction	Non covalent interaction with high specificity
Disulphide bridges	Covalent interaction resulting in formation of S-S covaalent bond

The installation of amino, hydroxyl, carboxyl and sulphydryl groups is the most widely used surface functionalization technique. The conjugation of these functional groups is mediated through the linker molecule which links between the nanomaterials and the biological molecule. The linker molecule is crucial as it aids in retaining the structure and hence the biological activity.

Nanostructures with carboxyl group covalently conjugate with biomolecules possessing primary amine groups through carbodiimide chemistry. Treatment of amine functionalised substrate with carbodiimide coupling agents like *N, N'* - dicyclohexylcarbodiimide (DCC) and 1-ethyl-3-(3-dimethylaminopropyl)-carbo diimide (EDC) in the presence of solvent dimethylformamide (DMF) and succinic anhydride. Similarly, hydroxyl groups possessing substrate can be converted to carboxy functional groups by treatment with chloroacetic acid. The thiolated substrates can be inducted with -COOH groups by S-carboxymethylation of the - SH (sulphydryl) groups in the presence of iodoacetate.

The addition of amino groups as a coupling strategy to conjugate the biomolecule is one of the widest used in surface chemistry. The amino groups can be added to carboxylated substrates by treatment with 1-ethyl-3-(3-dimethylaminopro-yl)-carbodiimide (EDC) and ethylenediamine. On the thiolated substrate, amino groups are added by treatment with ethyleneimine or 2-bromoethylamine. A wide range of substrates istreated with (3-aminopropyl) triethoxysilane (APTES) to induce amino groups. APTES are deposited on nanomaterials, electrode or solid materials and nanocomposites possessing active hydroxyl groups under optimal temperature, time, solvent and concentration conditions. Click chemistry is another approach used for immobilization of antibodies which involves cuprous catalysed terminal alkyne-azide cycloaddition reaction [21].

Immobilization of antibodies through affinity approach is carried out by the help of an anchor protein called protein A/G which binds through the Fc region of the antibody through non-covalent interactions [22, 23]. Metal nanoparticles like gold, silver, platinum have attracted interest in nanoparticle synthesis due to their superior three dimensional structure compared to their bulk materials. Metal nanoparticles can be modified by the above mentioned strategies (Table **1**). The orientation of the antibody molecule can be achieved by multipoint metal and sulphur bond formation, which allows the antibodies to orient in desired geometry. The antibodies through their free amines can also be conjugated to the metal nanoparticles through straight forward carboxylic acid-amine coupling. Metal nanoparticles can be synthesized with controlled size and shape with different surface coating. The metal nanoparticles exhibit characteristic surface plasmon resonance peaks arising from the interaction of electrons and the electromagnetic field, giving rise to absorption and scattering of light. The localised surface plasmon resonance is characteristic of the size and shape of the particular metal nanoparticles, which can be used for immunosensing applications [24, 25]. Amongst the noble metal nanoparticles, gold nanoparticles are widely used metals for immunosensor applications because of their superior properties like less toxicity, biocompatibility, amenability towards modification, optical and electronic properties. Gold nanoparticles form a powerful platform for detecting single molecules. For example, streptavidin can be detected upon their binding to the biotin modified gold-stars by their spectral shift in localised surface plasmon resonance (LSPR). The horseradish peroxidase (HRP) enzyme was immobilised on gold nanoparticles *via* biotin streptavidin linkage. HRP is an extensively used label in biosensing applications as it forms colored, redox molecules in the presence of H_2O_2 and colored substrate Diaminobenzidine (DAB). DAB is oxidised to an insoluble colored aggregate of polybenzimidazole which enables the shift of LSPR, allowing the detection of HRP attached to gold particles. The additional LSPR scattering of light is due to the immobilization of gold nanoparticles [25]. Similar to gold metal nanoparticles, magnetic nanoparticles are widely used in biosensing applications. Nanosized magnetic particles due to their nanosize, exhibit a reduced number of magnetic domains conferring the property of superparamagnetism. This has led to the development of magnetic labels which are used for the detection of microbial pathogens like *Eschrechia coli* or *Salmonella* in food [26].

Carbon nanomaterials (CNs) are potential candidates employed in biosensors due to their sensitivity and allowing detection in lower limits owing [27, 28]. This is due to their capability of immobilizing large quantities of bioreceptor components at a reduced volume. They also act as a transduction element. Carbon nanotubes (CNTs) can be tailored to functionalization and conjugation with metallic nanoparticles or organic compounds. The surface functionalization extends new physical, electrochemical, and optical properties to the nanomaterials. CNTs, graphene, and fullerenes [29 - 31] are the examples which serve as scaffolds for the immobilization of biomolecules at their surface and are also used as transducers for the conversion of signals associated with the recognition of biological analytes. Considering all these features, CNs are being extensively utilized in biosensor applications.

CONCLUSION

Tailoring of surfaces of nanoparticles provides versatile platforms for immunodiagnostic applications in clinical setup. A large number of antibody immobilization techniques are available that can be designed and tuned for specific applications. The advantages of fluorescent, optical, plasmonic and scattering properties of nanomaterials like gold, graphene or quantum dots are used in the development of biosensors. The crucial part in the design of immunosensor is the selection of a highly specific antibody or its derivative possessing the epitope which binds to a specific antigen and shows low cross reactivity with its cognate. Immunosensing technology offers many challenges like the issues of toxicity and biocompatibility related to the materials chosen as platforms. However, the technology offers great potential in the area of clinical diagnostics by offering a wide variety of platforms.

CONSENT FOR PUBLICATION

Not applicable.

CONFLICT OF INTEREST

The author declares no conflict of interest, financial or otherwise.

ACKNOWLEDGEMENT

Declared none.

REFERENCES

[1]   D.D. Chaplin, "Overview of the immune response", *J. Allergy Clin. Immunol.,* vol. 125, no. 2, suppl. Suppl. 2, pp. S3-S23, 2010.
        [http://dx.doi.org/10.1016/j.jaci.2009.12.980] [PMID: 20176265]

[2]   S. Akira, K. Takeda, and T. Kaisho, "Toll-like receptors: critical proteins linking innate and acquired immunity", *Nat. Immunol.,* vol. 2, no. 8, pp. 675-680, 2001.
        [http://dx.doi.org/10.1038/90609] [PMID: 11477402]

[3]   R.A. Sperling, and W.J. Parak, "Surface modification, functionalization and bioconjugation of colloidal inorganic nanoparticles", *Philos. Trans.- Royal Soc., Math. Phys. Eng. Sci.,* vol. 368, no. 1915, pp. 1333-1383, 2010.
        [http://dx.doi.org/10.1098/rsta.2009.0273] [PMID: 20156828]

[4] S. Sharma, H. Byrne, and R.J. O'Kennedy, "Antibodies and antibody-derived analytical biosensors", *Essays Biochem.*, vol. 60, no. 1, pp. 9-18, 2016.
[http://dx.doi.org/10.1042/EBC20150002] [PMID: 27365031]

[5] J.G. Manjunatha, "Electroanalysis of estriol hormone using electrochemical sensor", *Sens. Biosensing Res.*, vol. 16, pp. 79-84, 2013.
[http://dx.doi.org/10.1016/j.sbsr.2017.11.006]

[6] M.M. Charithra, and J.G. Manjunatha, "Poly (l-Proline) modified carbon paste electrode as the voltammetric sensor for the detection of Estriol and its simultaneous determination with Folic and Ascorbic acid", *Mater. Sci. Energy Technol.*, vol. 2, no. 3, pp. 365-371, 2020.
[http://dx.doi.org/10.1016/j.mset.2019.05.002]

[7] S. Muniandy, S. J. Teh, K. L. Thong, and I. G. Thiha, A. Dinshaw, C.W. Lai, F. Ibrahim, and B.F. Leo, *Carbon Nanomaterial-Based Electrochemical Biosensors for Foodborne Bacterial Detection.* Cri Revi Anal Chem, 2019.
[http://dx.doi.org/10.1080/10408347.2018.1561243]

[8] R. Chauhan, J. Singh, P.R. Solanki, T. Manaka, M. Iwamoto, T. Basu, and B.D. Malhotra, "Label-free piezoelectric immunosensor decorated with gold nanoparticles: Kinetic analysis and biosensing application", *Sens. Actuators B Chem.*, vol. 222, pp. 804-814, 2016.
[http://dx.doi.org/10.1016/j.snb.2015.08.117]

[9] S. Babacan, P. Pivarnik, and S. Letcher, "Evaluation of antibody immobilization methods for piezoelectric biosensor application, Biosens", *Bioelectron,* vol. 15, p. 615e621, 2000.

[10] E. Gizeli, and C.R. Lowe, *Biomolecular Sensors.* 1st ed. CRC Press, Taylor and Francis, 2002.
[http://dx.doi.org/10.1201/9780203212196]

[11] https://www.thegraphenecouncil.org/blogpost/1501180/Graphene-Updates?tag=Biosensor

[12] B.H. Kim, M.J. Hackett, J. Park, and T. Hyeon, "Synthesis, Characterization, and Application of Ultra small Nanoparticles", *Chem. Mater.*, vol. 26, pp. 59-71, 2014.
[http://dx.doi.org/10.1021/cm402225z]

[13] L.Y.T. Chou, K. Ming, and W.C.W. Chan, "Strategies for the intracellular delivery of nanoparticles", *Chem. Soc. Rev.*, vol. 40, no. 1, pp. 233-245, 2011.
[http://dx.doi.org/10.1039/C0CS00003E] [PMID: 20886124]

[14] Y. Gao, X. Xie, F. Li, Y. Lu, T. Li, S. Lian, Y. Zhang, H. Zhang, H. Mei, and L. Jia, "A novel nanomissile targeting two biomarkers and accurately bombing CTCs with doxorubicin", *Nanoscale,* vol. 9, no. 17, pp. 5624-5640, 2017.
[http://dx.doi.org/10.1039/C7NR00273D] [PMID: 28422250]

[15] B. Pelaz, C. Alexiou, R.A. Alvarez-Puebla, F. Alves, A.M. Andrews, S. Ashraf, L.P. Balogh, L. Ballerini, A. Bestetti, C. Brendel, S. Bosi, M. Carril, W.C. Chan, C. Chen, X. Chen, X. Chen, Z. Cheng, D. Cui, J. Du, C. Dullin, A. Escudero, N. Feliu, M. Gao, M. George, Y. Gogotsi, A. Grünweller, Z. Gu, N.J. Halas, N. Hampp, R.K. Hartmann, M.C. Hersam, P. Hunziker, J. Jian, X. Jiang, P. Jungebluth, P. Kadhiresan, K. Kataoka, A. Khademhosseini, J. Kopeček, N.A. Kotov, H.F. Krug, D.S. Lee, C.M. Lehr, K.W. Leong, X.J. Liang, M. Ling Lim, L.M. Liz-Marzán, X. Ma, P. Macchiarini, H. Meng, H. Möhwald, P. Mulvaney, A.E. Nel, S. Nie, P. Nordlander, T. Okano, J. Oliveira, T.H. Park, R.M. Penner, M. Prato, V. Puntes, V.M. Rotello, A. Samarakoon, R.E. Schaak, Y. Shen, S. Sjöqvist, A.G. Skirtach, M.G. Soliman, M.M. Stevens, H.W. Sung, B.Z. Tang, R. Tietze, B.N. Udugama, J.S. VanEpps, T. Weil, P.S. Weiss, I. Willner, Y. Wu, L. Yang, Z. Yue, Q. Zhang, Q. Zhang, X.E. Zhang, Y. Zhao, X. Zhou, and W.J. Parak, "Diverse Applications of Nanomedicine", *ACS Nano,* vol. 11, no. 3, pp. 2313-2381, 2017.
[http://dx.doi.org/10.1021/acsnano.6b06040] [PMID: 28290206]

[16] A.H. Sneharani, "In: Facile Chemical Fabrication of Designer Biofunctionalized Nanomaterials", In: *Functionalised Nanomaterials I; Taylor and Francis* CRC Press, 2020.
[http://dx.doi.org/10.1201/9781351021623]

[17]   M. Bugiel, H. Fantana, V. Bormuth, A. Trushko, F. Schiemann, J. Howard, E. Schäffer, and A. Jannasch, "Versatile microsphere attachment of GFP-labeled motors and other tagged proteins with preserved functionality", *J. Biol. Chem.,* vol. •••, p. 2, 2015.

[18]   K. Saha, S.S. Agasti, C. Kim, X. Li, and V.M. Rotello, "Gold nanoparticles in chemical and biological sensing", *Chem. Rev.,* vol. 112, no. 5, pp. 2739-2779, 2012.
[http://dx.doi.org/10.1021/cr2001178] [PMID: 22295941]

[19]   Y. Huang, P. Kannan, L. Zhang, T. Chen, and D-H. Kim, "Concave gold nanoparticle-based highly sensitive electrochemical IgG immunobiosensor for the detection of antibody–antigen interactions", *RSC Advances,* vol. 5, pp. 58478-58484, 2015.
[http://dx.doi.org/10.1039/C5RA10990F]

[20]   M. Brinkley, "A brief survey of methods for preparing protein conjugates with dyes, haptens, and cross-linking reagents", *Bioconjug. Chem.,* vol. 3, no. 1, pp. 2-13, 1992.
[http://dx.doi.org/10.1021/bc00013a001] [PMID: 1616945]

[21]   J. Bolley, E. Guenin, N. Lievre, M. Lecouvey, M. Soussan, Y. Lalatonne, and L. Motte, "Carbodiimide versus click chemistry for nanoparticle surface functionalization: a comparative study for the elaboration of multimodal superparamagnetic nanoparticles targeting αvβ3 integrins", *Langmuir,* vol. 29, no. 47, pp. 14639-14647, 2013.
[http://dx.doi.org/10.1021/la403245h] [PMID: 24171381]

[22]   B. Akerstrom, T. Brodin, and K. Reis, "Protein G: a powerful tool for binding and detection of monoclonal and polyclonal antibodies", *J. Immunol,* vol. 135, p. 2589e2592, 1985.

[23]   B.A. Björk, and J. Petersson, "Some physicochemical properties of protein A from Staphylococcus aureus", *Eur. J. Biochem,* vol. 29, p. 579e584, 1972.

[24]   S.K. Vashist, C.K. Dixit, and B.D. Maccraith, "Effect of antibody immobilization strategies on the analytical performance of a surface plasmon resonance-based immunoassay", *Analyst,* vol. 136, p. 4431e4436, 2011.

[25]   S.K. Dondapati, T.K. Sau, C. Hrelescu, T.A. Klar, F.D. Stefani, and J. Feldmann, "Label-free biosensing based on single gold nanostars as plasmonic transducers", *ACS Nano,* vol. 4, no. 11, pp. 6318-6322, 2010.
[http://dx.doi.org/10.1021/nn100760f] [PMID: 20942444]

[26]   H. Kuang, G. Cui, X. Chen, H. Yin, Q. Yong, L. Xu, C. Peng, L. Wang, and C. Xu, "A one-step homogeneous sandwich immunosensor for Salmonella detection based on magnetic nanoparticles (MNPs) and quantum Dots (QDs)", *Int. J. Mol. Sci.,* vol. 14, no. 4, pp. 8603-8610, 2013.
[http://dx.doi.org/10.3390/ijms14048603] [PMID: 23609493]

[27]   V. Vamvakaki, and N.A. Chaniotakis, "Carbon nanostructures as transducers in biosensors", *Sens. Actuators B Chem.,* vol. 126, pp. 193-197, 2007.
[http://dx.doi.org/10.1016/j.snb.2006.11.042]

[28]   J. G. Manjunatha, "Surfactant modified carbon nanotube paste electrode for the sensitive determination of mitoxantrone anticancer drug", *Journal of Electrochemical Science and Engineering,* vol. 7, no. 1, pp. 39-49, 2017.

[29]   J.G. Manjunatha, and M. Deraman, "Graphene paste electrode modified with sodium dodecyl sulfate surfactant for the determination of dopamine, ascorbic acid and uric acid", *Anal Bioanal Electrochem,* vol. 9, no. 2, pp. 198-213, 2017.

[30]   C. Raril, and J. Manjunatha, "G. A simple approach for the electrochemical determination of vanillin at ionic surfactant modified graphene paste electrode", *Microchem. J.,* vol. 154, p. 104575, 2020.
[http://dx.doi.org/10.1016/j.microc.2019.104575]

[31]   J.G. Manjunatha, M. Deraman, N.H. Basri, and I.A. Talib, "Selective detection of dopamine in the presence of uric acid using polymerized phthalo blue film modified carbon paste electrode", *Adv. Mat. Res.,* vol. 895, pp. 447-451, 2014.
[http://dx.doi.org/10.4028/www.scientific.net/AMR.895.447]

SUBJECT INDEX